Methods in Enzymology

Volume 188
HYDROCARBONS AND METHYLOTROPHY

METHODS IN ENZYMOLOGY

EDITORS-IN-CHIEF

John N. Abelson   Melvin I. Simon

DIVISION OF BIOLOGY
CALIFORNIA INSTITUTE OF TECHNOLOGY
PASADENA, CALIFORNIA

FOUNDING EDITORS

Sidney P. Colowick and Nathan O. Kaplan

Methods in Enzymology

Volume 188

Hydrocarbons and Methylotrophy

EDITED BY

Mary E. Lidstrom

KECK LABORATORIES
CALIFORNIA INSTITUTE OF TECHNOLOGY
PASADENA, CALIFORNIA

ACADEMIC PRESS, INC.
Harcourt Brace Jovanovich, Publishers
San Diego   New York   Boston
London   Sydney   Tokyo   Toronto

ACADEMIC PRESS, INC.
San Diego, California 92101

United Kingdom Edition published by
ACADEMIC PRESS LIMITED
24-28 Oval Road, London NW1 7DX

LIBRARY OF CONGRESS CATALOG CARD NUMBER: 54-9110

ISBN   0-12-182089-0   (alk. paper)

PRINTED IN THE UNITED STATES OF AMERICA
90 91 92 93   9 8 7 6 5 4 3 2 1

Table of Contents

CONTRIBUTORS TO VOLUME 188 xi

PREFACE xv

VOLUMES IN SERIES. xvii

Section I. Hydrocarbons and Related Compounds

A. Enzymes of Alkane and Alkene Utilization

1. Hydrocarbon Monooxygenase System of *Pseudo-* SHELDON W. MAY AND
   monas oleovorans   ANDREAS G. KATOPODIS   3

2. Assay Methods for Long-Chain Alkane Oxidation W. R. FINNERTY   10
   in *Acinetobacter*

3. Primary Alcohol Dehydrogenases from *Acineto-* W. R. FINNERTY   14
   bacter

4. Aldehyde Dehydrogenases from *Acinetobacter*   W. R. FINNERTY   18

5. Propane-Specific Alcohol Dehydrogenase from W. ASHRAF AND
   Rhodococcus rhodochrous PNKb1   J. C. MURRELL   21

6. Cell-Free Assay Methods for Enzymes of Propane J. C. MURRELL AND
   Utilization   W. ASHRAF   26

7. Quinoprotein Alcohol Dehydrogenase from *Pseu-* B. W. GROEN AND
   domonas aeruginosa and Quinohemoprotein Al- J. A. DUINE   33
   cohol Dehydrogenase from *Pseudomonas testos-*
   teroni

B. Enzymes Involved in Cyclic and Aromatic Hydrocarbon Utilization

8. Toluene Dioxygenase from *Pseudomonas putida* LAWRENCE P. WACKETT   39
   F1

9. Naphthalene Dioxygenase from *Pseudomonas* BURT D. ENSLEY AND
   NCIB 9816   BILLY E. HAIGLER   46

10. Benzene Dioxygenase from *Pseudomonas putida* PHILIP J. GEARY,
    ML2 (NCIB 12190)   JEREMY R. MASON, AND
                                                   CHRIS L. JOANNOU   52

11. Phthalate Dioxygenase   CHRISTOPHER J. BATIE
                                                   AND DAVID P. BALLOU   61

12. Cyclohexanone 1,2-Monooxygenase from *Acineto-* PETER W. TRUDGILL   70
    bacter NCIMB 9871

13. Cyclopentanone 1,2-Monooxygenase from *Pseudomonas* NCIMB 9872 — PETER W. TRUDGILL — 77

14. Protocatechuate 3,4-Dioxygenase from *Brevibacterium fuscum* — JAMES W. WHITTAKER, ALLEN M. ORVILLE, AND JOHN D. LIPSCOMB — 82

15. Protocatechuate 4,5-Dioxygenase from *Pseudomonas testosteroni* — DAVID M. ARCIERO, ALLEN M. ORVILLE, AND JOHN D. LIPSCOMB — 89

16. Protocatechuate 2,3-Dioxygenase from *Bacillus macerans* — SANFORD A. WOLGEL AND JOHN D. LIPSCOMB — 95

17. Gentisate 1,2-Dioxygenase from *Pseudomonas acidovorans* — MARK R. HARPEL AND JOHN D. LIPSCOMB — 101

18. Synthesis of ^{17}O- or ^{18}O-Enriched Dihydroxy Aromatic Compounds — ALLEN M. ORVILLE, MARK R. HARPEL, AND JOHN D. LIPSCOMB — 107

19. Catechol 2,3-Dioxygenases from *Pseudomonas aeruginosa* 2x — I. A. KATAEVA AND L. A. GOLOVLEVA — 115

20. Catechol and Chlorocatechol 1,2-Dioxygenases — KA-LEUNG NGAI, ELLEN L. NEIDLE, AND L. NICHOLAS ORNSTON — 122

21. Muconate Cycloisomerase — RICHARD B. MEAGHER, KA-LEUNG NGAI, AND L. NICHOLAS ORNSTON — 126

22. Muconolactone Isomerase — RICHARD B. MEAGHER, KA-LEUNG NGAI, AND L. NICHOLAS ORNSTON — 130

23. *cis*-1,2-Dihydroxycyclohexa-3,5-diene (NAD) Oxidoreductase (*cis*-Benzene Dihydrodiol Dehydrogenase) from *Pseudomonas putida* NCIB 12190 — JEREMY R. MASON AND PHILIP J. GEARY — 134

24. Hydroxybenzoate Hydroxylase — BARRIE ENTSCH — 138

25. Polycyclic Aromatic Hydrocarbon Degradation by *Mycobacterium* — CARL E. CERNIGLIA AND MICHAEL A. HEITKAMP — 148

26. Benzoate-CoA Ligase from *Rhodopseudomonas palustris* — JANE GIBSON, JOHANNA F. GEISSLER, AND CAROLINE S. HARWOOD — 154

C. Yeast and Fungal Enzymes of Hydrocarbon Degradation

27. Lignin Peroxidase from Fungi *Phanerochaete chrysosporium* — T. KENT KIRK, MING TIEN, PHILIP J. KERSTEN, B. KALYANARAMAN, KENNETH E. HAMMEL, AND ROBERTA L. FARRELL — 159

28. Long-Chain Alcohol Dehydrogenase of *Candida* MITSUYOSHI UEDA AND
    Yeast   ATSUO TANAKA   171

29. Long-Chain Aldehyde Dehydrogenase of *Candida* MITSUYOSHI UEDA AND
    Yeast   ATSUO TANAKA   176

Section II. Methylotrophy

A. Dissimilatory Enzymes of Methylotrophic Bacteria

30. Soluble Methane Monooxygenase from *Methylo-* SIMON J. PILKINGTON
    coccus capsulatus Bath   AND HOWARD DALTON   181

31. Methane Monooxygenase from *Methylosinus tri-* BRIAN G. FOX,
    chosporium OB3b   WAYNE A. FROLAND,
                                                  DAVID R. JOLLIE, AND
                                                  JOHN D. LIPSCOMB   191

32. Methanol Dehydrogenase from *Hyphomicrobium* J. FRANK AND
    X   J. A. DUINE   202

33. Methanol Dehydrogenase from *Methylobacterium* DARREN J. DAY AND
    extorquens AM1   CHRISTOPHER ANTHONY 210

34. Modifier Protein for Methanol Dehydrogenase of ANTONY R. LONG AND
    Methylotrophs   CHRISTOPHER ANTHONY 216

35. Methanol Dehydrogenase from Thermotolerant N. ARFMAN AND
    Methylotroph *Bacillus* C1   L. DIJKHUIZEN   223

36. Methylamine Oxidase from *Arthrobacter* P1   WILLIAM S. MCINTIRE 227

37. Methylamine Dehydrogenase from *Thiobacillus* JOHN E. VAN WIELINK,
    versutus   JOHANNES FRANK, AND
                                                  JOHANNIS A. DUINE   235

38. Methylamine Dehydrogenases from Methylo- VICTOR L. DAVIDSON   241
    trophic Bacteria

39. Methylamine Dehydrogenase from *Methyloba-* MICHAEL Y. KIRIUKHIN,
    cillus flagellatum   ANDREY Y. CHISTOSERDOV,
                                                  AND YURI D. TSYGANKOV 247

40. Trimethylamine Dehydrogenase from Bacterium WILLIAM S. MCINTIRE   250
    W3A1

41. Isolation, Preparation, and Assay of Pyrroloquino- R. A. VAN DER MEER,
    line Quinone   B. W. GROEN,
                                                  M. A. G. VAN KLEEF,
                                                  J. FRANK,
                                                  J. A. JONGEJAN, AND
                                                  J. A. DUINE   260

42. Blue Copper Proteins Involved in Methanol and CHRISTOPHER ANTHONY 284
    Methylamine Oxidation

43. Soluble Cytochromes c from *Methylomonas* A4   ALAN A. DiSPIRITO   289

44. Soluble Cytochromes c of Methanol-Utilizing Bac- DARREN J. DAY AND
    teria   CHRISTOPHER ANTHONY   298

45. Cytochrome c_L and Cytochrome c_H from *Hypho-* J. FRANK AND
    microbium X   J. A. DUINE   303

46. Electron-Transfer Flavoproteins from *Methylo-*   MAZHAR HUSAIN   309
    philus methylotrophus and Bacterium W3A1

47. Formaldehyde Dehydrogenases from Methylo-   MARGARET M. ATTWOOD   314
    trophs

48. NAD-Linked, Factor-Independent, and Glutathi-   J. A. DUINE   327
    one-Independent Aldehyde Dehydrogenase from
    Hyphomicrobium X

49. Formate Dehydrogenase from *Methylosinus tri-*   DAVID R. JOLLIE AND
    chosporium OB3b   JOHN D. LIPSCOMB   331

50. Glucose-6-phosphate Dehydrogenase and 6-Phos-   LUDMILA V. KLETSOVA,
    phogluconate Dehydrogenase from *Methyloba-*   MICHAEL Y. KIRIUKHIN,
    cillus flagellatum   ANDREY Y. CHISTOSERDOV,
                                                       AND YURI D. TSYGANKOV   335

51. Glucose-6-phosphate Dehydrogenase and 6-Phos-   A. P. SOKOLOV AND
    phogluconate Dehydrogenase from *Arthrobacter*   Y. A. TROTSENKO   339
    globiformis

52. Glucose-6-phosphate Dehydrogenase from *Pseu-*   DIETMAR MIETHE AND
    domonas W6   WOLFGANG BABEL   346

53. Citrate Synthases from Methylotrophs   GABRIELE MÜLLER-KRAFT
                                                       AND WOLFGANG BABEL   350

54. Dichloromethane Dehalogenase from *Hyphomi-*   THOMAS LEISINGER AND
    crobium DM2   DORIS KOHLER-STAUB   355

B. Assimilation of Formaldehyde in Methylotrophic Bacteria

55. Assay of Assimilatory Enzymes in Crude Extracts   PATRICIA M. GOODWIN   361
    of Serine Pathway Methylotrophs

56. Serine Hydroxymethyltransferases from *Methylo-*   MARY E. LIDSTROM   365
    bacterium organophilum XX

57. Hydroxypyruvate Reductase from *Methylobacte-*   CINDER KREMA AND
    rium extorquens AM1   MARY E. LIDSTROM   373

58. Malyl-CoA Lyase from *Methylobacterium extor-*   A. J. HACKING AND
    quens AM1   J. R. QUAYLE   379

59. Synthesis of L-4-Malyl Coenzyme A   PEGGY J. ARPS   386

60. 3-Hexulose-6-phosphate Synthase from Thermo-tolerant Methylotroph *Bacillus* C1 — N. ARFMAN, L. BYSTRYKH, N. I. GOVORUKHINA, AND L. DIJKHUIZEN   391

61. 3-Hexulose-6-phosphate Synthase from *Mycobacterium gastri* MB19 — NOBUO KATO   397

62. 3-Hexulose-6-phosphate Synthase from *Acetobacter methanolicus* MB58 — ROLAND H. MÜLLER AND WOLFGANG BABEL   401

63. Transaldolase Isoenzymes from *Arthrobacter* P1 — P. R. LEVERING AND L. DIJKHUIZEN   405

C. Methylotrophic Enzymes in Methanol-Utilizing Yeast

64. Cytochemical Staining Methods for Localization of Key Enzymes of Methanol Metabolism in *Hansenula polymorpha* — MARTEN VEENHUIS AND IDA J. VAN DER KLEI   411

65. Alcohol Oxidase from *Hansenula polymorpha* CBS 4732 — IDA J. VAN DER KLEI, LEONID V. BYSTRYKH, AND WIM HARDER   420

66. Amine Oxidases from Methylotrophic Yeasts — PETER J. LARGE AND GEOFFREY W. HAYWOOD   427

67. Dihydroxyacetone Synthase from *Candida boidinii* KD1 — LEONID V. BYSTRYKH, WIM DE KONING, AND WIM HARDER   435

68. Triokinase from *Candida boidinii* KD1 — LEONID V. BYSTRYKH, WIM DE KONING, AND WIM HARDER   445

69. Glycerone Kinase from *Candida methylica* — KLAUS H. HOFMANN AND WOLFGANG BABEL   451

70. Formaldehyde Dehydrogenase from Methylotrophic Yeasts — NOBUO KATO   455

71. Formate Dehydrogenase from Methylotrophic Yeasts — NOBUO KATO   459

72. Catalase from *Candida boidinii* 2201 — MITSUYOSHI UEDA, SABIHA MOZAFFAR, AND ATSUO TANAKA   463

AUTHOR INDEX 469

SUBJECT INDEX 483

Contributors to Volume 188

Article numbers are in parentheses following the names of contributors. Affiliations listed are current.

CHRISTOPHER ANTHONY (33, 34, 42, 44), *Department of Biochemistry, The University of Southampton, Southampton S09 5TU, England*

DAVID M. ARCIERO (15), *Department of Genetics and Cell Biology, University of Minnesota, St. Paul, Minnesota 55108*

N. ARFMAN (35, 60), *Department of Microbiology, University of Groningen, 9751 NN Haren, The Netherlands*

PEGGY J. ARPS (59), *Institute of Arctic Biology, University of Alaska at Fairbanks, Fairbanks, Alaska 99775*

W. ASHRAF (5, 6), *Department of Biological Sciences, University of Warwick, Coventry CV4 7AL, England*

MARGARET M. ATTWOOD (47), *Department of Molecular Biology and Biotechnology, University of Sheffield, Sheffield S10 2TN, England*

WOLFGANG BABEL (52, 53, 62, 69), *Institut für Biotechnologie der Akademie der Wissenschaften, Leipzig 7050, German Democratic Republic*

DAVID P. BALLOU (11), *Department of Biological Chemistry, University of Michigan, Ann Arbor, Michigan 48109*

CHRISTOPHER J. BATIE (11), *Department of Biochemistry and Molecular Biology, Louisiana State University Medical Center, New Orleans, Louisiana 70012*

LEONID V. BYSTRYKH (60, 65, 67, 68), *Institute of Biochemistry and Physiology of Microorganisms, U.S.S.R. Academy of Sciences, Pushchino, Moscow 142292, U.S.S.R.*

CARL E. CERNIGLIA (25), *Microbiology Division, National Center for Toxicology Research, Food and Drug Administration, Jefferson, Arkansas 72079*

ANDREY Y. CHISTOSERDOV (39, 50), *Institute of Genetics and Selection of Industrial Microorganisms, 1st Dorozhni proezd 1, Moscow 113545, U.S.S.R.*

HOWARD DALTON (30), *Department of Biological Sciences, University of Warwick, Coventry CV4 7AL, England*

VICTOR L. DAVIDSON (38), *Department of Biochemistry, University of Mississippi Medical Center, Jackson, Mississippi 39216*

DARREN J. DAY (33, 44), *Department of Biochemistry, The University of Southampton, Southampton, S09 5TU, England*

WIM DE KONING (67, 68), *Department of Biochemistry, B.C.P. Jansen Institute, University of Amsterdam, 1018 TV Amsterdam, The Netherlands*

L. DIJKHUIZEN (35, 60, 63), *Department of Microbiology, University of Groningen, 9751 NN Haren, The Netherlands*

ALAN A. DISPIRITO (43), *Department of Biology, University of Texas at Arlington, Arlington, Texas 76019*

J. A. DUINE (7, 32, 37, 41, 45, 48), *Kluyver Laboratory of Biotechnology, Delft University of Technology, 2628 BC Delft, The Netherlands*

BURT D. ENSLEY (9), *Envirogen Inc., Lawrenceville, New Jersey 08648*

BARRIE ENTSCH (24), *Department of Biochemistry, Microbiology and Nutrition, University of New England, Armidale, N.S.W. 2351, Australia*

ROBERTA L. FARRELL (27), *Repligen Sandoz Research Corporation, Lexington, Massachusetts 02173*

W. R. FINNERTY (2, 3, 4), *Finnerty Enterprises Inc., Athens, Georgia 30605*

BRIAN G. FOX (31), *Department of Biochemistry, Medical School, University of Minnesota, Minneapolis, Minnesota 55455*

J. Frank (32, 37, 41, 45), *Kluyver Laboratory of Biotechnology, Delft University of Technology, 2628 BC Delft, The Netherlands*

Wayne A. Froland (31), *Department of Biochemistry, Medical School, University of Minnesota, Minneapolis, Minnesota 55455*

Philip J. Geary (10, 23), *Shell Research Ltd., Sittingbourne Research Centre, Sittingbourne, Kent ME9 8AG, England*

Johanna F. Geissler (26), *Pharmacology Division, CIBA-GEIGY AG, CH-4056 Basel, Switzerland*

Jane Gibson (26), *Section of Biochemistry, Molecular and Cell Biology, Cornell University, Ithaca, New York 14853*

L. A. Golovleva (19), *Institute of Biochemistry and Physiology of Microorganisms, U.S.S.R. Academy of Sciences, Pushchino, Moscow 142292, U.S.S.R.*

Patricia M. Goodwin (55), *School of Cell and Molecular Biology, North East Surrey College of Technology, Ewell, Epson, Surrey KT17 3DS, England*

N. I. Govorukhina (60), *Institute of Biochemistry and Physiology of Microorganisms, U.S.S.R. Academy of Sciences, Pushchino, Moscow 142292, U.S.S.R.*

B. W. Groen (7, 41), *Kluyver Laboratory of Biotechnology, Delft University of Technology, 2628 BC Delft, The Netherlands*

A. J. Hacking (58), *Dextra Laboratories Ltd., Reading, Berks RG6 2BX, England*

Billy E. Haigler (9), *HQ AFESC/RDVW, Tyndall Air Force Base, Panama City, Florida 32403*

Kenneth E. Hammel (27), *Department of Chemistry, S.U.N.Y. College of Environmental Science and Forestry, Syracuse, New York 13210*

Wim Harder (65, 67, 68), *MT-TNO Delft, 2600 AE Delft, The Netherlands*

Mark R. Harpel (17, 18), *Department of Biochemistry, Medical School, University of Minnesota, Minneapolis, Minnesota 55455*

Caroline S. Harwood (26), *Department of*

Microbiology, University of Iowa, Iowa City, Iowa 52242*

Geoffrey W. Haywood (66), *County Analyst's Laboratory Fillingate, Wanlip, Leicester LE7 8PF, England*

Michael A. Heitkamp (25), *Environmental Sciences Center, Monsanto Company, St. Louis, Missouri 63167*

Klaus H. Hofmann (69), *Institut für Angewandte und Technische Mikrobiologie, Ernst-Moritz-Arndt-Universität, Greifswald, 2200, German Democratic Republic*

Mazhar Husain (46), *Department of Applied Immunochemistry, Diagnostics Division, Abbott Laboratories, Abbott Park, Illinois 60064*

Chris L. Joannou (10), *Department of Microbiology, Kings College London, London W8 7AH, England*

David R. Jollie (31, 49), *Department of Biochemistry, Medical School, University of Minnesota, Minneapolis, Minnesota 55455*

J. A. Jongejan (41), *Kluyver Laboratory of Biotechnology, Delft University of Technology, 2628 BC Delft, The Netherlands*

B. Kalyanaraman (27), *Department of Radiology, Medical College of Wisconsin, Milwaukee, Wisconsin 63226*

I. A. Kataeva (19), *Institute of Biochemistry and Physiology of Microorganisms, U.S.S.R. Academy of Sciences, Pushchino, Moscow 142292, U.S.S.R.*

Nobuo Kato (61, 70, 71), *Department of Biotechnology, Faculty of Engineering, Tottori University, Koyama-Cho, Tottori 680, Japan*

Andreas G. Katodis (1), *Department of Chemistry and Biochemistry, Georgia Institute of Technology, Atlanta, Georgia 30332*

Philip J. Kersten (27), *Biotechnology Center, University of Wisconsin—Madison, Madison, Wisconsin 53705*

Michael Y. Kiriukhin (39, 50), *Institute of Genetics and Selection of Industrial Microorganisms, 1st Dorozhni proezd 1, Moscow 113545, U.S.S.R.*

T. Kent Kirk (27), *Institute for Microbial and Biochemical Technology, U. S. Forest Products Laboratory, Madison, Wisconsin 53705*

Ludmila V. Kletsova (50), *Institute of Genetics and Selection of Industrial Microorganisms, 1st Dorozhni proezd 1, Moscow 113545, U.S.S.R.*

Doris Kohler-Staub (54), *Mikrobiologisches Institut, Eidgenössische Technische, Hochschule, ETH-Zentrum, LFV, CH-8092 Zürich, Switzerland*

Cinder Krema (57), *Department of Biology, University of Texas at Arlington, Arlington, Texas 76019*

Peter J. Large (66), *Department of Applied Biology, The University of Hull, Hull HU6 7RX, England*

Thomas Leisinger (54), *Mikrobiologisches Institut, Eidgenössische Technische, Hochschule, ETH-Zentrum, CH-8092 Zürich, Switzerland*

P. R. Levering (63), *Microbiological R & D Labs, Organon International B. V., 5340 BH Oss, The Netherlands*

Mary E. Lidstrom (56, 57), *Environmental Engineering Science, California Institute of Technology, Pasadena, California 91125*

John D. Lipscomb (14, 15, 16, 17, 18, 31, 49), *Department of Biochemistry, Medical School, University of Minnesota, Minneapolis, Minnesota 55455*

Antony R. Long (34), *Department of Biochemistry, The University of Southampton, Southampton S09 5TU, England*

Jeremy R. Mason (10, 23), *Department of Microbiology, Kings College London, London W8 7AH, England*

Sheldon W. May (1), *Department of Chemistry and Biochemistry, Georgia Institute of Technology, Atlanta, Georgia 30332*

William S. McIntire (36, 40), *Department of Veterans Affairs, Medical Center, Molecular Biology Division, San Francisco, California 94121*

Richard B. Meagher (21, 22), *C/P Genetics Department, University of Georgia, Athens, Georgia 30602*

Dietmar Miethe (52), *Institut für Biotechnologie der Akademie der Wissenschaften, Leipzig 7050, German Democratic Republic*

Sabiha Mozaffar (72), *Laboratory of Industrial Biochemistry, Department of Industrial Chemistry, Faculty of Engineering, Kyoto University, Yoshida, Sakyo-ku, Kyoto 606, Japan*

Roland H. Müller (62), *Institute of Biotechnology, Academy of Sciences of the German Democratic Republic, Leipzig 7050, German Democratic Republic*

Gabriele Müller-Kraft (53), *Institut für Biotechnologie der Akademie der Wissenschaften, Leipzig 7050, German Democratic Republic*

J. C. Murrell (5, 6), *Department of Biological Sciences, University of Warwick, Coventry CV4 7AL, England*

Ellen L. Neidle (20), *Department of Microbiology, University of Texas Medical School, Houston, Texas 77225*

Ka-Leung Ngai (20, 21, 22), *Department of Biochemistry, Molecular Biology and Cell Biology, Northwestern University, Evanston, Illinois 60208*

L. Nicholas Ornston (20, 21, 22), *Department of Biology, Yale University, New Haven, Connecticut 06511*

Allen M. Orville (14, 15, 18), *Department of Biochemistry, Medical School, University of Minnesota, Minneapolis, Minnesota 55455*

Simon J. Pilkington (30), *Department of Biological Sciences, University of Warwick, Coventry CV4 7AL, England*

J. R. Quayle (58), *University of Bath, Bath BA2 7AY, England*

A. P. Sokolov (51), *Institute of Biochemistry and Physiology of Microorganisms, U.S.S.R. Academy of Sciences, Pushchino, Moscow 142292, U.S.S.R.*

ATSUO TANAKA (28, 29, 72), *Laboratory of Industrial Biochemistry, Department of Industrial Chemistry, Faculty of Engineering, Kyoto University, Yoshida, Sakyo-ku, Kyoto 606, Japan*

MING TIEN (27), *Department of Molecular and Cell Biology, Pennsylvania State University, University Park, Pennsylvania 16802*

Y. A. TROTSENKO (51), *Institute of Biochemistry and Physiology of Microorganisms, U.S.S.R. Academy of Sciences, Pushchino, Moscow 142292, U.S.S.R.*

PETER W. TRUDGILL (12, 13), *Department of Biochemistry, University College of Wales, Aberystwyth, Dyfed SY23 3DD, Wales*

YURI D. TSYGANKOV (39, 50), *Institute of Genetics and Selection of Industrial Microorganisms, 1st Dorozhni proezd 1, Moscow 113545, U.S.S.R.*

MITSUYOSHI UEDA (28, 29, 72), *Laboratory of Industrial Biochemistry, Department of Industrial Chemistry, Faculty of Engineering, Kyoto University, Yoshida, Sakyo-ku, Kyoto 606, Japan*

IDA J. VAN DER KLEI (64, 65), *Department of Microbiology, Biological Center, University of Groningen, 9751 NN Haren, The Netherlands*

R. A. VAN DER MEER (41), *Kluyver Laboratory of Biotechnology, Delft University of Technology, 2628 BC Delft, The Netherlands*

M. A. G. VAN KLEEF (41), *Kluyver Laboratory of Biotechnology, Delft University of Technology, 2628 BC Delft, The Netherlands*

JOHN E. VAN WIELINK (37), *Kluyver Laboratory of Biotechnology, Delft University of Technology, 2628 BC Delft, The Netherlands*

MARTEN VEENHUIS (64), *Laboratory for Electron Microscopy, Biological Center, University of Groningen, 9751 AA Haren, The Netherlands*

LAWRENCE P. WACKETT (8), *Department of Biochemistry, Gray Freshwater Biological Institute, University of Minnesota, Navarre, Minnesota 55392*

JAMES W. WHITTAKER (14), *Department of Chemistry, Carnegie-Mellon University, Pittsburgh, Pennsylvania 15213*

SANFORD A. WOLGEL (16), *Department of Biochemistry, Medical School, University of Minnesota, Minneapolis, Minnesota 55455*

Preface

In the past decade there has been an explosion of interest in organisms that grow on one-carbon compounds and in those that grow on higher hydrocarbons. This is a result of the commercial interest in the unique enzymes involved in these specialty metabolic pathways and to a growing understanding of the important role these organisms play in carbon cycling in nature. More recent interest in the use of these organisms and their enzymes for detoxification of hazardous waste has once again put the spotlight on these unique microorganisms. It seems clear that the study of bacteria involved in the utilization of one-carbon compounds and hydrocarbons is only beginning to realize its potential.

As in all biological systems, the ultimate understanding and exploitation of properties are highly dependent on information concerning the enzymes involved. This volume is the first comprehensive compilation of methods for the assay and purification of the enzymes involved in the utilization of reduced one-carbon compounds and higher hydrocarbons. In many cases, improved assay and purification methods are presented, while in others purification is reported for the first time.

In organizing and editing this volume, I have attempted to produce a source that will have value for researchers in academia, government, and industry alike. The information presented should provide a quick and handy reference guide for anyone with interest in these enzymes.

MARY E. LIDSTROM

METHODS IN ENZYMOLOGY

VOLUME I. Preparation and Assay of Enzymes
Edited by SIDNEY P. COLOWICK AND NATHAN O. KAPLAN

VOLUME II. Preparation and Assay of Enzymes
Edited by SIDNEY P. COLOWICK AND NATHAN O. KAPLAN

VOLUME III. Preparation and Assay of Substrates
Edited by SIDNEY P. COLOWICK AND NATHAN O. KAPLAN

VOLUME IV. Special Techniques for the Enzymologist
Edited by SIDNEY P. COLOWICK AND NATHAN O. KAPLAN

VOLUME V. Preparation and Assay of Enzymes
Edited by SIDNEY P. COLOWICK AND NATHAN O. KAPLAN

VOLUME VI. Preparation and Assay of Enzymes (*Continued*)
  Preparation and Assay of Substrates
  Special Techniques
Edited by SIDNEY P. COLOWICK AND NATHAN O. KAPLAN

VOLUME VII. Cumulative Subject Index
Edited by SIDNEY P. COLOWICK AND NATHAN O. KAPLAN

VOLUME VIII. Complex Carbohydrates
Edited by ELIZABETH F. NEUFELD AND VICTOR GINSBURG

VOLUME IX. Carbohydrate Metabolism
Edited by WILLIS A. WOOD

VOLUME X. Oxidation and Phosphorylation
Edited by RONALD W. ESTABROOK AND MAYNARD E. PULLMAN

VOLUME XI. Enzyme Structure
Edited by C. H. W. HIRS

VOLUME XII. Nucleic Acids (Parts A and B)
Edited by LAWRENCE GROSSMAN AND KIVIE MOLDAVE

xvii

VOLUME XIII. Citric Acid Cycle
Edited by J. M. LOWENSTEIN

VOLUME XIV. Lipids
Edited by J. M. LOWENSTEIN

VOLUME XV. Steroids and Terpenoids
Edited by RAYMOND B. CLAYTON

VOLUME XVI. Fast Reactions
Edited by KENNETH KUSTIN

VOLUME XVII. Metabolism of Amino Acids and Amines (Parts A and B)
Edited by HERBERT TABOR AND CELIA WHITE TABOR

VOLUME XVIII. Vitamins and Coenzymes (Parts A, B, and C)
Edited by DONALD B. MCCORMICK AND LEMUEL D. WRIGHT

VOLUME XIX. Proteolytic Enzymes
Edited by GERTRUDE E. PERLMANN AND LASZLO LORAND

VOLUME XX. Nucleic Acids and Protein Synthesis (Part C)
Edited by KIVIE MOLDAVE AND LAWRENCE GROSSMAN

VOLUME XXI. Nucleic Acids (Part D)
Edited by LAWRENCE GROSSMAN AND KIVIE MOLDAVE

VOLUME XXII. Enzyme Purification and Related Techniques
Edited by WILLIAM B. JAKOBY

VOLUME XXIII. Photosynthesis (Part A)
Edited by ANTHONY SAN PIETRO

VOLUME XXIV. Photosynthesis and Nitrogen Fixation (Part B)
Edited by ANTHONY SAN PIETRO

VOLUME XXV. Enzyme Structure (Part B)
Edited by C. H. W. HIRS AND SERGE N. TIMASHEFF

VOLUME XXVI. Enzyme Structure (Part C)
Edited by C. H. W. HIRS AND SERGE N. TIMASHEFF

VOLUME XXVII. Enzyme Structure (Part D)
Edited by C. H. W. HIRS AND SERGE N. TIMASHEFF

VOLUME XXVIII. Complex Carbohydrates (Part B)
Edited by VICTOR GINSBURG

VOLUME XXIX. Nucleic Acids and Protein Synthesis (Part E)
Edited by LAWRENCE GROSSMAN AND KIVIE MOLDAVE

VOLUME XXX. Nucleic Acids and Protein Synthesis (Part F)
Edited by KIVIE MOLDAVE AND LAWRENCE GROSSMAN

VOLUME XXXI. Biomembranes (Part A)
Edited by SIDNEY FLEISCHER AND LESTER PACKER

VOLUME XXXII. Biomembranes (Part B)
Edited by SIDNEY FLEISCHER AND LESTER PACKER

VOLUME XXXIII. Cumulative Subject Index Volumes I–XXX
Edited by MARTHA G. DENNIS AND EDWARD A. DENNIS

VOLUME XXXIV. Affinity Techniques (Enzyme Purification: Part B)
Edited by WILLIAM B. JAKOBY AND MEIR WILCHEK

VOLUME XXXV. Lipids (Part B)
Edited by JOHN M. LOWENSTEIN

VOLUME XXXVI. Hormone Action (Part A: Steroid Hormones)
Edited by BERT W. O'MALLEY AND JOEL G. HARDMAN

VOLUME XXXVII. Hormone Action (Part B: Peptide Hormones)
Edited by BERT W. O'MALLEY AND JOEL G. HARDMAN

VOLUME XXXVIII. Hormone Action (Part C: Cyclic Nucleotides)
Edited by JOEL G. HARDMAN AND BERT W. O'MALLEY

VOLUME XXXIX. Hormone Action (Part D: Isolated Cells, Tissues, and Organ Systems)
Edited by JOEL G. HARDMAN AND BERT W. O'MALLEY

VOLUME XL. Hormone Action (Part E: Nuclear Structure and Function)
Edited by BERT W. O'MALLEY AND JOEL G. HARDMAN

VOLUME XLI. Carbohydrate Metabolism (Part B)
Edited by W. A. WOOD

VOLUME XLII. Carbohydrate Metabolism (Part C)
Edited by W. A. WOOD

VOLUME XLIII. Antibiotics
Edited by JOHN H. HASH

VOLUME XLIV. Immobilized Enzymes
Edited by KLAUS MOSBACH

VOLUME XLV. Proteolytic Enzymes (Part B)
Edited by LASZLO LORAND

VOLUME XLVI. Affinity Labeling
Edited by WILLIAM B. JAKOBY AND MEIR WILCHEK

VOLUME XLVII. Enzyme Structure (Part E)
Edited by C. H. W. HIRS AND SERGE N. TIMASHEFF

VOLUME XLVIII. Enzyme Structure (Part F)
Edited by C. H. W. HIRS AND SERGE N. TIMASHEFF

VOLUME XLIX. Enzyme Structure (Part G)
Edited by C. H. W. HIRS AND SERGE N. TIMASHEFF

VOLUME L. Complex Carbohydrates (Part C)
Edited by VICTOR GINSBURG

VOLUME LI. Purine and Pyrimidine Nucleotide Metabolism
Edited by PATRICIA A. HOFFEE AND MARY ELLEN JONES

VOLUME LII. Biomembranes (Part C: Biological Oxidations)
Edited by SIDNEY FLEISCHER AND LESTER PACKER

VOLUME LIII. Biomembranes (Part D: Biological Oxidations)
Edited by SIDNEY FLEISCHER AND LESTER PACKER

VOLUME LIV. Biomembranes (Part E: Biological Oxidations)
Edited by SIDNEY FLEISCHER AND LESTER PACKER

VOLUME LV. Biomembranes (Part F: Bioenergetics)
Edited by SIDNEY FLEISCHER AND LESTER PACKER

VOLUME LVI. Biomembranes (Part G: Bioenergetics)
Edited by SIDNEY FLEISCHER AND LESTER PACKER

VOLUME LVII. Bioluminescence and Chemiluminescence
Edited by MARLENE A. DELUCA

VOLUME LVIII. Cell Culture
Edited by WILLIAM B. JAKOBY AND IRA PASTAN

VOLUME LIX. Nucleic Acids and Protein Synthesis (Part G)
Edited by KIVIE MOLDAVE AND LAWRENCE GROSSMAN

VOLUME LX. Nucleic Acids and Protein Synthesis (Part H)
Edited by KIVIE MOLDAVE AND LAWRENCE GROSSMAN

VOLUME 61. Enzyme Structure (Part H)
Edited by C. H. W. HIRS AND SERGE N. TIMASHEFF

VOLUME 62. Vitamins and Coenzymes (Part D)
Edited by DONALD B. MCCORMICK AND LEMUEL D. WRIGHT

VOLUME 63. Enzyme Kinetics and Mechanism (Part A: Initial Rate and Inhibitor Methods)
Edited by DANIEL L. PURICH

VOLUME 64. Enzyme Kinetics and Mechanism (Part B: Isotopic Probes and Complex Enzyme Systems)
Edited by DANIEL L. PURICH

VOLUME 65. Nucleic Acids (Part I)
Edited by LAWRENCE GROSSMAN AND KIVIE MOLDAVE

VOLUME 66. Vitamins and Coenzymes (Part E)
Edited by DONALD B. MCCORMICK AND LEMUEL D. WRIGHT

VOLUME 67. Vitamins and Coenzymes (Part F)
Edited by DONALD B. MCCORMICK AND LEMUEL D. WRIGHT

VOLUME 68. Recombinant DNA
Edited by RAY WU

VOLUME 69. Photosynthesis and Nitrogen Fixation (Part C)
Edited by ANTHONY SAN PIETRO

VOLUME 70. Immunochemical Techniques (Part A)
Edited by HELEN VAN VUNAKIS AND JOHN J. LANGONE

VOLUME 71. Lipids (Part C)
Edited by JOHN M. LOWENSTEIN

VOLUME 72. Lipids (Part D)
Edited by JOHN M. LOWENSTEIN

VOLUME 73. Immunochemical Techniques (Part B)
Edited by JOHN J. LANGONE AND HELEN VAN VUNAKIS

VOLUME 74. Immunochemical Techniques (Part C)
Edited by JOHN J. LANGONE AND HELEN VAN VUNAKIS

VOLUME 75. Cumulative Subject Index Volumes XXXI, XXXII, and XXXIV–LX
Edited by EDWARD A. DENNIS AND MARTHA G. DENNIS

VOLUME 76. Hemoglobins
Edited by ERALDO ANTONINI, LUIGI ROSSI-BERNARDI, AND EMILIA CHIANCONE

VOLUME 77. Detoxication and Drug Metabolism
Edited by WILLIAM B. JAKOBY

VOLUME 78. Interferons (Part A)
Edited by SIDNEY PESTKA

VOLUME 79. Interferons (Part B)
Edited by SIDNEY PESTKA

VOLUME 80. Proteolytic Enzymes (Part C)
Edited by LASZLO LORAND

VOLUME 81. Biomembranes (Part H: Visual Pigments and Purple Membranes, I)
Edited by LESTER PACKER

VOLUME 82. Structural and Contractile Proteins (Part A: Extracellular Matrix)
Edited by LEON W. CUNNINGHAM AND DIXIE W. FREDERIKSEN

VOLUME 83. Complex Carbohydrates (Part D)
Edited by VICTOR GINSBURG

VOLUME 84. Immunochemical Techniques (Part D: Selected Immunoassays)
Edited by JOHN J. LANGONE AND HELEN VAN VUNAKIS

VOLUME 85. Structural and Contractile Proteins (Part B: The Contractile Apparatus and the Cytoskeleton)
Edited by DIXIE W. FREDERIKSEN AND LEON W. CUNNINGHAM

VOLUME 86. Prostaglandins and Arachidonate Metabolites
Edited by WILLIAM E. M. LANDS AND WILLIAM L. SMITH

VOLUME 87. Enzyme Kinetics and Mechanism (Part C: Intermediates, Stereochemistry, and Rate Studies)
Edited by DANIEL L. PURICH

VOLUME 88. Biomembranes (Part I: Visual Pigments and Purple Membranes, II)
Edited by LESTER PACKER

VOLUME 89. Carbohydrate Metabolism (Part D)
Edited by WILLIS A. WOOD

VOLUME 90. Carbohydrate Metabolism (Part E)
Edited by WILLIS A. WOOD

VOLUME 91. Enzyme Structure (Part I)
Edited by C. H. W. HIRS AND SERGE N. TIMASHEFF

VOLUME 92. Immunochemical Techniques (Part E: Monoclonal Antibodies and General Immunoassay Methods)
Edited by JOHN J. LANGONE AND HELEN VAN VUNAKIS

VOLUME 93. Immunochemical Techniques (Part F: Conventional Antibodies, Fc Receptors, and Cytotoxicity)
Edited by JOHN J. LANGONE AND HELEN VAN VUNAKIS

VOLUME 94. Polyamines
Edited by HERBERT TABOR AND CELIA WHITE TABOR

VOLUME 95. Cumulative Subject Index Volumes 61–74, 76–80
Edited by EDWARD A. DENNIS AND MARTHA G. DENNIS

VOLUME 96. Biomembranes [Part J: Membrane Biogenesis: Assembly and Targeting (General Methods; Eukaryotes)]
Edited by SIDNEY FLEISCHER AND BECCA FLEISCHER

VOLUME 97. Biomembranes [Part K: Membrane Biogenesis: Assembly and Targeting (Prokaryotes, Mitochondria, and Chloroplasts)]
Edited by SIDNEY FLEISCHER AND BECCA FLEISCHER

VOLUME 98. Biomembranes (Part L: Membrane Biogenesis: Processing and Recycling)
Edited by SIDNEY FLEISCHER AND BECCA FLEISCHER

VOLUME 99. Hormone Action (Part F: Protein Kinases)
Edited by JACKIE D. CORBIN AND JOEL G. HARDMAN

VOLUME 100. Recombinant DNA (Part B)
Edited by RAY WU, LAWRENCE GROSSMAN, AND KIVIE MOLDAVE

VOLUME 101. Recombinant DNA (Part C)
Edited by RAY WU, LAWRENCE GROSSMAN, AND KIVIE MOLDAVE

VOLUME 102. Hormone Action (Part G: Calmodulin and Calcium-Binding Proteins)
Edited by ANTHONY R. MEANS AND BERT W. O'MALLEY

VOLUME 103. Hormone Action (Part H: Neuroendocrine Peptides)
Edited by P. MICHAEL CONN

VOLUME 104. Enzyme Purification and Related Techniques (Part C)
Edited by WILLIAM B. JAKOBY

VOLUME 105. Oxygen Radicals in Biological Systems
Edited by LESTER PACKER

VOLUME 106. Posttranslational Modifications (Part A)
Edited by FINN WOLD AND KIVIE MOLDAVE

VOLUME 107. Posttranslational Modifications (Part B)
Edited by FINN WOLD AND KIVIE MOLDAVE

VOLUME 108. Immunochemical Techniques (Part G: Separation and Characterization of Lymphoid Cells)
Edited by GIOVANNI DI SABATO, JOHN J. LANGONE, AND HELEN VAN VUNAKIS

VOLUME 109. Hormone Action (Part I: Peptide Hormones)
Edited by LUTZ BIRNBAUMER AND BERT W. O'MALLEY

VOLUME 110. Steroids and Isoprenoids (Part A)
Edited by JOHN H. LAW AND HANS C. RILLING

VOLUME 111. Steroids and Isoprenoids (Part B)
Edited by JOHN H. LAW AND HANS C. RILLING

VOLUME 112. Drug and Enzyme Targeting (Part A)
Edited by KENNETH J. WIDDER AND RALPH GREEN

VOLUME 113. Glutamate, Glutamine, Glutathione, and Related Compounds
Edited by ALTON MEISTER

VOLUME 114. Diffraction Methods for Biological Macromolecules (Part A)
Edited by HAROLD W. WYCKOFF, C. H. W. HIRS, AND SERGE N. TIMASHEFF

VOLUME 115. Diffraction Methods for Biological Macromolecules (Part B)
Edited by HAROLD W. WYCKOFF, C. H. W. HIRS, AND SERGE N. TIMASHEFF

VOLUME 116. Immunochemical Techniques (Part H: Effectors and Mediators of Lymphoid Cell Functions)
Edited by GIOVANNI DI SABATO, JOHN J. LANGONE, AND HELEN VAN VUNAKIS

VOLUME 117. Enzyme Structure (Part J)
Edited by C. H. W. HIRS AND SERGE N. TIMASHEFF

VOLUME 118. Plant Molecular Biology
Edited by ARTHUR WEISSBACH AND HERBERT WEISSBACH

VOLUME 119. Interferons (Part C)
Edited by SIDNEY PESTKA

VOLUME 120. Cumulative Subject Index Volumes 81–94, 96–101

VOLUME 121. Immunochemical Techniques (Part I: Hybridoma Technology and Monoclonal Antibodies)
Edited by JOHN J. LANGONE AND HELEN VAN VUNAKIS

VOLUME 122. Vitamins and Coenzymes (Part G)
Edited by FRANK CHYTIL AND DONALD B. MCCORMICK

VOLUME 123. Vitamins and Coenzymes (Part H)
Edited by FRANK CHYTIL AND DONALD B. MCCORMICK

VOLUME 124. Hormone Action (Part J: Neuroendocrine Peptides)
Edited by P. MICHAEL CONN

VOLUME 125. Biomembranes (Part M: Transport in Bacteria, Mitochondria, and Chloroplasts: General Approaches and Transport Systems)
Edited by SIDNEY FLEISCHER AND BECCA FLEISCHER

VOLUME 126. Biomembranes (Part N: Transport in Bacteria, Mitochondria, and Chloroplasts: Protonmotive Force)
Edited by SIDNEY FLEISCHER AND BECCA FLEISCHER

VOLUME 127. Biomembranes (Part O: Protons and Water: Structure and Translocation)
Edited by LESTER PACKER

VOLUME 128. Plasma Lipoproteins (Part A: Preparation, Structure, and Molecular Biology)
Edited by JERE P. SEGREST AND JOHN J. ALBERS

VOLUME 129. Plasma Lipoproteins (Part B: Characterization, Cell Biology, and Metabolism)
Edited by JOHN J. ALBERS AND JERE P. SEGREST

VOLUME 130. Enzyme Structure (Part K)
Edited by C. H. W. HIRS AND SERGE N. TIMASHEFF

VOLUME 131. Enzyme Structure (Part L)
Edited by C. H. W. HIRS AND SERGE N. TIMASHEFF

VOLUME 132. Immunochemical Techniques (Part J: Phagocytosis and Cell-Mediated Cytotoxicity)
Edited by GIOVANNI DI SABATO AND JOHANNES EVERSE

VOLUME 133. Bioluminescence and Chemiluminescence (Part B)
Edited by MARLENE DELUCA AND WILLIAM D. MCELROY

VOLUME 134. Structural and Contractile Proteins (Part C: The Contractile Apparatus and the Cytoskeleton)
Edited by RICHARD B. VALLEE

VOLUME 135. Immobilized Enzymes and Cells (Part B)
Edited by KLAUS MOSBACH

VOLUME 136. Immobilized Enzymes and Cells (Part C)
Edited by KLAUS MOSBACH

VOLUME 137. Immobilized Enzymes and Cells (Part D)
Edited by KLAUS MOSBACH

VOLUME 138. Complex Carbohydrates (Part E)
Edited by VICTOR GINSBURG

VOLUME 139. Cellular Regulators (Part A: Calcium- and Calmodulin-Binding Proteins
Edited by ANTHONY R. MEANS AND P. MICHAEL CONN

VOLUME 140. Cumulative Subject Index Volumes 102–119, 121–134

VOLUME 141. Cellular Regulators (Part B: Calcium and Lipids)
Edited by P. MICHAEL CONN AND ANTHONY R. MEANS

VOLUME 142. Metabolism of Aromatic Amino Acids and Amines
Edited by SEYMOUR KAUFMAN

VOLUME 143. Sulfur and Sulfur Amino Acids
Edited by WILLIAM B. JAKOBY AND OWEN GRIFFITH

VOLUME 144. Structural and Contractile Proteins (Part D: Extracellular Matrix)
Edited by LEON W. CUNNINGHAM

VOLUME 145. Structural and Contractile Proteins (Part E: Extracellular Matrix)
Edited by LEON W. CUNNINGHAM

VOLUME 146. Peptide Growth Factors (Part A)
Edited by DAVID BARNES AND DAVID A. SIRBASKU

VOLUME 147. Peptide Growth Factors (Part B)
Edited by DAVID BARNES AND DAVID A. SIRBASKU

VOLUME 148. Plant Cell Membranes
Edited by LESTER PACKER AND ROLAND DOUCE

VOLUME 149. Drug and Enzyme Targeting (Part B)
Edited by RALPH GREEN AND KENNETH J. WIDDER

VOLUME 150. Immunochemical Techniques (Part K: *In Vitro* Models of B and T Cell Functions and Lymphoid Cell Receptors)
Edited by GIOVANNI DI SABATO

VOLUME 151. Molecular Genetics of Mammalian Cells
Edited by MICHAEL M. GOTTESMAN

VOLUME 152. Guide to Molecular Cloning Techniques
Edited by SHELBY L. BERGER AND ALAN R. KIMMEL

VOLUME 153. Recombinant DNA (Part D)
Edited by RAY WU AND LAWRENCE GROSSMAN

VOLUME 154. Recombinant DNA (Part E)
Edited by RAY WU AND LAWRENCE GROSSMAN

VOLUME 155. Recombinant DNA (Part F)
Edited by RAY WU

VOLUME 156. Biomembranes (Part P: ATP-Driven Pumps and Related Transport: The Na,K-Pump)
Edited by SIDNEY FLEISCHER AND BECCA FLEISCHER

VOLUME 157. Biomembranes (Part Q: ATP-Driven Pumps and Related Transport: Calcium, Proton, and Potassium Pumps)
Edited by SIDNEY FLEISCHER AND BECCA FLEISCHER

VOLUME 158. Metalloproteins (Part A)
Edited by JAMES F. RIORDAN AND BERT L. VALLEE

VOLUME 159. Initiation and Termination of Cyclic Nucleotide Action
Edited by JACKIE D. CORBIN AND ROGER A. JOHNSON

VOLUME 160. Biomass (Part A: Cellulose and Hemicellulose)
Edited by WILLIS A. WOOD AND SCOTT T. KELLOGG

VOLUME 161. Biomass (Part B: Lignin, Pectin, and Chitin)
Edited by WILLIS A. WOOD AND SCOTT T. KELLOGG

VOLUME 162. Immunochemical Techniques (Part L: Chemotaxis and Inflammation)
Edited by GIOVANNI DI SABATO

VOLUME 163. Immunochemical Techniques (Part M: Chemotaxis and Inflammation)
Edited by GIOVANNI DI SABATO

VOLUME 164. Ribosomes
Edited by HARRY F. NOLLER, JR., AND KIVIE MOLDAVE

VOLUME 165. Microbial Toxins: Tools for Enzymology
Edited by SIDNEY HARSHMAN

VOLUME 166. Branched-Chain Amino Acids
Edited by ROBERT HARRIS AND JOHN R. SOKATCH

VOLUME 167. Cyanobacteria
Edited by LESTER PACKER AND ALEXANDER N. GLAZER

VOLUME 168. Hormone Action (Part K: Neuroendocrine Peptides)
Edited by P. MICHAEL CONN

VOLUME 169. Platelets: Receptors, Adhesion, Secretion (Part A)
Edited by JACEK HAWIGER

VOLUME 170. Nucleosomes
Edited by PAUL M. WASSARMAN AND ROGER D. KORNBERG

VOLUME 171. Biomembranes (Part R: Transport Theory: Cells and Model Membranes)
Edited by SIDNEY FLEISCHER AND BECCA FLEISCHER

VOLUME 172. Biomembranes (Part S: Transport Membrane Isolation and Characterization)
Edited by SIDNEY FLEISCHER AND BECCA FLEISCHER

VOLUME 173. Biomembranes [Part T: Cellular and Subcellular Transport: Eukaryotic (Nonepithelial) Cells]
Edited by SIDNEY FLEISCHER AND BECCA FLEISCHER

VOLUME 174. Biomembranes [Part U: Cellular and Subcellular Transport: Eukaryotic (Nonepithelial) Cells]
Edited by SIDNEY FLEISCHER AND BECCA FLEISCHER

VOLUME 175. Cumulative Subject Index Volumes 135–139, 141–167 (in preparation)

VOLUME 176. Nuclear Magnetic Resonance (Part A: Spectral Techniques and Dynamics)
Edited by NORMAN J. OPPENHEIMER AND THOMAS L. JAMES

VOLUME 177. Nuclear Magnetic Resonance (Part B: Structure and Mechanism)
Edited by NORMAN J. OPPENHEIMER AND THOMAS L. JAMES

VOLUME 178. Antibodies, Antigens, and Molecular Mimicry
Edited by JOHN J. LANGONE

VOLUME 179. Complex Carbohydrates (Part F)
Edited by VICTOR GINSBURG

VOLUME 180. RNA Processing (Part A: General Methods)
Edited by JAMES E. DAHLBERG AND JOHN N. ABELSON

VOLUME 181. RNA Processing (Part B: Specific Methods)
Edited by JAMES E. DAHLBERG AND JOHN N. ABELSON

VOLUME 182. Guide to Protein Purification
Edited by MURRAY P. DEUTSCHER

VOLUME 183. Molecular Evolution: Computer Analysis of Protein and Nucleic Acid Sequences
Edited by RUSSELL F. DOOLITTLE

VOLUME 184. Avidin-Biotin Technology
Edited by MEIR WILCHEK AND EDWARD A. BAYER

VOLUME 185. Gene Expression Technology
Edited by DAVID V. GOEDDEL

VOLUME 186. Oxygen Radicals in Biological Systems (Part B: Oxygen Radicals and Antioxidents)
Edited by LESTER PACKER AND ALEXANDER N. GLAZER

VOLUME 187. Arachidonate Related Lipid Mediators
Edited by ROBERT C. MURPHY AND FRANK A. FITZPATRICK

VOLUME 188. Hydrocarbons and Methylotrophy
Edited by MARY E. LIDSTROM

VOLUME 189. Retinoids (Part A: Molecular and Metabolic Aspects) (in preparation)
Edited by LESTER PACKER

VOLUME 190. Retinods (Part B: Cell Differentiation and Clinical Applications) (in preparation)
Edited by LESTER PACKER

VOLUME 191. Biomembranes (Part V: Cellular and Subcellular Transport: Epithelial Cells) (in preparation)
Edited by SIDNEY FLEISCHER AND BECCA FLEISCHER

VOLUME 192. Biomembranes (Part W: Cellular and Subcellular Transport: Epithelial Cells) (in preparation)
Edited by SIDNEY FLEISCHER AND BECCA FLEISCHER

VOLUME 193. Mass Spectrometry (in preparation)
Edited by JAMES A. MCCLOSKEY

VOLUME 194. Guide to Yeast Genetics and Molecular Biology (in preparation)
Edited by CHRISTINE GUTHRIE AND GERALD R. FINK

VOLUME 195. Adenylyl Cyclase, G Proteins, and Guanylyl Cyclase (in preparation)
Edited by ROGER A. JOHNSON AND JACKIE D. CORBIN

VOLUME 196. Molecular Motors and the Cytoskeleton (in preparation)
Edited by RICHARD B. VALLEE

VOLUME 197. Phospholipases (in preparation)
Edited by EDWARD A. DENNIS

Section I

Hydrocarbons and Related Compounds

A. Enzymes of Alkane and Alkene Utilization
Articles 1 through 7

B. Enzymes Involved in Cyclic and Aromatic Hydrocarbon Utilization
Articles 8 through 26

C. Yeast and Fungal Enzymes of Hydrocarbon Degradation
Articles 27 through 29

[1] Hydrocarbon Monooxygenase System of
Pseudomonas oleovorans

By SHELDON W. MAY and ANDREAS G. KATOPODIS

The enzyme system from *Pseudomonas oleovorans* that catalyzes the hydroxylation of terminal methyl groups of alkanes and fatty acids was first shown by Coon and co-workers to consist of three protein components: rubredoxin; NADH: rubredoxin reductase; and a monooxygenase protein which they named "ω-hydroxylase."[1-7] Rubredoxin, a red iron-sulfur protein of molecular weight 19,000, which contains no labile sulfide and which exhibits an electron paramagnetic resonance (EPR) spectrum characteristic of high-spin ferric in a rhombic field, has been shown to function as an electron carrier in the system. The reductase is a 55,000 molecular weight flavoprotein, whereas the ω-hydroxylase has proved to be relatively insoluble and unstable, easily aggregated to a molecular weight of 2×10^6, and difficult to purify. It has been characterized as a nonheme iron protein, with one iron atom and one cysteine per 40,800 molecular weight polypeptide chain, containing a high concentration of phospholipid plus carbohydrate.

Work in this laboratory has established that the *P. oleovorans* hydrocarbon monooxygenase enzyme system very readily carries out stereospecific epoxidation of simple, aliphatic terminal olefins.[8-11] We examined the specificity of this reaction and carried out immobilization, metal substitu-

[1] E. J. McKenna and J. J. Coon, *J. Biol. Chem.* **245**, 3882 (1971).

[2] E. T. Lode and J. J. Coon, *J. Biol. Chem.* **246**, 791, (1971).

[3] T. Ueda, E. T. Lode, and J. J. Coon, *J. Biol. Chem.* **247**, 2109 (1972).

[4] A. Benson, K. Tomoda, J. Chang, G. Matsueda, E. T. Lode, M. J. Coon, and K. T. Yasunobu, *Biochem. Biophys. Res. Commun.* **42**, 640 (1971).

[5] T. Ueda and M. J. Coon, *J. Biol. Chem.* **247**, 5010 (1972).

[6] R. T. Ruettinger, G. R. Griffith, and M. J. Coon, *Biochem. Biophys. Res. Commun.* **57**, 1011 (1974).

[7] R. T. Ruettinger, G. R. Griffith, and M. J. Coon, *Arch. Biochem. Biophys.* **183**, 528 (1977).

[8] S. W. May and B. J. Abbott, *Biochem. Biophys. Res. Commun.* **48**, 1230 (1972).

[9] S. W. May and B. J. Abbott, *J. Biol. Chem.* **248**, 1725 (1973).

[10] S. W. May and R. D. Schwartz, *J. Am. Chem. Soc.* **96**, 4031 (1974).

[11] S. W. May, M. S. Steltenkamp, R. D. Schwartz, and C. J. McCoy, *J. Am. Chem. Soc.* **98**, 7856 (1976).

tion, chemical modification, and mechanistic studies.[12-19] More recently, we demonstrated that the catalytic competence of this nonheme iron monooxygenase system, the biological role of which is apparently terminal methyl hydroxylation, extends to additional chemical transformations:[19-21] (1) oxygenative O-demethylation of ethers; (2) oxygenative formation of aldehydes from terminal alcohols; and (3) stereoselective sulfoxidation of methyl thioether substrates.

Assay of Oxygenation Activity

Pseudomonas oleovorans monooxygenase is assayed by measuring the amount of 1,7-octadiene converted to 1,2-epoxy-7-octene under standard conditions. The assay mixture contains 50 μg rubredoxin, 30 μg *P. oleovorans* reductase, 0.5 mg NADH, and various amounts of the monooxygenase in a total volume of 1.2 ml buffer (50 mM Tris-Cl, 0.1% deoxycholate, pH 7.4). The bacterial reductase may be substituted with 20 μg of spinach ferredoxin reductase, which is commercially available, in which case NADPH must be used as the reductant. In cases where extended incubations are required, a NADPH recycling system may be added; this consists of 1 unit glucose-6-phosphate dehydrogenase and 2 mg glucose 6-phosphate. The assay is initiated by the addition of 20 μl 1,7-octadiene solution (0.25 ml in 5 ml acetone), and the mixture is incubated at 30° for 10 min at 300 rpm. At the end of this period a 1.0-ml aliquot is extracted with 0.200 ml hexane (10 μl 2-octanol in 50 ml hexane is used as an internal standard). The hexane layer is analyzed by gas chromatography (GC) using a 30-m bonded 5% phenyl methyl silicon capillary column, and peaks are quantitated relative to the internal standard. A similar protocol can be used for other monooxygenase substrates; sulfoxide products are extracted with chloroform instead of hexane before injection in the gas chromatograph.

[12] S. W. May, *Catal. Org. Synth. 4th, 1976,* 101 (1977).
[13] S. W. May, R. D. Schwartz, B. J. Abbott, and O. R. Zaborsky, *Biochim. Biophys. Acta* **403,** 245 (1975).
[14] S. W. May, M. S. Steltenkamp, K. R. Borah, A. G. Katopodis, and J. R. Thowsen, *J. Chem. Soc., Chem. Commun.,* 845 (1979).
[15] S. W. May and J. Y. Kuo, *J. Biol. Chem.* **252,** 2390 (1977).
[16] S. W. May and J. Y. Kuo, *Biochemistry* **17,** 3333 (1978).
[17] S. W. May, L. G. Lee, A. G. Katopodis, J. Y. Kuo, K. Wimalasena, and J. R. Thowsen, and *Biochemistry* **23,** 2187 (1984).
[18] S. W. May, S. L. Gordon, and M. S. Steltenkamp, *J. Am. Chem. Soc.* **99,** 2017 (1977).
[19] A. G. Katopodis, K. Wimalasena, J. Lee, and S. W. May, *J. Am. Chem. Soc.* **106,** 7928 (1984).
[20] S. W. May and A. G. Katopodis, *Enzyme Microb. Technol.* **8,** 17 (1986).
[21] A. G. Katopodis, H. A. Smith, and S. W. May, *J. Am. Chem. Soc.* **110,** 897 (1988).

Growth and Maintenance of *Pseudomonas oleovorans*

Pseudomonas oleovorans is grown under aseptic conditions in a 300-ml flask containing 100 ml of autoclaved P-1 medium[18] incubated at 28° with shaking at 400 rpm. Inoculation is performed by washing a slant of TF4-1L strain with P-1 medium and adding it to the flask. The culture is incubated for 26 hr. Octane (1 ml) is added at 12, 16, 20, and 24 hr, and the pH is adjusted to 7.3 by addition of the appropriate amount of 5 N NaOH at every feeding. At the end of the incubation the OD_{660} is usually about 30. A 50-ml portion of the culture is then used to inoculate 1 liter of P-1 medium containing 10 ml octane. The second flask is incubated under the same conditions for 30 hr with feedings of 10 ml octane and pH adjustment at similar intervals. The OD_{660} of the resulting culture is usually 25. The culture is then centrifuged, resuspended in 50 ml sterile P-1 medium containing 10% glycerol, and then stored in 1.0-ml sealed ampoules at $-70°$.

For large-scale growth, 1 liter of P-1 medium containing 10 ml octane is inoculated with a 1.0-ml vial of stored *P. oleovorans* and incubated as indicated above for 48 hr. The culture is fed with 10 ml octane at 12, 24, 34, 44, and 48 hr, and the pH is adjusted to 7.3 at every feeding. It is possible at this point to recognize when the cells need feeding since considerable foaming occurs when growth substrate is depleted; this foaming stops on addition to octane. At the end of 48 hr the master flask is used to inoculate a large-scale growth, and the leftover portion may be kept for up to 6 days at 4° for further inoculations. A fermentor charged with 10 liters of P-1 medium and 150 ml of octane is inoculated with 330 ml of cells, from either a fresh master flask or a chilled one which has been warmed up in the incubator for 5 hr. The fermentor temperature is set at 28°, the air supply at 0.7 liters/min, and agitation at 800 rpm. It is critical that the temperature not exceed 30° during fermentation, since above this temperature the culture turns yellowish and all enzymatic activities of interest are markedly diminished. The dissolved oxygen concentration in the fermentation solution rises sharply upon depletion of octane, at which point 50 ml octane is added and the pH adjusted to 7.3. Cell growth is monitored by measuring the increase in OD_{660}, and the air supply is steadily increased as long as the dissolved oxygen level stays below 10%. The cells are harvested by centrifugation at 16 hr or when the OD_{660} is approximately 35 and stored as a wet paste at $-70°$.

Isolation of Rubredoxin

The isolation described here is a modification of a previously published procedure[2] and affords much higher yields of rubredoxin in a shorter time. The entire procedure is carried out at 4°. Typically, 300 g of *P. oleovorans* wet paste is thawed overnight and then suspended in 1.5 liters of 20 mM Tris-Cl buffer, pH 7.4. The pH of the suspension is adjusted to pH 7.4 by the addition of 1.0 M Tris base, and the suspension is then stirred for 2 hr. The suspension is centrifuged at 20,000 g for 30 min. The pellet may be frozen and utilized later for the isolation of the hydroxylase. The pH of the supernatant is again adjusted to 7.4, and the solution is made 0.2% in streptomycin sulfate by the dropwise addition of 5% (w/v) buffered streptomycin sulfate solution. The mixture is stirred for 30 min and centrifuged at 20,000 g for 30 min.

The white pellet, containing nucleic acids, is discarded, and the supernatant is applied to a DEAE-cellulose column (4 × 45 cm) already equilibrated with 20 mM Tris-Cl buffer, pH 7.4. Loading is done at a flow rate of 150 ml/hr, and the whole sample is loaded in approximately 11 hr. The column is then washed with 50 ml of 20 mM Tris-Cl, pH 7.4, containing 70 mM KCl at a flow rate of 60 ml/hr. A linear gradient of the same buffer containing from 70 to 500 mM KCl, total volume 1 liter, is then applied at a flow rate of 50 ml/hr, and fractions of 15 ml each are collected. Rubredoxin elutes at about 300 mM KCl and is pooled by color, including in the pool all the tubes with reddish color. The rubredoxin pool is concentrated to about one-fourth its volume, diluted back up to volume with 20 mM Tris-Cl, pH 7.4, and then concentrated to about 30 ml. It is then applied to a second DEAE-cellulose column (2.5 × 30 cm) equilibrated with 20 mM Tris-Cl, pH 7.4. A linear gradient of 200 to 500 mM KCl is then applied at a flow rate of 30 ml/hr, and fractions of 15 ml are collected. Fractions containing rubredoxin with $A_{280}A_{497}$ ratios less than 10.0 are pooled, concentrated to approximately 5 ml, and applied to a Sephadex G-75 column (2.5 × 120 cm), equilibrated with 20 mM Tris-Cl, pH 7.4, 100 mM KCl buffer, and eluted at a flow rate of 15 ml/hr. The rubredoxin pool from this column typically has an A_{280}/A_{497} ratio of 6.9, and this material is concentrated and stored at $-70°$.

Isolation of Reductase

Typically, the *P. oleovorans* cells in 100 g of wet paste are lysed in 500 ml of 20 mM potassium phosphate (KP$_i$), 10% glycerol, pH 7.4, buffer, and the nucleic acids are precipitated as described above for the rubredoxin isolation. The lysate is then applied on the DEAE-cellulose column (2.5 × 30 cm), equilibrated with the lysis buffer. The column is then washed with 50 ml of the same buffer, and a gradient to 100 mM KCl is applied (total volume 500 ml) at a flow rate of 25 ml/hr, with 15-ml fractions being collected. Fractions with an A_{280}/A_{450} ratio of 16 or less are combined. The reductase pool is concentrated to about 5 ml and applied on a Sephadex G-75 column (2.5 × 120 cm) equilibrated with 20 mM

KP_i, 10% (v/v) glycerol, 50 mM KCl buffer, pH 7.4, and eluted at a flow rate of 15 ml/hr using the same buffer. Fractions of 5 ml are collected and assayed using the ferricyanide reduction method.[3] Fractions with more than 22 units of activity per 10 μl of sample are concentrated and applied on a blue agarose column (1.5 × 6 cm) equilibrated with 100 mM KP_i, 20% glycerol, pH 7.6, buffer. The column is washed with 100 ml of the same buffer, and the reductase is eluted with 500 mM KCl buffer at a flow rate of 8 ml/hr. The material is then pooled, concentrated, assayed, and stored at −70° (2.5 mg of protein/ml). The stored material has a specific activity of 355 μmol of ferricyanide reduced/mg/min.

Isolation of *Pseudomonas oleovorans* Monooxygenase

Typically, 100 g of cell membranes, obtained after the lysis step in the isolation of the soluble enzymes, is suspended in 500 ml of 20 mM Tris-Cl buffer, pH 7.4, and stirred for 1 hr at 4° The solution is centrifuged (20,000 g for 30 min at 4°), and the pellet is suspended in 300 ml of the same buffer containing 0.2% deoxycholate and stirred in the cold for 30 min. The suspension is split into three equal volumes and sonicated (instrument rated for 200 W set at 75% output) for a total of 3 min each in 1-min blasts. Care is taken never to allow the temperature to exceed 10°. The sonicated solution is diluted with 200 ml of the same buffer and is stirred for 20 min. After centrifugation (20,000 g for 1 hr at 4°) the supernatant is subjected to ammonium sulfate precipitation. The 30–40% pellet is kept and resuspended in 20 ml buffer (100 mM Tris-Cl, 20% glycerol, 0.1% deoxycholate, pH 7.4). The solution is then centrifuged (100,000 g for 30 min at 4°), and the supernatant is stored frozen in vials of 2.5 ml each or subjected to the next purification step.

The enzyme solution is then applied on a BioGel A-15 column (2.5 × 40 cm) equilibrated with 100 mM Tris-Cl, 20% glycerol, 0.1% deoxycholate buffer, pH 7.4, and eluted at a flow rate of 10 ml/hr. Fractions of 3.5 ml each are assayed for hydroxylase activity, pooled, concentrated using an Amicon YM30 membrane, and stored at −70°. Enzymatic activity drops about 30% after one freeze/thaw cycle and more sharply after subsequent cycles. For maximal activity it is best to freeze the enzyme after the ammonium sulfate step and use the hydroxylase immediately after the gel-filtration column.

Cell-Free Production of Oxygenated Products

Milligram quantities of hydroxylase products may be obtained by using sonicated cell extracts. This procedure is useful for the production of nonmetabolizable products such as epoxides and methyl sulfoxides. Control experiments in our laboratory with 1,2-epoxyoctane and methyl *n*-octyl sulfoxide indicated that both these products readily accumulate in the presence of crude cell extracts, and their optical purities remain unchanged even after long incubation times. Typically, 50 g of *P. oleovorans* cells is suspended in 200 ml of cold buffer (100 mM Tris-Cl, pH 7.4) and sonicated for 5 min in 1-min blasts. The solution is diluted with 500 ml of warm buffer, the pH is adjusted to 7.4, and 10 mg of NADH is added. The mixture is shaken at 300 rpm in an incubator at 30°, and the reaction is initiated by the addition of substrate solution (0.5 g of substrate in 10 ml acetone). Product formation is linear for several hours, and the mixture may be incubated overnight. The solution is then saturated with NaCl and extracted twice with 200 ml chloroform. After drying and concentration of the organic layer, products may be isolated by flash chromatography.[21]

Properties

Pseudomonas rubredoxin differs from those of the anaerobes in that it contains two metal-binding sites, and experiments with the cyanogen bromide-cleaved enzyme support the conclusion that each binding site is composed of cysteine residues from only one end of the molecule.[2] Iron bound in the N-terminal site is exceedingly labile, and rubredoxin is always isolated as the species containing only one iron which is bound in the C-terminal site. The sequence data established that a cluster of five cysteines exists near each terminus of rubredoxin,[4] and since the Fe atoms are clearly tetrahedrally sulfur-ligated, a single uncoordinated sulfhydryl group is presumably available near each metal-binding site or, alternatively, a diterminal disulfide exists. "Extra sulfhydryls" are not present in rubredoxins from anaerobic bacteria such as *Megasphaera elsdenii*, *Desulfovibrio gigas*, *Clostridium pasteuranium*, *Micrococcus aerogenes*, and *Desulfovibrio vulgaris*. It is striking that although anaerobic rubredoxins do accept electrons from the *P. oleovorans* reductase and have metal-binding sites and redox potentials analogous to those of *P. oleovorans* rubredoxin, they are still unable to support monooxygenation by the *P. oleovorans* system.

In order to ascertain whether the extra sulfhydryls of *P. oleovorans* rubredoxin are essential to monooxygenation activity, we carried out selective chemical modification experiments both with intact rubredoxin and with its C-terminal CNBr fragment.[16,17] First, we prepared and fully characterized cobalt rubredoxin, in which both metal-binding sites are occupied by cobalt atoms, each of which is bound in the normal rubredoxin-type ligation environment. The relative stability of cobalt binding allowed selective chemical modification with iodoacetamide under conditions which leave the ligation environment of cobalt intact. We were then able to remove the cobalt and restore iron, yielding rubredoxin with only the extra sulfhydryls blocked. In the case of the rubredoxin C-terminal peptide, the single extra sulfhydryl was modified with methylmethane thiosulfonate. Activity assays demonstrated that although chemical modification reduced epoxidative activity, both modified rubredoxin and modified C-peptide were still active, with the less bulky modification agent causing a significantly smaller activity loss. We therefore conclude that the extra sulfhydryls are not essential for the activity of *P. oleovorans* rubredoxin in supporting monooxygenation. We have also found that rubredoxin immobilized on CNBr-activated Sepharose exhibits normal spectral properties and redox potential and is active as an electron-transfer protein.[15]

Work in our laboratory has established that the *P. oleovorans* ω-hydroxylation" enzyme system readily carries out a number of oxygenation reactions quite distinct from hydroxylation at terminal methyl; among these oxygenation activities are stereospecific epoxidation of terminal olefins, sulfoxidation of methyl sulfides, and O-demethylation of methyl ethers.[19-21] We have demonstrated that these oxygenation activities exhibit regio- and stereospecificities which are highly intriguing from a biotechnological perspective.[20,21] We have also provided evidence for a chemical mechanism for this enzyme system.[18,19,22] Studies with *cis*- and *trans*-1-deutero-1-octene established 70% inversion of olefin configuration and no loss of deuterium during epoxidation. To explain these results and a preference for the production of 90% of the (*R*)-epoxide isomer of 1,7-octadiene, we have proposed a mechanism in which the oxygen of the putative active site iron-oxo species of the enzyme initially attacks the terminal carbon of the olefin substrate. Initial oxygen attack at the terminal carbon is supported by the exclusive formation of terminal alcohols and terminal epoxides by the monooxygenase system. Furthermore, the O-demethylation activity of the monooxygenase system and the observation of aldehyde products during epoxidation reactions indicate that the initial attack of oxygen occurs at the terminal carbon of the olefin. Closure to the epoxide occurs from the *si* side of C-2 of the olefin, giving rise to the 90% (*R*)-epoxide. The 70% inversion arises from preferential initial attack on the *re* face of the olefin followed by rotation before closure to the epoxide.

[22] J. E. Colbert, A. G. Katopodis, and S. W. May, *J. Am. Chem. Soc.* **112**, 3993 (1990).

[2] Assay Methods for Long-Chain Alkane Oxidation in *Acinetobacter*

By W. R. Finnerty

The oxidation of alkanes by microorganisms is an oxygen-dependent reaction catalyzed by monooxygenases and dioxygenases. An enzyme system consisting of rubredoxin, NADH-rubredoxin reductase, and ω-hydroxylase catalyzes short-chain alkane (C_6-C_8) and fatty acid hydroxylation in *Pseudomonas putida*.[1-5] This chapter describes methods to measure the oxidation of long-chain alkanes ($C_{12}-C_{18}$) in whole cells and cell-free preparations of *Acinetobacter* HO1-N.

Pathway of Alkane Oxidation in *Acinetobacter* HO1-N

$$RCH_3 + O_2 \rightarrow RCH_2OOH \rightarrow RCH_2OH \rightleftharpoons RCHO \rightleftharpoons RCOOH$$

Assay Method

Principle. The assay is based on the oxidation of [^{14}C]alkane to the stable oxygenated products ^{14}C-labeled fatty alcohol, ^{14}C-labeled fatty aldehyde, and ^{14}C-labeled fatty acid of the same carbon number as the alkane.

Reagents

Tris-Cl buffer, 50 mM, pH 7.8
Basal salts medium E containing (in g/liter): K_2HPO_4, 10; NaH_2PO_4, 5; $(NH_4)_2SO_4$, 2; $MgSO_4 \cdot 7H_2O$, 0.2; $CaCl_2 \cdot 2H_2O$, 0.001; $FeSO_4 \cdot 7H_2O$, 0.001; pH 7.8
$FeSO_4 \cdot 7H_2O$, 10 μM, freshly prepared
NAD and NADP, 50 mM
NADH and NADPH, 50 mM
Substrates. [1-^{14}C]Hexadecane (New England Nuclear Corporation, Boston, MA) is purified by silicic acid chromatography prior to use and diluted with carrier hexadecane to a specific activity of 25 μCi/3.42 mmol.

[1] J. A. Peterson, D. Basu, and M. J. Coon, *J. Biol. Chem.* **241**, 5162 (1966).
[2] J. A. Peterson, M. Kusunose, E. Kusunose, and M. J. Coon, *J. Biol. Chem.* **242**, 4334 (1967).
[3] J. A. Peterson and M. J. Coon, *J. Biol. Chem.* **243**, 329 (1968).
[4] E. J. McKenna and M. J. Coon, *J. Biol. Chem.* **245**, 3882 (1970).
[5] T. Ueda, E. T. Lode, and M. J. Coon, *J. Biol. Chem.* **247**, 2109 (1972).

Procedure

Whole cell assay. The reaction mixture contains 8.8 ml of basal salts medium E and 1 ml of early exponential phase hexadecane-grown cells (10 mg dry cell weight/ml). The cell suspension is washed once with basal salts medium E, adjusted to the proper cell density, and incubated at 28° with shaking at 250 rpm for 15 min. The reaction is initiated by the addition of 0.2 ml [1-^{14}C]hexadecane, and 50-μl samples are taken at 5-min intervals for 30–60 min. The samples are spotted directly onto an activated 0.4-mm-thick silica gel G thin-layer plate, air-dried, and developed twice in petroleum ether. The origin is collected in scintillation vials, and the total radioactivity is quantified by liquid scintillation spectrometry. Sequential development of the origin allows for the determination of radioactivity in the cellular lipid and protein. Neutral lipids and phospholipids are eluted in petroleum ether–diethyl ether–glacial acetic acid (70:30:1, v/v/v) and chloroform–methanol–5.0 M NH$_4$OH (65:30:5, v/v/v), respectively, and the radioactivity remaining at the origin is determined after each elution.

Alternatively, the sample is extracted with 3 ml of chloroform–methanol (1:2, v/v) and back-extracted with 3 ml of 2 M KCl, and 20 μl of the chloroform phase is spotted on silica gel G thin-layer plates and developed twice in petroleum ether. The origin is collected, and the radioactivity is quantified for the determination of total ^{14}C-labeled cellular lipid. Sequential development of the origin allows for the determination of radioactivity in the cellular neutral lipids and phospholipids. Neutral lipids and phospholipids are eluted from the origin in petroleum ether–diethyl ether–glacial acetic acid (70:30:1, v/v/v) and chloroform–methanol–5.0 M NH$_4$OH (65:30:5, v/v/v), respectively. Chloroform-soluble lipids are saponified in sealed tubes containing 0.2 N KOH in 80% methanol at 120° for 30 min and recovered in petroleum ether after acidification with 4 N HCl.

Cell-free assay. The reaction mixture contains 2 ml of 50 mM Tris-Cl buffer, pH 7.8, 10 μl of NAD and NADP, 10 μl of NADH and NADPH, 50 μl of FeSO$_4$·7H$_2$O, 0.1 ml of soluble enzyme fraction, and 0.1 ml of the membrane fraction. The reaction is initiated by the addition of 0.1 ml of [1-^{14}C]hexadecane. The reaction is incubated at 30° with shaking at 200 rpm, and 10-μl samples are taken at 5-min intervals for 30 min. The samples are spotted onto silica gel G thin-layer plates and developed twice in petroleum ether. The origin is collected, and the radioactivity is determined by liquid scintillation spectrometry. Sequential development of the origin (see above) allows for the determination of radioactivity in specific lipids.

Definition of Specific Activity. Specific activity is defined as total counts per minute (cpm) in radioactive lipid per milligram protein for cell-free assays and counts per minute in radioactive lipid or protein per total dry cell weight for whole cell assays. The protein content for derived cell fractions is determined by the method of Lowry *et al.*[6]

Preparation of Cell Extracts and Cell Fractionation. Cells are preinduced for growth on hexadecane as the sole carbon and energy source and subcultured 4 times at 6-hr intervals. Cells are harvested in the early exponential growth phase and used immediately. Cell-free extracts are prepared from 2 liters of early exponential phase cultures, washed once in 50 mM Tris-Cl buffer, pH 7.8, and suspended in the same buffer at a consistency of 1 g wet cell weight/ml. This cell suspension is broken by one passage through a well-chilled French pressure cell at 10,000 lb/square inch. The cell lysate is diluted 2-fold with cold 50 mM Tris-Cl buffer, pH 7.8, and centrifuged at 30,000 g at 5° for 15 min. The resulting supernatant fluid is centrifuged at 225,000 g at 5° for 60 min, yielding a supernatant fluid defined as the soluble enzyme fraction and a pellet. The pellet is suspended in 50 mM Tris-Cl buffer, pH 7.8, layered onto 2 ml of 80% sucrose, and centrifuged at 225,000 g at 5° for 30 min. The final membrane fraction is collected from the top of the sucrose cushion and used immediately. Such enzyme preparations are used within 30 min of final preparation because inactivation is rapid ($<$ 1 hr). Cell-free preparations contain less than 5 colony-forming units (cfu)/ml.

Properties: Whole Cells

Whole cell preparations prepared properly will convert radiolabeled long-chain alkanes into characteristic cellular lipid and protein at greater than 60% efficiency. An advantage in using hexadecane-grown cells is that greater than 98% of all cellular fatty acid is palmitate or palmitoleate.[7] The time course of alkane carbon distribution into cellular components is as follows: neutral lipid, 5 min; phospholipid, 5 – 10 min; and protein, 10 – 15 min. Cetyl palmitate and phosphatidylethanolamine represents the highest specific activity lipids present in the neutral lipids and phospholipids, respectively. The steady-state distribution of hexadecane carbon to cellular components is 20% lipid and 70% protein. The total incorporation of alkane carbon is highly dependent on the efficiency of the preincubation of whole cells, which reduces or eliminates the intracellular pools of hexade-

[6] O. H. Lowry, N. J. Rosebrough, A. L. Farr, and R. J. Randall, *J. Biol. Chem.* **193**, 265 (1951).
[7] R. A. Makula and W. R. Finnerty, *J. Bacteriol.* **95**, 2102 (1968).

TABLE I
CELL-FREE REACTION CONDITIONS

Conditions	Product formation (cpm/mg protein)
Complete system	275,000
Minus soluble enzyme fraction	0
Minus membrane fraction	0
Minus NAD^+ and $NADP^+$	0
Minus NADH and NADPH	0
Minus $FeSO_4 \cdot 7H_2O$	0
Minus oxygen (N_2 atmosphere)	0

cane characteristic of *Acinetobacter* HO1-N.[8] Optimal hexadecane oxidation occurs with early exponential phase cells, with a 70–80% loss of activity in the late exponential or stationary growth phase.

Stability. The whole cell preparation is stable for up to 4 hr.

pH Range. Hexadecane oxidation by whole cells occurs over a broad pH range of 7–10, with optimal activity between pH 7.8 and 8.0.

Substrate Specificity. Alkane oxidation by whole cells is optimal in homologous alkane-induced, alkane-substrate systems. Hexadecane-induced cells oxidize [^{14}C]tetradecane and [^{14}C]octadacane 20% less efficiently and do not oxidize [^{14}C]hexane, [^{14}C]heptane, or [^{14}C]octane.

Properties: Cell-Free Extracts

The cell-free oxidation of [1-^{14}C]hexadecane to stable ^{14}C-labeled lipid is less than 1%, with radioactivity present in fatty acid, fatty aldehyde, fatty alcohol, and wax ester. A significant percentage of total chloroform-soluble radioactivity is localized in cetyl palmitate, an end product of alkane metabolism by *Acinetobacter* HO1-N. Addition of homologous and heterologous rubredoxin or flavodoxin did not stimulate product formation. Optimal reaction conditions are shown in Table I. Representative incorporation of [^{14}C]hexadecane to cell components is shown in Table II.

Stability. The soluble enzyme fraction and the membrane fraction are stable for less than 1 hr. Sulfhydryl reagents such 2-mercaptoethanol, cysteine, dithiothreitol, or glutathione are ineffective as stabilizers, as are 4 M ammonium sulfate suspensions, 0.5 M glycerol, or storage at 5°, −20°, or −70°.

[8] C. C. L. Scott and W. R. Finnerty, *J. Bacteriol.* **127**, 481 (1976).

TABLE II

INCORPORATION OF [^{14}C] HEXADECANE INTO CELLULAR COMPONENTS

Components	Whole Cells (total cpm/60 min)		Cell-free system (total cpm/mg protein/30 min)	
	Unsaponified	Saponified	Unsaponified	Saponified
Cell system	7,260,000	—	275,000	—
Neutral lipid	1,070,000	1,701,000	235,000	210,000
Fatty alcohol	224,000	539,000	85,000	81,000
Fatty aldehyde	12,000	5,000	10,000	8,700
Fatty acid	353,000	1,102,000	57,000	91,000
Wax ester	471,000	—	83,000	—
Phospholipid	726,000	—	—	—
Protein	4,936,800	—	—	—

Optimum. The pH optimum is 7.8–8.5, with 90% loss of enzyme activity below pH 7.5 or above pH 8.5.

Inhibitors. The reaction is inhibited 100% by 10 mM p-chloromercuribenzoate or the absence of molecular oxygen.

Substrate Specificity. The enzyme system oxidizes [^{14}C]tetradecane and [^{14}C]octadecane but not [^{14}C]hexane, [^{14}C]heptane, or [^{14}C]octane.

[3] Primary Alcohol Dehydrogenases from *Acinetobacter*

By W. R. FINNERTY

Acinetobacter species oxidize short- and long-chain primary alcohols and aromatic alcohols by soluble and membrane-bound enzyme systems to the corresponding carboxylic acids.[1–5] This chapter describes the properties of NAD$^+$- and NADP$^+$-dependent soluble and membrane-bound primary alcohol dehydrogenases in *Acinetobacter* HO1-N.

[1] L. M. Fixter and M. N. Nagi, *FEMS Microbiol. Lett.* **22**, 297 (1984).
[2] M. Beardmore-Gray and C. Anthony, *J. Gen. Microbiol.* **129**, 2979 (1983).
[3] H. Tauchert, M. Grunow, H. Harnisch, and H. Aurich, *Acta Biol. Med. Ger.* **35**, 1267 (1976).
[4] H. Tauchert, M. Roy, W. Schopp, and H. Aurich, *Z. Allg. Mikrobiol.* **15**, 457 (1975).
[5] M. E. Singer and W. R. Finnerty, *J. Bacteriol.* **164**, 1017 (1985).

Soluble NAD⁺- and NADP⁺-Dependent Primary Alcohol
Dehydrogenases

$$RCH_2-OH + NAD^+ \rightleftharpoons RCHO + NADH + H^+$$

Assay Method

Principle. The assay is based on the rate of reduction of NAD⁺ and
NADP⁺ at 340 nm in the presence of an alcohol.

Reagents

Pyrophosphate–phosphate buffer, 20 mM, pH 9.0
Tris-Cl buffer, 10 mM, pH 7.8
NAD⁺ or NADP⁺, 2.5 mM

Substrates. Primary alcohols are added at the following final concentrations: ethanol, 0.52 M; n-propanol and n-butanol, 25 mM, n-pentanol through n-decanol, 0.5 mM in 2.5% dimethylformamide; n-dodecanol through n-hexadecanol, 50 μM in 2.5% dimethylformamide. Alcohol dehydrogenases are inhibited less than 1% by 2.5% dimethylformamide.

Procedure. The reaction mixture contains 0.2 ml of 20 mM pyrophosphate–phosphate buffer, pH 9.0; 0.1 ml of enzyme suitably diluted in 10 mM Tris-Cl buffer, pH 7.8; 10 μl of NAD⁺ or NADP⁺; and 0.50 ml of 10 mM Tris-Cl buffer, pH 7.8. The reaction is initiated by the addition of 0.2 ml of alcohol substrate to yield the final alcohol concentration. Alcohol dehydrogenase activities are determined by recording the rate of NAD⁺ or NADP⁺ reduction for 5 min at 340 nm using a Cary 219 spectrophotometer. Controls are run to determine the rate of NADH or NADPH oxidation.

Definition of Specific Activity. Specific activity is expressed in terms of nanomoles of NAD⁺ or NADP⁺ reduced per minute per milligram protein. The protein content of all cell fractions is determined by the method of Lowry *et al.*[6]

Preparation of Cell Extracts and Cell Fractionation. Cells are grown under conditions required for induction of specific alcohol dehydrogenases.[5] Cell lysis and fractionation are performed at 0°. Cells are suspended in 10 mM Tris-Cl, pH 7.8, to a consistency of 1 g wet cell weight/ml. DNase is added at 2 μg/ml, and cold cell suspensions are broken by three passages in a French pressure cell at 12,000 lb/square inch. The cell lysate is diluted 3-fold with 10 mM Tris-Cl buffer, pH 7.8, and is centrifuged for 10 min at 30,000 g. The resulting supernatant fluid is centrifuged at

[6] O. H. Lowry, N. J. Rosebrough, A. L. Farr, and R. J. Randall, *J. Biol. Chem.* **193**, 265 (1951).

225,000 g for 1.5 hr, yielding a supernatant fluid and a pellet. This supernatant fluid is centrifuged at 225,000 g for 1.5 hr and is designated the soluble enzyme fraction. The pellet is suspended in 10 mM Tris-Cl buffer, pH 7.8, and is centrifuged at 225,000 g for 1.5 hr; the resulting pellet is designated the membrane fraction.

Properties

Primary aliphatic alcohols (2–16 carbons) are oxidized by the soluble enzyme fraction which contains multiple alcohol dehydrogenase (ADH) enzyme activities designated ADH-A and ADH-B. ADH-A is a soluble, NAD$^+$-dependent, inducible ethanol dehydrogenase, ADH-B is a soluble, constitutive, NADP$^+$-dependent alcohol dehydrogenase active against medium-chain alcohols (octanol). Hexadecanol dehydrogenase (HDH) is a soluble, NAD$^+$-dependent, inducible enzyme that differs from either ADH-A or ADH-B. HDH is highly unstable and can be stabilized by precipitation at 50% ammonium sulfate saturation, dissolving the precipitate in a 15-fold reduced volume of 10 mM Tris-Cl buffer, pH 7.8, and freezing at −20°. ADH-A and ADH-B are stable for 2–3 weeks at −20°. The alcohol dehydrogenases show optimal activity between pH 7.8 and 9.0. ADH-A and ADH-B are major protein bands exhibiting alcohol dehydrogenase activity along with two to four other minor alcohol dehydrogenase protein bands. HDH is not visible as an enzymatically active protein band in zymograms.

Substrate Specificity. The soluble enzyme fraction exhibits a broad substrate spectrum ranging from 2 to 16 carbon atoms. Ethanol-induced NAD$^+$-dependent alcohol dehydrogenases oxidizes the following n-alcohols at the relative rates given: ethanol, 20; propanol, 60; butanol, 50; pentanol, 5; hexanol, 22; heptanol, 40; octanol, 60; nonanol, 100; decanol, 95; dodecanol, 30; tetradecanol, 20; pentadecanol, 15; hexadecanol, 10.

Electron Acceptor Specificity. The soluble enzyme fraction is unable to use dichlorophenol-indophenol, phenazine methosulfate, phenazine ethosulfate, 3-(4,5-dimethyl)thiazolyl-2,5-diphenyltetrazolium bromide or cytochrome c as final electron acceptors in place of NAD$^+$ or NADP$^+$. Pyridine nucleotide-independent oxidation of primary alcohols is not detected in the soluble enzyme fraction.

Inhibitors. The primary alcohol dehydrogenases are inhibited 100% by 1 mM p-chloromercuribenzoate but are not inhibited by 10 mM concentrations of azide, semicarbazide, hydroxylamine, arsenite, o-phenanthroline, and EDTA.

Kinetic Constants. The kinetic parameters for the alcohol substrates of ethanol dehydrogenase (EDH) and hexadecanol dehydrogenase following growth on different substrates are listed in Table I.

TABLE I
KINETIC CONSTANTS OF ETHANOL DEHYDROGENASE AND HEXADECANOL DEHYDROGENASE

	K_m (μM) (apparent), NAD$^+$		V_{max} (nmol/min), NAD$^+$		Specific activity (nmol/min/mg protein), NAD$^+$	
Growth substrate	EDH	HDH	EDH	HDH	EDH	HDH
Ethanol	512	3.0	138.0	1.0	92.6	0.8
Hexadecanol	255	1.6	7.8	5.8	1.8	4.6
Hexadecane	242	2.8	5.8	4.8	2.8	5.9
Palmitic acid	151	7.3	3.4	1.1	3.1	1.1

Membrane-Bound NAD$^+$-Dependent Hexadecanol Dehydrogenase

Assay Method

Principle. The assay is based on a radioisotopic method for the direct measurement of the oxidation of hexadecanol to hexadecanal and hexadecanoic acid.

Reagents

Tris-Cl buffer, 100 mM, pH 7.8
Tris-Cl buffer, 20 mM, pH 8.8
NAD$^+$, 2.5 mM
[1-^{14}C]hexadecanol (55.9 μCi/μmol) diluted to 3.95 μCi/μmol with *n*-hexadecanol in 0.5% dimethylformamide

Procedure. The reaction mixture contains 0.8 ml of 20 mM Tris-Cl buffer, pH 8.8; 0.1 ml of enzyme in 100 mM Tris-Cl buffer, pH 7.8; and 10 μl of 2.5 mM NAD$^+$. The reaction is initiated by the addition of 0.1 ml of [1-^{14}C]hexadecanol dissolved in 0.5% dimethylformamide. The reaction mixture is incubated at 30° for 30 min. The reaction is terminated by spotting 50 μl of the reaction mixture directly onto an activated 0.4-mm-thick silica gel G thin-layer plate, which is air-dried and developed in petroleum ether–diethyl ether–glacial acetic acid (70:30:1, v/v/v). The fatty acid and fatty aldehyde areas of the thin-layer chromatogram, determined by cochromatography with authentic standards, are collected in scintillation vials, and the radioactivity is quantified by liquid scintillation spectrometry.

Definition of Specific Activity. Specific activity is expressed as nanomoles product formed per 30 min per milligram protein.

TABLE II
LOCALIZATION OF HEXADECANOL DEHYDROGENASE

	Specific activity (nmol product formed/30 min/mg protein)	
Growth substrate	Soluble enzyme fraction	Membrane fraction
Hexadecanol	0.20	2.93
Hexadecane	1.11	2.37
Ethanol	0.28	1.53
Palmitic acid	0.59	2.69

Properties

Hexadecanol dehydrogenase is highly unstable, with greater than 50% loss of activity occurring in 24 hr. HDH is transiently induced 72-fold in hexadecanol-grown cells but remains constant in specific activity over an 8-hr time course in ethanol-, hexadecane-, and palmitate-grown cells. HDH induction is inhibited by rifampicin in hexadecanol-grown cells, whereas HDH activity remains at constitutive levels in hexadecane-, ethanol-, and palmitate-grown cells. The localization of HDH in the membrane fraction is shown in Table II.

[4] Aldehyde Dehydrogenases from *Acinetobacter*

By W. R. FINNERTY

Aldehyde dehydrogenases are described from a number of alkane-utilizing microorganisms. *Acinetobacter calcoaceticus* 69V contains a membrane-bound, $NADP^+$-dependent fatty aldehyde dehydrogenase induced by growth on hexadecanol or hexadecane.[1,2] *Acinetobacter* HO1-N exhibits inducible and constitutive NAD^+- and $NADP^+$-dependent short-chain and long-chain aldehyde dehydrogenases.[3,4] This chapter describes the properties of the soluble and membrane-bound NAD^+- and $NADP^+$-dependent aldehyde dehydrogenases in *Acinetobacter* HO1-N.

[1] H. Aurich and G. Eitner, *Z. Allg. Mikrobiol.* **17,** 263 (1977).
[2] H. Sorger and H. Aurich, *Z. Allg. Mikrobiol.* **18,** 587 (1978).
[3] M. E. Singer and W. R. Finnerty, *J. Bacteriol.* **164,** 1017 (1985).
[4] M. E. Singer and W. R. Finnerty, *J. Bacteriol.* **164,** 1011 (1985).

Soluble and Membrane-Bound NAD⁺- and NADP⁺-Dependent Aldehyde Dehydrogenases

$$\text{RCHO} + \text{NAD}^+ (\text{NADP}^+) \rightleftharpoons \text{RCOOH} + \text{NADH (NADPH)} + \text{H}^+$$

Assay Method

Principle. The assay is based on the rate of reduction of NAD^+ or NADP^+ at 340 nm in the presence of an aldehyde.

Reagents

Pyrophosphate–phosphate buffer, 20 mM, pH 9.0
Tris-Cl buffer, 10 mM, pH 7.8
NAD^+ or NADP^+, 2.5 mM

Substrates. The following substrates are used: Acetaldehyde, 330 mM; hexanal, 50 mM in 2.5% dimethylformamide; decanal, 50 mM in 2.5% dimethylformamide; dodecanal, 50 mM in 2.5% dimethylformamide. Aldehyde dehydrogenases are inhibited less than 1% by 2.5% dimethylformamide.

Procedure. The reaction mixture contains 0.2 ml of 20 mM pyrophosphate–phosphate buffer, pH 9.0; 0.1 ml of enzyme suitably diluted in 10 mM Tris-Cl buffer, pH 7.8; 10 μl of NAD^+ or NADP^+; and 0.6 ml of 10 mM Tris-Cl buffer, pH 7.8. The reaction is initiated by the addition of 0.1 ml of aldehyde substrate to yield the final concentration specified above. Aldehyde dehydrogenase activities are determined by recording the rate of NAD^+ or NADP^+ reduction for 5 min at 340 nm using a Cary 219 spectrophotometer.

Definition of Specific Activity. Specific activity is expressed in terms of nanomoles of NAD^+ (NADP^+) reduced per minute per milligram protein. The protein content of cell fractions is determined by the method of Lowry et al.[5]

Preparation of Cells Extracts and Cell Fractionation. Cells are grown under conditions required for induction of specific aldehyde dehydrogenases.[3,4] Cell lysis and fractionation are performed at 0°. Cells are suspended in 10 mM Tris-Cl buffer, pH 7.8, to a consistency of 1 g wet weight/ml. DNase is added at a concentration of 2 μg/ml, and the cell suspensions are broken by three passages in a French pressure cell at 12,000 lb/square inch. The cell lysate is diluted 3-fold with 10 mM Tris-Cl buffer, pH 7.8, and centrifuged for 10 min at 30,000 g. The resulting supernatant fluid is

[5] O. H. Lowry, N. J. Rosebrough, A. L. Farr, and R. J. Randall, *J. Biol. Chem.* **193**, 265 (1951).

centrifuged at 225,000 g for 1.5 hr, yielding a supernatant fluid and a pellet. This supernatant fluid is centrifuged at 225,000 g for 1.5 hr and is designated the soluble enzyme fraction. The pellet is suspended in 10 mM Tris-Cl buffer, pH 7.8, and is centrifuged at 225,000 g for 1.5 hr; the resulting pellet is designated the membrane fraction.

Properties

The soluble NAD$^+$-dependent acetaldehyde dehydrogenase is induced in ethanol-grown cells with a 56- to 117-fold increase in specific activity over cells grown on hexadecanol, hexadecane, or palmitate. The membrane-bound NAD$^+$-dependent fatty aldehyde dehydrogenase is constitutive in ethanol-, hexadecanol-, hexadecane-, and palmitate-grown cells and is induced 2-fold in dodecanal-grown cells. The NAD$^+$-dependent soluble and membrane-bound aldehyde dehydrogenase activities derived from ethanol-grown cells oxidize the following aldehydes at the relative rates given: acetaldehyde, 22; hexanal, 12; decanal, 98; dodecanal, 100. The NADP$^+$-dependent, membrane-bound fatty adlehyde dehydrogenase is indeuced in hexadecanol-, hexadecane-, and dodecanal-grown cells with 11-, 9-, and 15-fold higher specific activities, respectively, than present in palmitate-grown cells. Maximum induction occurs in 15 min for dodecanal- and hexadecanol-grown cells and within 3 hr in hexadecane-grown cells. The NAD$^+$-dependent aldehyde dehydrogenases are stimulated in activity 75% after a 6-hr incubation at 58°. The NADP$^+$-dependent fatty aldehyde dehydrogenase is unstable at 58°, with a 50% loss of enzyme activity after 3.5 hr.

TABLE I

KINETIC CONSTANTS OF NAD$^+$-DEPENDENT ACETALDEHYDE DEHYDROGENASE AND NADP$^+$-DEPENDENT FATTY ALDEHYDE DEHYDROGENASE

Cell type[a]	Acetaldehyde dehydrogenase			Fatty aldehyde dehydrogenase		
	K_m (μM) (apparent)	V_{max} (nmol/min)	Specific activity (nmol/min/mg protein)	K_m (μM) (apparent)	V_{max} (nmol/min)	Specific activity (nmol/min/mg protein)
A	0.05	10.0	183.0	18.0	25.0	14.2
B	11.00	13.0	10.2	13.0	500.0	331.7
C	11.00	5.0	6.9	5.0	537.0	237.6
D	45.00	18.3	1.9	18.3	38.0	29.2

[a] A, Ethanol-grown cells; B, hexadecanol-grown cells; C, hexadecane-grown cells; D, palmitate-grown cells.

TABLE II
ALDEHYDE DEHYDROGENASE ISOZYMES

Growth substrates	Ald-a	Ald-b	Ald-c	Ald-d
Ethanol	−	+	+	+
Hexadecanol	+	+	+	−
Hexadecane	+	+	+	−
Dodecanal	+	+	−	+
Palmitate	−	+	−	+

Kinetic Constants. The kinetic constants of the soluble NAD^+-dependent acetaldehyde dehydrogenase and the membrane-bound $NADP^+$-dependent fatty aldehyde dehydrogenase are listed in Table I.

Isozymes. The isozyme pattern of aldehyde dehydrogenases in *Acinetobacter* HO1-N is related to the growth substrate as shown in Table II.[4] The substrate-specific isozymes are present in both the soluble enzyme fraction and the membrane fraction and exhibit identical electrophoretic mobilities, indicating similar, if not identical, proteins.

[5] Propane-Specific Alcohol Dehydrogenase from *Rhodococcus rhodochrous* PNKb1

By W. ASHRAF and J. C. MURRELL

2-Propanol + NAD^+ ⇌ acetone + NADH + H^+

Propane-specific alcohol dehydrogenase catalyzes the NAD^+-dependent dehydrogenation of 1-propanol and 2-propanol in propane-grown *Rhodococcus rhodochrous* PNKb1. The purification described below is similar to other reports.[1,2] However, the enzyme is significantly different from other reported propane-associated alcohol dehydrogenases.

Assay Methods

Principle. Two quantitative and one qualitative assay can be used. Method 1 involves measuring alcohol dehydrogenase activity by monitoring the alcohol-dependent reduction of NAD^+ at 340 nm in a

[1] C. T. Hou, R. Patel, A. I. Laskin, I. Barist, and N. Barnabe, *Appl. Environ. Microbiol.* **46,** 98 (1983).

[2] J. P. Coleman and J. J. Perry, *J. Gen. Microbiol.* **131,** 2901 (1985).

spectrophotometer.[1-3] Method 2 utilizes gas chromatography in measuring and identifying the reaction products of the dehydrogenase reaction.[1,2] Method 3, which is qualitative, involves running samples on nondenaturing polyacrylamide gels and staining (by formazan production) for alcohol dehydrogenase proteins by using a combination of a dye and an artificial electron carrier.[4] Method 1 is used routinely in the purification and characterization of the propane-specific alcohol dehydrogenase from *R. rhodochrous* PNKb1.

Reagents for Assay Method 1

Tris–NaOH buffer, 20 mM, pH 10.0
NAD$^+$, 1.0 mM
1-Propanol or 2-propanol, 1 M
Alcohol dehydrogenase (purified from *R. rhodochrous* PNKb1)

Reagents for Assay Method 2

Tris–NaOH buffer, 20 mM, pH 10.0
NAD$^+$, 10 mM
1-Propanol or 2-propanol, 1 M
Alcohol dehydrogenase (purified from *R. rhodochrous* PNKb1)

Reagents for Assay Method 3

Tris–NaOH buffer, 20 mM, pH 10.0
NAD$^+$, 116 mg
Nitro blue tetrazolium, 20 mg
Phenazine methosulfate, 2.5 mg
1-Propanol or 2-propanol, 3 mg/ml Alcohol dehydrogenase (purified from *R. rhodochrous* PNKb1)

Procedure for Assay Method 1

The assay mixture contains, in a final volume of 1 ml, Tris–NaOH buffer, 940 μl; NAD$^+$, 10 μl; 1-propanol or 2-propanol, 50 μl; and alcohol dehydrogenase, 10 units. The reaction mixture is equilibrated for 1 min at 30°, and the reaction is initiated by addition of either 1-propanol or 2-propanol and monitored by the reduction of NAD$^+$ at 340 nm. The reaction is followed for at least 5 min to ensure measurement of a linear rate.

[3] N. R. Woods and J. C. Murrell, *J. Gen. Microbiol.* **135**, 2335 (1989).
[4] D. I. Stirling and H. Dalton, *J. Gen. Microbiol.* **107**, 19 (1978).

Procedure for Assay Method 2

The assay mixture contains, in a final volume of 1 ml, Tris–NaOH buffer, 940 μl; NAD$^+$, 10 μl; 1-propanol or 2-propanol, 50 μl; and alcohol dehydrogenase, 50 units. The reaction mixture is equilibrated for 1 min at 30° in a 5-ml Suba-sealed reaction flask before the alcohol is added using a syringe. The reaction is followed using a Pye-Unicam gas chromatograph fitted with a flame ionization detector to identify reaction products, which are quantified using a Hewlett Packard 3390 A Integrator. 1-Propanol, 2-propanol, propanal, and acetone solutions of 2.0 mM (freshly prepared daily) are used as standards. Samples (5 μl) taken from the reaction flask at 10-min intervals are injected onto a glass column (2 m × 2 mm) packed with Porapak R, run isothermally at 160° with nitrogen as the carrier gas (20 ml/min). Detector and injector temperatures are maintained at 225° throughout.

Procedure for Assay Method 3

A cell-free extract of *R. rhodochrous* PNKb1, containing 100 μg soluble protein, is electrophoresed at 4° on 4.5 to 15% (w/v) polyacrylamide gradient gels [containing no sodium dodecyl sulfate (SDS)]. After separation of proteins, the location of the alcohol dehydrogenase in the gel is visualized by means of an activity stain. The assay mixture in which the polyacrylamide gel is incubated contains, in a final volume of 100 ml, Tris–NaOH, 99 ml; nitro blue tetrazolium, 20 mg; NAD$^+$, 116 mg; phenazine methosulfate, 2.5 mg; and either 1-propanol or 2-propanol (1 ml of a 3 mg/ml solution). A gel incubated in the above mixture without substrate is used as control. The reaction is allowed to proceed in the dark after initiation by the addition of alcohol. The reaction is terminated by immersion in 10% (v/v) acetic acid. The gel may be stored in the dark in this solution.

Units. One unit of alcohol dehydrogenase is defined as that amount catalyzing the reduction or oxidation of 1 nmol of NAD$^+$ or NADH/min at 30°.

Purification Procedure

Growth of Organism. The *Rhodococcus rhodochrous* PNKb1 used originated from Woods and Murrell.[3] It is grown at 30° in a fermentor, sparged with propane–air (50:50, v/v), in a mineral salts medium containing, per liter, the following: Na$_2$HPO$_4$, 0.72 g; KH$_2$PO$_4$, 0.26 g; NH$_4$Cl, 1 g; MgSO$_4$·7H$_2$O, 1 g; CaCl$_2$, 0.2 g; Fe EDTA, 4 mg; Na$_2$MoO$_4$·2H$_2$O,

2.5 mg; EDTA, 0.5 mg; $FeSO_4 \cdot 7H_2O$, 0.2 mg; $ZnSO_4 \cdot 7H_2O$, 10 μg; $MnCl_2 \cdot 4H_2O$, 3μg; H_3BO_3, 30 μg; $CaCl_2 \cdot 6H_2O$, 0.2 mg; $CuCl_2 \cdot 2H_2O$, 1 μg; $NiCl_2 \cdot 6H_2O$, 2 μg; $Na_2MoO_4 \cdot 2H_2O$, 3 μg. *It should be noted that propane–air mixtures containing less than 10% propane are explosive.* The actively growing culture is harvested during late exponential growth by centrifugation at 20,000 g for 10 min at 4° and the washed cell suspension is frozen by dropwise addition into liquid nitrogen and stored at −70°.

Step 1: Preparation of Cell-Free Extract. Cell pellets are resuspended in 1/20 volume of 20 mM Tris-HCl, pH 6.8. The suspension is disrupted by three passages through a French pressure cell at 137 MPa. The lysed suspension is centrifuged at 48,000 g for 30 min at 4° to remove whole cells and cellubar debris including membranes. Other reports concerning the preparation of cell-free extracts from *Rhodococcus* have described the use of an X-press in the process.[5,6]

Step 2: DEAE-Cellulose Chromatography. The soluble cell-free extract is applied to a column (10 × 2.0 cm) of DEAE-cellulose (Sigma, St. Louis, MO) previously equilibrated with 20 mM Tris-HCl, pH 6.8, containing 1 mM dithiothreitol (buffer A). After washing the column with 2 column volumes of buffer A, the alcohol dehydrogenase is eluted with a step gradient (each 0.1 M increment being 1 column volume) from 0 to 0.5 M NaCl in buffer A. Fractions containing the highest alcohol dehydrogenase activity are combined and dialyzed overnight against 10 liters of buffer A.

Step 3: Butanol-Sepharose 4B Chromatography. The colored DEAE-cellulose alcohol dehydrogenase fraction is applied to a column (3 × 3 cm) of butanol-Sepharose 4B, obtained by coupling 4-amino-1-butanol to cyanogen bromide-activated Sepharose 4B (Pharmacia, Piscataway, NJ),[7] previously equilibrated with buffer A. After washing with 2 column volumes of buffer A, a step gradient containing 0.1 to 0.5 M NaCl in buffer A is used to elute the column. The fraction containing the highest alcohol dehydrogenase activity is dialyzed as above. This step could be omitted from our purification procedure. It is included because the attachment of other alcohols to activated Sepharose 4B may be of greater significance when purifying other alcohol dehydrogenases.

4: NAD-Agarose Chromatography. The colorless butanol-Sepharose 4B fraction is applied to a type III NAD-agarose (Sigma; NAD linked to agarose via the C-8 of the purine ring) column (bed volume, 1 ml).[2,8] The

[5] L. Eggeling and H. Sahm, *Arch. Microbiol.* **126,** 141 (1980).

[6] L. Eggeling and H. Sahm, *Environ. Biochem.* **150,** 129 (1985).

[7] *Pharmacia Affinity Chromatography Handbook.* Pharmacia LKB Biotechnology, Uppsala, Sweden, 1988.

[8] S. Barry and P. O'Carra, *Biochem. J.* **135,** 595 (1973).

column is washed with 10 ml of buffer A containing 0.5 M NaCl (buffer B) to remove any unbound proteins. Alcohol dehydrogenase is selectively eluted with 1 mM NAD in buffer B, and the active fraction is dialyzed as previously described. The alcohol dehydrogenase is stored at −70° after being frozen by dropwise addition to liquid nitrogen. A typical purification is summarized in Table I.

Properties

Purity. The enzyme yields only one major protein band after denaturing electrophoresis and staining with Coomassie blue. Only very minor traces of impurities were detected after careful visual examination of 20 μg of electrophoresed purified protein.

Substrate Specificity. Alcohol dehydrogenase activity is detected with a range of short-chain primary and secondary alcohols (C_2–C_8). The enzyme also catalyzes the reverse reductase reaction at pH 6.5 with acetone, butanone, and cyclohexanone as substrates.

Effect of Inhibitors. Alcohol dehydrogenase activity is inhibited by the metal-complexing reagents 2′,2′-bipyridyl and 1,10-phenanthroline and the thiol reagent iodoacetate. Metal chelators such as EDTA and cyanide do not inhibit enzyme activity.

pH Optimum. The alcohol dehydrogenase has optimum activity at pH 10.5 and pH 6.5 for the reverse reaction, that is, the formation of 2-propanol from acetone at pH 6.5.

Stability. The purified enzyme is stable for several months at −70° in the presence of 1 mM dithiothreitol.

Michaelis Constants. The apparent K_m for alcohol dehydrogenase is 1.2×10^{-2} and 1.8×10^{-2} M for 1-propanol and 2-propanol, respectively.

TABLE I

PURIFICATION OF ALCOHOL DEHYDROGENASE FROM PROPANE-GROWN
Rhodococcus rhodochrous PNKb1

Fraction	Total volume (ml)	Total protein (mg)	Total activity (units)	Specific activity[a] (units/mg protein)	Purification factor	Yield (%)
Step 1: Cell-free extract	3.8	26.6	520	19.5	1	100
Step 2: DEAE-cellulose	23	9.7	500	51.5	2.6	96
Step 3: Butanol-sepharose 4B	13	4.4	246	56.0	2.9	47
Step 4: NAD-agarose	4.6	0.05	100	2000	103	19

[a] Results obtained by Assay Method 1 using 2-propanol as a substrate for the alcohol dehydrogenases.

In the reverse reaction, the apparent K_m for acetone is 5.7×10^{-2} M. The apparent K_m for the cofactors is 6.25×10^{-5} and 1.0×10^{-4} M for NAD^+ and NADH, respectively.

Molecular Weight and Subunit Structure. The molecular weight of alcohol dehydrogenase in 20 mM Tris-HCl buffer, pH 6.8, containing 1 mM dithiothreitol, as measured by gel filtration through a column of Sephacryl S-200 (Pharmacia), corresponds to 160,000. A single polypeptide corresponding to as molecular weight of 40,000 was observed after denaturing electrophoresis of purified alcohol dehydrogenase. This suggests the enzyme is composed of four identical subunits.

[6] Cell-Free Assay Methods for Enzymes of Propane Utilization

By J. C. Murrell and W. Ashraf

The following assay methods are based on *in vitro* studies of propane metabolism in a *Rhodococcus rhodochrous* strain PNKb1, isolated by Woods and Murrell.[1] In addition to growing on propane, this organism grows on all the potential intermediates of propane oxidation, including 1-propanol, 2-propanol, propanal, acetone, acetol, and propanoate. There are a number of potential routes for propane oxidation by bacteria. The pathways of propane metabolism and the enzymatic steps that may be involved are outlined in Scheme 1. All the enzyme activities described below have been demonstrated in cell-free extracts of *Rhodococcus rhodochrous* PNKb1, except where stated.

Growth of *Rhodococcus* and Preparation of Cell-Free Extracts

Rhodococcus rhodochrous PNKb1 is grown in fermentor culture on a basic mineral salts medium sparged with propane and air as described in the previous chapter (Ashraf and Murrell, this volume [5]). The culture is harvested during late exponential growth, and cell pastes are either stored at $-70°$ or immediately used for preparation of cell-free extracts. Cell pellets are resuspended in 1/20 volume of 20 mM Tris-HCl, pH 6.8, and disrupted by three passages through a French pressure cell operated at 137 MPa. Owing to the extremely tough nature of the cell wall of this bacterium, the lysed cell suspension is centrifuged at 10,000 *g* for 10 min at 4° to remove any unbroken whole cells which may interfere with subsequent

---

[1] N. R. Woods and J. C. Murrell, *J. Gen. Microbiol.* **135**, 2335 (1989).

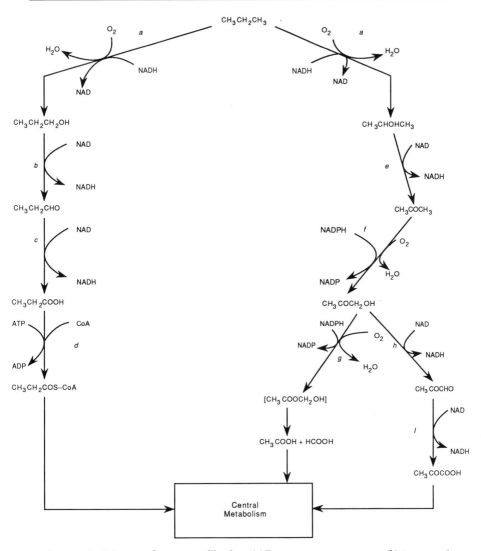

SCHEME 1. Pathways of propane utilization. *(a)* Propane monooxygenase, *(b)* 1-propanol dehydrogenase, *(c)* propanal dehydrogenase, *(d)* propionyl-CoA synthetase (propionate-CoA ligase), *(e)* 2-propanol dehydrogenase, *(f)* acetone monooxygenase, *(g)* acetol monooxygenase, *(h)* acetol dehydrogenase, and *(i)* methylglyoxal dehydrogenase.

assays on the membrane or particulate fractions of the cell-free extract. The resulting suspension is then centrifuged at 48,000 g for 30 min at 4° to yield a soluble cell-free extract, which is stored at $-70°$ after dropwise addition to liquid nitrogen, and a membrane pellet, which is resuspended in 20 mM Tris-HCl, pH 6.8, giving a particulate fraction which is also stored at $-70°$.

Propane Oxygenase Assay

$$\text{Propane} + O_2 + \text{NADH} \rightarrow \text{propanol} + H_2O + \text{NAD}^+$$

$$\text{Propene} + O_2 + \text{NADH} \rightarrow \text{1,2-epoxypropane} + H_2O + \text{NAD}^+$$

Principle. The propane oxygenase assay is based on the ability of this enzyme to convert propane to 1,2-epoxypropane in a NADH-dependent reaction. The epoxide formed is not further metabolized and may be quantified by gas chromatography. This indirect assay for propane oxygenase activity is more reliable than the measurement by gas chromatography of either 1-propanol or 2-propanol formed from the oxidation of propane because these metabolites may be further assimilated via alcohol dehydrogenases present in cell-free extracts of this and other propane-grown organisms. The presence of propane oxygenase activity in whole cells can be confirmed before the preparation of cell-free extracts by measuring the propane-stimulated oxygen uptake of whole cells in a standard oxygen electrode assay. The cell-free assay given is based on the method of Hou *et al.*[2]

Reagents

Tris-HCl buffer, 20 mM, pH 6.8

NADH, 100 mM (as many commercial preparations of NADH contain substantial amounts of ethanol, which may mask the 1,2-epoxypropane peak obtained after gas chromatography, it is essential that the ethanol be removed from an aqueous solution of NADH by ether extraction)

Propene (> 99.5% pure; British Oxygen Company, London)

Procedure. Assays are performed in 2-ml gas chromatography vials. The assay mixture contains, in a final volume of 0.25 ml, Tris-HCl buffer, 1–5 mg protein as a soluble cell-free extract, and 10 μl NADH. The vial is then sealed with a rubber cap and allowed to equilibrate for 1 min at 30° in a shaking water bath. The reaction is initiated by the removal of 0.9 ml air

[2] C. J. Hou, R. Patel, A. I. Laskin, N. Barnabe, and I. Barist, *Appl. Environ. Microbiol.* **46,** 171 (1983).

and the rapid addition of 0.9 ml propene. The vial is returned to the water bath, and 5-μl samples are removed every 5 min for analysis on a Pye Unicam gas chromatograph fitted with a 1.5 m $\times$ 2.3 mm glass column containing Porapak Q. The column is run isothermally at 180° with nitrogen (30 ml/min) as the carrier gas. The amount of 1,2-epoxypropane produced is quantified from the peak area measured using a reporting integrator (Hewlett Packard) that is calibrated with standard solutions. Rates are expressed as nanomoles 1,2-epoxypropane formed per minute per milligram protein. As this enzyme is usually labile, assays are routinely performed using freshly prepared soluble cell-free extracts. Whole cell epoxidation activity may also be measured using this method by omitting NADH from a buffered cell suspension (2–5 mg dry weight) of *Rhodococcus rhodochrous* PNKb1.

Alcohol and Aldehyde Dehydrogenase Assays

The enzymes alcohol and aldehyde dehydrogenases (EC 1.1.1.*x*) catalyze key steps in the assimilation of carbon during growth of bacteria on propane. In *Rhodococcus rhodochrous* all three enzymes are present and require NAD$^+$ for *in vitro* activity measurement (see Scheme 1). In other propane-utilizing bacteria these enzymes may exhibit dye-linked activities, and, for completeness, methods for these enzyme assays are also included below.

Principle. Two assay methods can be used. Method 1, based on that of Coleman and Perry,[3] measures spectrophotometrically the alcohol- or aldehyde-dependent reduction of NAD$^+$ at 340 nm. Method 2 involves measuring the phenazine methosulfate-linked activity of alcohol or aldehyde dehydrogenase spectrophotometrically and is a modification of a method of Van Den Tweel and deBont.[4]

Reagents for Method 1

Tris–NaOH buffer, 20 mM, pH 10.0
NAD$^+$, 10 mM
1-Propanol, 2-propanol, or propanal (of the purest grade available), 1 M

Reagents for Method 2

Tris–NaOH buffer, 20 mM, pH 9.0
Phenazine methosulfate (PMS), 20 mM

[3] J. P. Coleman and J. J. Perry, *J. Gen. Microbiol.* **131,** 2901 (1985).
[4] W. J. J. Van Den Tweel and J. A. M. deBont, *J. Gen. Microbiol.* **131,** 3155 (1985).

Dichlorophenolindophenol (DCPIP), 20 mM

1-Propanol, 2-propanol, or propanal (of the purest grade available), 1 M

Procedure for Method 1. The assay mixture contains, in a final volume of 1 ml, Tris–NaOH buffer, 20 μl NAD$^+$ and 0.1–1.0 mg of soluble cell-free extract (sufficient to give a linear rate for 3–5 min). The reaction cuvette is allowed to equilibrate at 30° for 2 min before initiate of the reaction with 50 μl of alcohol or aldehyde. The increase in absorbance at 340 nm is measured spectrophotometrically.

Procedure for Method 2. The assay mixture contains, in a final volume of 1.5 ml, Tris–NaOH buffer (previously sparged with oxygen-free nitrogen for 5 min), 7.5 μl PMS. 7.5 μl DCPIP, 15 μl NH$_4$Cl, and 0.1–1.0 mg soluble protein extract (sufficient to give a linear rate for 3 to 5 min). The reaction cuvette is allowed to equilibrate at 30° for 2 min before the assay is initiated by the addition of 50 μl of alcohol or aldehyde. The reduction of DCPIP by reduced PMS is measured spectrophotometrically by the change in absorbance at 600 nm. Specific activities can be estimated from the molar extinction coefficient for DCPIP of 16,100 (at pH 7.0).

Acetone and Acetol Monooxygenase Assays

Principle. Acetone and acetol monooxygenase activity (see Fig. 1) is measured as substrate-dependent stimulation of oxygen uptake in a Clark-type oxygen electrode, using a modification of the method by Hartmans and deBont.[5]

Reagents

Tris-HCl buffer, 20 mM, pH 6.8
NADPH, 100 mM
Acetone or acetol, 100 mM

Procedure. The assay mixture contains 2.81 ml of Tris-HCl buffer and 100 μl of soluble cell-free extract containing 0.1–1.0 mg of protein. This is allowed to equilibrate at 30° in the chamber of the oxygen electrode, and the endogenous oxygen uptake is recorded. Then 10 μl of NADPH is added and any change in rate measured. Subsequently, 80 μl of acetone or acetol is added, and the substrate-dependent rate of oxygen uptake is monitored. Acetone or acetol monooxygenase activity, corrected for endogenous activity, is expressed as nanomoles O$_2$ consumed per minute per milligram protein. The dissolved oxygen concentration of air-saturated buffer is calculated using the method of Robinson and Cooper.[6] It is worth

---

[5] S. Hartmans and J. A. M. deBont, *FEMS Microbiol Lett.* **36,** 151 (1986).
[6] J. Robinson and R. M. Cooper, *Anal. Biochem.* **33,** 390 (1970).

noting that no acetone monooxygenase activity can be detected in *Rhodococcus rhodochrous* PNKb1, despite the use of different buffer systems and substitution of NADPH with a variety of other electron donors. A similar situation is reported by Taylor *et al.*,[7] who were also wholly unsuccessful in measuring cell-free acetone oxygenase activity in a variety of acetone utilizers. It is thought that this enzyme is extremely labile in such organisms.

Acetol (1-hydroxyacetone) Dehydrogenase

Principle. Another possible fate of acetol in some propane utilizers (but not *Rhodococcus rhodochrous* PNKb1) may be conversion to methylglyoxal via an acetol dehydrogenase. This activity is assayed spectrophotometrically using the NAD-linked assay of Taylor *et al.*[7] The subsequent metabolism of methylglyoxal, if evidence of an acetol dehydrogenase is obtained, can be further investigated by the methods of Taylor *et al.*,[7] who detail procedures for elucidating the metabolism of acetone by several acetone-utilizing corynebacteria.

Reagents

Glycine–NaOH buffer, 50 mM, pH 10.0
NAD$^+$, 10 mM
Acetol, 100 mM

Procedure. The assay mixture contains, in a total volume of 1.0 ml, 0.90 ml glycine–NaOH buffer, 50 μl NAD$^+$, and 30 μl cell-free extract (containing 0.5–1.0 mg protein). The reaction cuvette is allowed to equilibrate at 30° for 2 min before the reaction is initiated by the addition of acetol (20 μl). The acetol dehydrogenase-dependent reduction of NAD$^+$ is measured at 340 nm. Acetol dehydrogenase activity, corrected for endogenous NAD$^+$ reduction, is expressed as nanomoles NAD$^+$ reduced per minute per milligram protein. It should also be borne in mind that this activity measurement may contain an element of methylglyoxal dehydrogenase activity.[7]

Propionyl-CoA Synthetase

Principle. Propionic acid resulting from the terminal oxidation of propane to 1-propanol and oxidation of 1-propanol via propanal dehydrogenase may be further metabolized via the enzyme propionyl-CoA synthetase (propionate-CoA ligase), the activity of which is considerably elevated

---

[7] D. G. Taylor, P. W. Trudgill, R. E. Cripps, and P. H. Harris, *J. Gen. Microbiol.* **118**, 159 (1980).

during growth of *Rhodococcus rhodochrous* on propane (compared with the activity of this enzyme found in cell-free extracts prepared from cells grown on pyruvate). The activity of this enzyme is measured by determination of propanoate-dependent sulfhydryl-CoA disappearance, based on the method of Grunert and Phillips.[8]

Reagents

Tris-HCl buffer, 10 mM, pH 7.4
Magnesium chloride, 250 mM
Potassium borate, 10 mM (freshly prepared)
Coenzyme A, sodium salt, 10 mM
ATP, 100 mM
Potassium propanoate, 10 mM, pH 8.0
Sodium cyanide/sodium carbonate (0.44 g NaCN and 21.2 g Na$_2$CO$_3$ in 100 ml water)
Sodium chloride (saturated solution in water)

Procedure. Assays are performed in small glass test tubes in a total volume of 0.2 ml. These contain 100 μl Tris-HCl, 10 μl magnesium chloride, 10 μl potassium borate, 10 μl ATP, 20 μl potassium propanoate, 40 μl coenzyme A, and 10 μl cell-free extract (containing 0.1–0.5 mg protein). Tubes are incubated at 30° for 15 min under a nitrogen atmosphere. A developing reagent is prepared in a second tube consisting of 2 ml saturated sodium chloride, 0.4 ml sodium cyanide/sodium carbonate, and 0.4 ml nitroprusside. This reagent is poured into the assay tube, mixed, and the absorbance read at 520 nm immediately against a blank lacking propanoate. Under these conditions, a change in sulfhydryl-CoA concentration of 0.1 μM is equivalent to a change in absorbance at 520 nm of 0.2 units. Units of activity are expressed as nanomoles CoA consumed per minute per milligram protein.

[8] R. R. Grunert and P. H. Phillips, *Arch. Biochem. Biophys.* **30**, 217 (1951).

[7] Quinoprotein Alcohol Dehydrogenase from *Pseudomonas aeruginosa* and Quinohemoprotein Alcohol Dehydrogenase from *Pseudomonas testosteroni*

By B. W. GROEN and J. A. DUINE

$$\text{Alcohol} + \text{dye} \rightarrow \text{aldehyde} + \text{dye} \cdot H_2$$

Pseudomonas species involved in the degradation of alkanes contain $NAP(P)^+$-independent alcohol dehydrogenases, but the nature of the cofactor is unknown. Similar types of dehydrogenases have been isolated and characterized, however, from alcohol-grown *Pseudomonas aeruginosa*,[1,2] *P. putida*[3,4] (the organism originally supposed to be *Acinetobacter calcoaceticus*[4] is now classified as a *P. putida* strain), and *Comamonas testosteroni* (*P. testosteroni*[5]). Although these are quinoproteins [enzymes with pyrroloquinoline quinone (PQQ) as a cofactor], two quite different types of enzymes appear to exist, namely, quinoprotein and quinohemoprotein alcohol dehydrogenase.

Quinoprotein Alcohol Dehydrogenase from *Pseudomonas aeruginosa*

Assay Method

Principle. Enzyme activity is determined by measuring the rate of ethanol-dependent reduction of Wurster's blue with ethylamine as an activator.

Reagents

Ethylamine-HCl, 1 M, pH 9.0
Ethanol, 10 mM
Wurster's blue,[6] 2 mM

---

[1] B. W. Groen, J. Frank, and J. A. Duine, *Biochem. J.* **223,** 921 (1984).

[2] M. Rupp and H. Görisch, *Biol. Chem. Hoppe-Seyler* **369,** 431 (1988).

[3] H. Görisch and M. Rupp, *in* "Proceedings of the First International Symposium on PQQ and Quinoproteins" (J. A. Jongejan and J. A. Duine, eds.), p. 111. Kluwer Academic Publ., Dordrecht, 1989.

[4] J. A. Duine and J. Frank, *J. Gen. Microbiol.* **122,** 201 (1981).

[5] B. W. Groen, M. A. G. van Kleef, and J. A. Duine, *Biochem. J.* **234,** 611 (1986).

[6] L. Michaelis and S. Granick, *J. Am. Chem. Soc.* **65,** 1747 (1943).

KCN, 1 M, adjusted to pH 9.0 with 2 M HCl

Pyrophosphate buffer, 0.1 M tetrasodium pyrophosphate adjusted to pH 9.0 with 2 M HCl

Calcium solution, 15 mg of $CaCl_2$/ml

Iron solution, 15 mg of $FeSO_4 \cdot 7H_2O$/ml

Trace element solution, in a total volume of 1 liter, 7.8 mg $CuSO_4 \cdot 5H_2O$, 10 mg H_3BO_3, 10 mg $MnSO_4 \cdot 4H_2O$, 70 mg $ZnSO_4$, and 10 mg MoO_3

Tris–Ca buffer, 10 mM Tris-HCl, pH 8.0, containing 1 mM $CaCl_2$

Procedure. To the cuvette (1 cm optical pathway) is added 0.9 ml pyrophosphate buffer, 10 μl ethylamine, 10 μl KCN, 20 μl ethanol, 50 μl Wurster's blue, and 20–50 μl of the enzyme solution. After mixing, the rate of absorption decrease is measured at 610 nm. The rate of Wurster's blue reduction is calculated from the linear part of the curve using a molar absorption coefficient of $12.4 \times 10^3 \, M^{-1} \, cm^{-1}$. One unit (U) is the amount of enzyme that reduces 2 μmol of Wurster's blue per minute at 20° under the indicated conditions. Protein concentrations are determined by the method of Lowry *et al.*[7]

Purification Procedure

Growth of the Organism. Pseudomonas aeruginosa LMD 80.53 is grown at 35° on a mineral medium. The basic medium has the following composition (per liter): 15.4 g $K_2HPO_4 \cdot 3H_2O$, 4.52 g KH_2PO_4, 0.5 g $MgCl_2 \cdot 6H_2O$, and 3 g $(NH_4)_2SO_4$. After autoclaving, 1 ml calcium solution, 1 ml iron solution, and 1 ml trace element solution (all sterilized by autoclaving) are added per liter of basic medium. The mineral medium is supplemented with 0.5% (v/v) ethanol (sterilized by filtration). Cultivation is carried out in a 2-liter flask with reciprocal shaking (250 rpm). Cells are harvested at the end of the exponential growth phase and washed with 20 mM potassium phosphate buffer, pH 7.0. The cell paste is stored at −40°.

Purification Steps. All operations are carried out at 4°, unless stated otherwise.

Step 1: Preparation of cell-free extract. Cell paste (9 g wet weight) is suspended in 13 ml of 50 mM Tris-HCl buffer, pH 7.5, containing 1 mM $CaCl_2$. The mixture is passed twice through a French pressure cell at 110 MPa. The viscosity of the suspension is lowered by adding a few milligrams of DNase. Cell debris is removed by centrifugation at 48,000 g for 20 min.

---

[7] O. H. Lowry, N. J. Rosebrough, A. L. Farr, and R. J. Randall, *J. Biol. Chem.* **193**, 265 (1951).

The supernatant is dialyzed overnight against 1 liter of Tris–Ca buffer, containing 0.02% (v/v) 2-mercaptoethanol.

Step 2: DEAE-Sepharose filtration. The dialyzed cell-free extract (22 ml) is applied to a DEAE-Sepharose fast flow (Pharmacia, Uppsala, Sweden) column (20 × 1.5 cm) equilibrated with Tris–Ca buffer. The enzyme passes through under these conditions. The column is washed with the same buffer, and the active fractions are pooled.

Step 3: Silica gel filtration. The pooled fractions (20 ml) are applied to a silica gel[8] column (10 × 1 cm) in Tris–Ca buffer. The enzyme passes through, and cytochrome c adheres to the column material. The column is washed with the same buffer, and active fractions are pooled and concentrated (to 2 ml) by using an immersible CX-30 ultrafiltration cartridge (Millipore, Bedford, MA). The concentrated enzyme solution is centrifuged at 48,000 g for 15 min.

Step 4: Gel-filtration chromatography. The concentrated enzyme solution is applied to a high-performance liquid chromatography (HPLC) gel-filtration column (Serva, Heidelberg, FRG, Si 200 polyol, 50 × 0.95 cm) equilibrated with 0.1 M sodium phosphate, pH 7.0. The enzyme is eluted with the same buffer, using a flow rate of 0.5 ml/min.

The result of a typical purification is summarized in Table I. A remarkable increase in activity is observed in the last purification step. The reason for this is unknown.

Properties

Purity. The final preparation is homogeneous, as judged by polyacrylamide gel electrophoresis and Coomassie Brilliant blue staining. An $A_{280 nm}/A_{340nm}$ ratio of 6 can be derived from the absorption spectrum of a homogeneous preparation.[1]

Enzyme Structure and Cofactor Content. The molecular weight of the native enzyme is 100,000, as determined by gel-filtration analysis. Gel filtration of denatured enzyme [incubation for 60 min at 95° in 0.1 M sodium phosphate buffer, pH 7.0, containing 1% (w/v) sodium dodecyl sulfate (SDS) and 1% (w/v) 2-mercaptoethanol] in 0.1 M sodium phosphate buffer, pH 7.0, containing 0.1% SDS, shows a molecular weight of 100,000, indicating that the enzyme is a monomer. On the other hand, studying a similar enzyme, Rupp and Görisch[2] have reported a molecular weight of 120,000 for the native enzyme and 61,000 for the denatured form (determined with polyamide gel electrophoresis), suggesting that it is a dimeric protein. On gel filtration of the denatured enzyme, a protein

---

[8] J. Frank and J. A. Duine, this volume [32].

TABLE I
PURIFICATION OF QUINOPROTEIN ALCOHOL DEHYDROGENASE

Fraction	Total protein (mg)	Specific activity (U/mg protein)	Purification (-fold)
Step 1: Cell-free extract	770	0.15	1
Step 2: DEAE filtration	80	1.1	7
Step 3: Silica gel filtration	30	2.1	14
Step 4: HPLC gel filtration	22	16	106

peak is observed containing no chromophoric substances, whereas a peak showing an absorption spectrum identical to that of PQQ elutes in the fractions containing the low molecular weight compounds. From this it appears that PQQ is the only cofactor in this enzyme, present in a ratio of 2.0 ± 0.2 molecules per enzyme molecule.

Catalytic Properties. Enzyme activity requires the addition of ammonium or primary amine salts as activators to the assay. Maximal activity is observed at pH 9, and the following electron acceptors appear to be active: Wurster's blue, phenazine methosulfate, and horse heart cytochrome *c*. The enzyme shows a broad substrate specificity: primary and secondary alcohols as well as aldehydes are substrates. Irreversible inactivation is observed by incubating the enzyme with cyclopropanol and electron acceptor. A ratio of 1.0 molecule of cyclopropanol per enzyme molecule is required to achieve complete inactivation. On the other hand, Rupp and Görisch[2] reported a ratio of 2 molecules of cyclopropanone ethyl hemiketal per enzyme molecule for inactivation of their enzyme preparation.

Quinohemoprotein Alcohol Dehydrogenase from *Comamonas testosteroni*

Assay Method

Principle. After reconstituting the apoenzyme with PQQ, the activity of the resulting holoenzyme is determined by measuring the butanol-dependent reduction of Wurster's blue.

Reagents

Tris–Ca buffer, 0.1 M Tris-HCl, pH 7.5, containing 3 mM CaCl$_2$
PQQ, 0.2 mM
Wurster's blue,[6] 2 mM
1-Butanol, 3 mM

Procedure. To the cuvette (1 cm optical pathway) are added 0.2 ml Tris–Ca buffer, 50 μl apoenzyme, and 20 μl PQQ solution. After a preincubation for 5 min at 20°, 0.6 ml Tris–Ca buffer and 100 μl Wurster's blue solution are added. The reaction is initiated by the addition of 30 μl butanol solution, followed by mixing. The rate of absorbance decrease is measured at 610 nm, using a molar absorbance coefficient of 12.4 $\times$ 10^3 M^{-1} cm^{-1}. One unit is the amount of enzyme that reduces 2 μmol of Wurster's blue per minute. Protein concentrations are determined by the method of Lowry *et al.*[7]

Purification Procedure

Growth of the Organism. Comamonas testosteroni ATCC 15667 officially does not grow on alcohols.[9] However, growth is observed after an adaptation time of about 3 days on primary alcohols (C$_2$–C$_5$).[5] High yields of enzyme are obtained when cultivation is performed in three phases. It is of crucial importance that growth occurs at a suboptimal temperature without the addition of antifoam agents. The same mineral medium is used as for growth of *P. aeruginosa.*

Phase 1. A 100-ml flask containing 35 ml of mineral medium supplemented with 0.3% ethanol is inoculated with cells from a nutrient broth agar slant. Growth occurs on a reciprocal shaker (250 rpm) at 30° to the end of the exponential phase (48–96 hr).

Phase 2. The entire contents of the first culture flask are added to 1 liter of fresh medium in a 2.5-liter flask. The culture is incubated for 24 hr at 30° on a reciprocal shaker.

Phase 3. A fermentor (20 liters) with 18 liters mineral medium containing 0.6% ethanol is inoculated with the culture from phase 2. The culture is agitated (250 rpm) at 23° for about 110 hr at an air flow of 3.5 liter/min. The pH is controlled and adjusted to pH 7.0 with 2 M NaOH. The cells are harvested by centrifugation at the end of the exponential phase, and the cell paste is stored frozen at −40°.

Purification Steps. Originally,[5] purification of enzyme required four chromatographic steps. A much simpler procedure has been adopted, as described here. If not indicated otherwise, operations are carried out at 4°.

Step 1: Preparation of cell-free extract. Cell paste (55 g) is thawed and suspended in 25 ml Tris–Ca buffer. The suspension is passed twice through a French pressure cell at 110 MPa. Intact cells and cell debris are removed by centrifugation at 48,000 g for 20 min.

---

[9] R. Y. Stanier, N. J. Palleroni, and M. Doudoroff, *J. Gen. Microbiol.* **43**, 159 (1966).

Step 2: CM-Sepharose chromatography. To the supernatant (60 ml) is added 600 ml of Tris–Ca buffer, and the mixture is applied to a CM-Sepharose fast flow (Pharmacia) column (15 × 4 cm), equilibrated with Tris–Ca buffer. After washing with the same buffer, elution of the enzyme is performed using a linear gradient of 0–0.3 M NaCl (400 ml) in 50 mM Tris-HCl buffer, pH 7.5, containing 3 mM CaCl$_2$. Enzyme-containing fractions have a rose-red color. Active fractions are pooled, dialyzed against 2 liters of 20 mM Tris-HCl buffer, pH 7.5, containing 1 mM CaCl$_2$. The dialysis residue is concentrated to 5 ml by pressure filtration using a Pellicon membrane (Millipore), Type PTGC 047.10).

Step 3: Cation-exchange chromatography on Mono S. The nearly homogeneous enzyme is further purified on a FPLC (fast performance liquid chromatography) Mono S 5/5 column (5.0 × 0.5 cm) (Pharmacia). The concentrated enzyme solution (100 μl) from Step 2 is applied to the column, equilibrated with Tris–Ca buffer. Elution of the apoenzyme occurs with a linear gradient of 0–0.3 M NaCl in Tris–Ca buffer (20 min at a flow rate of 0.5 ml/min), with monitoring of the eluate at 280 nm.

The purified enzyme has a specific activity of 20 U/mg protein under the indicated assay conditions. The result of a typical purification is summarized in Table II.[10]

Properties

Purity. Polyacrylamide gel electrophoresis of the final preparation shows a single band with Coomassie Brilliant blue as well as with activity staining.

Absorption Spectra. The final preparation shows a cytochrome c-like spectrum owing to the presence of heme c. The heme c becomes reduced on addition of substrate to the holoenzyme or sodium dithionite to the apoenzyme.

Enzyme Structure and Cofactor Content. The molecular weight of the apoenzyme is 70,000, as determined by gel filtration and polyacrylamide gel electrophoresis. Under denaturing conditions (in the presence of 1% SDS) a molecular weight of 65,000 is found, indicating that the enzyme is a monomer. It contains 1 heme c per apoenzyme molecule, and the addition of 1 molecule of PQQ is required to obtain maximal activity.

Activators and Inhibitors. The enzyme is not activated by amine or ammonium salts. Binding of PQQ requires the presence of calcium ions. Therefore, reconstitution is strongly inhibited by EDTA or zinc salts (the latter probably competing with the calcium-binding sites). Cyclopropanol is not an inhibitor for this enzyme.

[10] P. A. Poels, B. W. Groen, and J. A. Duine. *Eur. J. Biochem.* **166,** 575 (1987).

TABLE II
PURIFICATION OF QUINOHEMOPROTEIN ALCOHOL DEHYDROGENASE APOENZYME

Fraction	Total protein (mg)	Specific activity (U/mg protein)	Purification (-fold)
Step 1: Cell-free extract	2406	2.7[a]	1
Step 2: CM-Sepharose chromatography	390	7.5	2.8
Step 3: Mono S chromatography	290	10	3.7

[a] This value is probably too high since the product (butanal) formed in the assay is a substrate for molybdoprotein aldehyde dehydrogenase,[10] acting with Wurster's blue.

Catalytic Properties. The enzyme has a broad substrate specificity: primary and secondary alcohols as well as aldehydes are substrates. With Wurster's blue as the electron acceptor, maximal activity is found at pH 7.5. As aldehydes are excellent substrates, under certain conditions no free aldehyde is detected in the conversion of alcohols, but the alcohol is directly converted to the corresponding acid.

Electron Acceptors. The electron acceptor specificity is very broad: Wurster's blue, phenazine methosulfate, and potassium ferricyanide can be used effectively.

[8] Toluene Dioxygenase from *Pseudomonas putida* F1

By LAWRENCE P. WACKETT

Toluene dioxygenase is a three-component enzyme system that catalyzes the incorporation of both atoms from the same molecule of dioxygen into toluene to produce (+)-*cis*-1(*S*),2(*R*)-dihydroxy-3-methylcyclohexa-3,5-diene (Scheme 1).[1] The purification of each of the three components, which readily separate during chromatography, has been described previously. The reconstitution of dioxygenase activity requires a flavoprotein designated reductase$_{TOL}$,[2] an iron-sulfur protein denoted ferredoxin$_{TOL}$,[3] and a complex iron-sulfur protein known as ISP$_{TOL}$.[4]

[1] V. M. Kobal, D. T. Gibson, R. E. Davis, and A. Garza, *J. Am. Chem. Soc.* **95**, 4420 (1973).
[2] V. Subramanian, T.-N. Liu, W.-K. Yeh, M. Narro, and D. T. Gibson, *J. Biol. Chem.* **256**, 2723 (1981).
[3] V. Subramanian, T.-N. Liu, W.-K. Yeh, C. Serdar, L. P. Wackett, and D. T. Gibson, *J. Biol. Chem.* **260**, 2355 (1985).
[4] V. Subramanian, T.-N. Liu, W.-K. Yeh, and D. T. Gibson, *Biochem. Biophys. Res. Commun.* **91**, 1131 (1979).

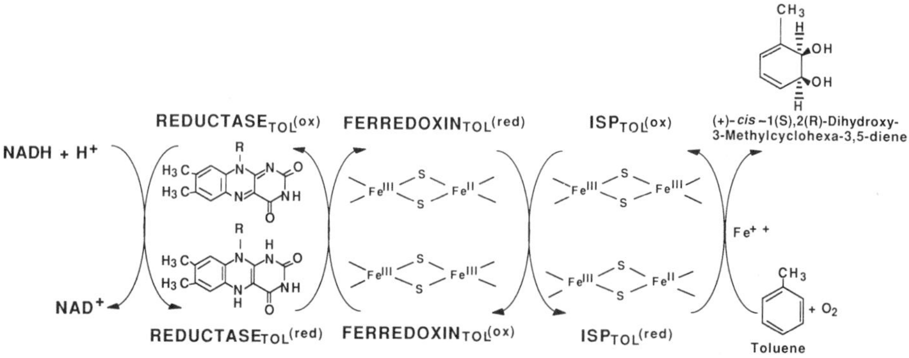

SCHEME 1. The three protein components of toluene dioxygenase, their redox-active groups, and the reaction catalyzed by the enzyme.

Assay Methods

Principle. Using crude extract or a reconstituted three-component system, dioxygenase activity is quantitatively determined with a sensitive radiometric assay. The standard protocol is an adaption of a published method[5] in which the polar product formed by the dioxygenation of toluene binds to silica gel and unreacted [^{14}C]toluene is removed by volatilization. The description of an alternative spectrophotometric assay for toluene dioxygenase using indole as the substrate has been made by Jenkins and Dalton.[6] The reductase$_{TOL}$ and ferredoxin$_{TOL}$ components can be assayed in the absence of ISP$_{TOL}$ by measuring the NADH-dependent reduction of cytochrome c spectrophotometrically. Crude cell extracts contain a number of proteins that reduce cytochrome c directly. However, following DEAE-cellulose chromatography, fractions containing reductase$_{TOL}$ and ferredoxin$_{TOL}$, which do not reduce cytochrome c each by themselves, can be combined to reconstitute cytochrome c reductase activity.

Reagents for Radiometric Assay

Tris-HCl, 50 mM, pH 7.5

Ferrous sulfate heptahydrate, 14.4 mM, freshly prepared in deionized water

NADH, 24 mM in Tris buffer, pH 7.5

[*methyl*-^{14}C]Toluene (Amersham, Arlington Heights, IL), 1.67 mM in N,N-dimethylformamide, 1.1×10^8 dpm/ml

[5] W.-K. Yeh, D. T. Gibson, and T.-N. Liu, *Biochem. Biophys. Res. Commun.* **78**, 401 (1977).

[6] R. O. Jenkins and H. Dalton, *FEMS Microbiol. Lett.* **30**, 227, (1985).

Reagents for Cytochrome c Reduction Assay

Tris-HCl, 50 mM, pH 7.5

Cytochrome c (Type V, from bovine heart; Sigma, St. Louis, MO), 4.36 mM in Tris buffer, pH 7.5

NADH, 12 mM in Tris buffer, pH 7.5

Procedure for Radiometric Assay. The assay mixture contains, in a final volume of 400 μl, Tris buffer, 320 μl; the three enzyme components, 30 μl; and ferrous sulfate, 10 μl. After incubation at room temperature for 2 min, 20 μl each of NADH and [^{14}C]toluene are added to the enzyme mixture with gentle shaking. The reaction is allowed to proceed for 5 min during which dioxygenase activity increases in a linear fashion with respect to time. A 10-μl aliquot is removed and spotted onto a 1 $\times$ 1 cm silica gel square cut from a plastic-backed thin-layer chromatography (TLC) sheet (100-μm-thick silica gel, Kodak chromagram sheet 13181). The TLC square is allowed to air dry in a fume hood for 25 min, which is sufficient for remaining [^{14}C]toluene to be removed by volatilization. The TLC square is then placed in a scintillation vial, 8 ml of Amersham OCS scintillation fluid is added, and the remaining radioactivity is determined in a scintillation counter. It is important to conduct nonenzyme controls to ascertain the background in the absence of toluene dioxygenase. The observed background radioactivity, typically 50–500 dpm, is variable with the purchased lot of [^{14}C]toluene and is probably reflective of different degrees of polar contaminants in the commercially prepared [^{14}C]toluene. A reliable measure of enzyme activity gives radioactivity that is at least 4 times higher than the control level.

Procedure for Cytochrome c Reduction Assay. The assay mixture contains, in a final volume of 1.0 ml, Tris buffer, 800 μl; water, 120 μl; cytochrome c, 20 μl; ferredoxin$_{TOL}$, 20 μl; reductase$_{TOL}$, 20 μl; and NADH, 20 μl, to initiate the reaction. The reduction of cytochrome c is followed continuously using a spectrophotometer operating at 550 nm. The activity can be quantitatively determined using the published extinction coefficient difference of 21 cm^{-1} mM^{-1} for the reduced minus oxidized forms of cytochrome c at 550 nm.[7] Significant levels of NADH-dependent cytochrome c reduction require both reductase$_{TOL}$ and ferredoxin$_{TOL}$ components, with electrons transferred with the ferredoxin to the cytochrome. However, old preparations of reductase$_{TOL}$ will catalyze low levels of NADH-dependent cytochrome c reduction in the absence of ferredoxin$_{TOL}$.

---

[7] T. Ueda, E. T. Lode, and M. J. Coon, *J. Biol. Chem.* **247**, 2109 (1972).

Purification Procedure

Growth of Organism. Pseudomonas putida F1 was isolated from soil in 1968 by David Gibson and co-workers.[8] It is grown in a 12-liter fermentor at 30° with toluene as the sole source of carbon in a mineral medium that has been described previously.[9] Toluene is supplied as a vapor by passing the air to be sparged into the fermentor vessel over liquid toluene. To obtain large amounts of cell paste, consecutive fermentations are conducted by draining all but 1 liter of the culture and topping with fresh medium. In this way, 50–100 g wet weight of cells is obtained for every growth cycle, which typically takes 4–8 hr. It is important for preparing highly active cell-free extracts to harvest the cells while still in exponential growth phase (absorbance at 600 nm ≤ 2.0). The cell paste is made by harvesting cells from fermentation cultures with a Sharples continuous centrifuge or by using a Pellicon cell concentrator (Amicon, Danvers, MA) coupled with a conventional preparative centrifuge. Cell pastes have been stored at −20 or −70° for at least 1 year without significant loss of toluene dioxygenase activity.

Preparation of Cell-Free Extract. It is crucial to disrupt frozen cells of *P. putida* F1 in buffers containing ethanol and glycerol to maintain a significant percentage of toluene dioxygenase activity during purification.[10] A good buffer in this regard is 50 mM potassium phosphate, pH 7.2, containing 10% (v/v) each of ethanol and glycerol (PEG). The cells are thawed at a ratio of 1 g cells per 3 ml PEG buffer and maintained at 0–5° during subsequent protocols. Cell disruption is accomplished by two passages through a French pressure cell at 20,000 psi. The lysate is then processed by centrifugation at 4° and 30,000 g for 30 min. The supernatant fluid is poured off and further centrifuged at 100,000 g for 60 min. The high-speed supernatant liquid, which is decanted through glass wool into a cold flask, serves as the starting point for chromatographic purification of toluene dioxygenase.

Affinity Chromatography. With all chromatographic procedures currently attempted, toluene dioxygenase separates into its component proteins, which must be reconstituted to obtain toluene dioxygenase activity. A protocol that revolves around affinity chromatography is shown as a flow diagram in Fig. 1. This procedure affords a rapid purification of the unstable ISP$_{TOL}$ component and a convenient purification of reductase$_{TOL}$ and ferredoxin$_{TOL}$.

[8] D. T. Gibson, J. R. Koch, and R. E. Kallio, *Biochemistry* **7**, 2653 (1968).
[9] R. Y. Stanier, N. J. Palleroni, and M. Doudoroff, *J. Gen. Microbiol.* **43**, 159 (1966).
[10] V. Subramanian and D. T. Gibson, unpublished data (1978).

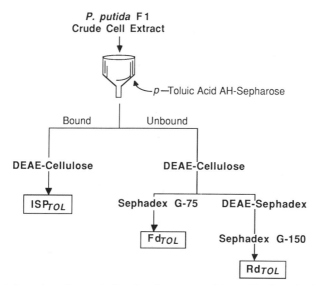

Fig. 1. A flow chart demonstrating the chromatographic purification of toluene dioxygenase components.

Synthesis of the *p*-toluic acid affinity matrix is as described by Subramanian *et al.*[4] The coupling of *p*-toluic acid to aminohexyl-Sepharose 4B can be spot tested by suspending an aliquot of the gel in 0.5 ml of a 0.1% (w/v) ninhydrin solution and heating to 100° for 5 min. The appearance of a deep blue color indicates the presence of the amide-linked ligand. The *p*-toluic acid-Sepharose gel is washed according to the manufacturer's instructions, packed into a glass column (20 × 1.6 cm for 2–10 g protein loaded), and equilibrated with N_2-saturated PEG buffer. It is necessary to transfer electrons to the ISP_{TOL} component for it to bind to the affinity column. Thus, an active cell-free extract is first incubated with 120 μM NADH for 15 min. Afterward, the extract is dialyzed against 20 volumes of N_2-saturated PEG buffer for 2 hr and then immediately loaded onto the affinity column at a flow rate of 120 ml/hr. In a typical purification with a 20 × 1.6 cm affinity matrix, the column is washed with 1 liter of N_2-saturated PEG buffer. Unbound protein through the first 400 ml of the wash is saved as a source of reductase$_{TOL}$ and ferredoxin$_{TOL}$ for subsequent purification. It is also used for assay of the ISP_{TOL} component. The material bound to the affinity column is eluted with a 400 ml linear gradient of 0.0–0.4 M KCl in PEG buffer. The dioxygenase protein generally elutes as a broad peak centered at a salt concentration that is somewhat variable from preparation to preparation. It is also worth noting that the affinity gel is most useful for two to four enzyme preparations if it is extensively washed between applications. The observed diminution in ISP_{TOL} binding thereafter might be reflective of ligand destruction by enzymatic oxidation.

Active fractions from the *p*-toluic acid-Sepharose column are pooled, diluted with 2 volumes of PEG buffer, dialyzed against 20 volumes of PEG for 3 hr, and then loaded onto a DEAE-cellulose column (24 × 2.4 cm) at a flow rate of 80 ml/hr. After washing with an equivalent volume of PEG buffer containing 0.1 M KCl, the ISP_{TOL} component is eluted with a linear gradient of 0.1–0.45 M KCl in PEG using a volume double that of the wash. The purification data for ISP_{TOL} are shown in Table I.

As shown in Fig. 1, the unbound material from the affinity column serves as the starting point for the purification of reductase$_{TOL}$ and ferredoxin$_{TOL}$. These two components can be completely resolved by DEAE-cellulose chromatography, and admixtures of appropriate fractions can be used to locate each protein using the cytochrome *c* reduction assay. In a representative purification, the affinity column wash through (600 ml, 3 g protein) is loaded onto a DEAE-cellulose column (32 × 2 cm) equilibrated in PEG buffer. After washing the column with 400 ml of PEG buffer, a 2 liter 0.0–0.4 M KCl gradient in the same buffer is used to elute the toluene dioxygenase components. Reductase$_{TOL}$ elutes first, with peak activity at about 0.10 M KCl, with ferredoxin$_{TOL}$ eluting at 0.25 M KCl.

The fractions containing ferredoxin are concentrated to approximately

TABLE I

PURIFICATION OF THE THREE COMPONENTS OF TOLUENE DIOXYGENASE[a]

Protein component	Purification step	Protein (mg)	Activity (units)	Specific activity (units/mg)	Recovery (%)
ISP_{TOL}	Crude extract	2380	760[b]	0.32[b]	100
	Affinity chromatography	114	588	5.14	77
	DEAE-cellulose	62	469	7.50	62
Reductase$_{TOL}$	DEAE-cellulose	673	466[c]	0.69[c]	100
	DEAE-Sephadex	106.5	336	3.15	72
	Sephadex G-200	18.2	224	12.3	48.4
Ferredoxin$_{TOL}$	DEAE-cellulose	283	568[c]	2.0[c]	100
	Sephadex G-75	15.3	272	17.8	47.9

[a] The starting material for reductase$_{TOL}$ and ferredoxin$_{TOL}$ purifications are the affinity column eluate from the ISP_{TOL} purification. The data shown are for purifications from independent extracts and have been published previously.[2-4]

[b] Dioxygenase activity determined by the radiometric assay; units are defined as nmol of product formed/min.

[c] NADH-dependent cytochrome-*c* reductase activity; units are defined as μmol of product formed/min.

5 ml by ultrafiltration using an Amicon UM10 filter. The protein concentrate is then loaded in two separate batches onto Sephadex G-75 superfine (85 × 2.5 cm) and eluted with PEG buffer. Ferredoxin$_{TOL}$ emerges in a homogeneous form as the middle peak of three proteins, which are detected by absorbance at 280 nm.

Reductase$_{TOL}$ fractions from DEAE-cellulose are diluted with 2 volumes of PEG buffer and applied to the top of a DEAE-Sephadex A-50 column (20 × 2 cm). The column is then washed with 500 ml of PEG buffer and further eluted with 500 ml of a linear 0.0–0.3 M KCl gradient in PEG. The reductase fractions, which elute at approximately 0.18–0.26 M KCl, are pooled, and the protein is precipitated with ammonium sulfate brought to 70% of saturation. The precipitate is collected by centrifugation, redissolved in 5 ml PEG, and applied to a Sephadex G-200 column (82 × 2.5 cm). The reductase elutes as the second major peak as determined by the cytochrome c reduction assay reconstituted with purified ferredoxin$_{TOL}$. Purification sequences for reductase$_{TOL}$ and ferredoxin$_{TOL}$ are shown in Table I.

Properties

Purity. The degree of purification is assessed by gel-filtration chromatography and polyacrylamide gel electrophoresis in the absence or the presence of sodium dodecyl sulfate as described previously.[2-4] The protocols described herein typically yield reductase$_{TOL}$ and ferredoxin$_{TOL}$ at 90% or greater purity. The affinity column purification shows greater variability, with preparations generally ranging from 50% to 95% homogeneous.

Stability. The reductase and ferredoxin components are generally amenable to storage, with most activity being retained after 6 months at −70°. In contrast, ISP$_{TOL}$ is more unstable, and activity is best maintained by storage of the enzyme as an ammonium sulfate precipitate.

Molecular Weights. Reductase$_{TOL}$ and ferredoxin$_{TOL}$ are single subunit proteins with molecular weights of 46,000 and 15,300, respectively. ISP$_{TOL}$ has a holoenzyme molecular weight of 151,000 ($\alpha_2\beta_2$) with subunit molecular weights of 52,500 and 20,800.

Cofactor Content. Reductase$_{TOL}$ is a flavoprotein,[2] and the other two components of toluene dioxygenase are [2Fe–2S] proteins.[3,4] Fully active reductase contains 1 mol of noncovalently bound FAD per mole protein. Purified reductase$_{TOL}$ is deficient in activity owing to loss of the flavin cofactor during chromatography and must be reconstituted with FAD to regain full activity.[2] Ferredoxin contains one [2Fe–2S] cluster which is thought to be of the Rieske type.[3] Fully active ISP$_{TOL}$ contains 4–6 iron atoms per holoenzyme,[10] but the detailed nature of the iron site(s) in this protein remains to be elucidated.

[9] Naphthalene Dioxygenase from *Pseudomonas* NCIB 9816

By BURT D. ENSLEY and BILLY E. HAIGLER

$$\text{NADH} + \text{H}^+ + \text{O}_2 + \text{Naphthalene} \rightarrow$$
$$cis\text{-}(1R,2S)\text{-dihydroxy-1,2-dihydronaphthalene} + \text{NAD}^+$$

Naphthalene dioxygenase (NDO) is a three-component enzyme system that catalyzes the initial step in naphthalene metabolism by some species of *Pseudomonas*. The first component of the enzyme system is a flavin- and iron-containing oxidoreductase (reductase$_{NAP}$) that transfers electrons from NADH to the second component in the system, a ferredoxin (ferredoxin$_{NAP}$), which in turn transfers the electrons to the iron-sulfur protein (ISP$_{NAP}$) terminal component in the system. The ISP$_{NAP}$ reacts with naphthalene and both atoms of molecular oxygen to form the oxidized product. The purification and assay of all three components in this enzyme system are described here. Purification of the three components of naphthalene dioxygenase has been the subject of previous reports.[1-3] A partial nucleotide sequence of the genes encoding naphthalene dioxygenase and the deduced amino acid sequence have been disclosed recently,[4] and the iron-sulfur protein (ISP$_{NAP}$) can also be purified from extracts of a recombinant *Escherichia coli*.[5]

Assay Methods

Although naphthalene oxygenase activity can be measured using a spectrophotometric assay with whole cells,[6] we have found an assay based on the conversion of [^{14}C]naphthalene to nonvolatile metabolites (J. E. Rogers, personal communication) to be the most convenient for measuring activity in crude cell extracts and purified enzyme systems. This assay is sensitive, specific, and reproducible.

[1] B. D. Ensley and D. T. Gibson, *J. Bacteriol.* **155**, 505 (1983).

[2] B. E. Haigler and D. T. Gibson, *J. Bacteriol.* **172**, 457 (1990).

[3] B. E. Haigler and D. T. Gibson, *J. Bacteriol.* **172**, 465 (1990).

[4] S. Kurkela, H. Lehvšlaiho, E. Palva, and T. Teeri, *Gene* **73**, 355 (1988).

[5] B. D. Ensley, T. D. Osslund, M. Joyce, and M. J. Simon, *in* "Microbial Metabolism and the Carbon Cycle" (S. Hagedorn, R. Hansen, and D. Kunz, eds.), p. 437. Harwood Academic, New York, 1988.

[6] K. M. Shamsuzzaman and E. A. Barnsley, *J. Gen. Microbiol.* **83**, 165 (1974).

Reagents

Tris-HCl buffer, 50 mM, pH 7.5

NADH (Sigma, St. Louis, MO), 50 mM in 1 ml Tris-HCl buffer, made up fresh

FAD, 5 mM in Tris-HCl buffer

[1(4,5,8)-^{14}C]Naphthalene (Amersham, Arlington Heights, IL), specific activity 5 mCi/mmol; 10 mM in dimethylformamide

Precoated thin-layer chromatography sheets (MC/B Manufacturing Chemists Inc.) ruled off into 1.5 $\times$ 2 cm rectangles.

Assay Procedures. To a 13 $\times$ 100 mm test tube, add Tris-HCl buffer to give a final volume of 0.4 ml and 20–100 μl of crude cell extract or appropriate amounts of partially purified components of naphthalene dioxygenase. Allow the reaction mixture to equilibrate at room temperature for 5 min. Add 20 μl of the FAD and NADH solutions, vortex, and start the reaction by adding 5–10 μl of dimethylformamide containing [^{14}C]naphthalene. Vigorous vortexing is requiring to obtain repeatable results. Incubate the reaction mixture for 5 min at room temperature and transfer 20 μl of the reaction mixture to the center of a 1.5 $\times$ 2 cm rectangle drawn on a precoated plastic-backed silica gel chromatography sheet. This sheet is immediately dried under a stream of air at room temperature for 15–20 min to remove water and unmetabolized [^{14}C]naphthalene. The square can be cut out of the sheet with a pair of a scissors, transferred to a liquid scintillation vial, mixed with 5–7 ml of liquid scintillation cocktail, vortexed, and remaining nonvolatile [^{14}C]naphthalene metabolites measured in a liquid scintillation counter.

Assay for ISP$_{NAP}$

Purified or partially purified ISP$_{NAP}$ activity can only be assayed in the presence of the other two components of the naphthalene dioxygenase system. Spinach ferredoxin and ferredoxin reductase are not acceptable substitutes. ISP$_{NAP}$ activity is measured in the presence of saturating amounts of the reductase$_{NAP}$ and ferredoxin$_{NAP}$ components. Saturating amounts of the other two components must be determined empirically because the activity of these components can vary considerably depending on the age, method of storage, and purification of the components. Typically, a reaction mixture will contain 100 μg of ferredoxin$_{NAP}$ and 50–100 μg of reductase$_{NAP}$ in 0.4 ml of Tris buffer. The ISP$_{NAP}$ is added with the other two components, allowed to equilibrate at room temperature for 5 min, and assayed as described above. One unit of ISP$_{NAP}$ is defined as the amount of enzyme that will catalyze the conversion of 1 nmol of [^{14}C]naphthalene to nonvolatile metabolites in 1 min. The reaction rate is linear with time for the 5-min incubation period, but it is linear with the amount of ISP_{NAP} added to the reaction only if the activities of the other two components are truly saturating at the concentrations added.

Assay for Reductase$_{NAP}$

Although the reductase component can be assayed using the naphthalene dioxygenase assay, a more convenient and rapid determination for this enzyme is accomplished by measuring cytochrome c reduction in the presence of NADH and FAD. More than one protein capable of reducing cytochrome c is present in crude cell extracts, but only one is capable of supporting naphthalene dioxygenase activity. Thus, only the cytochrome-c reductase activity obtained after the first chromatographic step is representative of reductase$_{NAP}$. A typical cytochrome-c reductase assay contains the following in 1.0 ml of 50 mM Tris-HCl buffer (pH 7.5): 88 nmol of cytochrome c, 300 nmol of NADH, 1.0 nmol of FAD, and appropriate amounts of reductase$_{NAP}$. The change in absorbance at 550 nm is measured in a recording spectrophotometer, and a molar extinction coefficient of $2.1 \times 10^4 \, M^{-1} \, cm^{-1}$ is used to calculate concentrations of cytochrome c reduced.[7] One unit of NADH–cytochrome-c reductase activity is the amount of protein required to catalyze the reduction of 1.0 μmol of cytochrome c per min.

Reagents for Reductase Assay

Tris-HCl buffer, 50 mM, pH 7.5
Cytochrome c, 4.4 mM in Tris-HCl buffer
NADH, 15 mM in Tris-HCl buffer, made up fresh
FAD, 0.1mM in Tris-HCl buffer

Procedures. The assay is performed in 1-ml cuvettes as follows. To the reference cuvette add 0.97 ml Tris-HCl buffer, 20 μl of cytochrome c solution, and 10 μl of the FAD solution. To the sample cuvette add sufficient Tris-HCl buffer to give a final volume to 1 ml, 20 μl of the cytochrome c solution, 10 μl of the FAD solution, and 10–20 μl of the enzyme preparation. Record the rate of background cytochrome c reduction at 550 nm. Add 20 μl of the NADH solution to both cuvettes and record the rate of cytochrome c reduction. Subtract the rate of background cytochrome c reduction from the rate after addition of the NADH.

---

[7] T. Ueda, E. T. Lode, and M. J. Coon, *J. Biol. Chem.* **247,** 2109 (1972).

Assay for Ferredoxin$_{NAP}$

Ferredoxin$_{NAP}$-dependent naphthalene dioxygenase activity is determined by measuring the rate of formation of [^{14}C]naphthalene dihydrodiol from [^{14}C]naphthalene as described above. Each assay contains, in a final volume of 0.4 ml of 50 mM Tris-HCl (pH 7.5), 1.0 μmol of NADH, 1 nmol FAD, 100 nmol [^{14}C]naphthalene, and saturating amounts of ISP$_{NAP}$ and reductase$_{NAP}$. The background activity present in control assays lacking ferredoxin$_{NAP}$ should be subtracted from all assays.

Purification Procedure

Growth of Organism

Pseudomonas species strain NCIB 9816 provided by W. C. Evans, University College of North Wales, Bangor, Wales, was used as a source of the enzyme. The organism is maintained on mineral salts agar plates,[6] with naphthalene supplied as vapor by placing crystals of naphthalene in the lid of the petri dish. Liquid cultures are grown in mineral salts medium containing 0.1% (w/v) naphthalene at 30° for 24–36 hr with shaking. Large quantities of cells can be prepared by inoculating a log-phase liquid culture into 8 liters of the same medium containing 0.2% (w/v) of naphthalene in a 10-liter fermentor. Air is supplied at 12 liters/min and 600 rpm agitation at 30°. Near the end of the logarithmic phase of growth (24–36 hr), the culture fluid is filtered through glass wool, and the cells are harvested by centrifugation. The cell paste is stored at −20° until used.

Preparation of Crude Cell Extracts

The frozen cell paste (50 g wet weight) is thawed in 150 ml of 50 mM Tris-HCl buffer, pH 7.8, containing 10% (v/v) ethanol, 10% (v/v) glycerol, and 0.5 mM dithiothreitol (TEG buffer). The cells are broken by two passages through a French press at 12,000 psi and 4°. The cell suspension is filtered through glass wool, incubated with 0.5 mg of DNase for 15 min at 4°, and centrifuged at 10,000 g for 15 min. The supernatant solution is centrifuged at 105,000 g for 1 hr and carefully decanted to give a clear crude extract solution.

Purification of Reductase$_{NAP}$

The purification of reductase$_{NAP}$ from crude cell extracts can be accomplished by the two-column procedure described below. All procedures should be carried out at 4°. This procedure employs a pseudoaffinity column containing the dye Cibacron Blue F3G-A bound to a cross-linked agarose gel such as Sepharose CL-6B. This dye has an affinity for enzymes requiring adenylyl-containing cofactors, including NAD. The gel matrix can be manufactured in the laboratory[8] or purchased from any of several commercial sources. Ferredoxin$_{NAP}$ and ISP$_{NAP}$ do not bind to this column and are eluted during the buffer wash. If further purification of these enzyme components is desired, these fractions should be assayed for the appropriate naphthalene dioxygenase activity. Purification of ferredoxin$_{NAP}$ and ISP$_{NAP}$ from these fractions is described later.

Apply crude cell extract (~7 g of protein in 150 ml of TEG buffer plus 1 mM dithiothreitol) to a blue dye-Sepharose column (2.6 × 28 cm) previously equilibrated with TEG buffer. Wash the column with sufficient buffer (0.5–0.8 liters) to bring the absorbance at 280 nm to 0.1 to 0.2. Because the reductase does not bind tightly to the blue dye-ligand column in crude cell extracts, the initial flow rate during application and the buffer wash should be kept low (<30 ml/hr). In addition, commercial preparations of blue dye-Sepharose often exhibit variability in binding capacity. Thus, conditions should be optimized prior to a large-scale purification. Elute the reductase$_{NAP}$ from the column with a continuous KCl gradient (0.0–220 mM) in 1 liter of TEG buffer. Collect fractions of 8–10 ml and assay for cytochrome-c reductase activity. Pool the appropriate fractions containing reductase$_{NAP}$-dependent cytochrome-c reductase activity. Concentrate and desalt the solution by ultrafiltration over an Amicon PM10 filter under N$_2$ (35 psi). Two to three equal volume buffer additions are sufficient to lower the salt concentration prior to DEAE-cellulose chromatography. The KCl can be removed by dialysis if the time is held to 3–5 hr; overnight dialysis results in a significant loss of activity. Apply the concentrated solution to the top of a DEAE-cellulose column (2.3 × 23 cm) previously equilibrated with TEG buffer. Wash the column with 500 ml of TEG buffer and elute the reductase with 500 ml of a cotinuous KCl gradient (0.0–120 mM). Collect fractions of 5–7 ml and assay for cytochrome-c reductast activity. Pool the fractions containing cytochrome-c reductase activity. Concentrate and dialyze the purified reductase$_{NAP}$ by ultrafiltration as described above. Store the purified protein at −20°.

Purification of Ferredoxin$_{NAP}$

Homogeneous ferredoxin$_{NAP}$ is obtained after three column chromatographic steps following separation from reductase$_{NAP}$ by blue dye-ligand chromatography. Ferredoxin$_{NAP}$ is located in column eluates by its ability to catalyze naphthalene dioxygenase activity in the presence of reduc-

[8] H. J. Bohme, G. Kopperschlager, J. Schulz, and E. Hofmann, *J. Chromatogr.* **69**, 209 (1972).

tase$_{NAP}$, ISP$_{NAP}$, FAD, and NADH. All procedures should be carried out at 4°.

Although ISP$_{NAP}$ and ferredoxin$_{NAP}$ do not bind to the blue dye-ligand column, their progress through the column is hindered owing to the molecular weight sieving action of the gel matrix. Thus, the fractions that elute in the void volume and buffer wash should be assayed for ferredoxin$_{NAP}$- and ISP$_{NAP}$-dependent naphthalene dioxygenase activity. Pool the fractions containing these activities and centrifuge at 10,000 g for 15 min to remove particulate matter. Owing to the low yield of ferredoxin$_{NAP}$ from this quantity of cell extract, we have found it convenient to store this fraction at $-20°$ and later to pool it with another such fraction. If this procedure is followed, the solution should be centrifuged as described above to remove particulates prior to the next column chromatographic step.

Apply the protein solution to a DEAE-cellulose column (4 × 12 cm) previously equilibrated with TEG buffer. Wash the column with 1 liter of TEG buffer. ISP$_{NAP}$ and ferredoxin$_{NAP}$ are eluted and separated with a continuous KCl gradient (0.0–0.4 M KCl) in 1 liter of TEG buffer. Collect fractions (10.0 ml) and assay for ISP$_{NAP}$- and ferredoxin$_{NAP}$-dependent naphthalene dioxygenase activity as described above. Pool the fractions which contain ferredoxin$_{NAP}$ and those containing ISP$_{NAP}$ activity separately. Fractions containing ISP$_{NAP}$ can be stored at $-20°$ and used in enzyme assays or later purified as described below. Concentrate the ferredoxin$_{NAP}$ solution over an Amicon filter (YM5 membrane) under N$_2$ (35 psi) to a final concentration of 5–7 ml. Apply the concentrated solution to the top of a column containing Sephadex G-75 superfine (2.6 × 90 cm) equilibrated with TEG buffer. Elute the protein with TEG buffer, collect fractions (2–4 ml), and assay for ferredoxin$_{NAP}$ activity. Pool the appropriate fractions containing ferredoxin$_{NAP}$ and apply the solution to a small column of DEAE-cellulose (1 × 8 cm) equilibrated with TEG buffer and wash with 50 ml of buffer. Elute the protein with a 100 ml continuous KCl gradient (0–800 mM) in TEG buffer. Collect fractions (2–3 ml) and assay as before. A single protein peak which coincides with ferredoxin$_{NAP}$ activity should be observed. Pool the appropriate fractions and dialyze against TEG buffer for 3 hr to remove KCl. Store the purified protein at $-20°$.

Purification of ISP$_{NAP}$

Fractions showing peak ISP$_{NAP}$ activity from the DEAE column described above are pooled and concentrated for the next step. The pooled fractions containing ISP$_{NAP}$ are concentrated over an Amicon XM50 membrane under N$_2$ at 30–40 psi to a final volume of 10–20 ml. The concentrated enzyme solution is then slowly brought to 40% saturation with ammonium sulfate by adding a saturated (NH$_4$)$_2$SO$_4$ solution in TEG buffer. The ammonium sulfate-saturated TEG should be prepared 2–3 days ahead of time and stored at 4°. Although it is possible to concentrate the enzyme solution by precipitating the enzyme with ammonium sulfate at 60–80% saturation and dissolving the precipitated enzyme in 40% ammonium sulfate-saturated TEG, concentrated enzyme solutions prepared in this manner are extremely difficult to elute in the next column step.

A slight precipitate forms on addition of ammonium sulfate to 40% saturation and can be removed by centrifugation at 10,000 g for 15 min. The supernatant solution is applied to a column (1.5 × 20.0 cm) of octyl-Sepharose 4B previously equilibrated with TEG buffer containing 40% ammonium sulfate. As the concentrated protein solution washes into the column, a dark red band can be seen forming at the top. The column is then washed with 50 ml of TEG buffer containing 40% ammonium sulfate, and the ISP_{NAP} is eluted with 200 ml of the same buffer containing 30% ammonium sulfate. Fractions of 8–10 ml each are collected and assayed for naphthalene dioxygenase activity. It is important to use a minimum volume of column fractions from this step when assaying for naphthalene dioxygenase activity since the ammonium sulfate in the buffer appears to interfere with the enzyme assay. Column fractions showing peak activity of ISP_{NAP} are pooled, dialyzed overnight against 2 liters of TEG buffer, concentrated to approximately 10 ml over an Amicon XM50 membrane, and stored at −20°.

[10] Benzene Dioxygenase from *Pseudomonas putida* ML2 (NCIB 12190)

By Philip J. Geary, Jeremy R. Mason, and Chris L. Joannou

Benzene + O_2 + NADH·H^+ → *cis*-1,2-dihydroxycyclohexa-3,5-diene + NAD^+

Benzene dioxygenase catalyzes the first step in the assimilation of benzene by *Pseudomonas putida*.[1] It comprises three components: a ferredoxin-NAD^+ reductase (A_2), an intermediate electron-transfer protein (B), and the terminal dioxygenase (A_1).[2] By analogy with similar systems it is proposed that these should be known respectively as reductase(ben), ferredoxin(ben), and ISP(ben). ISP(ben) comprises two subunits (α and β) in an $\alpha_2\beta_2$ configuration.[3]

[1] B. C. Axcell and P. J. Geary, *Biochem. J.* **146,** 173 (1975).
[2] S. E. Crutcher and P. J. Geary, *Biochem. J.* **177,** 393 (1979).
[3] M. Zamanian and J. R. Mason, *Biochem. J.* **244,** 611 (1987).

Assay Methods

Principle. Assay of benzene dioxygenase depends on the measurement of oxygen uptake, for which the preferred method uses a Clark-type oxygen electrode. A number of points should be considered when performing the assay. (1) The enzyme has a low turnover number; hence, to achieve realistic rates it is necessary to use milligram quantities of crude protein (cell-free extract). (2) The electron-transfer chain is saturated at relatively low concentrations (50–70 μg/ml) of the reductase component, suitably reactivated with FAD.[4] Under these conditions the reaction rate is directly proportional to the concentrations of both ISP(ben) and ferredoxin(ben) within the ranges investigated experimentally (2 and 1 mg/ml, respectively). As a result of this, assays using crude extract are nonlinear with respect to total protein concentration, and care must be taken to define and standardize conditions for the assay of individual components of the system. (3) The presence of catechol 1,2-dioxygenase (pyrocatechase) together with *cis*-1,2-dihydrobenzene-1,2,-diol dehydrogenase (*cis*-benzene glycol dehydrogenase)[5] in less purified preparations effectively doubles the reaction rate. (4) The reductase may be assayed independently using dichlorophenolindophenol (DCPIP)[1] and monitoring the reaction spectrophotometrically.

Reagents for Method 1 (Oxygen Electrode)
Potassium phosphate buffer, 25 mM, pH 7.2 (buffer A)
NADH, 3.3 mM in buffer A
FeSO$_4$·7H$_2$O, 2 mM freshly prepared in ice-cold distilled water and stored in ice
Benzene, 11.2 mM (1 μl/ml of distilled water)
Components of the benzene dioxygenase system and associated enzymes, namely, reductase(ben), 1 mg/ml in buffer A, reactivated with FAD;[4] ferredoxin(ben), 1 mg/ml in buffer A; and ISP(ben), 1 mg/ml in buffer A [ISP(ben) may be preactivated with Fe^{2+}, but the Fe^{2+} addition is still made to the reaction mixture[6]]

During purification, fractions may be located using combinations of the above pure proteins with column eluates, ammonium sulfate cuts, etc. In order that direct comparisons of activity may be made, the addition of catechol 1,2-dioxygenase and/or *cis*-benzene glycol dehydrogenase is often beneficial.[1] These may be obtained during the purification of benzene dioxygenase[2,4] or independently.[5] Addition is at the discretion of the researcher (100 μg each).

[4] P. J. Geary and D. P. E. Dickson, *Biochem. J.* **195**, 199 (1981).
[5] B. C. Axcell and P. J. Geary, *Biochem. J.* **136**, 927 (1973).
[6] P. J. Geary, F. Saboowalla, D. Patil, and R. Cammack, *Biochem. J.* **217**, 667 (1984).

Catechol 1,2-dioxygenase (pyrocatechase), 10 mg/ml in buffer A
cis-Benzene glycol dehydrogenase 10 mg/ml in buffer A

Reagents for Method 2 (Spectrophotometry)

Buffer A as above
NADH, 1 mM in buffer A
DCPIP, 2.5 mM in distilled water
Reductase(ben), 3 mg/ml in buffer A, reactivated with FAD[4]

Procedure for Method 1. The assay is performed in a Clark-type oxygen electrode cell with a total reaction volume of 500 μl, which obviates the need to concentrate column fractions. Reaction mixtures contain 100 μl (100 μg) of two out of the three components ISP(ben), ferredoxin(ben), or reductase(ben); 200 μl of a complementary column fraction (for example); Fe^{2+} solution (25 μl); and NADH solution (25 μl). The endogenous rate of oxygen uptake is followed for 3 min, following which the reaction is initiated by the addition of benzene solution (50 μl). The rate of oxygen uptake is recorded and corrected for the endogenous rate. (At 30° the oxygen available at saturation is ~0.12 μmol.[7])

Note that the addition of Fe^{2+} causes an apparent oxygen uptake, the effect of which may be minimized by allowing a 1-min incubation period prior to the addition of NADH. Incomplete activation by Fe^{2+} of ISP(ben) may cause the reaction rate to increase initially. This may be overcome by using ISP(ben) which has been preactivated under reducing conditions using Fe^{2+}, dithiothreitol, and mercaptoethanol.[6,8] A further cause of this phenomenon is the further degradation of cis-benzene glycol, catechol, etc., particularly in crude extracts. In these cases rates may be obtained from the linear portion of the oxygen uptake plot which occurs several minutes after initiation. It is advisable to use such rates qualitatively rather than quantitatively.

Procedure for Method 2. The assay mixture contains buffer A (0.87 − x ml) to which is added DCPIP solution (30 μl) and x ml of a solution of the reductase (typically 5 μl of pure reductase at 3 mg/ml or 20–50 μl of a column eluate) in a stoppered cuvette (10 mm light path) at 30°. The reaction is initiated by the addition of NADH solution (100 μl) and monitored by the change in extinction at 600 nm compared to a blank cuvette containing buffer A only. The rate of bleaching of DCPIP may be corrected for any endogenous bleaching observed prior to addition of NADH.

[7] S. Glasstone "Elements of Physical Chemistry," 1st Ed., p. 343. Van Nostrand, New York, 1946.
[8] E. Bill, F.-H. Bernhardt, and A. X. Trautwein, *Eur. J. Biochem.* **121,** 19 (1981).

Specific Activities/Units

Method 1. Specific activities are expressed at 30° as nanomoles of O_2 per minute per milligram of partially or completely purified fractions, *uncontaminated by other components,* in the presence of specific amounts of both other components added to the reaction mixture. Nonlinearity of crude extract activity with respect to protein concentration and cross-contamination of less purified components nullify attempts to produce a meaningful purification chart based on specific activity. However, protein recoveries may be quantified using an immunological technique described herein.

Method 2. Specific activity is expressed an nanomoles of DCPIP reduced per minute per milligram protein. A unit of reductase is that amount of protein reducing 1 nmol of DCPIP in 1 min at 30°.

Antibody Probing

Antibodies to the four peptides are used to detect the proteins during purification by the Western blotting technique. Antibodies are raised to purified protein preparations for all four subunits separated by preparative sodium dodecyl sulfate–polyacrylamide gel electrophoresis (SDS–PAGE) as described by Zamanian and Mason for ISP(ben) α and ISP(ben) β. Pure protein preparations are electrophoresed and stained with Coomassie blue for 5 min. The band is then cut out of the gel, electroeluted, and used to raise antibodies in rabbits.[3]

The rapid minigel systems (Bio-Rad, Richmond, CA) for electrophoresis and electroblotting are used. Protein samples are electrophoresed for 45 min [3 μg total protein per track for ISP(ben)α and reductase(ben) and 6 μg total protein per track for ferredoxin(ben) and ISP(ben)β, decreasing to 0.5 μg/track for pure preparations]. Constant amounts of protein are loaded onto gels from column fractions after OD_{280} estimations to determine both position and quantity. Approximate locations of components are determined by color and spectrum as described later.

After electrophoresis, gels are equilibrated in transfer buffer [25 mM Tris–192 mM glycine, pH 8.3, containing 20% (v/v) methanol] for 15 min. Peptides are electroblotted onto nitrocellulose for 1 hr at 100 V. The nitrocellulose membrane is equilibrated in Tris–NaCl buffer (20 mM Tris-HCl–500 mM NaCl, pH 7.5) for 5 min and then blocked with 5% (w/v) skim milk in the same buffer for 15 min. Antibody is then added [10 μg of purified IgG[3] in 1% (w/v) skim milk in the Tris–NaCl buffer] for 45 min at room temperature. The membranes are washed twice in Tris–NaCl buffer containing 0.05% (w/v) Tween 20 with gentle agitation for 10 min. Binding of second antibody and color development are performed by protocols given by Bio-Rad for either affinity purified goat anti-rabbit IgG horseradish peroxidase conjugate or alkaline phosphatase conjugate. The same blots can be reused and reprobed with the other antibodies. The purity of protein preparations can also be determined on blots by using amido black (sensitivity $0.5-1$ μg) or biotin-blot protein detection kit (Bio-Rad; $1-5$ ng sensitivity).

By labeling antibodies with ^{125}I it is possible to determine quantitatively the percent yield and recoveries during the purification. Antibodies are labeled with ^{125}I using the protocol for Iodo-Beads described by Pierce (Pierce & Warriner (UK) Ltd., Chester, UK). Samples containing unknown amounts of benzene dioxygenase subunits and known amounts of standard subunits are resolved by denaturing electrophoresis and blotted onto nitrocellulose. These are then probed with the appropriate ^{125}I-labeled antibody under the same conditions described above. The bands are visualized by using second antibody, excised, and counted in a γ counter. From the known amounts of a given subunit protein a standard curve can be obtained and used to determine the concentrations of unknown samples.

Purification Procedure*

Growth of Organism. Pseudomonas putida ML2 (NCIB 12190) was isolated from refinery soil using benzene as the sole carbon source. It is maintained on nutrient agar slopes, stored at 4°, prepared from benzene-grown shake cultures at regular (~ 3 monthly) intervals. Repeated transfer from agar slope to agar slope eventually results in loss of ability to grow on benzene.

Shake cultures are grown on a mineral salts medium containing, per liter: KH_2PO_4, 1.02 g; K_2HPO_4, 3.05 g; $(NH_4)_2SO_4$, 1.0 g; $MgSO_4 \cdot 7H_2O$, 0.4 g; $FeSO_4 \cdot 7H_2O$, 40 mg; Bacto-peptone, 40 mg; pH 7.2. To 50 ml of this medium in a 250-ml Erlenmeyer flask is added benzene (50 μl; 0.1%, v/v), and the flask is sealed using a sterilized rubber Suba-seal stopper. The culture is incubated at 30° with shaking at 200 rpm for 18 hr and then used to inoculate 9 liters of the same medium in a fermentor.

For fermentation similar levels of benzene are maintained by bubbling the input air through benzene at intervals in response to a preset level of this compound being detected in the off-gas.[5] (Benzene may also be me-

---

* Reprinted by permission from B. C. Axcell and P. J. Geary, *Biochem. J.* **136,** 927–934 and *Biochem. J.* **146,** 173–183; S. E. Crutcher and P. J. Geary, *Biochem. J.* **177,** 393–400; M. Zamanian and J. R. Mason, *Biochem. J.* **244,** 611–616. Copyright © 1973, 1975, 1979, 1987 The Biochemical Society, London.

tered by pumping, using appropriate detection and control.) The pH is maintained at 7.2 by the automatic addition of aqueous NH_4OH, which additionally provides nitrogen for growth. Cells are harvested in mid to late exponential growth phase (A_{660} 4.5–6), yielding 12–16 g/liter wet weight, which may be washed once in buffer A or stored immediately as a frozen paste at $-20°$ until required. All stages in the purification are performed at $4°$, under N_2, in 25 mM sodium–potassium phosphate buffer, pH 7.2, containing 0.1 mM dithiothreitol (buffer B).[2,3]

Note that because benzene dioxygenase (but not catechol 1,2-dioxygenase) is to a lesser extent expressed constitutively in *P. putida* ML2, reductase(ben) is advantageously purified from cells grown on succinate (using the same medium; pH control with H_2SO_4). Under these conditions the difficulty of separating reductase(ben) from catechol 1,2-dioxygenase is avoided.

Step 1: Preparation of Cell-Free Extract. Cell paste (100 g wet weight) is disrupted by two passes through an Aminco French pressure cell at 27.5 MPa followed by centrifuging for 30 min at 11,300 g. The pellet is resuspended in buffer B, disrupted further by ultrasonication 5 times for 30 sec each, at $0–4°$, in an MSE ultrasonic disintegrator (100 W), and centrifuged at 10,900 g for 30 min. The supernatants from both steps are pooled and centrifuged at 41,400 g for 1 hr to yield the cell-free extract as the supernatant.

Step 2: Protamine Sulfate Treatment. Nucleic acids are conveniently removed by precipitation with protamine sulfate.[5] Protamine sulfate (0.5 g/100 g wet weight of cells) is dissolved in 20 ml of buffer B and added dropwise with constant stirring to the cell-free extract. The suspension is stirred for a further 15 min and then clarified by centrifuging for 30 min at 41,400 g to yield the protamine sulfate-treated supernatant.

Step 3: Ammonium Sulfate Fractionation. The supernatant from Step 2 is brought to 50% saturation by the gradual addition of a saturated solution of ammonium sulfate in buffer B. After stirring for a further 15 min, the precipitate is collected by centrifuging at 41,400 g for 30 min and suspended in the minimum volume of buffer B. The suspension is then dialyzed for 18 hr against the same buffer (2 liters), during which time the suspended precipitate redissolves to yield a clear brown solution. A small amount of residual precipitate may be removed by centrifuging (41,400 g, 15 min).

Note. Protein fractions containing ISP(ben) and reductase(ben) are advantageously prepared using a 40–70% saturation ammonium sulfate fraction.[1,3] However, we have found that a 0–50% fraction retains more of the ferredoxin(ben) component,[2] which may be required for the assay.

Step 4: DEAE-Cellulose Chromatography. The dialyzed protein solution from Step 3 is applied to a column (4 cm² × 30 cm) of DEAE-cellulose (Whatman DE-52) previously equilibrated with buffer B. Proteins are eluted with a linear 0–6% (w/v) gradient of KCl in the same buffer (1000 ml total), collecting fractions of 9 ml at approximately 15-min intervals. Given good chromatography, it is possible to observe the locations of the proteins by their colors, ISP(ben) eluting first as a red band, shading into the yellow of the reductase (Area 1) and followed somewhat later by the darker red band characteristic of the ferredoxin plus *cis*-benzene glycol dehydrogenase (Area 2). The contents of tubes containing the two areas broadly defined above are separately combined and concentrated to 6–8 ml each by ultrafiltration through an Amicon (Danvers, MA) Diaflo PM10 membrane.

Note. Should large quantities of protein be required, it is advantageous to combine several similar preparations at this stage and further concentrate to 10–15 ml as above before proceeding with the separation.

ISP(ben) and Reductase(ben)

Step 5: Sephacryl S-200 Chromatography. ISP(ben) and reductase(ben) are separated by chromatography of the relevant fraction from Step 4 on a column (7 cm² × 80 cm) of Sephacryl S-200 superfine (Pharmacia, Uppsala, Sweden) equilibrated and eluted with buffer B. Red and yellow bands are easily observed. In addition, ISP(ben) may be located by its spectrum[2] and reductase(ben) by the use of assay Method 2. The fractions containing these two proteins are concentrated to small bulk as above (Amicon PM10 membrane).

Step 6a: Ultrogel AcA 34 Chromatography. ISP(ben) is further purified by chromatography in buffer B on a column (3 cm² × 80 cm) of Ultrogel AcA 34. The protein is located visually and spectrophotometrically and reconcentrated to yield pure ISP(ben).

*Step 6b: FAD Treatment of Reductase(ben) and Sephacryl S-300 Chromatography.** Reductase(ben) is further purified and reactivated by treatment with FAD (5 mg/100 mg protein) in 10% glycerol (v/v) followed by chromatography on a column (3 cm² × 80 cm) of Sephacryl S-300 eluted with buffer A. Location using Method 2 and reconcentration yields pure, reactivated reductase. Catechol 1,2-dioxygenase is partially separated from reductase(ben) at this stage if it is present and may be located by its red–brown color or polarographically using catechol as substrate (see Method 1), yielding small quantities for subsequent use in assays.

---

* Reprinted by permission from P. J. Geary and D. P. E. Dickson, *Biochem. J.* **195,** 199–203. Copyright © 1981 the Biochemical Society, London.

Ferredoxin(ben)

Step 7: Sephacryl S-200 Chromatography. Ferredoxin(ben) is further purified and separated from *cis*-benzene glycol dehydrogenase by chromatography on a column (7 cm^2 × 80 cm) of Sephacryl S-200 superfine equilibrated and eluted with buffer B. The ferredoxin is located spectrophotometrically[2] and by its color. Fractions are concentrated using an Amicon Diaflo UM2 membrane as the pure protein has been found to pass in significant amounts through larger pore sizes. *cis*-Benzene glycol dehydrogenase may be located in early fractions[5] and concentrated (Amicon PM10 membrane) for use in assays.

Step 8: Sephadex G-150 Chromatography. Ferredoxin(ben) is further purified by chromatography on a column of Sephadex G-150 superfine (3 cm^2 × 80 cm) equilibrated and eluted in buffer B. The protein is located and concentrated as Step 7 to yield pure ferredoxin(ben).

Preparation of Pure α and β Subunits of ISP(ben)

Purified ISP(ben) (480 μg) is separated into its α and β subunits by preparative SDS–PAGE in 8% (w/v) acrylamide gels. Proteins are detected by staining (5 min) in fresh Coomassie Brilliant Blue R-250 (0.25% w/v) in methanol–water–acetic acid (5:4:1, by volume). Partially stained gel slices containing α and β proteins are excised, and the proteins are extracted by electroelution into 25 mM Tris–glycine, pH 8.3, containing 0.1% (w/v) SDS. Electroelution is carried out at 50–60 V for 18 hr at room temperature. Extracted samples are concentrated to a final volume of about 1 ml in a Minicon B 15 concentration block at room temperature and stored at −20°.[3]

Properties

Purity. Proteins prepared as described above are substantially pure as judged by acrylamide gel electrophoresis.[9,10] ISP(ben) possesses catalase activity that so far has resisted all attempts to resolve and may be a feature of the protein itself. Reductase(ben) and ferredoxin(ben) may contain variable quantities of their respective apoenzymes, and reductase(ben) may additionally be contaminated by traces of catechol 1,2-dioxygenase unless specific steps are taken to avoid the latter.

Substrate Specificity. Benzene dioxygenase metabolizes lower alkyl-substituted benzenes (CH$_3$ > C$_2$H$_5$ ≫ C$_3$H$_7$) and halogenated benzenes

[9] D. E. Williams and R. A. Reisfeld, *Ann. N.Y. Acad. Sci.* **121,** 373 (1964).
[10] U. K. Laemmli, *Nature (London)* **227,** 680 (1970).

(F > Cl > Br). Trifluoromethylbenzene is converted to the corresponding diol. Benzoic acid, nitrobenzene, benzaldehyde, and phenol are not metabolized. The enzyme is specific for NADH.

Effect of Metal Ions. ISP(ben) requires the addition of Fe^{2+} for maximum activity, following purification, either by preactivation[6] or in the standard assay (see above). The protein has been shown to contain two Fe^{2+} ions per molecule in addition to two [2Fe–2S] iron-sulfur clusters.[2,6]

Stability. The enzyme, as purified and concentrated, is stable for 6 months to 1 year when stored frozen at $-20°$ or lower.

Kinetic Constants. For reductase(ben) the K_m for NADH is 7.82×10^{-6} M. The dissociation constant for FAD is 5.14×10^{-6} M.[11]

Molecular Weight and Subunit Structure. Molecular weights of components of benzene dioxygenase are as follows: reductase(ben), $82,000 \pm 2000$;[11] ferredoxin(ben), 12,300;[2] and ISP(ben) native, $168,000 \pm 4000$;[3] ISP(ben) α, 54,500; ISP(ben) β, 23,500.[3] ISP(ben) comprises four subunits in an $\alpha_2\beta_2$ configuration.[3] Reductase(ben) has two equal subunits of M_r $42,000 \pm 500$.

Iron-Sulfur Clusters and Redox Potentials.* Ferredoxin(ben) contains one [2Fe–2S] cluster and has a midpoint redox potential of -155 mV at pH 7.0. ISP(ben) contains two [2Fe–2S] clusters and has a midpoint redox potential of -112 mV at pH 7.0. Average electron spin resonance g values are 1.92 and 1.896, respectively,[6] the signal from ISP(ben) most closely resembling those of 4-methoxybenzoate monooxygenase[12] and the [2Fe–2S] Rieske proteins of the quinone–cytochrome c region of electron-transport chains of respiration and photosynthesis.[13] The presence of two [2Fe–2S] and two Fe^{2+} associated with an $\alpha_2\beta_2$ structure leads to the suggestion that ISP(ben) has two active centers.

Amino Acid Sequence of Ferredoxin(ben). The amino acid sequence of ferredoxin(ben), obtained directly using cleaved peptides, has been described.[14]

[11] S. E. Crutcher and P. J. Geary, unpublished observations (1977).

[12] H. Twilfer, F.-H. Bernhardt, and K. Gersonde, *Eur. J. Biochem.* **119**, 595 (1981).

[13] J. S. Rieske, R. E. Hansen, and W. S. Zaugg, *J. Biol. Chem.* **239**, 3017 (1964).

[14] N. Morrice, P. J. Geary, R. Cammack, A. Harris, F. Beg, and A. Aitken, *FEBS Lett.* **231**, 336 (1988).

* Reprinted by permission from P. J. Geary, F. Saboowalla, D. Patil, and R. Cammack, *Biochem. J.* **217**, 667–673. Copyright © 1984 The Biochemical Society, London.

[11] Phthalate Dioxygenase

By CHRISTOPHER J. BATIE and DAVID P. BALLOU

Introduction

Phthalate dioxygenase (PDO) catalyzes the reaction:

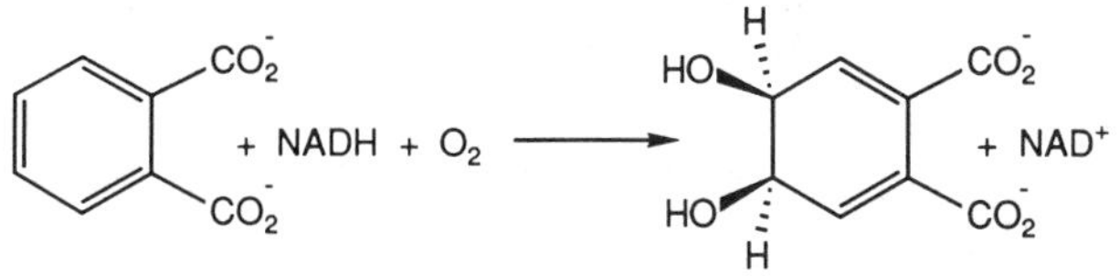

It is a member of a class of bacterial oxygenases which contain both nonheme mononuclear Fe^{2+} (presumably at the active site) and a Rieske-type [2Fe–2S] center.[1,2] These enzymes play an important role in the aerobic degradation of aromatic compounds by soil bacteria in that they attack the relatively stable benzene nucleus and activate it for oxygenative cleavage.[3,4]

Ribbons and co-workers first described a phthalate dioxygenase from *Pseudomonas cepacia*;[5] this chapter describes the purification and characterization of this two-component enzyme system. Two separate proteins are required for phthalate dioxygenase activity: phthalate dioxygenase and phthalate dioxygenase reductase (PDR). The former contains the substrate-binding site, the mononuclear Fe^{2+}, and a Rieske-type [2Fe–2S].[1] The reductase catalyzes electron transfer from NADH to the dioxygenase, and it contains one FMN and one plant ferredoxin (Fd)-[2Fe–2S] per protein.[1]

This system, which is induced at high levels by phthalate, has proved advantageous for structural studies that have utilized combinations of specific isotopic labeling and various spectroscopic techniques. These have included determination of the structure of the Rieske-type [2Fe–2S]

[1] C. J. Batie, E. LaHaie, and D. P. Ballou, *J. Biol. Chem.* **262,** 1510 (1987).
[2] J. S. Rieske, R. E. Hansen, and W. S. Zaugg, *J. Biol. Chem.* **239,** 3017 (1964).
[3] J. M. Wood, *Environ. Sci. Technol.* **16,** 291A (1982).
[4] C. Bull and D. P. Ballou, *J. Biol. Chem.* **256,** 12673 (1981).
[5] P. K. Keyser, B. G. Pujar, R. W. Eaton, and D. W. Ribbons, *Environ. Health Perspect.* **18,** 159 (1976).

center,[6-10] characterization of the ferrous-binding site in this class of protein,[11,12] and X-ray crystallographic studies of the structure of the reductase,[13] an iron-sulfur flavoprotein.

Assay Methods

Assay of Phthalate Dioxygenase. Phthalate dioxygenase is assayed by spectrophotometrically monitoring (at 340 nm) phthalate dioxygenase reductase-dependent NADH oxidation. The assay is performed in buffer A, $0.1 M$ HEPES, pH 8.0, at $25°$. The dioxygenase (10 nM to 1 μM) and ferrous ammonium sulfate (10 μM) are added to 1 ml of an assay cocktail consisting of buffer A, NADH (100 μM), and dipotassium phthalate (2000 μM). All concentrations are the final diluted concentrations. The enzymatic reaction is determined before and after addition of phthalate dioxygenase reductase (0.18 μM). One unit of enzyme activity is defined as that which causes a net oxidation of 1 μmol of NADH in 1 min. The specific activity of the most active purified dioxygenase is about 2 units/mg protein.[1]

The ferrous ammonium sulfate, which is added to reconstitute any enzyme that has lost its mononuclear iron, is prepared at 10 mM and stored under N_2 in a screw-top tube sealed with a septum; aliquots are removed by syringe. The concentrated reductase is stored on ice in order to stabilize it. The concentration of phthalate dioxygenase reductase used in the assay does not cause significant NADH oxidation in the absence of PDO. However, the assay is very sensitive to the amount of reductase because the concentration used is substantially less than the K_m value.

Assay of Phthalate Dioxygenase Reductase. Phthalate dioxygenase reductase is assayed by its cytochrome c reducing activity. Reductase (0.2 to 10 nM) is added to buffer A at $25°$ containing 50 μM horse heart cytochrome c and 100 μm NADH: the reaction is monitored at 550 nm. One

[6] J. F. Cline, B. M. Hoffman, E. LaHaie, D. P. Ballou, and J. A. Fee, *J. Biol. Chem.* **260,** 3251 (1985).

[7] R. J. Gurbiel, C. J. Batie, M. Sivaraja, A. E. True, J. A. Fee, B. M. Hoffman, and D. P. Ballou, *Biochemistry* **28,** 4861 (1989).

[8] H.-T. Tsang, C. J. Batie, D. P. Ballou, and J. E. Penner-Hahn, *Biochemistry* **28,** 7233 (1989).

[9] D. P. Ballou and C. J. Batie, *in* "Oxidases and Related Redox Systems" (T. S. King, H. S. Mason, and M. Morrison, eds.), p. 211. Alan R. Liss, New York, 1988.

[10] D. Kuila, J. A. Fee, J. R. Schoonover, W. H. Woodruff, C. J. Batie, and D. P. Ballou, *J. Am. Chem. Soc.* **109,** 1559 (1987).

[11] C. J. Batie, W. R. Dunham, and D. P. Ballou, unpublished work (1989).

[12] H.-T. Tsang, C. J. Batie, D. P. Ballou, and J. E. Penner-Hahn, unpublished work.

[13] C. C. Correll, C. J. Batie, D. P. Ballou, and M. L. Ludwig, *J. Biol. Chem.* **260,** 14633 (1985).

unit of phthalate dioxygenase reductase will oxidize 1 μmol of NADH/min (hence reduce 2 μmol cytochrome c/min). Purified reductase has a specific activity of 145 units/mg. If phthalate dioxygenase is present at more than 1 μM, much higher rates of cytochrome c reducing activity are observed.

Purification

Growth of Bacteria. The bacteria are grown aerobically at pH 6.8, 37°, on a phthalate–minimal salts medium containing, per liter: 3 g $(NH_4)_2HPO_4$; 1.2 g K_2HPO_4; 3 g potassium hydrogen phthalate; 0.6 g NaCl; 60 mg NaFeEDTA; 1 ml of micronutrients; 25 mg yeast extract; 0.2 g $MgSO_4 \cdot 7H_2O$; and, in 200-liter fermentations, 0.5 ml Antifoam A. The micronutrient solution consists of 10 mM HCl, 0.5 mM $CuSO_4$, 1.0 mM $Zn(C_2H_3O_2)_2 \cdot 2H_2O$, 0.9 m$M$ $MnCl_2$, and 2.3 mM $CaCl_2$.

The primary cultures of *Pseudomonas cepacia* DB01 are stored at $-20°$ in 50% glycerol.[4] Cultures for enzyme preparation are started by inoculating plates of solid medium (2% agar plus phthalate–minimal medium). The culture is then transferred to liquid medium. The culture is scaled up in stages from 5 ml to 1 liter to 200 liters; 15–18 hr is sufficient for full growth in each stage. During the 200-liter fermentation pH is maintained by adding, as required, phosphoric acid and dipotassium phthalate in a 2:1 molar ratio; about 1 kg of phthalate is required. Once the bacteria have reached full growth (600–800 Klett units) they are harvested with a Sharples centrifuge and stored at $-70°$. Prolonged storage at $-20°$ results in proteolysis of PDO and lower yields of both proteins. The yield of cell paste from 200 liters of culture is about 1 kg.

Purification Conditions. All operations are carried out at 4°. Potassium phosphate buffer, pH 6.9 (KP$_i$), is used throughout; the concentration is indicated at each point. All buffers used in purifying the dioxygenase component also contain 5 mM potassium phthalate (KP$_i$–phthalate); "20 mM KP$_i$–phthalate buffer" indicates 20 mM KP$_i$, 5 mM phthalate, pH 6.9. Phthalate stabilizes the dioxygenase and improves the specific activity of the final preparation. Tables I and II list the steps in the purification of phthalate dioxygenase and phthalate dioxygenase reductase, respectively.

Preparation of Cell Extract. Frozen cells are chilled in liquid N_2, powdered, then suspended in a buffer consisting of 20 mM KP$_i$–phthalate, 100 μM EDTA, and 100 μM phenylmethylsulfonyl fluoride (PMSF) (500 g cell paste:1 liter of buffer). The EDTA and PMSF are necessary to prevent proteolysis of the dioxygenase and reductase. The cells are lysed by 4 passages through a precooled Manton-Gaulin homogenizer. After each passage the homogenate warms to 15–18°; therefore, it is cooled to below 4° between passages.

TABLE I
PURIFICATION OF PHTHALATE DIOXYGENASE[a]

Step	Volume (ml)	Total protein (mg)	Total activity (units)	Specific activity (units/mg)	Yield (%)
Crude extract	1,120	21,800	7,310	0.335	—
DEAE-Sepharose	435	2,620	4,173	1.59	57.0
Sephadex G-100	22.5	1,340	1,026	0.77	14.0
Sephacryl S-300	174	870	1,141	1.31	15.6

[a] Starting material: 429 g cell paste.

In a typical preparation 500 g of cell paste is homogenized; nucleic acids are then precipitated by addition of protamine sulfate. The protamine sulfate (9 mg/g of cell paste) is added dropwise in the form of a 5% solution in water. After addition, the mixture is stirred for 30 min. Centrifugation at 100,000 g for 1 hr yields the clear cell extract.

DEAE-Sepharose I. The cell extract is pumped onto a DEAE Fast-Flow Sepharose (Pharmacia, Piscataway, NJ) column (5 × 40 cm) which has been equilibrated with 20 mM KP$_i$–phthalate; the flow rate is approximately 20 ml/min. The column is washed with 2 volumes of 20 mM KP$_i$–phthalate, and the enzymes are eluted by an 8 liter gradient from 20 to 350 mM KP$_i$–phthalate. Phthalate dioxygenase and phthalate dioxygenase reductase coelute in a broad, red–brown band at about 200 mM KP$_i$. Fractions which include over 90% of the reductase activity are combined to give Fraction I.

Protocatechuate dioxygenase, 4,5-dihydrodihydroxyphthalate dehydrogenase, and 4,5-dihydroxyphthalate decarboxylase all elute early in the

TABLE II
PURIFICATION OF PHTHALATE DIOXYGENASE REDUCTASE[a]

Step	Volume (ml)	Total protein (mg)	Total activity (units)	Specific activity (units/mg)	Yield (%)
Crude extract	1120	21,800	92,000	2.1	—
DEAE-Sepharose	435	2,620	47,800	9.1	52.0
Sephadex G-100	380	160	41,420	130	45.0
DEAE-Sepharose	102	58.1	21,200	181	23.0

[a] Starting material: 429 g cell paste.

gradient in incompletely resolved bands. The protocatechuate dioxy-genase-containing fractions have the characteristic burgundy color of the enzyme. These enzymes can be purified further as detailed elsewhere.[1,4]

Separation of PDO and PDR: Sephadex G-100. The reductase is separated from the oxygenase by gel filtration on Sephadex G-100 (Pharmacia). The fraction containing PDO and PDR is concentrated to about 50 mg/ml by ultrafiltration over an Amicon PM30 membrane or by dialysis versus a KP_i–phthalate buffer (50 mM) containing at least 20% polyethylene glycol (w/v). The concentrate is equilibrated with 50 mM KP_i–phthalate, 500 mM KCl by dialysis; the KCl is necessary to prevent complex formation between the two proteins. The sample is then clarified by centrifugation at 20,000 g for 10 min and applied to a Sephadex G-100 column (5 × 90 cm) which is equilibrated with the same buffer. The flow rate is 1.5 ml/min. Phthalate dioxygenase reductase activity elutes in an orange band. It is then concentrated by ultrafiltration over an Amicon PM30 membrane and equilibrated with 45 mM KP_i by gel filtration over Sephadex G-25. After removal of phthalate, the A_{462}/A_{280} ratio should be at least 0.15.

Ammonium Sulfate Precipitation of PDO. Phthalate dioxygenase elutes in the void volume of the G-100 column; it is concentrated by precipitation with ammonium sulfate. Solid $(NH_4)_2SO_4$ (226 g/liter) is added to the enzyme; NH_4OH (2 M) is added as necessary to maintain constant pH. The solution is then stirred for 30 min. The precipitate, which contains the dioxygenase, is collected by centrifugation at 13,000 g for 30 min. The further purification of PDO is described below.

Purification of Phthalate Dioxygenase Reductase

Table II lists the steps in the purification of PDR.

DEAE-Sepharose II. Phthalate dioxygenase reductase is applied to a DEAE-Sepharose column (2.5 × 10 cm) equilibrated with 45 mM KP_i. The enzyme binds at the top of the column but is eluted by a 500 ml linear gradient from 45 to 100 mM KP_i. Those fractions that have an A_{462}/A_{280} ratio over 0.32 are pure and are combined. On occasion it is necessary to repeat the Sephadex G-100 step to achieve purity; if so, phthalate is omitted from the buffer. The purified reductase is concentrated by ultrafiltration and dialyzed versus 50 mM HEPES buffer, pH 6.8, containing 20% glycerol. Reductase is stored in a liquid N_2 freezer.

In some preparations (e.g., those using cell paste that is several months old) a fraction of the reductase remains bound to the top of the DEAE-Sepharose column at the end of the gradient; it is eluted by 250 mM KP_i buffer. The properties of this form of the reductase have not been characterized.

Purification of Phthalate Dioxygenase

Sephacryl S-300. The dioxygenase pellet is resuspended and dialyzed overnight versus 50 mM KP$_i$–phthalate. It is then applied to an S-300 Sephacryl column (5 × 90 cm) that is equilibrated with the same buffer. The enzyme is eluted at 1.5 ml/min. Fractions for which A_{464}/A_{410} ratios exceed 1.4 are pooled. At this stage the dioxygenase should be over 90% pure as judged by polyacrylamide gel electrophoresis (PAGE) in the presence of sodium dodecyl sulfate (SDS).

DEAE-Sepharose. Final purification of the dioxygenase is achieved by a second ion-exchange step. This step is not always necessary to achieve sufficient purity for many studies; it does, however, remove residual catalase and a blue copper protein. The enzyme is applied to a DEAE-Sepharose column (5 × 20 cm) which is equilibrated with 50 mM KP$_i$–phthalate buffer. The enzyme sticks in a tight band at the top of the column and is eluted with a 4 liter gradient from 50 to 150 mM KP$_i$–phthalate buffer. The fractions with the highest A_{464}/A_{410} ratios are pooled. This ratio should be approximately 1.62. The final pool of dioxygenase is concentrated to approximately 50 mg/ml by ultrafiltration over an Amicon PM30 membrane. The concentrated dioxygenase is dialyzed versus 50 mM HEPES, pH 8.0, 5 mM phthalate, 20% glycerol, then frozen and stored in a liquid N$_2$ freezer.

Alternate Purification Method

The preparative method described above is optimized for best yield of the reductase. An alternate method is to separate PDO and PDR after DEAE-Sepharose I by ammonium sulfate fractionation. This method is better adapted to purification of large quantities of dioxygenase, but it is less reliable for preparation of reductase. In some preparations the reductase activity is largely destroyed in the $(NH_4)_2SO_4$ fractionation.

The dioxygenase is precipitated by 40% saturated $(NH_4)_2SO_4$, as described above. The reductase, which remains in the supernatant solution, is precipitated by addition of $(NH_4)_2SO_4$ to 75% saturation (212 g/liter supernatant). The oxygenase is purified as described above, beginning with the Sephacryl S-300 gel filtration. The reductase is resuspended, equilibrated with 50 mM KP$_i$, 500 mM KCl and purified by the Sephadex G-100 and DEAE-Sepharose steps described above.

Physical Properties

Stability. The components of the phthalate dioxygenase system are sufficiently stable to lend themselves to spectroscopic and kinetic studies.

The dioxygenase is very stable; it retains essentially 100% activity after 12 hr of incubation at room temperature at either pH 6.8 or 8.0. It also survives 1 hr at 42° and 5 min at 60° without significant loss of activity. At 4° phthalate dioxygenase appears to be stable for more than 1 week.[1]

The reductase is much less stable than dioxygenase. It does retain full activity during 24-hr incubation at 4° in 100 mM HEPES, pH 8.0. However, prolonged incubation at 4°, especially in low ionic strength KP$_i$ buffer, results in irreversible loss of activity. The denaturation is accompanied by loss of [2Fe–2S] absorbance and the release of free flavin. At higher temperatures the reductase denatures more quickly; for example, at 37° in HEPES, pH 8.0, the half-life ($t_{1/2}$) is 2 min. However, addition of glycerol to 100 mM HEPES, pH 8.0 [20% (v/v)] stabilizes PDR; in this buffer the reductase is stable at 4° for 1 week and at 25° for 4 hr.

Spectral Properties. Phthalate dioxygenase has the UV–visible absorbance spectra shown in Fig. 1A. That of the oxidized dioxygenase has peaks at 330 and 464 nm and a shoulder at 560 nm; the absorbancy at 464 nm is 7800 M^{-1} cm^{-1}.[1] This spectrum is very similar to the spectrum of the oxidized Rieske protein from *Thermus thermophilus*.[1,14] When reduced, the enzyme has less absorbance, with peaks at 382 and 508 nm; electron paramagnetic resonance (EPR) spectra of PDO closely resemble those of the respiratory Rieske proteins.[1,6] Titrations indicate that the enzyme takes up one electron equivalent per three irons. The absorbance spectrum of phthalate dioxygenase reductase (Fig. 1B) has peaks at 345 and 462 nm; the absorbancy at 462 nm is 24,000 M^{-1} cm^{-1}. Shoulders at 410 and 540 nm are consistent with the presence of a plant Fd-type [2Fe–2S] center.[1]

Molecular Weight and Prosthetic Groups. Analysis by denaturing gel electrophoresis and gel filtration indicates that phthalate dioxygenase is found as a tetramer of identical 48-kDa monomers.[1] Amino-terminal sequencing confirms the presence of only one kind of peptide. Each monomer appears to be an active catalytic unit, and each has three irons and two acid-labile sulfides. Two iron atoms are bound in a Rieske-type [2Fe–2S] center, and one Fe^{2+} is bound at a mononuclear site.[1]

The reductase is present as a 34-kDa monomer. It has one FMN, two Fe^{2+}, and two S^{2-} per peptide. EPR spectra indicate the presence of a plant Fd-type [2Fe–2S]; EPR data also indicate that the two centers are too far apart for significant magnetic interaction.[1,13] Phthalate dioxygenase reductase has been crystallized; its structure should aid our understanding of the structural requirements for efficient flavin–[2Fe–2S] electron transfer.[13]

---

[14] J. A. Fee, K. L. Findling, T. Yoshida, R. Hille, G. E. Tarr, D. O. Hearshen, W. R. Dunham, E. P. Day, T. A. Kent, and E. Munck, *J. Biol. Chem.* **259**, 124 (1984).

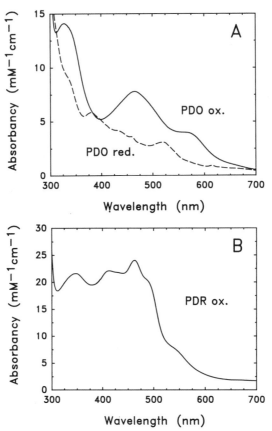

FIG. 1. Spectra of phthalate dioxygenase and phthalate dioxygenase reductase. (A) Phthalate dioxygenase in the oxidized and reduced forms. The enzyme was reduced by addition of excess glucose 6-phosphate in the presence of catalytic quantities of glucose-6-phosphate dehydrogenase and PDR. (B) Phthalate dioxygenase reductase in the oxidized form.

Structure of Rieske [2Fe–2S] Center. The abundant yield and stability of PDO have allowed us to utilize various spectroscopic and isotopic labeling studies to determine the structure of the [2Fe–2S] center in PDO and, by analogy, of all Rieske-type centers. These studies indicate that one iron of the [2Fe–2S] center has two histidine ligands.[6–8,10] Two cysteines ligate the other iron.[12,14]

Spectroscopic Investigation of Mononuclear Site.[11,12,15] Although the mononuclear Fe^{2+} will dissociate from the enzyme in the absence of substrate, the apoenzyme readily binds 1 equivalent of Fe^{2+} per [2Fe–2S]. Reconstitution of activity is accompanied by perturbations in the EPR and visible spectra of the Rieske-type [2Fe–2S] center. Although only Fe^{2+} restores activity, the enzyme will also bind several other divalent cations including Zn^{2+}, Co^{2+}, Ni^{2+}, Mn^{2+}, Cd^{2+}, and Cu^{2+}. EXAFS (X-ray absorption fine structure) spectra of PDO reconstituted with Zn^{2+} and Co^{2+} indicate that the site is penta- or hexacoordinate with all low-Z ligands (O or N). The bond lengths determined by EXAFS spectroscopy are most consistent with predominantly O-ligation. The structure of the site is sensitive to substrate binding and to the redox state of the [2Fe–2S]; XANES (X-ray absorption near edge structure) spectra suggest that substrate binding may be associated with a shift from hexacoordinate to pentacoordinate ligation. EPR spectra of $^{63}Cu^{2+}$-reconstituted, [^{15}N]histidine-labeled PDO are most consistent with the presence of two histidyl ligands.

Catalytic Properties

Specificity. Efficient product formation by phthalate dioxygenase requires phthalate dioxygenase reductase; incubation of PDO with other electron-transfer proteins does not yield product.[1,9] PDO will accept a variety of phthalate analogs. The vicinal carboxyls are required. However, a variety of substitutions at the 3 and 4 positions are tolerated; these include methyl, chloro, and fluoro groups. In addition, both the pyridine analogs 1,2-dicarboxypyridine and 2,3-dicarboxypyridine are effectors. The extent of coupling of NADH oxidation to oxygenation varies widely; only phthalate stimulates tightly coupled oxygenation.[1,9,15]

Phthalate dioxygenase reductase is specific for NADH; no activity is observed with NADPH. However, the reductase efficiently reduces many electron acceptors. These include PDO, cytochrome c, dichlorophenolindophenol (DCPIP), ferricyanide, methylene blue, and nitro blue tetrazolium; PDR reduces the first four substrates at similar rates.[1] This indicates that some step(s) involving electron transfer from NADH to FMN is rate-limiting.[1,15]

Steady-State Kinetics. Phthalate dioxygenase has maximal activity from pH 6.8 to 8.2. The activity of PDR as measured with PDO has a similar pH dependence, but the cytochrome c reducing activity of PDR is maximal from pH 7.5 to 9.5. Optimal PDO activity is found at ionic strength of 80 mM; activity drops off rapidly below 40 mM and above 100 mM. These data suggest that the relative strength of PDO–PDR complexes are important in achieving optimal activity. PDR activity is less sensitive to KCl, reaching 50% maximal activity at 1 M salt.[1,15]

The steady-state kinetic patterns of reductase activity indicate a ternary complex mechanism. At 4°, V_{max} is 17 NADH oxidized/sec, K_m(cytochrome c) is 2.2 μM, and K_m(NADPH) 5 μM. Kinetic analysis of activity with phthalate dioxygenase indicates a V_{max} of 12 NADH oxidized/sec; K_m (NADH) is approximately 10 μM; K_m(PDO) is 1.4 μM. NAD (1 mM) did not inhibit.[1,15]

Steady-state kinetic analysis of the phthalate dioxygenase reaction is consistent with the following model. Reduced PDR first interacts with PDO and reduces the Rieske [2Fe–2S] center. Phthalate and O_2 then bind reversibly and PDR transfers one more electron to the enzyme. Now the oxygenation occurs, probably by reduction of O_2 bound to the mononuclear Fe, followed by nucleophilic attack by aromatic ring. At 25°, V_{max} is 20/sec, K_m(PDR) is 2.3 μM, and K_m(phthalate) is 300 μM. At 4°, V_{max} is 0.5/sec, K_m(PDR) is approximately 1 μM, K_m(phthalate) is 500 μM, $K_m(O_2)$ is 125 μM.[15]

Acknowledgment

This work and the unpublished work referenced herein was supported by a grant from the National Institutes of Health (GM 20877).

[15] C. J. Batie and D. P. Ballou, unpublished work (1989).

[12] Cyclohexanone 1,2-Monooxygenase from *Acinetobacter* NCIMB 9871

By Peter W. Trudgill

Cyclohexanone 1,2-monooxygenase is a member of a class of monooxygenases that mimic the chemical Baeyer–Villiger reaction[1] and catalyze the insertion of an oxygen atom *between* two carbon atoms, one of which carries a keto group, with the formation of a lactone. These monooxygenases are widely distributed in nature and play an important role in the degradation of keto compounds by microorganisms.[2-5] A number have been purified to homogeneity, and all of them are flavoproteins. 2,5-Diketocamphane 1,2-monooxygenase from *Pseudomonas putida* is the most structurally complex,[6] consisting of two dissimilar polypeptides which interact through FMN bound firmly to one of them. 2-Oxo-Δ^3-4,5,5-trimethylcyclopentenylacetyl-CoA monooxygenase[7] from the same organism

[1] A. Baeyer and V. Villiger, *Ber. Dtsch. Chem. Ges.* **32**, 3625 (1899).
[2] L. N. Britton and A. J. Markovetz, *J. Biol. Chem.* **252**, 8561 (1977).
[3] P. W. Trudgill, *in* "Microbial Degradation of Organic Compounds" (D. T. Gibson, ed.), p. 131. Dekker, New York, 1984.
[4] P. W. Trudgill, *in* "The Bacteria: Volume 10" (J. R. Sokatch, ed.), p. 483. Academic Press, New York, 1986.
[5] P. W. Trudgill, this volume [13].
[6] D. G. Taylor and P. W. Trudgill, *J. Bacteriol.* **165**, 489 (1986).
[7] H. J. Ougham, D. G. Taylor, and P. W. Trudgill, *J. Bacteriol.* **153**, 140 (1982).

and androstene-3,17-dione 13,17-monooxygenase[8,9] from *Cyclindrocarpon radicicola* ATCC 11011 are both homodimers that bind one molecule of FAD.

In contrast, cyclohexanone 1,2-monooxygenase from *Nocardia globerula* CL1 and *Acinetobacter* NCIMB 9871 are single polypeptide chains binding a single molecule of FAD.[10] Unfortunately, stock cultures of *Nocardia globerula* CL1 have been lost and the organism is, therefore, no longer available for study. The mechanism of this biological Baeyer–Villiger reaction has been thoroughly investigated and shown to mimic the chemical reaction.[1] Biological oxygen insertion is achieved through a 4a-hydroperoxyflavin intermediate acting as a nucleophile.[11] In addition, the gene coding for the enzyme from *Acinetobacter* NCIMB 9871 has been isolated and sequenced, and the complete amino acid sequence of 542 residues has been derived by translation of the nucleotide sequence.[12]

The enzyme from each organism catalyzes the following reaction:

Cyclohexanone + NADPH + H$^+$ + O$_2$ → ϵ-caprolactone + NADP$^+$ + H$_2$O

This reaction is the first of two steps involved in cleavage of the carbocyclic ring of cyclohexanone and related compounds by bacteria growing with cyclohexanol or cyclohexanone as sole source of carbon. The caprolactone formed is potentially unstable and is cleaved to form 6-hydroxyhexanoate by the action of an induced lactone hydrolase.

Assay Methods

Principle. Two assay methods can be used that are also of more general application to this class of oxygenases. Method 1 involves following the cyclohexanone-stimulated oxidation of NADPH at 340 nm. Method 2 depends on the measurement of cyclohexanone-stimulated oxygen consumption in a suitably sensitive closed system fitted with a Clark-type oxygen electrode. Although Method 1 is the more convenient, and may be used routinely, it has the disadvantage that it does not distinguish between NADPH-dependent monooxygenase and any contaminating NADPH-linked dehydrogenase activities. Method 2 should always be used intermittently as a check that it is oxygenase rather than dehydrogenase activity that is being measured by Method 1.

[8] E. Itagaki, *J. Biochem.* **99**, 815 (1986).
[9] E. Itagaki, *J. Biochem.* **99**, 825 (1986).
[10] N. A. Donoghue, D. B. Norris, and P. W. Trudgill, *Eur. J. Biochem.* **63**, 175 (1976).
[11] C. C. Ryerson, D. P. Ballou, and C. T. Walsh, *Biochemistry* **21**, 2644 (1982).
[12] Y.-C. J. Chen, O. P. Peoples, and C. T. Walsh, *J. Bacteriol.* **170**, 781 (1988).

Reagents for Method 1

Glycine – NaOH buffer, 0.1 M, pH 9.0 (preincubated at 30°)
NADPH, 1.5 mM in buffer
Cyclohexanone, 20 mM in buffer

Reagents for Method 2

Glycine – NaOH buffer, 0.1 M, pH 9.0 (preincubated at 30°)
NADPH, 10 mM in buffer
Cyclohexanone, 20 mM in buffer

Procedure for Method 1. The assay mixture contains, in a final volume of 1 ml (semimicrocuvette), NADPH, 0.1 ml; extract or enzyme fraction containing 0.01 – 0.1 unit of activity; and glycine – NaOH buffer to 0.9 ml. The mixture is preincubated at 30° in the thermostatic cell holder of the spectrophotometer for 1 – 2 min and the endogenous rate of NADPH oxidation followed for 1 min. After preincubation, cyclohexanone (0.1 ml) is added and the stimulated rate of NADPH oxidation followed. This rate, minus the endogenous rate, is taken to be the cyclohexanone 1,2-monooxygenase activity.

Procedure for Method 2. The assay mixture contains, in a volume of 3 ml in a Yellow Springs Oxygen Monitor (Yellow Springs Instruments, Yellow Springs, OH) (volumes of components can be varied *pro rata* to suit systems of different working capacity), NADPH, 0.1 ml; 0.1 – 0.3 unit of cyclohexanone monooxygenase; and glycine – NaOH buffer to 2.9 ml. After temperature equilibration (30°) and insertion of the electrode, endogenous oxygen consumption is followed for 1 – 2 min. Cyclohexanone (0.1 ml) is then added, stimulated oxygen consumption followed for an appropriate period, and the substrate-stimulated rate calculated.

Units. One unit of cyclohexanone 1,2-monooxygenase is defined as that amount of enzyme catalyzing the cyclohexanone-stimulated oxidation of 1 μmol of NADPH or consumption of 1 μmol of oxygen per minute at 30°. Specific activity is expressed as units per milligram of protein.

Growth and Purification Procedure

Growth of Organism. *Acinetobacter* NCIMB 9871 (available from NCIMB, Torry Research Station, P.O. Box 31, 135 Abbey Road, Aberdeen AB9 8DG, Scotland) is grown at 30° in a mineral salts medium containing, per liter: KH_2PO_4, 2 g; Na_2HPO_4, 4 g; $(NH_4)_2SO_4$, 2 g; and trace element solution,[13] 4 ml. The composition of the trace element solution is shown in Table I. Large crops of cells, suitable for enzyme purification, are grown

---

[13] R. F. Rosenberger and S. R. Elsden, *J. Gen. Microbiol.* **118**, 159 (1960).

TABLE I
TRACE ELEMENT SOLUTION[a]

Compound[b]	Amount
Magnesium oxide (g)	10.1
Calcium carbonate (g)	2.0
Concentrated hydrochloric acid (ml)	53
Ferrous sulfate heptahydrate (g)	5.6
Zinc sulfate heptahydrate (g)	1.4
Manganese sulfate tetrahydrate (g)	1.1
Cupric sulfate pentahydrate (g)	0.25
Cobalt sulfate heptahydrate (g)	0.28
Boric acid (g)	0.06

[a] From Ref. 13.
[b] Compounds should be added to 500 ml of distilled water with continued stirring and in the order listed. Once all components are dissolved the volume is made up to 1 liter.

according to the following regime: four 25-ml samples of sterile mineral salts medium in 100-ml conical flasks are inoculated from a nutrient agar slant, cyclohexanone is added to diffusion tubes,[14] and the cultures are grown at 30° on an orbital incubator at 150 rpm. After 48 hr of growth these cultures are used to inoculate four 1-liter batches of sterile medium in 2-liter conical flasks to which 1 ml of cyclohexanol (sterilized by membrane filtration) is added directly. After 16 hr of growth these batch cultures are used to inoculate 40 liters of medium in a suitable fermentor, stirred at approximately 400 rpm and supplied with air at 2 liters/min. This regime may, of course, be modified *pro rata* according to the fermentor capacity available. The culture is harvested, by continuous flow centrifugation, when A_{580} reaches about 2, and the cell paste is washed by resuspension in phosphate buffer (KH_2PO_4, 2 g/liter; Na_2HPO_4, 4 g/liter; pH 7.1) followed by centrifugation at 15,000 g and 3°, for 15 min. The cell paste (~ 100 g) is then resuspended in 1.5 volumes of the sodium potassium phosphate buffer and used immediately or stored at $-20°$, at which temperature it can be maintained for more than 1 year without appreciable loss of oxygenase activity.

Step 1: Preparation of Cell-Free Extract. The resuspended cell paste is thawed and disrupted by a single passage through a French pressure cell with a pressure difference at the orifice of 20,000 psi (148 MPa), incubated with 1–2 mg of DNase for 15 min, and centrifuged at 25,000 g and 3° for 30 min to yield the cell-free extract as supernatant.

[14] D. Claus and N. Walker, *J. Gen. Microbiol.* **36,** 107 (1964).

Step 2: Ammonium Sulfate Fractionation. The cell extract is fractionated with neutral saturated (at 3°) aqueous $(NH_4)_2SO_4$. The fraction precipitating between 50 and 80% saturation is retained, redissolved in 20 mM sodium potassium phosphate buffer, and dialyzed overnight against 2 liters of the same buffer to remove $(NH_4)_2SO_4$.

Step 3: DEAE-Cellulose Chromatography. The dialyzed protein solution from Step 2 is applied to a column (18 × 5 cm) of DEAE-cellulose (Whatman DE-52), previously equilibrated with 20 mM sodium potassium phosphate buffer, pH 7.1, and eluted with a linear gradient composed of 1 liter of the buffer and 1 liter of the buffer containing 0.7 M KCl. Fractions of 15 ml are collected, the yellow oxygenase band (fractions 50–63) is assayed for protein and oxygenase activity, and fractions with a specific activity over 6.0 are pooled. Solid $(NH_4)_2SO_4$ is added to give 90% saturation, and the protein pellet is harvested by centrifugation at 25,000 g and 3° for 15 min and redissolved in 7–10 ml of the sodium potassium phosphate buffer.

Step 4: Sephadex G-150 Chromatography. The redissolved protein from Step 3 is loaded onto a column (60 × 2.5 cm) of Sephadex G-150 that has been equilibrated with 1 mM sodium potassium phosphate buffer (pH 7.1) and eluted by passage of the same buffer through the column. Fractions of 5 ml are collected, and the yellow oxygenase band (fractions 34–37) with a specific activity greater than 7.0 is pooled.

Step 5: Hydroxylapatite Chromatography. The pooled fractions from Step 4 are loaded directly onto a column of hydroxylapatite (BioGel HTP; 8 × 2.5 cm) equilibrated with 1 mM sodium potassium phosphate buffer, pH 7.1, and the enzyme eluted with a linear gradient formed from 250 ml of the above buffer and 250 ml of 100 mM sodium potassium phosphate buffer of the same pH. Fractions of 5 ml are collected and yellow fractions containing cyclohexanone 1,2-monooxygenase pooled. The enzyme may be concentrated either by $(NH_4)_2SO_4$ precipitation or by ultrafiltration and stored at −20° until required.

A typical purification is summarized in Table II.

Properties

Purity. The enzyme (specific activity ~21) is homogeneous as judged by electrophoresis on nondenaturing polyacrylamide gels of varied pore sizes. In all cases a single protein band is visualized by staining with Coomassie Brilliant Blue G-250 (CI 42655).

Molecular Weight and Subunit Structure. The molecular weight of the native enzyme, obtained from ultracentrifugal analysis ($s^{\circ}_{20,w}$ 4.0; diffusion coefficient 5.69 × 10^{-7} cm²/sec) and the Svedberg equation, is 63,700;

TABLE II

PURIFICATION OF CYCLOHEXANONE 1,2-MONOOXYGENASE FROM *Acinetobacter* NCIMB 9871[a]

Fraction	Total volume (ml)	Total protein (mg)	Total activity (units)	Specific activity (units/mg)	Yield (%)
Step 1: Cell-free extract	198	4554	2732	0.60	100
Step 2: (NH$_4$)$_2$SO$_4$ fractionation	100	1500	1920	1.28	70
Step 3: Pooled concentrated fractions after DEAE-cellulose	8	170	1094	6.44	40.1
Step 4: Pooled fractions from Sephadex G-150	20	33	492	14.8	18
Step 5: Pooled fractions from hydroxylapatite before concentration	45	9	189	21	6.9

[a] Adapted from N. A. Donoghue, D. B. Norris, and P. W. Trudgill, *Eur. J. Biochem.* **63,** 180 (1976) with permission.

from the Archibald approach to equilibrium experiments it is 59,000. Sodium dodecyl sulfate – polyacrylamide gel electrophoresis, together with molecular weight markers, gives a subunit molecular weight of 59,000, showing that the holoenzyme is a single polypeptide chain.[10]

Absorption Spectrum. The enzyme is bright yellow with absorbance maxima at 274, 382, and 435 nm. The ratio of absorbances at 274 and 435 nm of 10.8 : 1, for the pure enzyme, is a useful guide to enzyme purity.

Prosthetic Group. The prosthetic group is dissociated from the holoenzyme either by placing 1–2 mg of protein, dissolved in 20 mM sodium potassium phosphate buffer (pH 7.1), in a boiling water bath in the dark for 5 min or by adjusting the holoenzyme in (NH$_4$)$_2$SO$_4$ (90% saturation) to pH 3 with 0.25 M HCl and incubating at 0° for 30 min. Protein is then removed by centrifugation at 20,000 g for 10 min. The isolated prosthetic group has absorbance maxima at 373 and 450 nm and cochromatographs with authentic FAD. Quantitative analysis of the FAD dissociated from known amounts of protein shows that one molecule of FAD is bound to each molecule of apoenzyme.[10]

Preparation of Native Apoenzyme and Reconstitution of Active Holoenzyme. The native apoenzyme can be obtained by the acidic (NH$_4$)$_2$SO$_4$ precipitation procedure[10] but is best prepared as follows.[11] A few milligrams of holoenzyme in 0.5–1 ml sodium potassium phosphate buffer (pH 7.1) is dialyzed against 1 liter of 0.1 M sodium potassium phosphate buffer (pH 7.1), containing 2 M KBr, 0.3 mM EDTA, and 1 mM dithiothreitol, at 3° for 48 hr. Dialysis is then continued for a further 24 hr, three 1-liter changes of buffer *without the KBr* being used. This apoenzyme preparation is almost inactive but can be reconstituted with FAD, or FAD analogs, although FMN is ineffective. A typical reconstitution system contains, in a 1-ml-capacity spectrophotometer cuvette, 0.1 M glycine – NaOH buffer (pH 9.0), 0.16 mM NADPH, 0.2 mg of apoenzyme, and 0.6 μM FAD or FAD analog. The cuvette is incubated at 0° for 30 min, equilibrated at 30° for 1 – 2 min, and the reaction started by addition to 1 μmol of cyclohexanone. Apoenzyme that has been stored at 4° for up to 1 month can be fully reconstituted. A variety of FAD analogs are competent, and reactivated enzyme is obtained with 7-chloro-8-demethyl-FAD, 9-aza-FAD, and 6-methyl-FAD. 5-Deaza-FAD, which is catalytically inactive in oxygenase reactions,[15] inhibits FAD-reconstituted activity in a competitive manner (K_I 62 μM).

Substrate Specificity. The enzyme is active toward a variety of structurally related ketones including 2- and 4-methylcyclohexanones, cyclobutanone, cyclopentanone, cycloheptanone, and cyclooctanone and structurally more complex cyclic ketones such as norcamphor, (+)-camphor, and dihydrocarvone. Apparent K_m values for cyclic ketones range from 4 μM for cyclohexanone to 3.6 mM for cyclopentanone. It is absolutely specific for NADPH as electron donor, with an apparent K_m of 20 μM.

Reaction Mechanism. The insertion of a single oxygen atom into cyclohexanone and related ketones with the formation of the corresponding lactones involves a two-electron oxidation of the ketone substrate, a two-electron oxidation of the NADPH, and a four-electron reduction of the oxygen. Steady-state kinetics demonstrate a Ter – Ter mechanism, and a stable 4a-hydroperoxyflavin carries out a nucleophilic attack on the ketone; H_2O_2 is not involved in the oxygenation reaction.[11] A novel feature of this enzyme is that, in addition to its ketolactonizing activity, it also carries out S-oxygenation of thiane to thiane 1-oxide and of 4-tolylethyl sulfide to 4-tolylethyl sulfoxide,[12] a nucleophilic displacement by the sulfur on the terminal oxygen of the 4a-hydroperoxide.[16]

pH Optimum. The enzyme has a pH optimum of 9.0 in glycine – NaOH and Tris-HCl buffers. The pH – activity curve is fairly broad, and activity varies by less than 35% over a pH range from 8 to 10.

Inhibitors. The enzyme is sensitive to sulfhydryl reactive agents. Titration with *p*-hydroxymercuribenzoate, in which the formation of the cysteine mercaptide is followed at 250 nm,[10] reveals three susceptible sulfhydryl groups per molecule, and this is confirmed by titration with

[15] R. Spencer, J. Fisher, and C. T. Walsh, *Biochemistry* **15,** 1043 (1976).
[16] D. R. Light, D. J. Waxman, and C. T. Walsh, *Biochemistry* **21,** 2490 (1982).

5,5'-dithiobis(2-nitrobenzoate). When inhibition of oxygenating activity is used as the indicator of p-hydroxymercuribenzoate binding 50% inhibition is obtained at an inhibitor/enzyme molar ratio of approximately 0.5 : 1, indicative of a single reactive sulfhydryl group at the catalytic center. This is reinforced by the observed protection of the enzyme against p-hydroxymercuribenzoate inhibition that may be obtained by prior addition of NADPH and suggests a possible role for a catalytic center cysteine in binding NADPH.

[13] Cyclopentanone 1,2-Monooxygenase from *Pseudomonas* NCIMB 9872

By PETER W. TRUDGILL

Cyclopentanone 1,2-monooxygenase catalyzes the first of two steps involved in the cleavage of the carbocyclic ring of cyclopentanone and related compounds by *Pseudomonas* NCIMB 9872 growing with them as sole sources of carbon according to the following equation:

$$\text{Cyclopentanone} + \text{NADPH} + \text{H}^+ + \text{O}_2 \rightarrow \text{5-valerolactone} + \text{NADP}^+ + \text{H}_2\text{O}$$

The purification described herein has been the subject of a previous report,[1] and enzymes catalyzing similar oxygen-insertion reactions with ketone substrates have been purified to homogeneity from a number of sources.[2-5]

Assay Methods

Principle. Two assay methods can be used. Method 1 involves following the cyclopentanone-stimulated oxidation of NADPH at 340 nm. Method 2 depends on the measurement of cyclopentanone-stimulated oxygen consumption in a suitably sensitive closed system fitted with a Clark-type oxygen electrode. Although Method 1 is the more convenient, and may be used routinely, it has the disadvantage that it does not distinguish between monooxygenase and NADPH-linked dehydrogenase activities. Method 2

[1] M. Griffin and P. W. Trudgill, *Eur. J. Biochem.* **63**, 199 (1976).
[2] N. A. Donoghue, D. B. Norris, and P. W. Trudgill, *Eur. J. Biochem.* **63**, 175 (1976).
[3] P. W. Trudgill, this volume [12].
[4] E. Itagaki, *J. Biochem.* **99**, 815 (1986).
[5] L. N. Britton and A. J. Markovetz, *J. Biol. Chem.* **252**, 8561 (1977).

METHODS IN ENZYMOLOGY, VOL. 188

should always be used intermittently as a check that oxygenase activity is indeed being assayed by Method 1.

Reagents for Method 1

Sodium potassium phosphate buffer, 40 mM, pH 7.7 (preincubated at 30°)
NADPH, 1.5 mM in buffer
Cyclopentanone, 20 mM in buffer

Reagents for Method 2

Sodium potassium phosphate buffer, 40 mM, pH 7.7 (preincubated at 30°)
NADPH, 10 mM in buffer
Cyclopentanone, 20 mM in buffer

Procedure for Method 1. The assay mixture contains, in a final volume of 1 ml (semimicrocuvette), NADPH, 0.1 ml; extract or enzyme fraction containing 0.01–0.1 unit of activity; and sodium potassium phosphate buffer to 0.9 ml. This mixture is preincubated at 30° in the thermostatic cell holder of the spectrophotometer for 1–2 min and the endogenous rate of NADPH oxidation then followed for 1 min. After preincubation, cyclopentanone (0.1 ml) is added and the stimulated rate of NADPH oxidation followed. This rate, minus the cyclopentanone-independent rate, is taken to be due to cyclopentanone monooxygenase activity.

Procedure for Method 2. The assay mixture contains, in a volume of 3 ml in a Yellow Springs Oxygen Monitor (Yellow Springs Instruments, Yellow Springs, OH) (volumes of components can be varied *pro rata* to suit systems of different working capacity), NADPH, 0.1 ml; 0.1–0.3 unit of cyclopentanone monooxygenase; and sodium potassium phosphate buffer to 2.9 ml. After temperature equilibration (30°) and insertion of the electrode, endogenous oxygen consumption is followed for 1–2 min. Cyclopentanone (0.1 ml) is then added, substrate-stimulated oxygen consumption followed for an appropriate period, and the substrate-stimulated rate then calculated.

Units. One unit of cyclopentanone 1,2-monooxygenase activity is defined as that amount of enzyme catalyzing the cyclopentanone-stimulated oxidation of 1 μmol of NADPH or consumption of 1 μmol of oxygen per minute at 30°.

Growth and Purification Procedure

Growth of Organism. *Pseudomonas* NCIMB 9872 (available from NCIMB, Torry Research Station, P.O. Box 31, 135 Abbey Road, Aberdeen

AB9 8DG, Scotland) is grown at 30° in a mineral salts medium containing, per liter: KH_2PO_4, 2 g; Na_2HPO_4, 4 g; $(NH_4)_2SO_4$, 2 g; and trace element solution, 4 ml.[3,6] Large crops of cells, suitable for enzyme purification, are grown according to the following regime: 25 ml of sterile mineral salts medium in a 100-ml conical flask is inoculated from a nutrient agar slant, cyclopentanone is supplied from a diffusion tube,[7] and the flask is incubated on an orbital incubator at 30°. After 48 hr of growth this is used to inoculate 1 liter of medium in a 2-liter conical flask to which 2 g of cyclopentanone is added directly after sterilization. After 24 hr of growth the whole of this is used to inoculate a fermentor containing 15 liters of mineral salts medium to which 45 g of cyclopentanone is added after sterilization. The culture is aerated (4 liters/min), stirred (150 rpm), and harvested when the A_{580} reaches 2–2.6 by continuous flow centrifugation. Cells are washed by resuspension in phosphate buffer (KH_2PO_4 2 g/liter, Na_2HPO_4 4 g/liter, pH 7.1) followed by centrifugation at 15,000 g and 3°, for 15 min. The cell paste (~90 g) is then resuspended in an equal volume of the phosphate buffer and used immediately or stored at −20°.

Step 1: Preparation of Cell-Free Extract. The resuspended cell paste is thawed and disrupted by a single passage through a French pressure cell with a pressure difference at the orifice of 20,000 psi (148 MPa), incubated with 1–2 mg of DNase for 15 min, and centrifuged at 20,000 g and 3° for 45 min to yield the cell-free extract as supernatant.

Step 2: Ammonium Sulfate Fractionation. The cell-free extract is fractionated with a neutral saturated (at 3°) solution of $(NH_4)_2SO_4$. The fraction precipitating between 38 and 50% saturation is retained, redissolved in 20 mM sodium potassium phosphate buffer, pH 7.1, and dialyzed overnight against 2 liters of the same buffer to remove $(NH_4)_2SO_4$.

Step 3: DEAE-Cellulose Chromatography. The dialyzed protein solution from Step 2 is applied to a column (15 × 5 cm) of DEAE-cellulose (Whatman DE-52), previously equilibrated with the 20 mM sodium potassium phosphate buffer, and eluted with a linear gradient composed of 1 liter of 20 mM sodium potassium phosphate buffer and 1 liter of the buffer containing 0.5 M KCl. Fractions of 15 ml are collected, the yellow cyclopentanone monooxygenase band (fractions 55–70) are assayed for protein and activity, and those with a specific activity over 1.2 are pooled. Solid $(NH_4)_2SO_4$ is added to the pooled fractions to give a 65% saturated solution, and the protein pellet is harvested by centrifugation at 20,000 g and 3° for 15 min and redissolved in 5 ml of the 20 mM sodium potassium phosphate buffer.

[6] R. F. Rosenberger and S. R. Elsden, *J. Gen. Microbiol.* **22,** 726 (1960).
[7] D. Claus and N. Walker, *J. Gen. Microbiol.* **36,** 107 (1964).

Step 4: Sephadex G-200 Chromatography. The redissolved protein from Step 3 is applied to a column (90 × 2.5 cm) of Sephadex G-200 that has been equilibrated with the 20 mM sodium potassium phosphate buffer and is eluted by passage of this buffer through the column. Fractions of 5 ml are collected (the enzyme can be recognized by the bright yellow color of the bound flavin prosthetic group), assayed for enzyme and protein, and those with specific activity exceeding 2 pooled and dialyzed overnight against 1 mM sodium potassium phosphate buffer, pH 6.8.

Step 5: Hydroxylapatite Chromatography. The dialyzed pooled fraction from Step 4 is loaded onto a column of hydroxylapatite (7 × 2.5 cm) and eluted with a linear gradient formed from 300 ml of 1 mM sodium potassium phosphate buffer, pH 6.8, and 300 ml of 100 mM sodium potassium phosphate buffer at the same pH. Fractions of 5 ml are collected and assayed for protein and enzyme; fractions with a specific activity over 4.0 constitute enzyme of greater than 90% purity. The enzyme may be concentrated either by $(NH_4)_2SO_4$ precipitation or by ultrafiltration.

A typical purification is summarized in Table I.

Properties

Purity. The enzyme is almost homogeneous, containing a trace of a single impurity when analyzed by nondenaturing polyacrylamide gel electrophoresis and stained with Coomassie Brilliant Blue G-250 (CI 42655).

Molecular Weight and Subunit Structure. The molecular weight of the native enzyme, measured by gel permeation through a column of Sephadex G-200, corresponds to 200,000, close to the value of 194,000 obtained from ultracentrifugal analysis ($s_{20,w}^{\circ}$ 10.92; diffusion coefficient 4.64 × 10^{-7} cm^2/sec at 20°). Sodium dodecyl sulfate–polyacrylamide gel electrophoresis together with molecular weight markers gives a subunit molecular weight of 54,000, suggesting that the holoenzyme may be composed of four identical subunits.

Absorption Spectrum. The enzyme is bright yellow with absorbance maxima at 273, 379, and 437 nm. The ratio of absorbance at 273 and 437 nm is a useful guide to purity; the enzyme, purified according to this regime, has a ratio of 3.5:1.

Prosthetic Group. The prosthetic group is dissociated from 1–2 mg of enzyme in 20 mM sodium potassium phosphate buffer (pH 7.1) by the addition of an equal volume of absolute ethanol followed by incubation in the dark for 5 min at 60° and removal of the denatured apoprotein by centrifugation at 20,000 g and 3°, for 10 min. The isolated prosthetic group has absorbance maxima at 373 and 450 nm, and it cochromatographs with authentic FAD. Assay of the FAD/protein ratio gives values of between 2.2 and 3.4 molecules of FAD bound to each tetrameric enzyme molecule.

TABLE I
PURIFICATION OF CYCLOPENTANONE 1,2-MONOOXYGENASE
FROM *Acinetobacter* NCIMB 9872[a]

Fraction	Total volume (ml)	Total protein (mg)	Total activity (units)	Specific activity (units/mg)	Yield (%)
Step 1: Cell-free extract	80	4100	451	0.11	100
Step 2: 38–50% (NH₄)₂SO₄ fraction after dialysis	35	1225	514	0.42	114
Step 3: Pooled concentrated fractions after DEAE-cellulose	7	85	127	1.50	28
Step 4: Pooled fractions from Sephadex G-200	21	34	83	2.45	19
Step 5: Pooled fractions from hydroxylapatite before concentration	20	7.8	33	4.30	7

[a] Adapted from M. Griffin and P. W. Trudgill, *Eur. J. Biochem.* **63**, 202 (1976) with permission.

Substrate Specificity. The enzyme is active toward a variety of structurally related ketones including cyclobutanone, cyclohexanone, methylcyclohexanones, cycloheptanone, cyclooctanone, and norbornanone. It is absolutely specific for NADPH as electron donor, and its absolute requirement for molecular oxygen is shown by the reaction stoichiometry cyclopentanone:NADPH:O_2 of 1:0.97:0.91.

pH Optimum. In both universal barbitone or sodium potassium phosphate buffer the enzyme has a pH optimum of 7.7. The pH–activity curve is broad, and in the barbitone buffer the activity varies by less than 10% over a pH range from 6.5 to 8.5.

Michaelis Constants. Detailed kinetic studies of this three-substrate system have not been performed, but the apparent K_m value for cyclopentanone at 0.15 mM NADPH and saturating levels of oxygen is less than 1 μM. Studies with NADPH as the variable component are also hampered by similar considerations, measurable changes in reaction rate only being observed below 3 μM NADPH.

Inhibitors. The enzyme is inhibited by sulfhydryl-active compounds and, in contrast to cyclohexanone monooxygenase,[2,3] displays sensitivity to metal ion-chelating agents, especially those with a preferential affinity for copper.[1] However, analysis of enzyme for Fe and Cu both give metal atom/protein ratios of less than unity, casting doubt on a role for either element in catalysis.

[14] Protocatechuate 3,4-Dioxygenase from *Brevibacterium fuscum*

By James W. Whittaker, Allen M. Orville, and John D. Lipscomb

Protocatechuate   β-Carboxy-*cis,cis*-muconate

Protocatechuate is an intermediate in the bacterial biodegradation pathways of a wide variety of aromatic compounds.[1] Protocatechuate dioxygenases catalyze the pivotal aromatic ring-opening steps in these pathways.[2] Enzymes which catalyze cleavage between the 2–3, 3–4, and 4–5 carbons of the aromatic ring of protocatechuate have been reported.[3-5] Preparation procedures for each of these enzymes are described in this and other chapters in this volume.[6,7] The enzymes are separated into two subclassifications termed intradiol or extradiol based on the site of cleavage relative to the vicinal hydroxyl functions of the substrate. The intradiol dioxygenase protocatechuate 3,4-dioxygenase (EC 1.13.11.3), has been purified from at least eight different bacterial strains.[8] The enzyme from *Pseudomonas aeruginosa*[9,10] is the best characterized and the crystal structure is known.[11] The enzyme from *Brevibacterium fuscum* is also well characterized and exhibits optical, electron paramagnetic resonance

[1] D. T. Gibson and V. Subramanian, *in* "Microbial Degradation of Organic Compounds" (D. T. Gibson, ed.), p. 181. Dekker, New York and Basel, 1984.

[2] P. Chapman, *in* "Degradation of Synthetic Organic Molecules in the Biosphere," p. 17. National Academy of Sciences, Washington D.C., 1972.

[3] R. L. Crawford, J. W. Bromley, and P. E. Perkins-Olson, *Appl. Environ. Microbiol.* **37,** 614 (1979).

[4] R. Y. Stanier and J. L. Ingraham, *J. Biol. Chem.* **210,** 799 (1954).

[5] S. Trippitt, S. Dagley, and D. A. Stopher, *Biochem. J.* **76,** 9p (1960).

[6] D. M. Arciero, A. M. Orville, and J. D. Lipscomb, this volume [15].

[7] S. A. Wolgel and J. D. Lipscomb, this volume [16].

[8] See, for example, C. Bull and D. Ballou, *J. Biol. Chem.* **256,** 12673 (1981); D. R. Durham, L. A. Stirling, N. L. Ornston, and J. J. Perry, *Biochemistry,* **19,** 149 (1980).

[9] H. Fujisawa and O. Hayaishi, *J. Biol. Chem.* **243,** 2673 (1968).

[10] H. Fujisawa, this series, Vol. 17A, p. 526.

[11] D. H. Ohlendorf, J. D. Lipscomb, and P. C. Weber, *Nature (London)* **336,** 403 (1988).

(EPR), Mössbauer, and resonance Raman spectra which are significantly better resolved than those of the other highly purified protocatechuate 3,4-dioxygenases.[12] The preparation shown here is modified from that previously reported.[12]

Assay Method

Principle. The enzyme activity is measured by either monitoring the disappearance of protocatechuate optically at 290 nm or by monitoring the utilization of O_2 polarographically. The former method has been described in an earlier volume of this series.[10] The latter method is preferred and is described here.

Reagents

MOPS buffer: 50 mM 3-(N-morpholino)propanesulfonic acid adjusted to pH 7.0 at 23°; allow to equilibrate fully with air ($\sim$ 250 μM O_2)

Argon or nitrogen gas passed over a column of BASF Inc. copper catalyst (Kontes, Vineland, NJ, #655960) to remove traces of contaminating O_2

Protocatechuate, 0.1 M stock solution: add 46.2 mg recrystallized and desiccated protocatechuate (Sigma, St. Louis, MO) to 3 ml MOPS buffer in a septum-sealed vial, neutralize the solution with one equivalent of 6 N NaOH, and evacuate the vial and refill with N_2 or Ar at least 6 times to achieve anaerobiosis

Protocatechuate 3,4-dioxygenase for polarograph calibration (either purified from *P. aeruginosa*[10] or *B. fuscum,* or a commercial preparation; Sigma)

Procedure

Oxygen Polarograph. Commercial Clark-type oxygen polarograph instruments (Gilson Medical Electronics, Middleton, WI, or Yellow Springs Instruments, Yellow Springs, OH) are adequate for routine assay. The polarograph circuit shown in Fig. 1 has proved to be more sensitive and is preferred for kinetic investigations of the enzyme.

Calibration of Oxygen Polarograph. The total volume of the stoppered oxygen polarograph chamber is determined. The chamber is completely filled with buffer ($\sim$ 1.5 ml), and the capillary-bored stopper is inserted so that excess buffer is displaced into the bore. After the temperature has equilibrated at 23°, 1 μl of protocatechuate stock solution (0.1 μmol) is

[12] J. W. Whittaker, J. D.Lipscomb, T. A. Kent, and E. Münck, *J. Biol. Chem.* **259,** 4466 (1984).

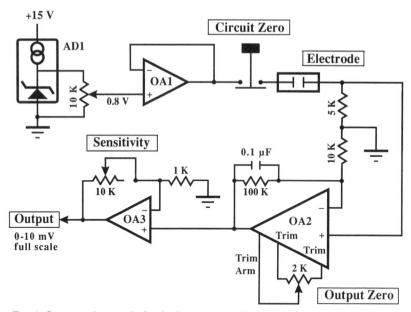

FIG. 1. Oxygen polarograph circuit. Components: AD1, precision voltage reference source (Analog Devices, Norwood, MA, AD580); OA1,OA3, pretrimmed operational amplifier (Precision Monolithics, Santa Clara, CA, OP-07); OA2, FET input instrumentation amplifier (Analog Devices, 52J). The boxed labels indicate front panel controls. All front panel controls are 10-turn precision resistors. A stable ±15 V modular power supply such as the Analog Devices 904 is required. The oxygen electrode (Yellow Springs Instruments, 5331) is contained in a capillary-bore stoppered, water-jacketed 1.5 ml chamber (Gilson OX-15253). The solution in the chamber is stirred using a small magnetic stirring plate (Cole Parmer #04805). It is important to eliminate ground loops in the circuit or significant instability will occur. It is recommended that a single ground point be used and that only this point be connected to the electrode ground. Avoid routing AC power lines near the circuit components. The output of the circuit can be monitored with strip-chart recorder or other recording device.

added using a precision Hamilton syringe. A base line is established on the recording device, and then approximately 20 units of purified protocatechuate 3,4-dioxygenase is added. The substrate in the chamber is rapidly and completely utilized under these conditions. The sensitivity of the polarograph circuit is adjusted to give approximately full-page displacement on the recording device. The amplitude of the displacement in any convenient units is termed the calibration number (C) and is used in the activity calculation shown below. Calibration is required whenever the O_2 porous membrane of the oxygen electrode is changed, and at frequent intervals during the lifetime of the membrane (~ 2 days).

Definition of a Unit of Activity. One unit of enzyme is defined as the amount that catalyzes the uptake of 1 μmol of O_2 per minute at 23° in buffer equilibrated with air.

Assay. The oxygen polarograph chamber is filled and the temperature equilibrated as described above. Then 30 μl of protocatechuate stock solution (~ 2 mM) is added and the recording device allowed to stabilize. The reaction is initiated by addition a known volume (V, in μl) of the enzyme solution to be determined. The initial velocity of O_2 utilization per minute (R) is determined in the same units as the calibration number, C.

Activity Calculations. Calculation of activity is as follows:

$$\text{Units/ml} = (R/10C)(1000/V)$$

Specific activity is defined as units per milligram protein. Protein concentrations of fractions obtained during the isolation are determined using the procedure of Bradford.[13] The purified enzyme concentration is determined using molar absorptivities spectrophotometrically [ϵ_{280} 465 mM^{-1} cm^{-1} or ϵ_{435} 2.9 mM^{-1} cm^{-1} (per iron, five iron atoms per molecule[12])] that are referenced to quantitative amino acid analysis.

Purification Procedure

Reagents

MOPS buffer, as above
Phenylmethylsulfonyl fluoride (PMSF) dissolved in ethanol (Sigma, *caution, very toxic*)
Ribonuclease and deoxyribonuclease (Sigma)
Enzyme-grade ammonium sulfate (Schwarz Mann Chemical Co., Cleveland, OH)

Growth of Organism. Brevibacterium fuscum (ATCC 15933) is cultured in medium containing (per liter) 3.0 g $(NH_4)_2SO_4$, 1.6 g K_2HPO_4, 2.5 g NaCl, 0.27 g $MgSO_4 \cdot 7H_2O$, 50 mg $Fe(NH_4)_2(SO_4)_2 \cdot 6H_2O$, and 0.5 g yeast extract (Difco, Detroit, MI) with 3.0 g of *p*-hydroxybenzoate (Pfaltz and Bauer Inc., Stamford, CT) as a growth substrate. The magnesium sulfate is autoclaved separately, and the iron salt is filter-sterilized for addition to the medium after cooling. The medium is adjusted to pH 7.0 with 6 N NaOH. Cells are transferred from an agar slant containing this medium to a 200-ml flask containing 100 ml of medium. After 12 hr of growth on a rotary shaker at 300 rpm and 30°, 5 ml of the culture is transferred to a 2-liter flask containing 1 liter of medium. This is grown for 12 hr on the rotary shaker under the same conditions. The entire culture is then transferred to an 18-liter carboy containing 15 liters of culture medium. The culture is vigorously aerated (~ 1 liter/sec) at 30° until approximately 6 hr after the end of log phase growth (~ 30 hr) with the addition of 3.0 g/liter of neutralized *p*-hydroxybenzoate after 16 and 24 hr. The cells

[13] M. Bradford, *Anal. Biochem.* **72**, 248 (1976).

are harvested with a water-cooled continuous-flow centrifuge, frozen rapidly in powdered dry ice, and stored at $-80°$. The yield is approximately 10 g/liter of medium.

Step 1: Preparation of the Cell-Free Extract. Frozen cells (200 g) are hammered into a fine meal and rapidly suspended in 175 ml of buffer containing 2 mM PMSF, 10 mM EDTA, and 1 mg each ribonuclease and deoxyribonuclease. Cells are disrupted using a Branson Model 350 sonicator with a 3/4-inch tip at 80% of full power for 50 min. The temperature is maintained at $5-10°$. Particulates are removed by centrifugation for 30 min at 22,000 g and $4°$. A second batch of frozen cells is similarly processed, and the supernatants are combined.

Step 2: Ammonium Sulfate Fractionation. The cell-free extract is diluted to 1800 ml with buffer containing 2 mM PMSF and 10 mM EDTA. The solution is brought to 40% of saturation in ammonium sulfate by adding finely ground solid ammonium sulfate while maintaining the pH at 7.0 with NaOH. After 15 min of equilibration, the precipitate is removed by centrifugation at 7500 g for 45 min at $4°$. The protocatechuate 3,4-dioxygenase activity is precipitated by increasing the ammonium sulfate concentration to 55% of saturation, followed by centrifugation for 30 min at 7500 g and $4°$.

Step 3: Ion-Exchange Chromatography. The pellet is dissolved in 300 ml of buffer and centrifuged for 20 min at 22,000 g and $4°$ to remove any insoluble protein. The conductivity of the solution is adjusted to that of a 0.2 M solution of NaCl (~ 11 mS/cm) by dilution with 5 mM MOPS buffer, pH 7.0. The protein is loaded at 450 ml/hr onto a 3.7 × 25.7 cm DEAE-Sepharose CL-6B fast-flow (Pharmacia, Piscataway, NJ) column equilibrated with buffer plus 0.2 M NaCl at $4°$. After washing the column with 2 bed volumes of buffer plus 0.2 M NaCl, the flow is decreased to 100 ml/hr, and the column is developed with a linear gradient from 0.2 to 0.4 M NaCl (750 and 750 ml) in buffer. The protocatechuate 3,4-dioxygenase elutes at approximately 0.3 M NaCl in a total volume of about 10% of the gradient volume. Fractions with activity over 15% of the peak fraction are pooled.

Step 4: Hydrophobic Interaction Chromatography. Solid ammonium sulfate is added to the pooled ion-exchange step fractions to bring the total concentration to 0.9 M. After adjusting the pH to 7.0, the enzyme solution is loaded onto a 4 × 16.5 cm Phenyl-Sepharose CL-4B (Pharmacia) column at $4°$ that had been equilibrated with buffer, plus 1 M ammonium sulfate, pH 7.0. After washing with 2 bed volumes of 1 M ammonium sulfate, in buffer, the column is developed with a decreasing linear gradient of 1 to 0 M ammonium sulfate in buffer (1000 and 1000 ml) at 100 ml/hr. Homogeneous protocatechuate 3,4-dioxygenase is eluted at approximately

0.45 M ammonium sulfate in about 10% of the gradient volume. Fractions exhibiting over 15% of the activity of the peak fraction are pooled.

Step 5: Concentration and Storage. The pooled fraction from the Phenyl-Sepharose colume is concentrated to approximately $\frac{1}{10}$ of the original volume on an ultrafiltration membrane (Amicon PM30, Danvers, MA). The residual ammonium sulfate is removed either by size-exclusion chromatography on Sephadex G-25 (Bio-Rad, Richmond, CA) equilibrated in buffer or by dialysis against buffer. Then the enzyme is concentrated further to approximately 30 mg/ml protein for storage. The protein is frozen quickly by immersion in liquid N_2. The purification procedure is summarized in Table I.

Properties

Purity and Stability. No impurities are detected in the preparation by denaturing or nondenaturing polyacrylamide gel electrophoresis when stained with Coomassie blue. Crystallization of the enzyme has been achieved, and the activity of the preparation is not increased.[12] The enzyme can be stored for several years at $-20°$ without significant loss of activity or changes of spectral properties.

Subunit Structure and Molecular Weight. Like all known protocatechuate 3,4-dioxygenases, the enzyme from *B. fuscum* is composed of equal numbers of two subunit types (α, M_r 22,500, and β, M_r 40,000). The overall molecular weight determined by analytical ultracentrifugation is 315,000, suggesting that the quaternary structure is $(\alpha\beta)_5$. Iron analyses by inductively coupled plasma emission spectroscopy suggests that there is one iron per $\alpha\beta$ pair or five irons per molecule. There is no evidence for cooperativity between the five active sites.

Kinetic Properties. The enzyme is maximally active in a broad plateau between pH 7 and 8.75. The kinetic properties are dependent on the concentration and nature of the salt ions in solution.[12] Sodium sulfate acts as a nonessential activator and moderates other salt effects when present at 100 mM. The K_m values at pH 7.0 in 50 mM MOPS buffer plus 100 mM Na_2SO_4 for protocatechuate and O_2 determined individually with the other substrate in saturation are 125 and 800 μM, respectively. The turnover number is 25,000 min^{-1} per active site. At the concentration of dissolved O_2 resulting from equilibration of buffer in air, the effective turnover number is 4,170 min^{-1} per site. Despite the relatively high K_m value for O_2, the irreversibility of the reaction drives it essentially to completion. Thus, this enzyme can be used to remove effectively all oxygen from solutions if sufficient protocatechuate is supplied.

Substrate Range. The enzyme demonstrates an absolute requirement

TABLE I
PURIFICATION OF PROTOCATECHUATE 3,4-DIOXYGENASE FROM *Brevibacterium fuscum*

Step	Volume (ml)	Activity (units/ml)	Protein (mg)	Specific activity (units/mg)	Yield (%)
1. Cell-free extract[a]	750	88.0	33,150	2.00	100
2a. 40% $(NH_4)_2SO_4$ supernatant	1800	32.7	15,900	3.71	89
2b. 55% $(NH_4)_2SO_4$ precipitate	300	193	6,810	8.50	88
3. DEAE-Sepharose CL-6B	150	304	2,120	21.5	69
4. Phenyl-Sepharose HIC	200	238	1,160	41.0	73

[a] From 400 g of cells, wet weight.

for vicinal hydroxyl groups in the 3- and 4-positions. Substrate analogs in which the carboxyl group is replaced by H, CHO, CH_2COOH, or $(CH_2)_2COOH$ are turned over, but at rates decreased at least 100-fold relative to protocatechuate.[9,14] Most substituted catechols bind to the enzyme and act as competitive inhibitors.

Spectroscopic Properties. The enzyme exhibits a burgundy red color (λ_{max} 435 nm, ϵ 2.9 mM^{-1} cm^{-1} per iron) owing to charge transfer from two endogenous tyrosine ligands to the active site iron.[12,15] The spectrum shifts to 474 nm for the anaerobic enzyme substrate complex. The EPR and Mössbauer spectra of the enzyme are characteristic of high-spin Fe(III).[12] The spectral properties of the enzymes of this type have been reviewed in detail,[15] and proposals for the mechanism of action have been put forward.[14,16]

Acknowledgment

This work was supported by a grant from the National Institutes of Health (GM 24689).

[14] J. W. Whittaker and J. D. Lipscomb, *J. Biol. Chem.* **259,** 4487 (1984).
[15] L. Que, Jr., *Adv. Inorg. Biochem.* **5,** 167 (1983).
[16] L. Que, Jr., J. D. Lipscomb, E. Münck, and J. M. Wood, *Biochim. Biophys. Acta* **485,** 60 (1977).

[15] Protocatechuate 4,5-Dioxygenase from *Pseudomonas testosteroni*

By DAVID M. ARCIERO, ALLEN M. ORVILLE, and JOHN D. LIPSCOMB

Protocatechuate   α-Hydroxy-γ-carboxymuconic semialdehyde

The extradiol cleaving protocatechuate 4,5-dioxygenase (EC 1.13.11.8) is induced in some nonfluorescent pseudomonads by growth on a variety of aromatic compounds.[1,2] The enzyme catalyzes the critical ring-opening step in the metabolism of protocatechuate, which is itself an intermediate in the biodegradation of many aromatic compounds.[3] The preparation described here has been the subject of a previous report.[4] A preparation of the enzyme from another, unidentified strain of *Pseudomonas* has also been reported.[5]

Assay Method

Principle. The enzyme activity is measured by monitoring (1) the rate of appearance of the yellow ring fission product optically at 410 nm or (2) the rate of utilization of O_2 polarographically.

Reagents

MOPS buffer, 50 mM 3-(N-morpholino)propanesulfonic acid, pH 8.2, 23°, plus 10% glycerol, 2 mM cysteine, and 100 μM ferrous ammonium sulfate; the cysteine and iron are prepared as a 10-fold concentrated solution and then added to the buffer; the buffer must be prepared daily to avoid oxidation of the iron

[1] S. Trippett, S. Dagley, and D. A. Stopher, *Biochem. J.* **76,** 9p (1960).

[2] M. L. Wheelis, N. J. Palleroni, and R. Y. Stanier, *Arch. Mikrobiol.* **59,** 302 (1967).

[3] S. Dagley, *in* "Degradation of Synthetic Organic Molecules in the Biosphere," p. 1. National Academy of Sciences, Washington, D.C., 1972.

[4] D. M. Arciero, J. D.Lipscomb, B. H. Huynh, T. A. Kent, and E. Münck, *J. Biol. Chem.* **258,** 14981 (1983).

[5] K. Ono, M. Nozaki, and O. Hayaishi, *Biochim. Biophys. Acta* **220,** 224 (1970).

Argon or nitrogen gas passed over a column of BASF Inc. copper catalyst (Kontes, Vineland, NJ, #655960) to remove traces of contaminating O_2

Protocatechuate, 0.1 M stock solution: add 46.2 mg recrystallized and desiccated protocatechuate (Sigma, St. Louis, MO) to 3 ml MOPS buffer in a septum-sealed vial, neutralize the solution with one equivalent 6 N NaOH, and evacuate the vial and refill with N_2 or Ar at least 6 times to achieve anaerobiosis

Procedure

Optical Method. Buffer (1 ml) is added to a cuvette and allowed to equilibrate at 23° in a water-jacketed cell holder of a spectrophotometer adjusted to record at 410 nm. Then 10 μl of protocatechuate stock solution (final concentration, 1 mM) is added. The reaction is initiated by adding a known volume of the solution to be assayed, and the time course of the reaction is recorded.

Oxygen Utilization Method. The assay using the oxygen polarograph can be initiated more rapidly than the optical method described above. Since enzymes of this type inactivate during turnover, a better estimate of the activity is obtained by making measurements as soon as possible after the reaction is initiated. For this reason, the polarographic method is preferred. This method is described in detail elsewhere in this volume.[6]

Definition of Unit of Enzyme. One unit of enzyme is defined as the amount that oxidizes 1.0 μmol of protocatechuate or utilizes 1 μmol of O_2 per minute at 23° in buffer equilibrated with air (~ 250 μM O_2).

Protein Determination. The method of Lowry et al.[7] is used for routine protein determination. Purified enzyme in buffer free of iron, glycerol, and cysteine used as stabilizers has an absorbance at 280 nm of 1.77 ml/mg · cm as determined by quantitative amino acid analysis.[4]

Purification Procedure

Reagents

MOPS buffer, as described above

Ribonuclease and deoxyribonuclease (Sigma)

Enzyme-grade ammonium sulfate (Schwarz Mann Chemical Co., Cleveland, OH)

Protamine sulfate solution (Sigma, grade X), 9% by weight in 50 mM MOPS buffer

---

[6] J. W. Whittaker, A. M. Orville, and J. D. Lipscomb, this volume [14].

[7] O. H. Lowry, N. J. Rosebrough, A. L. Farr, and R. J. Randall, *J. Biol. Chem.* **193**, 265 (1951).

Growth of the Organism. *Pseudomonas testosteroni* Pt-L5 (ATCC 49249) is cultured in medium containing (per liter) 2.5 g NaCl, 1.6 g K$_2$HPO$_4$, 3.0 g (NH$_4$)$_2$HPO$_4$, 0.27 g MgSO$_4$·7H$_2$O, 50 mg Fe(NH$_4$)$_2$(SO$_4$)$_2$·6H$_2$O, 0.5 g yeast extract (Difco, Detroit, MI) with 3.0 g of *p*-hydroxybenzoate (Pfaltz and Bauer, Waterbury, CT) as a growth substrate. The magnesium sulfate is autoclaved separately, and the iron salt is filter-sterilized for addition to the medium after it is cool. The medium is adjusted to pH 7.0 with 6 *N* NaOH. Cells are transferred from an agar slant containing this medium to a 200-ml flask containing 100 ml of medium. After 12 hr of growth on rotary shaker at 300 rpm and 28°, 5 ml of the culture is transferred to a 2-liter flask containing 1 liter of medium. This is grown for 12 hr on the rotary shaker under the same conditions. The entire culture is then transferred to an 18-liter carboy containing 15 liters of culture medium. The culture is vigorously aerated ($\sim$ 1 liter/sec) at 28° until approximately 6 hr past the end of log phase growth ($\sim$ 27 hr) with the addition of 3.0 g/liter of neutralized *p*-hydroxybenzoate after 9 and 16 hr. The cells are harvested with a water-cooled continuous-flow centrifuge, frozen rapidly in powdered dry ice, and stored at $-$ 80°. The yield is approximately 10 g/liter of medium.

Step 1: Preparation of Cell-Free Extract. All centrifugation steps in the preparation are carried out at 4°, 17,700 *g*. It has been found that the inclusion of the stabilizing agents glycerol, Fe(II), and cysteine are essential for the maintenance of activity. In the absence of these agents the enzyme is inactivated on all types of column resins examined. Frozen cells (210 g) are allowed to slowly thaw at 4° for 12 hr. Then they are suspended in 500 ml of buffer containing 2 mg each ribonuclease and deoxyribonuclease. Cells are disrupted in two batches with a Branson Model 350 sonicator with a ¾-inch tip at 80% power for 18 min at 5 – 10°. Particulates are removed by centrifugation for 25 min.

Step 2: Heat Precipitation. The cell-free extract is heated to 50° in a water bath for 2 min with constant agitation. It is then rapidly cooled to 4° in an 2-propanol – dry ice bath. The solution is centrifuged for 45 min, and the supernatant is separated. The pellet is washed with 115 ml of buffer and centrifuged again. The pellets are discarded and the supernatants combined.

Step 3: Protamine Sulfate Precipitation. To the rapidly stirred combined supernatants, 25 ml of protamine sulfate solution is added over the course of 20 min. After 10 min of additional stirring, the solution is centrifuged for 30 min and the pellet discarded. The effect of protamine sulfate is quite dependent on the procedure used in its manufacture as well as the composition of the protein solution to which it is added. It is suggested that the exact amount of protamine sulfate solution to be added in this step be experimentally determined for each preparation.

Step 4: Ammonium Sulfate Precipitation. The solution is brought to 48% of saturation in ammonium sulfate by adding finely ground solid ammonium sulfate while maintaining the pH at 8.2 with 6 N NaOH. After 15 min of equilibration, the solution is centrifuged for 30 min. The protocatechuate 4,5-dioxygenase activity is found in the pellet, which is then redissolved in 300 ml of buffer containing 1 M ammonium sulfate.

Step 5: Hydrophobic Interaction Chromatography. The solution is loaded onto a 2 × 20 cm column of Phenyl-Sepharose CL-4B (Pharmacia, Piscataway, NJ) at 4° equilibrated with buffer plus 1 M ammonium sulfate. After washing the column with 400 ml of the equilibration buffer, the enzyme is eluted in a tight band with buffer containing no ammonium sulfate. Fractions containing greater than 60 units/ml are pooled.

Step 6: Ion-Exchange Chromatography. The pooled fractions are diluted to 450 ml with buffer and applied to a 2.2 × 24 cm column of DE-52 (Whatman) equilibrated with buffer at 4°. After washing the column with 600 ml each of buffer and buffer containing 0.1 M KCl, the enzyme is eluted with buffer containing 0.5 M KCl. Fractions with greater than 100 units/ml are pooled.

Step 7: Affinity Chromatography. An affinity column (2.2 × 15 cm) is prepared by coupling vanillic acid (Sigma) to AH-Sepharose (Pharmacia) using a water-soluble carbodiimide (ethyldimethylaminopropylcarbodiimide) according to the manufacturer's procedures. The pooled fractions are diluted to 200 ml with buffer and loaded onto the column equilibrated in buffer at 4°. The column is washed with 200 ml each of buffer and buffer containing 0.1 M KCl before the enzyme is eluted with buffer containing 0.5 M KCl. Fractions with activity greater than 100 units/ml are pooled.

Step 8: Size-Exclusion Chromatography. The pooled fractions are concentrated to approximately 5 ml on an Amicon YM5 ultrafilter and applied to a column (4 × 80 cm) of Fractogel TSK-55F equilibrated in buffer at 4°. After elution, the fractions with specific activity greater than 200 are pooled.

Step 9: Concentration and Storage. The pooled fraction is concentrated at 4° to approximately 30 mg/ml on an ultrafiltration membrane (Amicon YM5) or by pulse elution from a short Phenyl-Sepharose CL-4B column. The protein is frozen quickly by immersion in liquid N_2. The purification procedure is summarized in Table I.

Properties

Purity and Stability. No impurities are detected in the preparation by denaturing or nondenaturing polyacrylamide gel electrophoresis (PAGE)

TABLE I
PURIFICATION OF PROTOCATECHUATE 4,5-DIOXYGENASE FROM
Pseudomonas testosteroni

Step	Volume (ml)	Activity (units/ml)	Protein (mg)	Specific activity (units/mg)	Yield (%)
1. Cell-free extract[a]	670	260	29,480	5.9	100
2. Heat treatment supernatant	555	360	16,650	12.0	116
3. Protamine sulfate supernatant	530	374	11,240	17.6	115
4. 48% $(NH_4)_2SO_4$ fractionation	315	572	4,915	36.7	104
5. Phenyl-Sepharose chromatography	225	735	2,200	75.1	96.0
6. DE-52 ion-exchange fractionation	200	619	1,100	113	71.8
7. Affinity fractionation	55.5	1,410	550	142	45.4
8. Fractogel HW-55F fractionation	56.0	1,510	400	212	50.3

[a] From 210 grams of cells, wet weight.

when stained with Coomassie blue. Crystallization of the enzyme has been achieved, and the specific activity of the preparation is not increased.[4] The enzyme can be stored at liquid N_2 temperature without significant loss of activity or spectral properties for several years. Dialysis of the enzyme can be performed only at temperatures at or below 4°. The enzyme is rapidly inactivated by oxidizing agents such as H_2O_2 or ferricyanide. After oxidation, it can be partially reactivated by ascorbate.

Metal Requirement. The metal-free enzyme is inactive. Inductively coupled plasma emission spectroscopy shows that iron is the only metal present at a significant mole fraction of the enzyme concentration. Mössbauer spectroscopy shows that the iron is present in active enzyme in the high-spin ferrous state.[4] Owing to the iron included as a stabilizer in the preparative procedure, a variable amount of high-spin ferric iron is often present and bound to the enzyme in ferromagnetic clusters. These are readily removed by treatment of the enzyme with the chelator Tiron (4,5-dihydroxy-1,3-benzenedisulfonic acid, Sigma) under anaerobic conditions, followed by chromatography on DEAE or Phenyl-Sepharose to separate the protein from the chelator–iron complex, as well as excess Tiron.

Subunit Structure and Molecular Weight. Electrophoresis in the presence of sodium dodecyl sulfate (SDS) shows that the enzyme consists of approximately equal concentrations of two types of subunits with apparent molecular weights of 17,700 and 33,800. Quantitative N-terminal peptide sequencing of the enzyme and of its resolved subunits is consistent with this observation. Nevertheless, the overall molecular weight determined by analytical gel filtration is 142,000, which does not support any integer combination of subunits in the native structure. However, there are several indications that the enzyme readily forms dissociable polymers, which would complicate the determination.[4] Iron analysis of enzyme from which adventitiously bound iron has been removed without decrease in specific activity contains 1.0 ± 0.1 iron bound per 103 kDa. Thus the native quaternary structure may be $(\alpha\beta)_2 Fe$. Most enzymes of this type elaborate an $(\alpha_4)Fe_4$ quaternary structure.

Kinetic Properties. As is the case for all enzymes of this type, kinetic determinations are complicated because the enzyme is inactivated slowly during turnover. At a typical assay concentration of approximately 0.1 μM enzyme, and under the assay conditions described above, the observed K_m values for protocatechuate and O_2 are 80 and 125 μM, respectively. The turnover number under these conditions is approximately 21,000 min^{-1}. The pH optimum for stability and turnover is between 8.0 and 8.5 in all buffers tested. The stability but not the activity decreases sharply at pH 7.0.

Substrate Range. The enzyme demonstrates an absolute requirement for vicinal hydroxyl groups in the 3- and 4-positions. Sulfonylcatechol is the only known substrate analog with a modification in the carboxylate position which is turned over. Many groups including OH, OCH_3, CH_3, and Cl can occupy the 5-position without loss of turnover.[4,8,9] These alternate substrates are turned over at appreciable rates. For example, 5-methylprotocatechuate retains 67% of the rate of protocatechuate. The products from the 5-methoxylated and 5-halogenated substrate analogs undergo spontaneous ring closure, resulting in loss of the substituent.[10,11] Most other substituted catechols bind to the enzyme and act as competitive inhibitors.

Spectroscopic Properties. The enzyme as isolated exhibits no optical spectrum in the visible range, and it is essentially electron paramagnetic resonance (EPR) silent. A small, axial EPR signal from ferric ion exhibited by the enzyme is eliminated by ascorbate reduction with a concomitant increase in specific activity. The complex of the enzyme with nitric oxide is EPR active, exhibiting a signal characteristic of an $S = 3/2$ spin system.[4] Such signals have been observed from many proteins containing mononuclear iron centers and are thought to be diagnostic of the presence of ferrous ion with ligand sites accessible to exogenous ligands. This signal has

[8] R. Zabinski, E. Münck, P. M. Champion, and J. M. Wood, *Biochemistry* **11**, 3212 (1972).

[9] D. M. Arciero, Ph.D. thesis, University of Minnesota, Minneapolis, 1985.

[10] P. J. Kersten, S. Dagley, J. W. Whittaker, D. M. Arciero, and J. D. Lipscomb, *J. Bacteriol.* **152**, 1154 (1982).

[11] P. J. Kersten, P. J. Chapman, and S. Dagley, *J. Bacteriol.* **162**, 693 (1985).

[12] D. M. Arciero, A. M. Orville, and J. D. Lipscomb, *J. Biol. Chem.* **260**, 14035 (1985).

been exploited in studies of the iron ligation and its interaction with substrates.[12,13] A proposal for the mechanism of action has been suggested based on the spectroscopic studies.[13]

Acknowledgment

This work was supported by a grant from the National Institutes of Health (GM 24689).

[13] D. M. Arciero and J. D. Lipscomb, *J. Biol. Chem.* **261**, 2170 (1986).

[16] Protocatechuate 2,3-Dioxygenase from *Bacillus macerans*

By SANFORD A. WOLGEL and JOHN D. LIPSCOMB

Protocatechuate   α-Hydroxy-δ-carboxymuconic semialdehyde

The extradiol-cleaving protocatechuate 2,3-dioxygenase has been detected only in aerobic spore-forming bacteria. The enzyme was first identified in extracts of *Bacillus circulans*[1] and was partially purified from *Bacillus macerans*.[2] The preparation of apparently homogeneous enzyme from the latter organism described here exhibits approximately 6-fold higher specific activity.[3] The enzyme catalyzes the critical ring-cleaving step in the biodegradation pathways of many complex aromatic compounds.[4]

Assay Method

Principle. The enzyme activity is measured by monitoring (1) the rate of appearance of the yellow ring fission product optically at 350 nm or (2)

[1] R. L. Crawford, *J. Bacteriol.* **121**, 531 (1975); R. L. Crawford, *J. Bacteriol.* **127**, 204 (1976).
[2] R. L. Crawford, J. W. Bromley, and P. E. Perkins-Olson, *Appl. Environ. Microbiol.* **37**, 614 (1979).
[3] S. A. Wolgel, M.S. thesis, University of Minnesota, Minneapolis, 1986.
[4] D. T. Gibson and V. Subramanian, *in* "Microbial Degradation of Organic Molecules" (D. T. Gibson, ed.), p. 181. Dekker, New York, 1984.

the rate of utilization of O_2 polarographically. The optical method is less satisfactory because the product is enzymatically decarboxylated by an enzyme present throughout most of the preparative procedure. The decarboxylated product absorbs at 375 nm with a greater extinction coefficient. Consequently it is difficult to correlate the change in absorbance with activity. The polarographic method is unaffected by the decarboxylase and consequently is the method of choice. Enzymes of this type inactivate slowly during turnover. A better estimate of the enzyme activity is obtained by making measurements as soon as possible after the reaction is initiated. The polarographic method can be initiated and measurements begun significantly more rapidly than by hand-mixed optical methods.

Reagents

MOPS buffer: 50 mM 3-(N-morpholino)propane sulfonate (Sigma, St. Louis, MO), adjusted to pH 7.0 with 6 N NaOH at 23°, fully equilibrated with air

Argon or nitrogen gas, passed over a column of BASF Inc. copper catalyst (Kontes, Vineland, NJ, #655960) to remove traces of contaminating O_2

Protocatechuate, 0.1 M stock solution: add 46.2 mg recrystallized and desiccated protocatechuate (Sigma) to 3 ml MOPS buffer in a septum-sealed vial, neutralize the solution with 1 equivalent 6 N NaOH, and evacuate the vial and refill with N_2 or Ar at least 6 times to achieve anaerobiosis.

Procedure

Method. The polarographic assay method is performed as described elsewhere in this volume.[5]

Definition of Unit of Activity. One unit of enzyme is defined as the amount of enzyme that consumes 1.0 μmol of protocatechuate per minute at 23° in buffer equilibrated with air ($\sim$250 μM O_2). Specific activity is expressed as units per milligram of protein.

Determination of Protein Concentration. Routine protein concentration determinations during the course of the preparation are made using the method of Bradford versus a standard of bovine serum albumin.[6] The protein concentration of homogeneous enzyme is based on a determination made for one sample by quantitative amino acid analysis. This protein sample is then used to construct a standard curve for the Bradford protein assay, from which subsequent protein concentrations are determined. Purified protein, when freed of cysteine and iron employed as stabilizers, has an absorbance of 1.4 ml/mg·cm at 280 nm in water.

[5] J. W. Whittaker, A. M. Orville, and J. D. Lipscomb, this volume [14].
[6] M. Bradford, *Anal. Biochem.* **72,** 248 (1976).

Purification Procedure

Reagents

Water, deionized and glass-distilled

Purified agar (Difco, Detroit, MI)

Pasteurized soil extract: one part soil is mixed with 2 parts water and autoclaved for 1 hr; the solution is filtered on filter paper and autoclaved again for 1 hr

Soil Extract Medium. The soil extract medium contains, per liter, 5 g peptone, 5 g yeast extract, and 250 ml pasteurized soil extract. After the pH of the solution is adjusted to 7.0 with NaOH, the solution is autoclaved.

Growth Medium. The growth medium contains three components, minimal medium, salts, and substrate, each of which is prepared and autoclaved separately. The pH specified for each component is established before autoclaving. The minimal medium contains, per liter, 12.75 g $K_2PO_4 \cdot 3H_2O$, 3.0 g $NaH_2PO_4 \cdot H_2O$, and 6.0 g NH_4Cl, adjusted to pH 7.4 with NaOH. The salts mixture contains, per liter, 0.3 g nitrilotriacetic acid, 0.6 g $MgSO_4 \cdot 7H_2O$, 4.5 mg $MnSO_4 \cdot H_2O$, 5.55 mg $CoCl_2 \cdot 6H_2O$; 9 mg $ZnSO_4 \cdot 7H_2O$, and 30 mg $Fe(NH_4)_2(SO_4)_2 \cdot 6H_2O$, adjusted to pH 7.4 with NaOH. The substrate mixture contains 3.0 g *p*-hydroxybenzoic acid per liter, adjusted to pH 7.1 with NaOH. Growth medium is prepared by mixing equal volumes of each component. Each component can be prepared as a $10 \times$ concentrate.

Growth of Microorganism. Bacillus macerans strain JJ1b (ATCC 35889) used in the preparation of protocatechuate 2,3-dioxygenase has proved to be very sensitive to growth conditions. The following procedures will ensure viable cultures with an adequate supply of enzyme. A loopful of bacteria is streaked onto a plate of 1.5% purified agar containing growth medium adjusted to pH 7.4 (prior to autoclaving) with NaOH. The plate is incubated for 36–40 hr at 45° Incubation for longer periods results in cell sporulation. A single colony from this plate is then used to inoculate another plate, which is similarly incubated. A loopful of cells from the freshly prepared single-colony isolate is transferred into 60 ml of soil extract medium in a 250-ml flask and shaken at 300 rpm at 30° on a gyrorotary shaker. After 11 hr, the flask is transferred into 600 ml of soil extract medium in a 2-liter flask, which is then cultured under the same conditions. After approximately 6 hr, when the cells are rapidly dividing, they are transferred into 5.4 liters of growth medium in a 9.5-liter carboy and grown at 30° with forced aeration with filter-sterilized air. Substrate depletion in the culture is monitored by following the absorbance decrease at 248 nm of medium freed of cells by centrifugation. When the substrate is exactly half-depleted (~7 hr), the culture is transferred equally into four

20-liter carboys, each containing 12 liters of growth medium, and the growth continued. *Immediately* on depletion of the substrate (~ 12 hr), the carboys are harvested using an HPK 40 Pellicon tangential-flow cell-harvesting system equipped with a 0.2-μm HVLP membrane (Millipore, Bedford, MA). The cells are washed with cold 50 mM sodium MOPS buffer, pH 7.0, and centrifuged at 9000 g for 20 min. The cell paste is frozen in liquid N_2 and stored at $-27°$.

The bacteria are damaged by very high rates of aeration early in their growth cycle. Consequently, the aeration rate in the 20-liter carboy is maintained at a moderate level (~ 10 liters/min) until after the initial growth phase, which is indicated by the media turning slightly green. At this point the aeration can safely be increased to about 20 liters/min. Additionally, *B. macerans* appears to be rather sensitive to all antifoam agents except Mazu DF 60P organic antifoam (Mazer Chemicals, Gurnee, IL). This agent is not added unless there is considerable foaming. Less than 1 ml per carboy is sufficient.

Purification

Reagents

50 mM MOPS buffer, pH 6.9, containing 100 μM ferrous ammonium sulfate and 2 mM cysteine; the iron salt and cysteine are added from a freshly prepared 100× stock solution, and the solution is prepared daily.

Ribonuclease and deoxyribonuclease (Sigma)

Step 1: Cell-Free Extract. All steps are carried out at 4°, and all centrifugations are 35,000 g for 20 min at 4°. Frozen cell paste (60 g) is allowed to thaw at room temperature. The thawed cells are suspended in approximately 180 ml of buffer. The cells are disrupted by sonication (Branson Model 350 sonicator with a $\frac{3}{4}$-inch flat tip). Complete disruption is effected by three 30-sec sonic bursts at 80% power with 15-sec intervening pauses. Two milligrams each of ribonuclease, deoxyribonuclease, and $MgCl_2$ are then added. After stirring for 5 min, the cellular homogenate is centrifuged.

Step 2:DEAE-Sepharose Fast-Flow Chromatography. The supernatant is made 100 μM in ferrous ammonium sulfate and 2 mM in cysteine and diluted with buffer until the ionic strength is below that of 25 mM NaCl in buffer. This solution is loaded onto a DEAE-Sepharose Fast Flow (Pharmacia) column (5 × 20 cm) that had been washed overnight with 200 mM MOPS buffer, pH 6.9, containing the iron and cysteine stabilizers and then equilibrated with the usual buffer. The column is washed with 800 ml of buffer, and then a linear gradient (1000 and 1000 ml) from 0 to 0.6 M

NaCl in buffer is applied. The enzyme is eluted in a sharp peak approximately halfway through the gradient. The fractions exhibiting at least 50% and 10% of the activity of the peak fraction on the low- and high-salt sides of the peak, respectively, are pooled and concentrated to 40 ml by ultrafiltration under N_2 on a YM10 membrane (Amicon, Danvers, MA).

Step 3: Hydrophobic Interaction Chromatography. The concentrate is made 1.5 M in NaCl and then applied to a Phenyl-Sepharose (Pharmacia, Piscataway, NJ) column (3 × 10.5 cm), preequilibrated with 1.5 M NaCl in buffer. The column is washed with 150 ml of 1.5 M NaCl in buffer. Homogeneous enzyme is eluted as a broad peak near the middle of a linear gradient (200 and 200 ml) decreasing from 1.5 to 0 M NaCl in buffer. Fractions exhibiting at least 5% of the activity of the peak fraction are pooled and concentrated by ultrafiltration as described above. The enzyme is frozen quickly and stored in liquid N_2.

Removal of Adventitiously Bound Fe(III). Removal of adventitious Fe(III) will decrease the stability of the preparation and is required only for spectroscopic studies which are sensitive to nonactive site iron. For this step, the buffer is 50 mM MOPS, pH 6.9, and all column buffers are bubbled with N_2. A 1 mg/ml solution of enzyme in buffer is anaerobically incubated with 0.3 mM Tiron (4,5-dihydroxy-1,3-benzenedisulfonic acid, Sigma) for 60 min. The solution is made 1.5 M in NaCl and loaded onto a Phenyl-Sepharose column (1 × 5 cm) preequilibrated with a buffer solution of 1.5 M NaCl and 2 mM cysteine. The Tiron–adventitiously bound metal complex is removed by extensively washing the column with 50 mM MOPS buffer at pH 7.0 containing 1.5 M NaCl and 2 mM cysteine. The enzyme is eluted by the same buffer solution without NaCl. The purification procedure is summarized in Table I.

Properties

Purity and Stability. No impurities are detected in the preparation by denaturing polyacrylamide gel electrophoresis (PAGE). Purified enzyme is stable for at least 2 years at concentrations greater than 7.5 mg/ml when stored at liquid N_2 temperature. In the presence of the stabilizing agents ferrous ion and cysteine, the enzyme is stable for 1 day at 4°. The enzyme is inactivated by H_2O_2, $K_3Fe(CN)_6$, NaN_3, and long-term exposure to air. This inactivation appears to be due to oxidation of the active site Fe(II). Full activity is restored by reduction with ascorbate (10:1 molar ratio to enzyme). The enzyme is completely unstable to dialysis and is destabilized by potassium ion. Maximum activity and stability is observed at pH 7.0.

Subunit Structure and Molecular Weight. The holoenzyme molecular weight is determined to be 140,000 by calibrated gel-filtration chromatog-

TABLE I
PURIFICATION OF PROTOCATECHUATE 2,3-DIOXYGENASE
FROM *Bacillus macerans*[a]

Step	Volume (ml)	Activity (units/ml)	Protein (mg)	Specific activity (units/mg)	Yield (%)
1. Cell-free extract	212	123	3220	8.10	100
2. DEAE-Sepharose[b]	42	477	725	25.9	72
3. Phenyl-Sepharose[b]	9	1850	83	201	64

[a] Based on 60 g wet weight cells.
[b] After concentration by pressure dialysis.

raphy, 149,000 by sedimentation velocity ultracentrifugation, and 146,000 by amino acid analysis. The enzyme appears to have four identical subunits as demonstrated by a single protein band on electrophoresis in the presence of sodium dodecyl sulfate (SDS) (M_r 35,500) and a single N-terminal amino acid sequence. Reduced, Tiron-treated enzyme contains four Fe atoms per mole as determined by inductively coupled plasma emission spectroscopy, suggesting a quaternary structure of $\alpha_4 Fe_4$. Mössbauer spectroscopy shows that the iron is high-spin ferrous.

Kinetic Constants. Under the assumption that the enzyme follows an ordered Bi–Uni kinetic reaction mechanism, the K_m values for protocatechuate and oxygen are 24 and 132 μM, respectively. The turnover·number is 18,000 min^{-1} per active site at saturating protocatechuate and oxygen concentrations at 23°. The high K_m for oxygen implies that the standard assay condition ($\sim 250 \mu M$ oxygen) is not saturating with respect to oxygen.

Substrate Range. The enzyme has a broader substrate range than any of the well-studied protocatechuate and catecholic dioxygenases. Compounds that serve as substrates include many 3- or 4-substituted catechols, many 2- or 5-substituted protocatechuates, esters of protocatechuate, as well as protocatechuate analogs in which the carboxylate function is replaced by propionate, mandelate, or acetate. The substrate analog 4-amino-3-hydroxybenzoate is also turned over. This is the only example of turnover of a nonorthodiol benzenoid ring by an extradiol dioxygenase. All substrates are cleaved in the proximal extradiol manner.

Inhibitors. The enzyme is inhibited by Fe(II) chelators including *o*-phenathroline; α,α-dipyridyl, nitrilotriacetic acid, and EDTA. Numerous benzenoid substrate analogs are weakly binding inhibitors, including 4-substituted 3-hydroxybenzoates, 3-substituted 4-hydroxybenzoates, and monohydroxybenzoates. Three compounds, 2-hydroxyisonicotinic *N*-

oxide, 6-hydroxynicotinic acid *N*-oxide, and 4-nitrocatechol, are tight binding inhibitors.

Spectroscopic Properties. The enzyme as isolated exhibits no optical spectrum in the visible range. Preparations exhibit a weak electron paramagnetic resonance (EPR) spectrum from contaminating Mn^{2+} often found associated with enzymes isolated from *Bacillus*. This metal is removed without activity loss by the same procedure which removes adventitious iron. The active site Fe(II) reversibly binds nitric oxide to produce an EPR-active complex that exhibits a spectrum characteristic of an $S = 3/2$ system. The spectrum is altered by substrate and inhibitor binding, and direct coordination of substrate to the iron has been demonstrated in this complex through the use of specific isotopic labeling.[3,7]

Acknowledgment

This work was supported by a grant from the National Institutes of Health (GM 24689).

[7] S. A. Wolgel and J. D. Lipscomb, *Fed. Proc., Fed. Am. Soc. Exp. Biol.* **45,** 1521 (1986).

[17] Gentisate 1,2-Dioxygenase from *Pseudomonas acidovorans*

By MARK R. HARPEL and JOHN D. LIPSCOMB

Gentisic Acid   Maleylpyruvic Acid

Gentisate 1,2-dioxygenase[1] catalyzes the cleavage of gentisate and related structures in the environment. Gentisate serves as the focal point in the biodegradation of a large number of simple and complex aromatic

[1] EC 1.13.11.4, gentisate: oxygen oxidoreductase

compounds.[2,3] Consequently, the gentisate pathway is distributed throughout the microbial world.[4-8] A purification of gentisate 1,2-dioxygenase from *Moraxella osloensis* has been published.[9] The preparation from *Pseudomonas acidovorans* described here exhibits approximately a 40-fold higher specific activity and does not require postpurification activation.[10]

Assay Method

Principle. Gentisate 1,2-dioxygenase activity can be conveniently measured by monitoring the time course of either reactant (oxygen) depletion or product (maleylpyruvate) formation. The former technique involves polarographic measurement of enzyme-stimulated oxygen consumption in the presence of gentisate. The latter technique involves optical measurement of the increase of absorbance at 330 nm arising from the formation of maleylpyruvate. The polarographic assay has been found to be more sensitive and better suited for kinetic studies. A description of the optical assay can be found elsewhere.[11]

Reagents

MOPS[12] buffer, 50 mM, containing 10% glycerol, adjusted to pH 7.4 with NaOH
Gentisic acid, charcoal-treated and recrystallized from hot water
Gentisate stock solution, 0.1 M, prepared anaerobically using recrystallized gentisic acid neutralized with 1 equivalent of NaOH and stored under argon in a sealed septum vial

Procedure. Gentisate 1,2-dioxygenase activity is measured by the calibrated oxygen uptake assay using a Clark-type oxygen electrode as described elsewhere in this volume.[13] Gentisate is added to the assay chamber to a final concentration of 1.25 mM from the 0.1 M stock solution. Stan-

---

[2] P. J. Chapman, *in* "Degradation of Synthetic Organic Molecules in the Biosphere," (I.C. Gunsalus, ed.), p. 17. National Academy of Sciences, Washington, D.C., 1972.

[3] R. C. Bayly and M. G. Barbour, *in* "Microbial Degradation of Organic Molecules" (D. T. Gibson, ed.), p. 253. Dekker, New York, 1978.

[4] L. Lack, *Biochim. Biophys. Acta* **34,** 117 (1959).

[5] D. J. Hopper, P. J. Chapman, and S. Dagley, *Biochem. J.* **122,** 29 (1971).

[6] R. L. Crawford, *J. Bacteriol.* **127,** 204 (1976).

[7] J. N. Ladd, *Nature (London)* **194,** 1099 (1962).

[8] R. C. Crawford, *J. Bacteriol.* **167,** 818 (1986).

[9] R. C. Crawford, S. W. Hutton, and P. J. Chapman, *J. Bacteriol.* **121,** 794 (1975).

[10] M. R. Harpel and J. D. Lipscomb, *J. Biol. Chem.* **265,** 6301 (1990).

[11] M. L. Wheelis, N. J. Palleroni, and R. Y. Stanier, *Arch. Mikrobiol.* **59,** 302 (1967).

[12] MOPS, 3-(*N*-Morpholino)propanesulfonic acid.

[13] J. W. Whittaker, A. M. Orville, and J. D. Lipscomb, this volume [14].

dard dioxygenase assays are carried out at 23° and atmospheric oxygen concentrations. The assay reaction is initiated by addition of enzyme. Gentisate and its analogs are prone to autoxidation resulting in a small background level of oxygen consumption that must be accounted for.

Units and Enzyme Concentration. One unit is defined as the consumption of 1 μmol of O_2 per minute as determined above. This corresponds to turnover of 1 μmol of gentisate per minute. Specific activity is given as units per milligram of protein. Enzyme concentration is determined colorimetrically[14] versus bovine serum albumin.

Purification Procedure

Bacterial Growth. Pseudomonas acidovorans (ATCC 17438) is grown under inducing conditions for gentisate 1,2-dioxygenase and the immediate enzymes of the gentisate pathway. The bacterium is cultured in the minimal medium of Stanier[15] supplemented with 0.2 g/liter yeast extract and 3 g/liter neutralized 3-hydroxybenzoic acid (free acid, Eastman Kodak Co., Rochester, NY). Per liter, the minimal medium contains 2.78 g Na_2HPO_4, 2.78 g KH_2PO_4, 1 g $NH_4(SO)_4$, and 20 ml Hutner's vitamin-free concentrated mineral base.[16] The Hutner's base is autoclaved separately from the rest of the medium, then aseptically added to it. The 3-hydroxybenzoic acid is neutralized by aseptically adding 3.6 ml of 6 N NaOH per liter of the autoclaved medium. The bacteria are maintained on agar containing the same medium.

Large-scale growth experiments are carried out in 18-liter carboys containing 15 liters of medium. A 100 ml liquid culture is inoculated from the agar maintenance culture and grown overnight at 30° with controlled shaking at 250 rpm in an environmental incubator. Then 1000 ml of liquid medium in a 2000-ml shake flask is inoculated with 10 ml of the overnight culture. This is grown under the same conditions. When the optical density of the culture reaches 0.15–0.2 at 650 nm, indicating mid–late log phase of growth, the entire culture is used to inoculate one carboy of medium. This culture is incubated at room temperature with forced aeration from filter-sterilized air at a flow rate of approximately 95 liter/min. An antifoaming agent (2 ml Antifoam A Emulsion, Sigma, St. Louis, MO) is added to the carboy at this time, and as necessary through-

[14] M. Bradford, *Anal. Biochem.* **72**, 248 (1976).
[15] R. Y. Stanier, N. J. Palleroni, and M. Doudoroff, *J. Gen. Microbiol.* **43**, 236 (1966).
[16] G. Cohen-Bazire, W. R. Sistrom, and R. Y. Stanier, *J. Cell. Comp. Physiol.* **49**, 25 (1957); R. M. Smibert and N. R. Krieg, *in* "Manual of Methods for General Bacteriology," (P. Gerhardt, R. G. E. Murray, R. N. Costilow, E. W. Nester, W. A. Wood, N. R. Krieg, and G. B. Phillips, eds.), p. 434. American Society for Microbiology, Washington, D.C., 1981.

out the growth to avoid excessive foaming of the culture. The concentration of 3-hydroxybenzoate is monitored during the growth by measuring the absorbance of centrifuged samples at 288 nm. When the growth substrate is 70–80% depleted (after ∼24 hr), the culture is supplemented with 45 g of neutralized 3-hydroxybenzoate. When growth substrate becomes depleted for the second time (∼12 hr additional time) the culture is harvested with a Sharples continuous-flow centrifuge. Growth yields are typically 10 g/liter wet weight. Cell paste is stored at −20°.

Purification of gentisate 1,2-dioxygenase is performed at 4°. Buffers are 50 mM MOPS, 10% glycerol (EM Science, Cherry Hill, NJ, ACS grade), pH 7.4, containing freshly prepared 100 μM ferrous ammonium sulfate and 2 mM L-cysteine. The reduced iron is essential for retaining full gentisate 1,2-dioxygenase activity throughout the purification. Table I shows a typical purification scheme.

Step 1: Cell-Free Extract. Cell paste (150 g) is resuspended in 300 ml of buffer. Approximately 1 mg each of ribonuclease and deoxyribonuclease are added. The cells are disrupted by sonication (Branson Sonifier, 14 kHz output) for 25 min, maintaining the solution temperature at 4–10° with a 2-propranol–dry ice bath. The suspension is then centrifuged for 30 min at 8875 g.

Step 2: Heat Treatment. Gentisate 1,2-dioxygenase is stable in the centrifuged cell-free extract to heating to 60°. The decanted supernatant from Step 1 is heated slowly with constant stirring in a 65° bath. When the protein solution reaches 60°, it is removed from the bath and immediately cooled to about 15° in a dry ice–2-propanol slurry. The denatured protein is removed by centrifugation at 8875 g for 45 min. This step irreversibly denatures approximately 80% of the total protein from the cell-free extract, with an 80% recovery of gentisate 1,2-dioxygenase activity in the supernatant.

Step 3: Ammonium Sulfate Fractionation. The heat denaturation supernatant is further fractionated by ammonium sulfate precipitation. The pH of the heat step supernatant is first readjusted to pH 7.4 by dropwise addition of 6 N NaOH. The solution is then brought to 33% saturation (196 g/liter) with ammonium sulfate (Ultrapure, Schwarz/Mann Biotech) by the addition of ammonium sulfate powder over a period of 30 min with constant stirring. NaOH (6 N) is added dropwise as necessary to maintain the pH at 7.4. The suspension is equilibrated for an additional 30 min, then centrifuged for 30 min (8875 g). The supernatant contains the gentisate 1,2-dioxygenase activity. This is decanted and brought to 40% ammonium sulfate saturation (additional 43 g/liter) in the same manner as described above, then centrifuged. The supernatant is free of gentisate 1,2-dioxygenase activity and is discarded.

Step 4: Fast-Flow DEAE-Sepharose CL-6B Chromatography. The 40%

TABLE I
PURIFICATION OF GENTISATE 1,2-DIOXYGENASE FROM
Pseudomonas acidovorans

Step	Volume (ml)	Protein (mg)	Units (μmol O$_2$/min)	Recovery (%)	Specific activity (units/mg)	(-fold) Purification
Cell-free extract[a]	375	4460	35,020	100	7.9	1
60° Supernatant	257	810	28,140	80	34.8	4.4
33–40% (NH$_4$)$_2$SO$_4$ precipitation	242	200	27,190	78	136	17.2
Fast-flow DEAE-Sepharose CL-6B	182	44.0	20,120	58	461	58.4
Phenyl-Sepharose CL-4B (PM30 concentration)	19	22.0	10,670	31	484	61.3

[a] From 146 g wet weight cell paste.

ammonium sulfate pellet is resuspended in buffer to a conductivity of no more than 3 mS (Radiometer CDM83 Conductivity Meter, 3.16 mS/cm cell constant) and loaded onto a column of Fast-Flow DEAE-Sepharose CL-6B (Pharmacia, 4.5 × 14 cm) that had been previously equilibrated in buffer containing Fe^{2+} and cysteine as described above. The column is washed with 200 ml buffer, then 500 ml of 0.1 M NaCl in buffer. The gentisate 1,2-dioxygenase activity is eluted with a gradient (1.8 liter) from 0.1 to 0.22 M NaCl in buffer. Fractions containing activities 30% or greater of the peak fraction are pooled. The flow rate is maintained at 31.5 cm/hr.

Step 5: Phenyl-Sepharose CL-4B Chromatography. The pooled fractions from Step 4 are brought to 0.5 M in ammonium sulfate and loaded onto a column of Phenyl-Sepharose CL-4B (Pharmacia, 2.5 × 8 cm) that had been preequilibrated in buffer containing 1 M ammonium sulfate. The column is washed with 100 ml of 0.4 M ammonium sulfate in buffer, and then a reverse salt gradient (300 ml) is run from 0.4 to 0 M ammonium sulfate in buffer. The gentisate 1,2-dioxygenase activity begins to elute approximately halfway through this gradient, and fractions with activity of 30% or greater of the peak fraction are pooled. The pool is concentrated by ultrafiltration under N$_2$ (Amicon PM30 membrane), with several cycles of resuspension in buffer containing no iron or cysteine. Any remaining adventitiously bound iron can be removed by anaerobic treatment with the Fe^{3+}-specific chelator Tiron (4,5-dihydroxy-1,3-benzenedisulfonic acid).[17] The purified and concentrated gentisate 1,2-dioxygenase is stored at $-80°$

[17] D. M. Arciero, J. D. Lipscomb, B. H. Huynh, T. A. Kent, and E. Münck, *J. Biol. Chem.* **258,** 14981 (1983).

with minimal loss of activity, retaining at least 85% of full activity for 1 year.

Properties

Physical Properties. Amino-terminal sequencing of purified gentisate 1,2-dioxygenase shows only a single high-yield polypeptide. A single band of protein is also seen in native gel electrophoresis. Denaturing electrophoresis (SDS–PAGE) reveals a major band at 39,800 molecular weight with two diffuse bands at slightly greater relative mobility, which related studies suggest arise from a microheterogeneity within the single subunit type of this enzyme. The holoenzyme molecular mass determined by gel-permeation chromatography on BioGel A0.5 M (Bio-Rad, Richmond, CA) is 164 kDa. This indicates an α_4 subunit structure. Isoelectric focusing of the protein reveals a pI of 4.55–4.8. Inductively coupled plasma emission analysis and colorimetric assays indicate that one equivalent of iron per subunit is present in fully active protein. The enzyme as isolated is essentially electron paramagnetic resonance (EPR) silent. However, the anaerobic complex with nitric oxide (NO) is EPR active ($S = 3/2$). Quantitation of this signal shows that a 1 : 1 complex of NO to the iron is formed. The appearance of this type of signal is diagnostic of Fe^{2+} in the enzyme as isolated.[17,18] The signal is perturbed by substrate binding.

Kinetic Properties and Stability. The enzyme exhibits a broad maximum of activity from pH 7 to 9. It is most stable near pH 7.4, showing a sharp drop in stability below pH 6 or above pH 8. It is relatively unstable in dilute solutions not containing the stabilizers glycerol, Fe^{2+}, and cysteine. At atmospheric oxygen tension ($\sim 250 \ \mu M$), the apparent Michaelis constant for gentisate equals 80 μM. The turnover number under these conditions is 19,300 min^{-1} (per active site). The $K_{m,app}$ for O_2 is determined to be 55 μM. A $K_{m,app}$ of 57 μM for gentisate at saturating [O_2] is derived from intercept replots with O_2 as the fixed changing substrate. The estimated turnover number at saturating oxygen concentration is approximately 24,000 min^{-1} (per active site). Under turnover conditions, gentisate 1,2-dioxygenase is rapidly inactivated at oxygen concentrations above 800 μM.

Substrate Specificity and Inhibition. Gentisate 1,2-dioxygenase from *P. acidovorans* catalyzes the turnover of a range of alkyl- and halo-substituted gentisates. Somewhat reduced rates are observed with gentisate containing substitutions (methyl, ethyl, 2-propyl, bromo, fluoro) at the ring C-3 position. Analogs of gentisate with similar substitutions in either the C-4 or in

---

[18] J. C. Salerno and J. N. Siedow, *Biochim. Biophys. Acta* **579**, 246 (1979).

both the C-3 and C-4 positions are also turned over, but at dramatically reduced rates ($<10\%$ that of gentisate). All three main functional groups of gentisate appear to be required for efficient turnover, in that little or no turnover is observed with compounds in which any one of these groups is substituted with another functionality. Additionally, monosubstituted benzoates such as salicylate, 3-hydroxybenzoate, and thiosalicylate are not turned over but are relatively good inhibitors of the enzyme. Gentisate 1,2-dioxygenase is inactivated in a time- and concentration-dependent manner by oxidants such as H_2O_2 and $K_3Fe(CN)_6$. Partial reactivation is effected by ascorbate.

Acknowledgment

This work was supported by a grant from the National Institutes of Health (GM 24689).

[18] Synthesis of [17]O- or ^{18}O-Enriched Dihydroxy Aromatic Compounds

By ALLEN M. ORVILLE, MARK R. HARPEL, and JOHN D. LIPSCOMB

Specific isotopic labeling of the oxygen atoms of phenolic and catecholic compounds is of use in spectroscopic and kinetic investigations of a variety of enzymes that bind these molecules as substrates or inhibitors.[1,2] A convenient and versatile synthetic procedure is depicted in Scheme 1. The position of incorporation of the isotopically enriched oxygen atom is determined by the source of the enriched oxygen (H_2O or O_2) and the position of substituents in the starting aromatic amine. This methodology is illustrated here with the synthesis of isotopically enriched homoprotocatechuate (3,4-dihydroxyphenylacetate, HPCA[3]) as an example. This procedure has appeared in a previous report.[2] These techniques have been shown to be applicable to a wide variety of phenolic and catecholic aromatic compounds (Table I).

---

[1] D. M. Arciero and J. D. Lipscomb, *J. Biol. Chem.* **261,** 2170 (1986).
[2] A. M. Orville and J. D. Lipscomb, *J. Biol. Chem.* **264,** 8791 (1989).
[3] Abbreviations used are given in Table I.

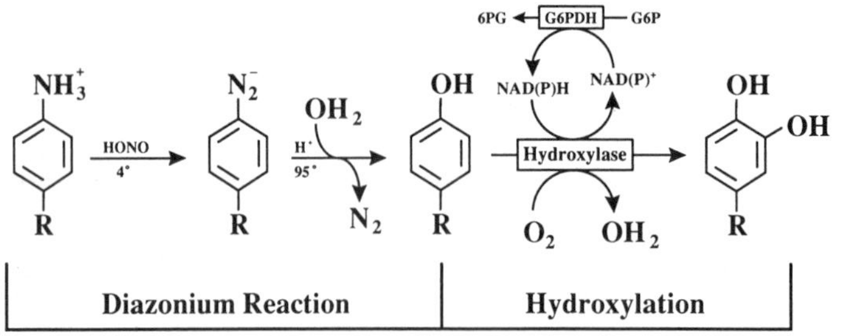

SCHEME 1

Chemical Methods

Diazonium Reaction: Synthesis of [^{17}O]- or [^{18}O]Hydroxyl-Enriched 4-Hydroxyphenylacetate

Reagents

4-Aminophenylacetate (4APA), recrystallized from hot water
Distilled ^{17}O- or ^{18}O-enriched water
Desiccated NaNO$_2$

Procedure. In this procedure an aromatic amine is converted to the corresponding diazonium salt and then hydrolyzed in ^{17}O- or ^{18}O-enriched water to form the labeled phenol. The method was adapted from that of Robertson and Jacobs.[4]

Formation of the diazonium salt. Add 50 mg of 4APA (1.0 molar equivalent) to a small vial containing 600 μl ^{17}O- or ^{18}O-enriched water. Cool the mixture in an ice bath with stirring and add 2.2 molar equivalents of 12 N HCl ($\sim$60 μl). In a separate vial, add approximately 100 μl ^{17}O- or ^{18}O-enriched water to 1.1 molar equivalents of NaNO$_2$ (25 mg) and cool on ice. Add the NaNO$_2$ solution to the amine in 5-μl aliquots from a cold Hamilton syringe over 30 min. Five minutes after the last addition of the NaNO$_2$ add about 50 mg urea to remove the excess HONO.

Hydrolysis of the diazonium salt. Heat, with continuous stirring ^{17}O- or ^{18}O-enriched water (600 μl) containing 2.2 molar equivalents 12 N HCl ($\sim$60 μl) to 90–100° in a two-neck pear-shaped reaction flask (10 ml) immersed in an oil bath. The vertical neck of the flask (♈14/20) is connected to a condenser, and the side neck (♈10/18) is sealed with a rubber septum. Add 25-μl aliquots of the cold diazonium salt solution through the septum with a cold syringe. Allow the evolution of N$_2$ to subside between additions. Continue heating for 15 min after the final addition. Partial

[4] G. R. Robertson and T. L. Jacobs, *in* "Laboratory Practice of Organic Chemistry," 4th Ed. Macmillan, New York, 1962.

TABLE I
Isotopically Enriched Dihydroxyaromatics Produced Enzymatically *in Vitro*[a]

Carbon source	Organism	Hydroxylase induced	Dioxygenase induced	Hydroxylase dihydroxy product	^{17}O Source	Ref.
4HBA	Pseudomonas testosteroni PtL-5 ATCC 49249	4HBA 3-hydroxylase	4,5-PCD	[4-^{17}O] Hydroxyl-PCA	$^{17}OH_2$	1
3HBA	P. testosteroni PtL-5 ATCC 49249	3HBA 4-hydroxylase	4,5-PCD 1,2-GTD	[3^{17}O] Hydroxyl-PCA	$^{17}OH_2$	1
4HPA	Brevibacterium fuscum ATCC 15993	4HPA 3-hydroxylase	2,3-HPCD 3,4-PCD	[4-^{17}O] Hydroxyl-HPCA [3-^{17}O] Hydroxyl-HPCA [3,4^{17}O] Dihydroxyl-HPCA	$^{17}OH_2$ $^{17}O_2$ $^{17}OH_2$, $^{17}O_2$	2 2 b
3HBA	Pseudomonas acidovorans ATCC 17438	3HBA 6-hydroxylase	1,2-GTD	[2-^{17}O] Hydroxyl-gentisate [5-^{17}O] Hydroxyl-gentisate [^{17}O] Carboxyl-gentisate	$^{17}O_2$ $^{17}OH_2$ $^{17}OH_2$	c c c

[a] 4HBA, 4-Hydroxybenzoate; 3HBA, 3-hydroxybenzoate; 4HPA, 4-hydroxyphenylacetate; 3HPA, 3-hydroxyphenylacetate; PCA, protocatechuate (3,4-dihydroxybenzoate); HPCA, homoprotocatechuate (3,4-dihydroxyphenylacetate); 4,5-PCD, protocatechuate 4,5-dioxygenase; 3,4-PCD, protocatechuate 3,4-dioxygenase; 2,3-HPCD, homoprotocatechuate 2,3-dioxygenase; 1,2-GTD, gentisate (2,5-dihydroxybenzoate) 1,2-dioxygenase.

[b] A. M. Orville and J. D. Lipscomb, unpublished.

[c] M. R. Harpel and J. D. Lipscomb, unpublished.

exchange of the carboxyl oxygens of some aromatic acids with water may occur during this step of the procedure. This exchange can be reversed as discussed in the carboxyl exchange section below.

Recovery of ^{17}O- or ^{18}O-enriched water. Cool the reaction to 0°, drain the condenser, and connect the reaction flask to a short-arm vacuum distillation curve. Connect the other end of the curve to a 10-ml pear-shaped collection flask. Cool both the reaction flask and the collection flask in liquid nitrogen, draw a vacuum, and then isolate the system from the vacuum line. On warming the reaction flask to 23°, the ^{17}O- or ^{18}O-enriched water is removed by distillation and is condensed in the cold collection flask. The enriched water recovered in this way must be neutralized, treated with activated charcoal, and distilled again before reuse.

Isolation of [^{17}O] or [^{18}O]hydroxyl-enriched 4-hydroxyphenylacetate. Dissolve the lyophilized hydrolysis product in 1 ml of water containing 1.2 molar equivalents of 6 N NaOH. The mixture is then treated with activated charcoal and filtered into a vial containing 5 equivalents of HCl. Saturate the filtrate with NaCl and extract it several times with 1 ml of anhydrous diethyl ether. Combine the ether extracts and dry them with anhydrous $MgSO_4$. After filtering, evaporate the solvent under a stream of argon or by vacuum distillation. The product is pure, with an isotopic enrichment essentially the same as that of the enriched water used in the hydrolysis.

This reaction proceeds readily with aniline and *m*- or *p*-amino-substituted benzoates or phenylacetates. The diazonium reaction does not produce the desired phenolic products in appreciable yields from aromatic compounds with substituents ortho to the amine.

Carboxyl Oxygen Exchange: Synthesis of [^{17}O] or [^{18}O]Carboxyl-Enriched Homoprotocatechuate

Reagents

Recrystallized HPCA
Distilled ^{17}O- or ^{18}O-enriched water
Argon or nitrogen gas passed over a column of BASF copper catalyst (Kontes, Vineland, NJ, 655960) to remove traces of contaminating O_2

Procedure. The carboxylate oxygens of aromatic acids are exchanged with ^{17}O- or ^{18}O-enriched water via an acid-catalyzed reaction. To HPCA (50 mg) in a thick-walled hydrolysis tube (bulb volume ~ 1 ml) add enriched or unenriched water (550 µl) and 12 N HCl (50 µl). Draw a vacuum and refill the tube with oxygen-free argon or nitrogen several times to achieve anaerobiosis. Freeze the sample, then evacuate and seal the tube. The sealed tube is placed in an oven at 110° for 24 hr during which time the exchange occurs. After cooling, the sample is frozen and the tube opened. The ^{17}O- or ^{18}O-enriched water is recovered by distillation as discussed above. Mass spectral analysis of the [^{17}O] or [^{18}O]carboxyl-enriched HPCA shows complete exchange. Benzoate, phenylacetate, and their m- and/or p-hydroxyl-substituted derivatives yield carboxyl oxygen-enriched compounds with this procedure. In the case of o-hydroxyl-substituted benzoates and phenylacetates, the incorporation is not limited exclusively to the carboxyl oxygens under the conditions described here.

Conditions that promote the carboxyl oxygen exchange are briefly achieved during the hydrolysis of the diazonium salt discussed above. Undesired carboxyl oxygen enrichment can be eliminated without affecting the enrichment of the hydroxyl oxygen atom by carrying out the exchange reaction in *unenriched* water.

Enzymatic Methods

Assay Methods

Catecholic dioxygenase and aromatic hydroxylase activity assays have been described in this and earlier volumes of this series.[5-8]

Microbial Induction

Monooxygenases of the type required for the right half of Scheme 1 are expressed by many bacteria when these organisms are cultured with the monohydroxy aromatic compound as the sole source of carbon and energy (Table I).[1,2,8-11] Several of these have been purified to homogeneity, and at least one is commercially available.[12]

Growth of the Organism. Brevibacterium fuscum (ATCC 15933) is cultured in the medium described elsewhere in this volume[5] with alteration in the carbon source as indicated below. Transfer a loop filled with cells from an agar plate containing medium plus 3 g/liter 4HBA to a 300-ml flask containing 100 ml of the same medium and carbon source. After

[5] J. W. Whittaker, A. M. Orville, and J. D. Lipscomb, this volume [14].

[6] H. Fujisawa, this series, Vol. 17A, p. 526.

[7] H. Kita and S. Senoh, this series, Vol. 17A, p. 645.

[8] M. Husain, L. M. Schopfer, and V. Massey, this series, Vol. 53, p. 543.

[9] J. L. Michalover and D. W. Ribbons, *Biochem. Biophys. Res. Commun.* **55**, 888 (1973).

[10] E. R. Blakley, W. Kurz, H. Halvorson, and F. J. Simpson, *Can. J. Microbiol.* **13**, 147 (1967).

[11] L-H. Wang, R. Y. Hamzah, Y. Yu, and S-C. Tu, *Biochemistry* **26**, 1099 (1987).

[12] 4-Hydroxybenzoate 3-hydroxylase is available from Sigma Chemical Co., St. Louis, MO.

14 hr of growth on a rotary shaker at 30°, transfer 5 ml of the culture to a 2-liter flask containing 1 liter of medium with a carbon source of 2.0 g 4HBA and 1.0 g 4HPA. After 14 hr of growth, transfer the culture to an 18-liter carboy containing 15 liters of medium with 2 g/liter 4HPA as the only carbon source. The culture is vigorously aerated (~ 1 liter/sec). until the end of log phase growth (~ 24 – 30 hr) with the addition of 2.0 g/liter of neutralized 4HPA after 12 hr. The cells are harvested with a water-cooled-continuous-flow centrifuge, frozen rapidly in powdered dry ice, and stored at −80°. The yield is approximately 5 g of cells (wet weight) per liter of medium. Under these conditions, the 4HPA 3-hydroxylase and the catecholic dioxygenases protocatechuate 3,4-dioxygenase (3,4-PCD) and homoprotocatechuate 2,3-dioxygenase (2,3-HPCD) are induced.

Purification of 4-Hydroxyphenylacetate 3-Hydroxylase

Reagents

MOPS buffer: 50 mM 4-morpholinopropanesulfonic acid, adjusted to pH 7.0 with 6 N NaOH

Phenylmethylsulfonyl fluoride (PMSF) (Sigma, St. Louis, MO), *toxic*

Ribonuclease and deoxyribonuclease (Sigma)

Enzyme-grade ammonium sulfate (Schwarz Mann Chemical Co., Cleveland, OH)

Flavin adenine dinucleotide (FAD) (Sigma)

A partial purification of 4HPA 3-hydroxylase is described here. The primary objective of this preparation procedure is to separate the hydroxylase from the catecholic dioxygenases that would degrade the reaction product during the extended *in vitro* enzymatic hydroxylation reaction.

Step 1: Preparation of Cell-Free Extract. Frozen cells (50 g) are hammered into a fine meal and rapidly suspended in 100 ml of MOPS buffer containing 2 mM PMSF, 10 mM EDTA, 1 mM 2-mercaptoethanol (2-ME), 10 μM FAD, and 2 mg each ribonuclease and deoxyribonuclease. Sonicate the cells for 45 min at 80% power with a Branson Model 350 sonicator equipped with a 3/4-inch flat tip. The temperature is maintained at 5 – 10°with a 2-propanol–dry ice bath. Remove the cell debris by centrifugation for 30 min at 22,000 g and 4°.

Step 2: Ammonium Sulfate Fractionation. The cell-free extract is made 55% saturated in ammonium sulfate, the pH is adjusted to 7.0 with 6 N NaOH, and the solution is centrifuged as above. All of the 3,4-PCD and approximately half of the 2,3-HPCD is found in the pellet. Bring the supernatant to 70% saturation in ammonium sulfate (pH 7.0) to precipitate the 4HPA 3-hydroxylase and residual 2,3-HPCD and centrifuge.

Step 3: Ion-Exchange Chromatography. Dissolve the pellet in 10 ml of MOPS buffer containing 2 mM PMSF and centrifuge at 22,000 g and 4° for 15 min to remove any insoluble protein. Dilute the supernatant to 200 ml and load it onto a column of Whatman DE-52 (3 × 18 cm) equilibrated with buffer. After washing with 75 mM ammonium sulfate in buffer, elute the residual 2,3-HPCD with a linear gradient (100 and 100 ml) from 75 to 150 mM ammonium sulfate in buffer. Elute the 4HPA 3-hydroxylase activity with a linear gradient (250 and 250 ml) from 150 to 300 mM ammonium sulfate in buffer. Fractions with more than 0.5 units/ ml hydroxylase activity are pooled and reactivated (~ 400%) by incubation with 1 mM 2-ME and 10 μM FAD for 20 min.

Step 4: Concentration and Storage. Precipitate the reactivated hydroxylase by bringing the solution to 70% saturation in ammonium sulfate (pH 7.0). After centrifugation, redissolve the pellet in 3 ml of MOPS buffer containing 1 mM 2-ME and 10 μM FAD. Typically this preparation has approximately 70 units/ml and a specific activity of about 20 units/mg protein. The enzyme is frozen with liquid N_2 and can be stored at 77 K for several months without loss of activity.

Synthesis of [^{17}O]- or [^{18}O]Hydroxyl-Enriched Homoprotocatechuate

Reagents

Buffer, 0.25 M MOPS, pH 7.0, containing 1 mM 2-ME and 10 μM FAD
Glucose 6-phosphate (G6P), 50 mM
4HPA, 25 mM, unenriched or [^{17}O]- or [^{18}O]hydroxyl-enriched
NADPH, 0.6 mM
^{17}O- or ^{18}O-enriched or unenriched O_2
Glucose-6-phosphate dehydrogenase (G6PDH) (Sigma)
4HPA 3-hydroxylase

Monitoring Hydroxylase Reaction Progress. The extended *in vitro* hydroxylase reaction progress can be followed in most cases by monitoring the loss of the monohydroxyaromatic substrate and/or increase in dihydroxyaromatic product with a UV spectrophotometer. However, the electronic spectra of 4HPA (λ_{max} 278 nm, ϵ 1400 M^{-1} cm^{-1}) and HPCA (λ_{max} 282 nm, ϵ 2500 M^{-1} cm^{-1}) do not permit simple independent quantitations in a complex mixture. Determination of the product concentration using HPLC is a suitable alternative.

HPLC Determinations. The column is a Beckman reversed-phase ODS, 4.6 × 150 mm and the mobile phase isocratic, 89% H_2O, 10% 2-propanol, 1% glacial acetic acid, v/v/v (pH ~2.5). Aliquots (10 μl) of the hydroxylation reaction mixture are taken at various intervals and acidified with 200 μl of 6 N HCl. Saturate the sample with NaCl and extract it several times with 0.5 ml of anhydrous diethyl ether. Evaporate the ether with a stream of argon and dissolve the residual material in 100 μl of the HPLC mobile phase. After filtering, inject 10 μl of this solution directly onto the analytical column for comparison with appropriate standards. For accurate quantitations, the partition coefficients of 4HPA and HPCA into the organic phase during the extractions and their respective molar absorptivity at the HPLC monitoring wavelength must be taken into account.

Procedure. The enzymatic conversion of monohydroxy aromatics to specific dihydroxy aromatics is depicted in the right half of Scheme 1. NADPH is regenerated using the standard system. Method A results in the conversion of [^{17}O]- or [^{18}O]hydroxyl- or carboxyl-enriched monohydroxyaromatic compound to the dihydroxy aromatic analog using unenriched O_2. Method B results in the incorporation of the oxygen from ^{17}O- or ^{18}O-enriched O_2 into the product.

Method A: Synthesis of [4-^{17}O]- or [^{18}O]hydroxyl-enriched homoprotocatechuate. Dissolve the [4-^{17}O]- or [^{18}O]hydroxyl-enriched 4HPA (~ 20 mg) from the diazonium reaction in 5 ml of buffer and adjust the pH to 7.0 with 6 N NaOH. Add G6P (70.5 mg) and NADPH (2.5 mg) and readjust the pH to 7.0. Add catalase and G6PDH (10 units each) from 100–1000× concentrated stock solutions. Initiate the reaction by the addition of 10 units 4HPA 3-hydroxylase from a concentrated stock solution. The O_2 concentration is maintained by directing a stream of compressed O_2 over the stirred reaction mixture. Aliquots (10 μl) are taken every 30 min to monitor the reaction progress as discussed above. Stop the reaction when complete (~ 4 hr) by acid precipitation of the enzymes. The isolation of the [4-^{17}O]- or [^{18}O]hydroxyl-enriched HPCA from the reaction mixture is discussed below.

Method B: Synthesis of [3-^{17}O]- or [^{18}O]hydroxyl-enriched homoprotocatechuate. The reaction is carried out in a glass manifold with a reaction well and a removable storage vessel equipped with a high-vacuum stopcock for recovery of the labeled O_2. Add the buffer, G6P, NADPH, and 4HPA to the reaction well. Adjust the pH as above, and seal the apparatus with a rubber septum. Draw a vacuum and refill with oxygen-free argon or nitrogen several times to achieve anaerobiosis. After freezing the mixture with a dry ice–2-propanol bath, draw a full vacuum and refill with 1 atmosphere of ^{17}O- or ^{18}O-enriched O_2. Thaw the solution and stir to ensure full O_2 equilibration. In a separate vial combine the catalase, G6PDH, and 4HPA 3-hydroxylase (10 units each) as a concentrated stock solution and make the solution anaerobic. Initiate the hydroxylation reaction by the addition of the enzymes to the reaction well through the rubber septum using a gas-tight syringe. Aliquots of the stirred solution are taken every 30 min and analyzed as described above to monitor the reaction progress. When the reaction is complete, acid-precipitate the enzymes and freeze the mixture in a dry ice–2-propanol bath. The ^{17}O- or ^{18}O-enriched O_2 is condensed in the previously evacuated storage flask with liquid N_2. After isolating the storage flask, the reaction mixture is thawed and the HPCA is isolated as described below.

Isolation of [3- or 4-^{17}O]- or [^{18}O]Hydroxyl-Enriched Homoprotocatechuate. Transfer the acidified reaction mixture to a microdistillation apparatus, reduce the volume to approximately 1 ml, and saturate the solution with NaCl. Add anhydrous diethyl ether and, if an emulsion forms, centrifuge. Separate the solvent phase from the aqueous phase. The extraction is repeated several times. The combined solvent phase is dried with anhydrous $MgSO_4$, filtered into a test tube, and evaporated under a stream of argon. Dissolve the residue in 1 ml of water containing 1.2 molar equivalents of NaOH and treat it with activated charcoal. Filter the solution into a test tube containing 5 molar equivalents of HCl, saturate with NaCl, extract with ether, dry, and evaporate as above. The product is pure, and the enrichment is essentially that of the starting labeled compound.

Acknowledgements

This work was supported by a grant from the National Institutes of Health (GM 24689).

[19] Catechol 2,3-Dioxygenases from *Pseudomonas aeruginosa* 2x

By I. A. KATAEVA and L.A. GOLOVLEVA

Cleavage of the aromatic ring is a key reaction in the oxidation of aromatic compounds. Dioxygenases capable of cleaving the aromatic ring widely occur in bacteria, but they differ in their mode of aromatic ring cleavage, specific inductions, and substrate specificity.[1] Certain bacteria of the genus *Pseudomonas* synthesize several aromatic ring-cleaving enzymes that enable them to oxidize various aromatic compounds.[2-4] For example, *Pseudomonas aeruginosa* 2x which grows on a wide set of aromatic sub-

[1] L. N. Ornston, *Bacteriol. Rev.* **35**, 87 (1971).
[2] L. N. Ornston, *J. Biol. Chem.* **241**, 3800 (1966).
[3] E. A. Barnsley, *J. Bacteriol.* **124**, 404 (1976).
[4] L. N. Ornston, *in* "Current Topics in Cellular Regulation" (B. L. Horecker and E. R. Stadtman, eds.), Vol. 12, p. 209. Academic Press, New York, 1977.

strates, including *P*-xylene, synthesizes four dioxygenases:[5-7] pyrocatechase (EC 1.13.11.1, catechol 1,2-dioxygenase), metapyrocatechase 1 and 2 (EC 1.13.11.2, catechol 2,3-dioxygenase) and protocatechuate 3,4-dioxygenase (EC 1.13.11.3). Metapyrocatechase 1 and metapyrocatechase 2 differ from each other in their specific inductions and other properties.[6,7] These enzymes catalyze the metacleavage of aromatic rings to form hydroxymuconic semialdehyde.[8]

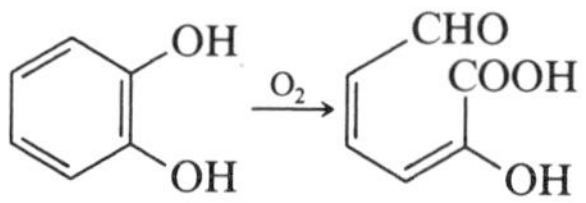

Determination of Enzymatic Activity

The activity of metapyrocatechase is determined spectrophotometrically from the rate of product accumulation.[8]

Reagents

Tris-HCl buffer, 50 mM, pH 7.5
Substrate solution: 10^{-4} M catechol in Tris-HCl buffer
Enzyme solution: 50–100 μg/ml metapyrocatechase 1 or 2

Procedure. Buffer (1.7 ml) is added to 0.2 ml of substrate solution and 0.2 ml of enzyme solution. The reaction mixture is incubated in a thermostatted Specord M-40 cell at 29°, and the variation of optical density is measured at 375 mm. The value of ϵ for the reaction product is 13,800.[8] One unit of enzyme activity was defined as the number of micromoles of substrate cleared per minute.

Purification

Metapyrocatechase 1

Metapyrocatechase 1 is isolated from cells grown on benzoate. All purification steps are conducted at 4° in 50 mM Tris-HCl buffer (pH 7.5) containing 10% acetone (acetone buffer).[9]

Step 1. Cells (100 g) are crushed in the acetone buffer as described for the purification of pyrocatechase.[5]

[5] N. V. Gorlatova and L. A. Golovleva, *Mikrobiologiya* **60**, 1002 (1981).
[6] L. A. Golovleva, I. A. Kataeva, V. Ya. Ilchenko, and P. Kh. Belyaeva, *Biokhimia* **48**, 565 (1983).
[7] I. A. Kataeva and L. A. Golovleva, *Mikrobiologiya J.* **46**, 30 (1984).
[8] J. Kojima, N. Itada, and O. Hayaishi, *J. Biol. Chem.* **242**, 2223 (1967).
[9] M. Nozaki, H. Kagamiyama, and O. Hayaishi, *Biochem. Biophys. Res. Commun.* **11**, 65 (1963).

TABLE I
PURIFICATION OF METAPYROCATECHASE 1

Step	Volume (mL)	Activity (U)	Protein (mg)	Specific Activity (U/mg)	Yield (%)	Purification (-fold)
Cell-free extract	400	6600	6005	1.1	100	—
Precipitation with acetone	53	6355	3180	2	95	1.8
DEAE-cellulose	500	5529	251	22	84	20
DEAE-cellulose	140	2851	102	28	43	25
Sephadex G-200	25	544	6.1	90	8	82

Step 2. Acetone cooled to $-10°$ is added to the supernatant to reach a concentration of 48%. The precipitate is discarded. Additional acetone is added to give a final concentration of 70%. The precipitate is collected, dissolved in a minimum volume of acetone buffer, and dialyzed against the same buffer for 24 hr.

Step 3. The preparation from Step 2 is applied to a column (1.5 × 25 cm) packed with DEAE-cellulose and equilibrated with acetone buffer. The column is washed with 1 liter of acetone buffer, then the elution is carried out with a linear $(NH_4)_2SO_4$ gradient [500 ml of acetone buffer in a mixer and 500 ml of the same buffer with 5% $(NH_4)_2SO_4$ in a storage bottle]. Metapyrocatechase is eluted at an $(NH_4)_2SO_4$ concentration of $1-2.5$%. The fractions showing specific enzyme activity over 20 units/mg are collected, concentrated in an Amicon (Danvers, MA) cell with a UMIO ultrafilter and dialyzed against acetone buffer.

Step 4. The preparation from Step 3 is applied to a column (1.5 × 12 cm) packed with DEAE-cellulose and equilibrated with acetone buffer. The enzyme is eluted with a stepped gradient of $(NH_4)_2SO_4$. The fractions exhibiting enzymatic activity are collected, concentrated to a 4 ml volume, and dialyzed.

Step 5. The preparation is passed through a column (2.5 × 75 cm) of Sephadex G-200. The fractions containing metapyrocatechase 1 with activity above 87 units/mg are collected, concentrated, and stored in acetone buffer under nitrogen at 4° for several weeks or in the lyophilized state at $-20°$ for several months. The purification procedure is summarized in Table I.

Metapyrocatechase 2

Metapyrocatechase 2 is isolated from cells grown on *p*-toluilate. To purify the enzyme, use is made of a slightly modified procedure for metapyrocatechase 1. Metapyrocatechase 2 is precipitated with acetone in the concentration of 48–75%. Chromatography on DEAE-cellulose is performed in one run. Metapyrocatechase is eluted at an $(NH_4)_2SO_4$ concentration of 3–4.5%. In the final step of enzyme purification, gel filtration is performed on Acrylex A-150 (Reanal, Hungary) column (2.5 × 120 cm). The fractions showing activity above 34 units/mg are collected, concentrated, and stored under the same conditions as for metapyrocatechase 1. The purification procedure is summarized in Table II.

Properties of Metapyrocatechases 1 and 2

Purity. Preparations of metapyrocatechases 1 and 2 produce identical symmetrical peaks during ultracentrifugation. They are also homogeneous during isoelectrical focusing and electrophoresis in sodium dodecyl sulfate (SDS)–polyacrylamide gels.

Physicochemical Properties. The molecular masses of metapyrocatechases 1 and 2 can be determined by several methods (Table III). The difference in molecular mass between the two enzymes is not high (10%). The molecular masses of metapyrocatechase 1 and 2 determined from electrophoresis in SDS–polyacrylamide gels are four times lower than those determined by other methods. Thus, both enzymes are composed of four identical subunits.

Amino Acid Analysis. Metapyrocatechases 1 and 2 differ essentially in their amino acid composition, in particular, with regard to the contents of cysteine, lysine, histidine, glutamine, and alanine (Table IV).

Spectral Characteristics. Concentrated solutions of metapyrocatechases 1 and 2 are colorless and show no absorption in the visible spectral range because they contain bivalent iron as a cofactor.[10]

Substrate Specificity. Metapyrocatechases 1 and 2 bring about the meta cleavage of catechol, 4-methylcatechol, 3-methylcatechol, 3,4-dihydroxybenzoate and 3-chlorocatechol. The rates of cleavage of these substrates by metapyrocatechases 1 and 2 differ (Table V). Metapyrocatechase 1 shows the maximum cleavage rate for catechol followed by 3-chlorocatechol and low rates for other substrates. Metapyrocatechase 2 is most effective in the cleavage of 4-methylcatechol but produces no effect on 3,4-dihydroxybenzoate.

Kinetics. Values of K_m for various substrates are also presented in Table V. Metapyrocatechase 2 is less sensitive to the inhibiting effect of ions of heavy metals and 4-chloromercuribenzoic acid (PCMB) compared to me-

[10] M. Nozaki, K. Ono, T. Nakazawa, S. Kotani, and O. Hayaishi, *J. Biol. Chem.* **243**, 2682 (1968).

TABLE II
Purification of Metapyrocatechase 2

Step	Volume (mL)	Activity (U)	Protein (mg)	Specific Activity (U/mg)	Yield (%)	Purification (-fold)
Cell-free extract	50	7495	4293	1.75	100	—
Precipitation with acetone	56	2113	558	3.79	28	2.16
DEAE-cellulose	350	1799	211	8.5	24	4.9
Sephadex G-200	60	998	67	14.9	13	8.5
Acrylex A-150	23	420	12	35	5.6	20

TABLE III
Molecular Masses of Metapyrocatechases 1 and 2

Method	Molecular mass	
	Metapyrocatechase 1	Metapyrocatechase 2
Gel filtration	134,000	143,000
Sedimentation equilibrium	—	144,600
Amino acid analysis	132,000	143,800
Electrophoresis in SDS–polyacrylamide gels	33,500	36,000

TABLE IV
Amino Acid Composition of Metapyrocatechases 1 and 2

Amino acid	Metapyrocatechase 1	Metapyrocatechase 2
Lys	12	61
His	60	25
Arg	76	72
Asp	144	147
Thr	64	62
Ser	44	59
Glu	120	161
Pro	65	39
Gly	100	113
Ala	97	138
Cys	32	15
Val	80	89
Met	24	27
Ile	40	59
Leu	128	121
Tyr	44	30
Phe	56	44
Total	1186	1262

TABLE V
SUBSTRATE SPECIFICITY OF METAPYROCATECHASES 1 AND 2

Substrate	Metapyrocatechase 1		Metapyrocatechase 2	
	Activity (%)	K_m (M)	Activity (%)	K_m(M)
Catechol	100	4.1×10^{-6}	85	2×10^{-6}
4-Methylcatechol	11	2×10^{-5}	100	8×10^{-6}
3-Methylcatechol	7.5	6×10^{-5}	28	10×10^{-6}
3,4-Dihydroxybenzoate	6.4	—	0	—
3-Chlorocatechol	3.6	—	6	—

tapyrocatechase 1. The latter is also activated by ferrous ions and is not inhibited by ferric ions (Table VI).

Induction of Metapyrocatechases 1 and 2

Metapyrocatechase 1 is synthesized mainly in cells induced by benzoate, whereas metapyrocatechase 2 is synthesized chiefly in cells induced by p-xylene, p-toluic alcohol, p-toluilate, or m-toluilate. As both enzymes are capable of cleaving catechol and methylcatechols, the substrate specificity of metapyrocatechase changes in cell-free extracts (Table VII).[7]

Physiological Role of Dioxygenases

Dioxygenases of *P. aeruginosa* 2x capable of cleaving the aromatic ring differ in their specific inductions and substrate specificity. Protocatechuate

TABLE VI
INFLUENCE OF POTENTIAL EFFECTORS ON ACTIVITY OF METAPYROCATECHASES 1 AND 2

Effector	Concentration (M)	Incubation time (min)	Activity (%)	
			Metapyrocatechase 1	Metapyrocatechase 2
CuCl	10^{-4}	5	3.5	32
CuSO$_4$	10^{-4}	5	45	72
AgNO$_3$	10^{-4}	5	10	65
FeSO$_4$	10^{-2}	5	97	191
FeCl$_2$	10^{-2}	5	95	164
FeCl$_3$	10^{-4}	5	81	100
PCMB	10^{-4}	30	6.5	20
EDTA	10^{-2}	60	96	95
o-Phenanthroline	10^{-2}	60	0	4.3
α,α'-Dipyridyl	10^{-2}	60	49	53

TABLE VII
EFFECT OF AROMATIC INDUCER STRUCTURE ON SUBSTRATE SPECIFICITY OF
METAPYROCATECHASES 1 AND 2 IN CELL-FREE EXTRACT

	Activity (%)				
Substrate	Benzoate	p-Xylene	p-Toluic alcohol	p-Toluilate	m-Toluilate
Catechol	100	100	100	100	100
4-Methylcatechol	20	107	87	105	115
3-Methylcatechol	6	23	20	26	33
Protocatechuate	6	3	3	4	2

3,4-dioxygenase is induced by p-hydroxybenzoate and protocatechuate and shows a high rate of cleavage only for protocatechuate. Pyrocatechase is induced by catechol and benzoate, and it cleaves catechol, 4- and 3-methylcatechols, 2-hydroxyaniline, and 3-chlorocatechol. The rate of 2-hydroxyaniline cleavage is markedly lower than that of pyrocatechol cleavage, but it proves to be sufficient for maintaining growth of cultures on this substrate. *Pseudomonas aeruginosa* capable of splitting 3-chlorocatechol can participate in the oxidation of this substrate only in mixed cultures, preventing the accumulation of toxic metabolites.[11,12]

Compared to metapyrocatechase 2, metapyrocatechase 1 has a narrower substrate specificity and effects a high rate of cleavage only on catechol. In addition, metapyrocatechase 2 produces a high rate of cleavage of 4- and 3-methylcatechols; its broad specificity resembles an enzyme from *Pseudomonas arvilla*.[13] Peculiarities of induction of metapyrocatechases 1 and 2 as well as differences in their substrate specificity and K_m values suggest that these enzymes are involved in the oxidation of different substrates. Metapyrocatechase 1 takes part in the oxidation of aromatic compounds, producing catechol as an intermediate, whereas metapyrocatechase 2 is involved in the oxidation of methyl-substituted aromatic substrates.

Thus, *P. aeruginosa* is capable of synthesizing a set of four enzymes for aromatic ring cleavage that are induced by various compounds and show differences in substrate specificity. These particular features are an essential factor enabling this organism to utilize a wide range of aromatic compounds as the sole source of carbon and energy.

[11] P. J. Chapman, *Biochem. Soc. Trans.* **4**, 452 (1970).
[12] W. Reineke and H.-J. Knackmuss, *Biochim. Biophys. Acta* **542**, 412 (1978).
[13] M. Nozaki, S. Kotani, K. Ono, and S. Senoh, *Biochim. Biophys. Acta* **220**, 213 (1970).

[20] Catechol and Chlorocatechol 1,2-Dioxygenases

By Ka-Leung Ngai, Ellen L. Neidle, and L. Nicholas Ornston

Introduction

Catechol and chlorocatechol 1,2-dioxygenases (EC 1.13.11.1; Scheme 1) incorporate molecular oxygen into catechol as they cleave between its hydroxylated carbons to produce *cis,cis*-muconic acid which is metabolized via the β-ketoadipate pathway.[1] Most broadly distributed in nature is catechol 1,2-dioxygenase, formerly named catechol oxygenase I, which exhibits little or no activity with chlorocatechol. This enzyme normally represents several percent of soluble protein in fully induced cells, and the oxygenase has been purified from a number of bacterial sources.[2-6] The purified enzyme possesses a specific activity of about 20 units/mg protein.

Chlorocatechol 1,2-dioxygenase, formerly known as catechol oxygenase II, acts on chlorocatechols to produce the corresponding muconic acid derivatives. This broad specificity, first observed with the enzyme from *Brevibacterium fuscum*,[7] was noted with a dioxygenase from *Pseudomonas* B13.[3] The purified enzyme exhibits relatively low k_{cat} values with catechol or chlorocatechols. The specific activity of the purified enzyme is about 2 units/mg protein, and it represents about 16% of the soluble protein in fully induced cells.[8]

Some *Pseudomonas* isolates express catechol oxygenase from the chromosomal *catA* gene and chlorocatechol oxygenase from the plasmid-borne *clcA* gene.[3] Plasmids carrying the *clcA* gene have been isolated,[9] and the chlorocatechol oxygenase gene has been sequenced.[10] The chromosomal

---

[1] R. Y. Stanier and L. N. Ornston, *Adv. Microb. Physiol.* **9**, 89 (1973).

[2] Y. P. Chen, A. R. Glenn, and M. J. Dilworth, *Arch. Microbiol.* **141**, 225 (1985).

[3] E. Dorn and H. J. Knackmuss, *Biochem. J.* **174**, 73 (1978).

[4] Y. Kojima, H. Fujisawa, A. Nakazawa, T. Nakazawa, F. Kanetsuna, H. Toniuchi, M. Nozaki, and O. Hayaishi, *J. Biol. Chem.* **242**, 3270 (1967).

[5] M. M. Nozaki, M. Iwaki, C. Nakai, Y. Saeki, K. Horiike, H. Kagamiyama, T. Nakazawa, Y. Ebina, S. Inouye, and A. Nakazawa, *in* "Oxygenases and Oxygen Metabolism" (M. Nozaki, S. Yamamoto, Y. Ishimura, M. J. Coon, L. Ernster, and R. W. Estabrook, eds.). Academic Press, New York, 1982.

[6] R. N. Patel, S. Mazumdar, and L. N. Ornston, *J. Biol. Chem.* **250**, 6567 (1975).

[7] H. Nakagawa, H. Inoue, and Y. Takeda, *J. Biochem. (Tokyo)* **54**, 65 (1963).

[8] K.-L. Ngai and L. N. Ornston, *J. Bacteriol.* **170**, 2412 (1988).

[9] D. Ghosal, I. S. You, D. K. Chatterjee, and A. M. Chakrabarty, *Proc. Natl. Acad. Sci. U.S.A.* **82**, 1638 (1985).

[10] B. Frantz and A. M. Chakrabarty, *Proc. Natl. Acad. Sci. U.S.A.* **84**, 4460 (1987).

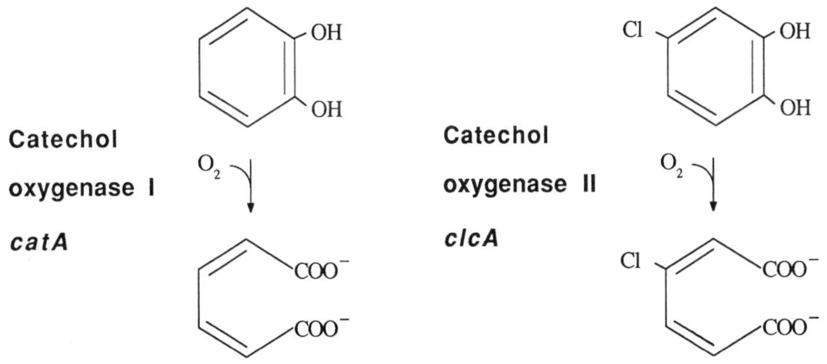

SCHEME 1. Reactions of catechol and chlorocatechol dioxygenases.

catechol oxygenase gene from *Acinetobacter calcoaceticus* has been cloned[11] and sequenced.[12] Because catechol and chlorocatechol oxygenases are expressed at high levels in induced cells, they are relatively amenable to purification. We describe here procedures that allow *Acinetobacter* catechol oxygenase and *Pseudomonas* chlorocatechol oxygenase, enzymes for which amino acid sequences can be deduced from DNA sequences, to be purified by high-performance liquid chromatography (HPLC). Our experience suggests that the procedures may be generally applicable to the purification of catechol and chlorocatechol oxygenases from diverse biological origins.

Assay

Muconate and chloromuconates absorb strongly at 260 nm, a wavelength at which the absorbance of catechol or chlorocatechol is slight. Therefore the activity of the oxygenases can be determined by measuring the increment in absorbance at 260 nm in the presence of catechol. Muconate cycloisomerase, the enzyme that acts on muconate, requires Mn^{2+} and, unlike the catechol oxygenases, is inhibited by EDTA. Therefore quantitative muconate accumulation from catechol is assured by adding EDTA to the assay mixture.

The assay mixture contains, in a final volume of 3.0 ml within a quartz cuvette with 1-cm light path, 0.1 mM catechol, 1 mM EDTA, 33 mM Tris-HCl buffer, pH 8.0, and from 1 to 10 milliunits of enzyme activity. An enzyme unit is the amount required for transformation of 1.0 μmol of substrate to product per minute at 25°. Conversion of 1.0 μmol of sub-

[11] E. L. Neidle and L. N. Ornston, *J. Bacteriol.* **168**, 815 (1986).
[12] E. L. Neidle, C. Hartnett, S. Bonitz, and L. N. Ornston, *J. Bacteriol.* **170**, 4874 (1988).

strate corresponds to an increment of 5.63 absorbance units under these conditions. The reaction is initiated by addition of enzyme, and activity is monitored spectrophotometrically at 260 nm. The presence of chlorocatechol oxygenase activity is assessed by substitution of chlorocatechol for catechol.

It is convenient to prepare mixtures sufficient for several enzyme assays. Stocks of 1.0 M Tris-HCl, pH 8.0, and 0.3 mM sodium EDTA can be prepared in advance and stored at room temperature. Concentrated solutions of 3 mM catechol or chlorocatechol should be prepared on the day of the assay and stored in the cold. Ten assay mixtures can be prepared by mixing 1.0 ml each of the reagent Tris, EDTA, and catechol solutions with 27 ml of distilled water.

Purification of Catechol 1,2-Oxygenase

Growth of Cells. Catechol 1,2-oxygenase can be induced by growing *A. calcoaceticus* ADP1 in mineral medium containing benzoate as the sole carbon and energy source.[6,11] High concentrations of benzoate are toxic to the organism, and the compound should be added periodically so that its concentration is always below 5 mM. Alternatively, the *catA* gene, cloned within pUC19 in *Escherichia coli* JM101(pIB1343), can be expressed to a level of 1.5 units/mg protein by growing the cells with L broth in the presence of 1 mM IPTG (isopropyl-β-D-thiogalactopyranoside).[11] Cells are grown to about 5 mg wet weight/ml of culture; this corresponds to a turbidity of about 300 Klett units or an apparent absorbance of about 1 at 600 nm in a Gilford spectrophotometer. The cells are harvested, washed with 50 mM sodium ethylenediamine (EDA) buffer, pH 7.3, and frozen if their storage is desired prior to extraction.

Preparation of Crude Extract. Cells are suspended in EDA buffer to a final concentration of about 2.5 g wet weight/10 ml and disrupted by sonication. The extract is clarified by centrifugation at 15,000 g for 15 min, treated with 0.25 M NaCl in order to dissociate enzyme aggregates, and dialyzed against several changes of 10 mM EDA buffer, pH 7.3. The dialyzate is ultracentrifuged at 100,000 g for 1 hr, and the supernatant liquid is filtered through a 0.45 μm Nalgene cellulose acetate membrane filter unit. The resulting solution is the crude extract.

High-Performance Liquid Chromatography. A binary gradient HPLC system with a TosoHaas (formerly Toya Soda) TSK DEAE-5PW (21.5 mm × 15 cm) semipreparative anion-exchange column is used. HPLC is performed at room temperature; with this exception, enzyme preparations are maintained at or below 4°. A flow rate of 5 ml/min is set, and about 10 ml of crude extract containing 200–250 mg of protein from the crude extract is injected into the HPLC column which has been equilibrated with 10 mM EDA buffer, pH 7.3. The column is washed with the same buffer containing 0–0.25 M NaCl in a linear gradient over 120 min. Protein in the eluant is monitored at 280 nm; fractions containing catechol oxygenase are readily detected because they have the characteristic red color of the enzyme. Catechol oxygenase is the predominant protein in these fractions,[11] which can be concentrated with a Centricon-30 ultrafilter unit (Amicon, 30,000 MW cutoff). Removal of contaminating proteins can be achieved by hydrophobic interaction chromatography as described below for chlorocatechol oxygenase.

Purification of Chlorocatechol 1,2-Oxygenase

Growth of Cells, Preparation of Crude Extracts, and High-Performance Liquid Chromatography. Cultures of *Pseudomonas* B13 are preadapted to growth with chlorobenzoate by selection for growth on 5 mM 3-chlorobenzoate plates. It may take 1 week for large colonies to appear, and these should be maintained by streaking on chlorobenzoate plates. Cultures are grown in mineral medium by addition of chlorobenzoate as the sole carbon source so that its concentration does not exceed 5 mM.

Procedures for preparation of crude extracts and HPLC DEAE chromatography are the same as those described above for catechol oxygenase. Crude extracts of chlorobenzoate-grown *Pseudomonas* B13 contain both catechol oxygenase and chlorocatechol oxygenase, and the presence of the latter enzyme can be established by monitoring activity with 3-chlorocatechol as the substrate.

Hydrophobic interaction chromatography (HIC) can be used to purify the oxygenases to homogeneity from fractions emerging from DEAE chromatography. A binary gradient system with a TosoHaas TSK Phenyl-5PW (7.5 mm × 7.5 cm) phenyl column is used. HPLC is performed at room temperature; the flow rate is set at 1 ml/min. A volume of 100 μl of concentrate from DEAE chromatography is injected into the HPLC column, which has been equilibrated with 10 mM EDA buffer, pH 7.3, containing 1.0 M ammonium sulfate. The enzyme is eluted with the same buffer containing ammonium sulfate in concentrations extending from 1.0 to 0.0 M in a linear gradient over 60 min. Oxygenase elutes toward the end of the gradient. Fractions containing the enzyme are pooled, desalted, and concentrated to 100 μl with a Centricon-30 ultrafilter unit.

Yield, Storage, Stability, and Properties

The purification procedures should yield several milligrams of pure enzyme. The oxygenases are stable when stored in 10 mM EDA buffer, pH 7.3, at 4°. The purified enzymes are burgundy red in color because they contain ferric ion, which participates in catalysis.[13] The amino acid sequences of catechol[12] and chlorocatechol[10] oxygenases, deduced from the DNA sequences, are evolutionarily homologous with those of *P. putida* protocatechuate 3,4-dioxygenase, for which the crystal structure has been determined.[14] The structural basis for the differences in substrate specificity and catalytic efficiency of the oxygenases is yet to be established. *Acinetobacter* catechol oxygenase[6] and *Pseudomonas* chlorocatechol oxygenase are dimers with respective subunit sizes of 34,351 and 28,922. Each subunit is associated with ferric ion, and pairs of tyrosyl and histidyl residues that ligate the ion appear to have been conserved by evolution at corresponding positions within the primary sequences of the intradiol oxygenases.[12]

[13] L. Que, Jr., *Struct. Bonding (Berlin)* **40**, 38 (1980).
[14] D. H. Ohlendorf, J. D. Lipscomb, and P. C. Weber, *Nature (London)* **336**, 403 (1988).

[21] Muconate Cycloisomerase

By RICHARD B. MEAGHER, KA-LEUNG NGAI, and
L. NICHOLAS ORNSTON

Introduction

Muconate cycloisomerase,   *catB*

Muconate cycloisomerase (EC 5.5.1.1) forms muconolactone from *cis,cis*-muconate during the metabolism of mandelate and benzoate via the β-ketoadipate pathway.[1] The crystal structure of the enzyme from *Pseudomonas putida* has been determined[2] and has been shown to be similar to the structure of mandelate racemase (Chen, Neidhart, and Petsko, personal communication). Comparison of the amino acid sequences of the race-

[1] R. Y. Stanier and L. N. Ornston, *Adv. Microb. Physiol.* **9**, 89 (1973).
[2] A. Goldman, D. L. Ollis, K.-L. Ngai, and T. A. Steitz, *J. Mol. Biol.* **182**, 353 (1985).

mase and the cycloisomerase reveals that the enzymes, catalyzing metabolically removed steps in the same pathway, share common ancestry (Ransom and Gerlt, personal communication). Another evolutionary relative of muconate cycloisomerase is chloromuconate cycloisomerase, an enzyme that removes chloride as it gives rise to a dienelactone.[3-6] It thus seems likely that *P. putida* muconate cycloisomerase will serve as reference for a number of comparative studies. These will be facilitated by procedures, presented here, for large-scale purification of the enzyme.[7]

Assay

cis,cis-Muconate absorbs strongly at 260 nm, and (+)-muconolactone, the product of the reaction, does not. Therefore the enzyme assay simply determines removal of substrate by measuring the decrement in absorbance at 260 nm. The enzyme requires divalent metal ions for activity, and this requirement is met by addition of Mn^{2+} to the assay mixture. A possible difficulty with the assay is isomerization of *cis,cis*-muconate to the cis,trans isomer, a spontaneous reaction that occurs if the cis,cis isomer is heated. The two isomers differ only slightly in spectral characteristics; however, the cis,trans isomer is a very poor substrate for the enzyme, and, in its presence, the assay yields unrealistically low activities that are nonlinear with respect to time even at initial substrate concentrations. Another possible concern is sensitivity of the cycloisomerase to polyanions such as phosphate which can serve as inhibitors of activity.

Assays are conducted at 25° in a volume of 3.0 ml in quartz cuvettes with a 1.0-cm light path. Stock reagents are 1.0 M Tris-HCl, pH 8.0; 10 mM $MnCl_2$; and 3 mM *cis,cis*-muconate. The muconate solution can be prepared in 20 mM Tris-HCl, pH 8.0, stored frozen, and thawed at room temperature. Addition of 1.0 ml of the three stock reagents to 27 ml of distilled water yields a reaction mixture for 10 assays. The assays are initiated by addition of 2 to 20 milliunits of enzyme. Disappearance of 1.0 μmol of substrate would correspond to a decrement in A_{260} of 5.75. A unit of enzyme is the amount that would remove 1.0 μmol of substrate per minute under standard assay conditions.

[3] T. L. Aldrich, B. Frantz, B. Gill, J. F. Kilbane, and A. M. Chakrabarty, *Gene* **52**, 185 (1987).
[4] B. Frantz and A. M. Chakrabarty, *Proc. Natl. Acad. Sci. U.S.A.* **84**, 4460 (1987).
[5] K.-L. Ngai and L. N. Ornston, *J. Bacteriol.* **170**, 2412 (1988).
[6] E. Schmidt and H.-J. Knackmuss, *Biochem. J.* **192**, 339 (1980).
[7] R. B. Meagher and L. N. Ornston, *Biochemistry* **12**, 3523 (1973).

Growth of Cells

Cultures of *P. putida* PRS2000 (or ATCC 12633, the biotype strain for the species) are grown in mineral medium[7] supplemented with mandelate so that the concentration ranges between 5 and 10 mM. This substrate, metabolized via benzoate, is less toxic then benzoate and allows growth to a level of 5 mg wet weight/ml of culture. During growth, the pH is maintained between 6.5 and 7.5 by addition of H_2SO_4. Cells are harvested by centrifugation, washed once with buffer A (10 mM ethylenediamine dihydrochloride, 1 μM $MnCl_2$, adjusted to pH 7.3 with NaOH), and frozen. Successful growth in a 200-liter fermentor should yield almost a kilogram wet weight of cells.

Purification

Preparation of Crude Extract. All steps[7] are conducted at 4° or below. The results are summarized in Table I. About 800 g wet weight of cells is mixed with 1 liter of buffer A and disrupted by sonication; during sonication the temperature is maintained below 4° by addition of cracked ice. Particulate matter is removed by centrifugation at 15,000 g for 1 hr. The pellet is resuspended in 1 liter of buffer A and centrifuged again. The combined supernatant liquids are termed the crude extract (Table I).

DEAE-Cellulose Chromatography. Salt treatment is required to dissociate enzyme aggregates prior to DEAE-cellulose chromatography.[8] Solid ammonium sulfate is added to a final concentration of 0.4 M, and the resulting material is dialyzed against changes of buffer A until the dialysis procedure no longer increases the conductivity of the buffer. The dialyzed material is centrifuged at 12,000 g for 30 min, and the supernatant liquid is applied to a 5 × 100 cm DEAE-cellulose column that has been equilibrated with buffer A. The sample is washed into the column with 500 ml of buffer A and eluted with 6 liters of buffer A containing NaCl in a linear gradient increasing from 0 to 200 mM. This step completely separates muconate cycloisomerase from muconolactone isomerase,[9] the enzyme that catalyzes the next reaction in the β-ketoadipate pathway. Fractions containing muconate cycloisomerase or muconolactone isomerase activity are pooled and termed the DEAE-cellulose eluate.

Ammonium Sulfate Fractionation and Crystallization. The DEAE-cellulose eluate containing muconate cycloisomerase is crystallized by treat-

[8] R. B. Meagher, G. M. McCorkle, and L. N. Ornston, *J. Bacteriol.* **111**, 465 (1972).

[9] R. B. Meagher, K.-L. Ngai, and L. N. Ornston, this volume [22].

TABLE I
PURIFICATION OF *cis,cis*-MUCONATE CYCLOISOMERASE

Step	Volume (ml)	Total activity (units × 10⁻³)	Total protein (g)	Specific activity (units/mg)	Recovery (%)	Purification (-fold)
1. Crude extract	3700	43.8	90.0	0.487	100	1.0
2. Dialyzate	4000	52.0	85.0	0.612	118	1.25
3. DEAE-cellulose eluate	486	40.0	5.0	8.0	91	16.4
4. 30–40% saturated ammonium sulfate fraction	40	32.5	1.8	18.0	74	37.0
5. First crystallization	10.5	21.9	0.252	87.0	50	178.0
6. Second crystallization	10.0	20.0	0.222	90.0	46	182.0
7. Third crystallization	10.0	19.0	0.201	90.0	44	182.0

ment with ammonium sulfate. Initially, ammonium sulfate is added to 30% of saturation and the precipitated material is removed. Ammonium sulfate is added to 40% of saturation. The precipitated material is separated by centrifugation and dissolved at room temperature in 20 mM Tris-HCl, 10 μM MnCl$_2$ (buffer B). This material, within a dialysis sack, is placed in 1 liter of 30% saturated ammonium sulfate, pH 7.0, containing 10 μM MnCl$_2$, and the entire preparation is placed at 4°. Crystals of the enzyme form on the side of the dialysis tubing, and crystallization generally is completed within 3 days. The crystals are collected by centrifugation, washed with the ammonium sulfate solution at 4°, and dissolved in buffer B at room temperature. Crystallization is repeated by the preceding procedure altered only in that the ammonium sulfate solution is at 20% of saturation. If desired, a third crystallization can be achieved by dialysis against 10% saturated ammonium sulfate. The crystalline enzyme is stable when stored in 30% saturated ammonium sulfate, 10 μM MnCl$_2$, 20 mM Tris-HCl, pH 7.4, at 4°.

Properties of Purified Muconate Cycloisomerase

Pseudomonas putida muconate cycloisomerase is formed by association of eight identical protein subunits with a molecular weight of about 40,000. Each subunit contains Mn^{2+} ion near the active site.[2] The protein has three main domains. The N-terminal domain, which is about 100 amino residues long, is a three-stranded antiparallel β sheet, followed by three α helices. The central 220 residues form a parallel-stranded $\alpha-\beta$ barrel with seven α helices linking eight parallel β strands and the small

C-terminal domain. The N-terminal domain of each subunit hangs over the N termini of the α helices in the barrel, close to the active site Mn^{2+}.

The lactonization catalyzed by muconate cycloisomerase involves syn addition of a proton. An enolate intermediate is likely to be involved, and convergent evolution may account for $\alpha-\beta$ barrels found in other enzymes that achieve metabolic transformations through enolates.[2] Sequence evidence shows that common ancestry accounts for similar structures in mandelate racemase (Ransom and Gerlt, personal communication) and chloromuconate cycloisomerase.[3,4] An intriguing question, yet to be answered, is the relationship of these enzymes to carboxymuconate cycloisomerase, an enzyme that catalyzes a reaction analogous to cycloisomerization of muconate in the β-ketoadipate pathway.[10] Whereas the proton addition catalyzed by muconate cycloisomerase is syn, the seemingly analogous addition catalyzed by carboxymuconate cycloisomerase is anti.[11]

[10] R. N. Patel, R. B. Meagher, and L. N. Ornston, *Biochemistry* **12**, 3531 (1973).
[11] R. V. J. Chari, C. P. Whitman, J. W. Kozarich, K.-L. Ngai, and L. N. Ornston, *J. Am. Chem. Soc.* **109**, 5514 (1987).

[22] Muconolactone Isomerase

By RICHARD B. MEAGHER, KA-LEUNG NGAI, and
L. NICHOLAS ORNSTON

Introduction

Muconolactone isomerase, *catC*

Muconolactone isomerase (EC 5.3.3.4) catalyzes the endocyclic migration of a double bond to form β-ketoadipate enol-lactone. The enzyme is physically unusual in that it is a decamer formed by association of identical subunits containing only 96 amino acid residues.[1,2] The primary sequence of the enzyme is of genetic interest because it offered the first evidence that DNA information transfer contributed to the evolutionary divergence of

[1] B. Katz, D. L. Ollis, and H. W. Wyckoff, *J. Mol. Biol.* **184**, 311 (1985).
[2] S. Katti, B. Katz, and H. W. Wyckoff, *J. Mol. Biol.* **205**, 557 (1988).

structural genes.[3,4] Muconolactone is formed by muconate cycloisomerase (EC 5.5.1.1), and substantial quantities of this enzyme and muconolactone isomerase can be purified from a single extract of an appropriately induced *Pseudomonas putida* culture.

Assay

The molar extinction coefficients of muconolactone and β-ketoadipate enol-lactone are equal at 230 nm. β-Ketoadipate, the product formed by hydrolysis of β-ketoadipate enol-lactone, does not absorb significantly at this wavelength. Therefore the rate of enzymatic conversion of muconolactone to β-ketoadipate enol-lactone is most conveniently measured by determining the decrement in absorbance at 230 nm in the presence of muconolactone, limiting amounts of muconolactone isomerase, and excess β-ketoadipate enol-lactone hydrolase which removes the product of the isomerase reaction.

(+)-Muconolactone, the chiral substrate of muconolactone isomerase, is formed enzymatically by the action of muconate cycloisomerase[5] on *cis,cis*-muconate.[6] Solutions of muconolactone are stable while frozen. Procedures for purification of β-ketoadipate enol-lactone hydrolase have been published,[4,7] but satisfactory preparations of the enol-lactone hydrolase, free from muconolactone isomerase activity, can be obtained from extracts of *Escherichia coli* cells in which the cloned hydrolase gene has been expressed.[8,9] The hydrolase from such extracts does not require extensive purification but should be filtered through DEAE-cellulose in the presence of 0.3 M NaCl to remove nucleic acids and other materials that absorb strongly at 230 nm. Enol-lactone hydrolase is stable when stored at 4° in 20 mM Tris-HCl, 50% saturated ammonium sulfate, and 1 mM dithiothreitol, pH 8.0.

Assays are conducted at 25° in a volume of 3.0 ml in quartz cuvettes with a 1.0-cm light path. A unit of activity is the amount of enzyme required to convert 1 μmol of substrate to product under standard assay conditions. Stock reagents are 1.0 M Tris-HCl, pH 8.0; 100 mM (+)-mu-

[3] L. N. Ornston and W. K. Yeh, *Proc. Natl. Acad. Sci. U.S.A.* **76,** 3996 (1979).

[4] W. K. Yeh, P. Fletcher, and L. N. Ornston, *J. Biol. Chem.* **255,** 6347 (1980).

[5] R. B. Meagher, K.-L. Ngai, and L. N. Ornston, this volume [21].

[6] L. N. Ornston and R. Y. Stanier, *J. Biol. Chem.* **241,** 3776 (1966).

[7] G. M. McCorkle, W. K. Yeh, P. Fletcher, and L. N. Ornston, *J. Biol. Chem.* **255,** 6335 (1980).

[8] R. C. Doten, K.-L. Ngai, D. J. Mitchell, and L. N. Ornston, *J. Bacteriol.* **169,** 3168 (1987).

[9] J. Hughes, M. Shapiro, J. Houghton, and L. N. Ornston, *J. Gen. Microbiol.* **134,** 2877 (1988).

conolactone; and β-ketoadipate enol-lactone hydrolase. One milliliter each of the Tris and the muconolactone stock solutions are mixed with 10 units of β-ketoadipate enol-lactone hydrolase and brought to a final volume of 30 ml in order to prepare sufficient mixture for 10 assays. This preparation is stable for several hours at room temperature. The assays are initiated by addition of 5 to 50 milliunits of enzyme. Disappearance of 1.0 μmol of substrate would correspond to a decrement in A_{230} of 0.476.

Purification

Growth of Cells, Preparation of Crude Extracts, and DEAE-Cellulose Chromatography. Early steps in purification of muconolactone isomerase are identical to those described for muconate cycloisomerase as described in the preceding article[5] and as summarized for muconolactone isomerase in Table I.[10] Muconolactone isomerase elutes after muconate cycloisomerase during DEAE-cellulose chromatography, and we here refer to fractions containing this enzyme as the DEAE-cellulose eluate.

Ammonium Sulfate Fractionation and Crystallization. The DEAE-cellulose eluate containing muconolactone isomerase is subjected to ammonium sulfate fractionation, and material precipitating between 50 and 60% of saturation is separated by centrifugation. The precipitated material is dissolved in 20 mM Tris-HCl, pH 7.4, at room temperature and placed within a dialysis sack which is added to 1 liter of 30% saturated ammonium sulfate, pH 7.0. The entire preparation is placed at 4°, and after 3 days crystals are collected from the material within the dialysis sack by centrifugation. The crystals are dissolved in 20 mM Tris-HCl, pH 7.4. After insoluble material has been removed by centrifugation, the enzyme is crystallized within a dialysis sack by the preceding procedure. If necessary, solubilization and crystallization of the enzyme can be repeated until it is pure.

Properties of Muconolactone Isomerase

The crystal structure of muconolactone isomerase has been solved at 3.3 Å resolution.[2] Facing pentameric arrays of subunits from the decameric enzyme. Emerging from each subunit is a C-terminal fragment that intertwines with the facing subunit to form the pocket comprising the active site. The protein subunits undergo reversible dissociation,[11] and so factors controlling their physical interaction can be explored.

---

[10] R. B. Meagher and L. N. Ornston, *Biochemistry* **12**, 3523 (1973).
[11] R. B. Meagher, Ph. D. dissertation, Yale University, New Haven, Connecticut, 1975.

TABLE I
PURIFICATION OF MUCONOLACTONE ISOMERASE

Step	Volume (ml)	Total activity (units $\times 10^{-3}$)	Total protein (g)	Specific activity (units/mg)	Recovery (%)	Purification (-fold)
1. Crude extract	3700	243	90.0	2.70	100	1.0
2. Dialyzate	4000	260	85.0	3.05	106	1.15
3. DEAE-cellulose eluate	1000	221	11.0	23.2	91	8.6
4. 50–60% saturated ammonium sulfate fraction	74	215	4.1	52.4	88	19.4
5. First crystallization	10	213	0.350	609.0	87	225.0
6. Second crystallization	6	208	0.240	860.0	85	318.0
7. Third crystallization	5	180	0.212	856.0	74	317.0

The absolute stereochemical course of enzymatic isomerization of muconolactone has been established.[12] A single univalent base serves as both proton acceptor and donor via a syn mechanism. Contrasting stereochemistry[12] is exercised by 4-carboxymuconolactone decarboxylase (EC 4.1.1.44), an enzyme that catalyzes a reaction formally analogous to that mediated by muconolactone isomerase.[13]

[12] R. V. J. Chari, C. P. Whitman, J. W. Kozarich, K.-L. Ngai, and L. N. Ornston, *J. Am. Chem. Soc.* **109**, 5520 (1987).
[13] D. Parke, *Biochim. Biophys. Acta* **578**, 145 (1979).

[23] cis-1,2-Dihydroxycyclohexa-3,5-diene (NAD) Oxidoreductase (cis-Benzene Dihydrodiol Dehydrogenase) from *Pseudomonas putida* NCIB 12190

By JEREMY R. MASON and PHILIP J. GEARY

$$\text{cis-Benzene dihydrodiol} + \text{NAD}^+ \xrightarrow{\text{Fe}^{2+}} \text{dihydroxybenzene} + \text{NADH} + \text{H}^+$$

The ability of microorganisms to catalyze the enzymatic fission of the benzenoid nucleus depends on the presence of at least two hydroxyl substituents on the aromatic ring.[1] The dihydroxylation reaction is frequently carried out by dioxygenase enzymes that incorporate two atoms of molecular oxygen into the substrate to form *cis*-dihydrodiols. The latter are then the substrates for dehydrogenation to form a catechol. The initial reactions of the latter mechanism may be represented by the general sequence, shown in Scheme 1.

Clearly, dihydrodiol dehydrogenases are a group of bacterial enzymes that play an important role in the biodegradation of many aromatic compounds. Dehydrogenases that catalyze the oxidation of *cis*-diols from benzoic acid,[2] benzene,[3] naphthalene,[4] toluene,[5] chloridazon,[6] and phenanthrene[7] have been purified. A comparison of these enzymes is summarized in Table I. In most cases they exhibit a high affinity to *cis*-dihydrodiols and NAD$^+$ and have pH optima between 7.9 and 9.8. Many of the dehydrogenases isolated from *Pseudomonas* species have been shown to be immunologically related.[8] The following gives an account of the purification and characterization of the *cis*-benzene dihydrodiol dehydrogenase.

Assay Method

Principle. *cis*-Benzene dihydrodiol dehydrogenase activity is assayed by the spectrophotometric measurement of NAD$^+$ reduction at 340 nm.

[1] D. T. Gibson and V. Subramanian, in "Microbial Degradation of Organic Compounds" (D. T. Gibson, ed.), p. 181. Dekker, New York, 1984.
[2] A. M. Reiner, *J. Bacteriol.* **108**, 89 (1971).
[3] B. C. Axcell and P. J. Geary, *Biochem. J.* **136**, 927 (1973).
[4] R. T. Patel and D. T. Gibson, *J. Bacteriol.* **119**, 879 (1974).
[5] J. E. Rogers and D. T. Gibson, *J. Bacteriol.* **130**, 1117 (1977).
[6] J. Eberspächer and F. Lingens, *Hoppe-Seyler's Z. Physiol. Chem.* **359**, 1323 (1978).
[7] K. Nagao, N. Takizawa, and H. Kiyohara, *Agric. Biol. Chem.* **10**, 2621 (1988).
[8] T. R. Patel and D. T. Gibson, *J. Bacteriol.* **128**, 842 (1976).

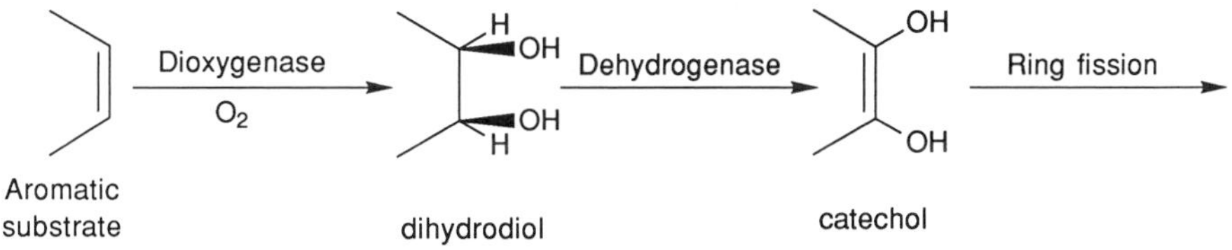

SCHEME 1. General sequence of the initial reactions in the bacterial oxidation of benzene.

Reagents

Potassium phosphate buffer, pH 7.4, 25 mM
NAD$^+$, 10 mM
FeSO$_4$, 3 mM
Glutathione (GSH), 3 mM
cis-Benzene dihydrodiol (Sigma, St. Louis, MO), 1.5 mM

Procedure. To a cuvette with a 1.0-cm light path is added 2.5 ml of buffer, 0.1 ml of enzyme or cell-free extract, and 0.1 ml each of FeSO$_4$, GSH, and NAD$^+$. The reaction is initiated by the addition of 0.1 ml of *cis*-benzene dihydrodiol. The NADH generated in the reaction is measured by the increase in extinction at 340 nm compared with that of a further

TABLE I
COMPARISON OF *cis*-DIHYDRODIOL DEHYDROGENASES

| | | Properties of purified enzyme | | | | | | |
| | | | Subunits | | K_m (μM) | | | |
Substrate	Organism	$M_r \times 10^{-3}$	No.	$M_r \times 10^{-3}$	Diol	NAD$^+$	pH optimum	Ref.
cis-Benzene dihydrodiol	*Pseudomonas putida*	440	4	110	286	43.5	7.9	3
cis-Toluene dihydrodiol	*P. putida*	104	4	27	<2	660	9.6	5
cis-Naphthalene dihydrodiol	*P. putida*	102	4	25.5	29	880	9.0	4
cis-Phenanthrene dihydrodiol	*Alcaligenes faecalis*	~100	4	27.5	21	58	9.0	7
cis-Toluene dihydrodiol	*Bacillus*	172	6	29.5	92	80	9.8	*a*
Chloridazo-dihydrodiol	—	220	4	50	250 1000[b]	200 250[b]	9.5 7.0[b]	6
cis-1,2-Benzoic acid diol	*Alcaligenes eutrophus*	94.6	4	24	200	150	—	2

[a] H. D. Simpson, J. Green, and H. Dalton, *Biochem. J.* **244,** 585 (1987).
[b] Two isoenzymes present.

3 ml of reaction mixture without substrate. The extinction of *cis*-benzene dihydrodiol at this wavelength and concentration is 0.001.

Growth of Microorganism

Growth of *Pseudomonas putida* NCIB 12190 is performed as described in [10] of this volume.

Purification Procedure*

Step 1: Preparation of Cell Extract. Washed cells (100 g) are thawed and resuspended in 100 ml of buffer A (25 mM potassium phosphate containing 0.1 mM dithiothreitol, pH 7.4). The suspension is cooled on ice and subjected to the maximum output from a Ultrasonic disrupter. Batches (50 ml) are treated for 6 times, 30 sec each. Cell debris and unbroken cells are removed by centrifugation at 40,000 g_{av} for 1 hr. The supernatant solution is decanted. These and all following purification operations are performed between 0 and 4°.

Step 2: Protamine Sulfate Removal of Nucleic Acids. Protamine sulfate (0.5 g/100 g wet weight of cells) is dissolved in the minimum volume of buffer A and added dropwise with constant stirring to the crude cell-free extract. The suspension is stirred for a further 15 min and the precipitate removed by centrifugation at 40,000 g for 30 min.

Step 3: Ammonium Sulfate Fractionation. The supernatant is brought to 40% saturation with a saturated solution of $(NH_4)_2SO_4$ which has been adjusted to pH 7.4 with aqueous NH_3. After the mixture has stirred for 15 min, the precipitate is collected by centrifugation at 40,000 g for 30 min and dissolved in a minimum volume of buffer A.

Step 4: DEAE-Cellulose Chromatography. Ammonium sulfate is removed by dialysis for 18 hr against buffer A, and the fraction is then applied to the top of a column of DEAE-cellulose (Whatman DE-52) previously equilibrated with buffer A. Proteins are eluted by means of a 0–6% (w/v) linear KCl gradient. The enzyme is eluted as a single peak between 4 and 5% KCl. Fractions containing the highest specific activity are pooled and concentrated using a Diaflo ultrafiltration system (Amicon, Danvers, MA) with an XM50 membrane.

Step 5: Sephadex G-200 Chromatography. The concentrated fraction from Step 4 is applied to the top of a column of Sephadex G-200 previously equilibrated with 100 mM potassium phosphate buffer containing 0.1 mM dithiothreitol, pH 7.4, and eluted with the same buffer. *cis*-Benzene dihydrodiol dehydrogenase activity is found to coincide exactly with the second protein peak eluted from the column.

Results of a typical purification are summarized in Table II.

* Reprinted by permission from B. C. Axcell and P. J. Geary, *Biochem. J.* **136**, 927. Copyright © 1973 by the Biochemical Society, London.

TABLE II
PURIFICATION OF *cis*-BENZENE DIHYDRODIOL DEHYDROGENASE[a]

Fraction	Total protein (mg)	Enzyme (total units)	Specific activity (units/mg protein)	Yield (%)	Purification (-fold)
Crude extract	7780	72,800	9.3	100	1.0
Protamine sulfate supernatant	7100	79,000	11.2	108	1.2
Ammonium sulfate fraction (dialyzed)	1960	56,500	28.8	78	3.1
DEAE-cellulose	168	18,200	108.0	25	11.6
Sephadex G-200	75	14,300	191.0	20	20.5

[a] Modified from Ref. 3.

Properties of *cis*-Benzene Dihydrodiol Dehydrogenase

Molecular Weight. From ultracentrifugation measurements the native molecular weight of the protein was estimated to be 440,000. Analysis of the denatured protein by polyacrylamide gel electrophoresis in the presence of sodium dodecyl sulfate indicated that the subunit molecular weight was 110,000.

Activation by Fe^{2+}. A loss of activity was experienced after ammonium sulfate precipitation of the enzyme and subsequent dialysis. The addition of 0.1 mM Fe^{2+} restored activity. No other metal ions were effective in restoring activity. The residual Fe^{2+} present in purified enzyme preparations before reactivation, determined by the 1,10-phenanthroline method, varied between 1 and 2 mol of Fe^{2+} per mole of enzyme.

Specificity. The enzyme was specific for the cis form of benzene dihydrodiol and showed no generation of NADH when incubated with *trans*-benzene dihydrodiol. $NADP^+$ could not substitute for NAD^+ as a hydrogen acceptor.

Kinetics. The enzyme gave K_m values for *cis*-benzene dihydrodiol and NAD^+, determined from double-reciprocal plots, of 286 and 43.5 μM, respectively. One mole of NADH was generated for each mole of *cis*-benzene dihydrodiol converted to catechol.

X-Ray Crystallography. X-Ray analysis of crystals of *cis*-benzene dihydrodiol dehydrogenase shows that they belong to the tetragonal space group I422, with cell dimensions $a = b = 133.1$ Å, $c = 273.8$ Å.[9] The enzyme molecule is composed of four identical subunits of 110,000 molecular weight, one of which comprises the asymmetric unit of the crystal. It is not clear whether the complete molecule has tetrahedral or square planar symmetry.

[9] P. J. Artymiuk, C. C. F. Blake, and P. J. Geary, *J. Mol. Biol.* **111**, 203 (1977).

[24] Hydroxybenzoate Hydroxylase

By BARRIE ENTSCH

There are three isomers of hydroxybenzoate, and each is hydroxylated by a separate enzyme in microorganisms. These enzymes are intracellular flavoprotein monooxygenases which require reduced pyridine nucleotide for the reaction. They channel aromatic compounds into the metabolic pathways which open the benzene ring to form common intermediates of energy metabolism.

The enzyme which hydroxylates 4-hydroxybenzoate (4-hydroxybenzoate 3-monooxygenase, EC 1.14.13.2; *p*-hydroxybenzoate hydroxylase) has been studied extensively over the past 20 years and is the subject of this chapter. The enzyme which hydroxylates 2-hydroxybenzoate [salicylate monooxygenase (hydroxylase), EC 1.14.13.1] is increasingly the subject of investigation at present at the molecular level.[1,2] Two enzymes hydroxylate 3-hydroxybenzoate (classified under EC 1.14.13.23 and 24). The enzyme which hydroxylates para to the substrate hydroxyl group is now under investigation.[3]

The biological reaction catalyzed by *p*-hydroxybenzoate hydroxylase is represented by the following equation:

$$HO-\hspace{-4pt}\langle\!\!\bigcirc\!\!\rangle\!\!-COO^- + NADPH + H^+ + O_2 \rightarrow HO-\hspace{-4pt}\langle\!\!\overset{\displaystyle HO}{\bigcirc}\!\!\rangle\!\!-COO^- + NADP^+ + H_2O$$

A wide range of soil microorganisms seem to be capable of this reaction or the same reaction with NADH. However, it is the enzyme from the fluorescent pseudomonads which has been studied in detail. The enzyme is not normally produced except in the presence of its substrate as a source of carbon.

A previous report on this enzyme from *Pseudomonas fluorescens* appeared in this series in 1978.[4] Since that time, much information has appeared in the literature, and the enzyme has become even more established as a model for flavoprotein oxygenases. This chapter presents an improved method of preparation, a selected summary of properties established since the previous contribution in this series, and some information on the gene for the enzyme.

[1] H. Kamin, R. H. White-Stevens, and R. P. Presswood, this series, Vol. 53, p. 527.
[2] G. H. Einarsdottir, M. T. Stankovich, and S.-C. Tu, *Biochemistry* 27, 3277 (1988).
[3] L.-H. Wang, R. Y. Hamzah, Y. Yu, and S.-C. Tu, *Biochemistry* 26, 1099 (1987).
[4] M. Husain, L. M. Schopfer, and V. Massey, this series, Vol. 53, p. 543.

Assay Method

The assay method has remained essentially unchanged in its present form since it was developed by Entsch *et al.*[5] from a knowledge of the pH and buffer ion dependence of the enzyme. The assay is reliable even with crude cell extracts.

Principle. Activity is measured spectrophotometrically by following the rate of *p*-hydroxybenzoate-dependent oxidation of NADPH at 340 nm in air-saturated buffer at 25°.

Procedure. Assays are conducted at 25° in 1-cm cells in a recording spectrophotometer. The reaction solution of 3 ml (or less) contains 33 mM Tris as sulfate salt, 0.33 mM EDTA as sodium salt, 0.33 mM sodium *p*-hydroxybenzoate, 0.23 mM NADPH, 3.3 μM FAD, and 0.26 mM oxygen, all at pH 7.9 to 8.0. The reaction is initiated by the addition of enzyme, and the NADPH oxidation rate is measured. Alternatively, the rate of reduction of oxygen can be measured in an oxygen electrode chamber. It is important to purify *p*-hydroxybenzoic acid for the reaction by recrystallization from hot water. FAD is necessary to keep the enzyme saturated with cofactor at the low enzyme concentrations in the assay.

Units. One unit of enzyme is the amount that oxidizes 1 μmol of NADPH or reduces 1 μmol of oxygen per minute under the assay conditions. Specific activity is expressed as units of enzyme per milligram of protein.

Purification

The procedure presented below has been evolved from the methods of Entsch *et al.*[5] and Müller *et al.*[6] and it is effective in the purification of enzyme from *P. fluorescens* and *Pseudomonas aeruginosa* rapidly in high yields. A recent report[7] on enzyme from *Corynebacterium cyclohexanicum* suggests that the affinity step below is not effective in the purification of enzyme from all sources. An alternative procedure is thus presented in that reference.

Although enzyme has been studied in the past from *P. fluorescens,* the procedure below is written for enzyme from *P. aeruginosa,* since we have found that the enzyme from the latter bacterium is almost identical to that from *P. fluorescens* (see Genetics section) but is induced to higher concen-

[5] B. Entsch, D. P. Ballou, and V. Massey, J. Biol. Chem. **251,** 2550 (1976).
[6] F. Müller, G. Voordouw, W. J. H. van Berkel, P. J. Steennis, S. Visser, and P. J. van Rooijen, *Eur. J. Biochem.* **101,** 235 (1979).
[7] T. Fujii and T. Kaneda, *Eur. J. Biochem.* **147,** 97 (1985).

trations in cells of *P. aeruginosa*. In addition, the biology of *P. aeruginosa* has been studied extensively because of its medical significance.

Growth of Bacteria. Pseudomonas aeruginosa PAO1C (ATCC 15692) is maintained for extended periods when 1 volume of culture in the medium below is added to 2 volumes of glycerol and stored at −70°. Cells are revived for cultures by streaking on plates containing the medium below plus 1.5% agar at 30° (allow 2 days). Colonies of cells are used to start liquid cultures as soon as the cells grow up. Plates are not stored. Cells can be grown in continuous culture or in batch culture. Only batch culture is described here. Any volume of culture can be grown, depending on the facilities for aeration. Growth rates are normally limited by oxygen supply.

The growth medium contains the following in 1.0 liter: 3.5 g of *p*-hydroxybenzoic acid, 4.0 g of NH_4NO_3, 1.0 g of NH_4Cl, 0.22 g $MgSO_4 \cdot 7H_2O$, and 3.5 g K_2HPO_4. The pH is adjusted to 7.0 with NaOH. Iron (0.5 mg) is added as the EDTA complex (1 mol of iron to 3 mol of EDTA), and micronutrients are as follows: 0.25 mg of $CuSO_4 \cdot 5H_2O$, 0.35 mg of $ZnSO_4 \cdot 7H_2O$, 0.35 mg of $MnCl_2 \cdot 2H_2O$, 2.0 mg of $CaCl_2 \cdot 2H_2O$, 0.45 mg of $Co(NO_3)_2 \cdot 6H_2O$, and 0.25 mg of ammonium molybdate. The iron and micronutrient solutions are sterilized separately and added when the macronutrient solution is cool.

Cultures are grown in two stages at 32 to 35°. A starter culture is grown from plates, and this provides an inoculum of about 5% in the main culture. Cells are harvested before the substrate is completely exhausted. Growth ceases when aeration is halted, and the culture is cooled. A test for substrate is easily achieved by sampling. Cells are removed from samples by centrifugation, a small aliquot of supernatant is added to a solution of 50 m*M* NaOH, then the absorbance at 282 nm is measured. The concentration of *p*-hydroxybenzoate is calculated from its extinction at 282 nm ($16.3 \text{ m}M^{-1} \text{ cm}^{-1}$). A yield of 5 to 6 g of cells (wet weight) per liter is obtained. The cells can be stored as a frozen paste at −70°.

Procedure. The procedure below has been arranged for amounts up to 100 g wet weight of cells. However, scale-up for larger quantities should not be difficult. With organization, procedure can be completed in 2 days. This time scale is recommended.

On Day 1 cells are thawed with 2 volumes of extraction buffer (50 m*M* potassium phosphate, 0.5 m*M* EDTA, 0.5 m*M* *p*-hydroxybenzoate, pH 7.0), and 5 mg of deoxyribonuclease I is added. The slurry is sonicated on ice with the temperature maintained under 16° until enzyme activity reaches a maximum (usually within 3 min). Approximately 13,000 units of enzyme should be obtained from 100 g of cells (at least 200 mg of enzyme). After the addition of a further 2 volumes of cold extraction buffer, the mixture is centrifuged at 30,000 *g* for 30 min at 2°. The yellow supernatant is siphoned off, and solid ammonium sulfate is dissolved in it at the rate of 24.5 g per 100 ml. The pH is adjusted to 7.0 with ammonia, and the mixture is stirred until it reaches 15°. Then the mixture is centrifuged at 30,000 g for 20 min at 15°. From here, chromatographic steps are most easily run at room temperature without loss of enzyme.

Hydrophobic chromatography. The supernatant is loaded onto a column of DEAE-cellulose (Whatman DE-52, bed volume of 1.5 times the original cell weight converted to milliliters) equilibrated with running buffer (25 mM potassium phosphate, 1.0 mM EDTA, pH 7.0) containing 45% saturation of ammonium sulfate at room temperature (about 20°). The column is then washed with 2 volumes of the same solution, and fractions containing enzyme are eluted with running buffer containing 32% saturation of ammonium sulfate. With a slow flow rate, most of the enzyme is eluted in about 2 bed volumes. The enzyme fractions only are pooled, and protein is precipitated by adding solid ammonium sulfate to a final concentration of 75% saturation, with pH in the range of 6.5 to 7.0. The precipitate is recovered by centrifugation at 20,000 g for 10 min. The protein is dissolved in ice-cold affinity buffer (10 mM Tris–maleate, 0.3 mM EDTA, pH 7.0) by the addition of a volume of about 0.5 of cell weight. The solution is dialyzed against affinity buffer at 2 to 4° to remove ammonium sulfate (overnight).

Affinity chromatography. On Day 2, the dialyzed solution is centrifuged to remove protein precipitate (after warming to 20°), then loaded onto a column of Blue Sepharose (Pharmacia Blue Sepharose CL-6B) equilibrated with affinity buffer at 20° (bed volume approximately the same as the original cell weight). The column is washed with 1 volume of affinity buffer. Protein is eluted with affinity buffer containing a gradient of 0 to 0.5 M KCl (gradient volume about 4 bed volumes). The fractions containing the enzyme peak are combined.

Hydroxylapatite chromatography. The solution above is loaded directly onto a column of hydroxylapatite equilibrated with 7 mM potassium phosphate buffer, pH 7.0 at 20° (bed volume about 35% of the original cell mass, if the Bio-Rad product BioGel HT is used). The column is washed with 2 volumes of 8 mM potassium phosphate, pH 7.0, then the yellow enzyme product is eluted by flushing the column with 48 mM potassium phosphate, pH 7.0. Enzyme elutes in about 1 bed volume. Fractions containing enzyme with an A_{280}/A_{450} absorbance ratio of less than 10 can be used for most experimental purposes. Yields of enzyme are in the range of 50 to 60% of the enzyme in the original cell-free extract.

Crystallization. Crystallization may be used to obtain enzyme for analytical studies of the molecule. To the material from hydroxylapatite, EDTA is added to 1.0 mM. Solid ammonium sulfate is added slowly until the solution is faintly turbid. The sample is quickly centrifuged, and the supernatant is allowed to stand in a cold room to slowly form crystals. Pure enzyme has an A_{280}/A_{450} ratio of 8.5 and a specific activity of 62 to 64. This is equivalent to a molecular activity in the standard assay of 2900 to 3000 per minute, when an extinction value of 11.3 mM^{-1} cm^{-1} is used for the enzyme at 450 nm.

Storage. Large quantities of enzyme can be stored indefinitely as a precipitate under a solution of 50 mM potassium phosphate and 0.5 mM EDTA, pH 6.5 to 7.0, with 70% saturated ammonium sulfate at 0 to 4° (away from light). Smaller quantities of enzyme can also be stored in solution (without ammonium sulfate) in the same buffer at −70° (≥2 mg/ml). If enzyme has partly degraded for some reason, intact molecules can usually be purified to full specific activity by running the sample through a small column of hydroxylapatite as above.

Properties

Stability. A number of reports have appeared since the previous chapter in this series on the heterogeneity of preparations of enzyme from *P. fluorescens.* The definitive paper from van Berkel and Müller[8] showed that multiple forms of the enzyme are artifacts of purification and storage. A single cysteine residue (Cys-116) at the surface of the molecule is susceptible to oxidation by oxygen. As a result, sulfinate and sulfonate derivatives can form which confer charge heterogeneity on the molecule. In addition, weak covalent disulfide cross-links between molecules can occur,[9] resulting in higher order molecular aggregates which disrupt ordered crystallization of the enzyme. However, catalytic activity is not influenced by these changes. Van Berkel and Müller[8] recommend a different final step in purification and storage of the enzyme from *P. fluorescens* with dithiothreitol to overcome these chemical changes. We have not used dithiothreitol in the storage of enzyme from *P. aeruginosa* as prepared by the method described here, although Cys-116 should be in the molecule.[10] Even after long periods of storage in solution, the preparations of enzyme from *P. aeruginosa* showed only small amounts of oxidized enzyme molecules compared to the native molecule, as determined by ion-exchange high-performance liquid chromatography (HPLC). Van Berkel and Müller[11] report that the enzyme has a tolerance to temperatures as high as 60°, but only in

[8] W. J. H. van Berkel and F. Müller, *Eur. J. Biochem.* **167**, 35 (1987).

[9] J. M. van der Laan, M. B. A. Swarte, H. Groendijk, W. G. J. Hol, and J. Drenth, *Eur. J. Biochem.* **179**, 715 (1989).

[10] B. Entsch, N. Yang, K. Weaich, and K. F. Scott, *Gene* **71**, 279 (1988).

[11] W. J. H. van Berkel and F. Müller, *Eur. J. Biochem.* **179**, 307 (1989).

the pH range of 5.5 to 6.5. The enzyme can be used with impunity in the pH range 5 to 8 if the temperature is kept below 40°.

Molecular Properties. In solution at physiological concentrations, the native enzyme is a dimer of identical subunits[8] (with independent active sites) held together by noncovalent forces.[12] Each polypeptide is 394 amino acids long and contains one molecule of FAD.[10,12] The native subunit molecular weight is 45,100 (from the known structure), but the dimer behaves as a molecule of 75 to 80,000 in solution, because of its hydrodynamic shape.[9]

Owing to the efforts of Drenth's laboratory, there is now a large body of knowledge on the 3-dimensional structure of the enzyme from *P. fluorescens,* and this information is equally applicable to the enzyme from *P. aeruginosa.*[10,13] The complex between *p*-hydroxybenzoate and oxidized enzyme was first described at 0.25-nm resolution by Wierenga *et al.*[14] Then the full amino acid sequence and its integration with the crystallographic data were reported in 1983.[12] Since then, the structure has been refined to 0.19-nm resolution.[15] A surprising discovery was that the tricyclic isoalloxazine ring is slightly twisted in the oxidized enzyme toward the configuration necessary to accommodate a flavin–C4a derivative. Other forms of the enzyme involved in catalysis have been reported at various resolutions: the complex between *p*-hydroxybenzoate and reduced enzyme,[18] the complex between 3,4-dihydroxybenzoate and oxidized enzyme,[16] enzyme without substrates,[17] and a model of flavin–C4a–peroxide in the active site with *p*-hydroxybenzoate.[18] In the crystalline state, the enzyme is a dimer, just as in free solution. This structural knowledge has greatly stimulated interest in further study of the enzyme.

Substrate Dependence. There has not been much new information about range of substrates for this enzyme. There was a major report in 1980[19] on aromatic fluorine derivatives of *p*-hydroxybenzoate (reprints containing the correct methods can be obtained from Vincent Massey).

[12] W. J. Weijer, J. Hofsteenge, J. J. Beintema, R. K. Wierenga, and J. Drenth, *Eur. J. Biochem.* **133,** 109 (1983).

[13] B. Entsch and D. P. Ballou, *Biochim. Biophys. Acta* **999,** 313 (1989).

[14] R. K. Wierenga, R. J. de Jong, K. H. Kalk, W. G. J. Hol, and J. Drenth, *J. Mol. Biol.* **131,** 55 (1979).

[15] H. A. Schreuder, P. A. J. Prick, R. K. Wierenga, G. Vriend, K. S. Wilson, W. G. H. Hol, and J. Drenth, *J. Mol. Biol.* **208,** 679 (1989).

[16] H. A. Schreuder, J. M. van der Laan, W. G. J. Hol, and J. Drenth, *J. Mol. Biol.* **199,** 637 (1988).

[17] J. M. van der Laan, Doctoral dissertation, Groningen, 1986.

[18] H. A. Schreuder, Doctoral dissertation, Groningen, 1988.

[19] M. Husain, B. Entsch, D. P. Ballou, V. Massey, and P. J. Chapman, *J. Biol. Chem.* **255,** 4189 (1980).

Fluorine can replace any or all of the ring hydrogens in p-hydroxybenzoate, and an effective substrate results. When fluorine is in the 3-position, it can be eliminated as fluoride. When this occurs, oxygenation of substrate becomes the rate-determining step.

Other potentially important analogs of p-hydroxybenzoate are known:[20] 5-hydroxypicolinate is a powerful effector (i.e., it stimulates rapid oxidation of NADPH by the enzyme without being a substrate), whereas 4-aminosalicylate and 6-aminonicotinate are excellent competitive inhibitors which do not stimulate NADPH oxidation. Analogs of NADPH have generally not been useful in studying this enzyme. Structural analyses have shown that the enzyme does not have a conventional Rossman fold to bind NADPH.[21] It has been reported on a number of occasions that some anions act as competitive inhibitors of NADPH binding. For example, Cl$^-$, I$^-$, CNS$^-$, and N$_3^-$ are all effective inhibitors in millimolar concentrations.[22] However, the catalytic effect of these anions is much more complex than competitive inhibition.[5] Inorganic anions such as sulfate and phosphate are safe to use for enzyme kinetic studies.

The first full description of the steady-state substrate kinetics of the enzyme from $P.$ $fluorescens$ was reported with p-mercaptobenzoate as substrate.[23] This has subsequently been followed by a full analysis with the native substrate, p-hydroxybenzoate, at $4°$,[24] for comparison to studies of enzyme transient species in the reaction. A complete collection of steady-state parameters for the enzyme from $P.$ $aeruginosa$ at pH 8.0 (the pH optimum) and $25°$ has been published.[13] The enzyme has a turnover number at these standard conditions of 3750 min^{-1}. The kinetic parameters reported in this paper should be close to the actual values for the enzyme from $P.$ $fluorescens$ under standard conditions.

Flavin Cofactor. As the component of the enzyme reaction at the center of catalysis, and the component readily studied in isolation (owing to electronic transitions in the visible), the flavin has received considerable attention. The perturbations of catalysis caused by analogs of the isoalloxazine ring have been studied by Massey's laboratory. Two important practical considerations surround this work. First, the natural nucleotide of the chemically modified riboflavin must be generated. The most favored method utilizes a preparation of flavokinase and FAD synthase from *Brevibacterium ammoniagenes* to generate the dinucleotide, as described

[20] B. Entsch, unpublished observations (1975) and (1989).
[21] R. K. Wierenga, J. Drenth, and G. E. Schulz, *J. Mol. Biol.* **167,** 725 (1983).
[22] H. Shoun, K. Arima, and T. Beppu, *J. Biochem.* **93,** 169 (1983).
[23] B. Entsch, D. P. Ballou, M. Husain, and V. Massey, *J. Biol. Chem.* **251,** 7367 (1976).
[24] M. Husain and V. Massey, *J. Biol. Chem.* **254,** 6657 (1979).

originally by Spencer *et al.*[25] Second, the natural FAD must be removed from the enzyme before binding the new analog. The simplest procedures involve dialysis or ammonium sulfate precipitation at low pH, as described in the papers by Ghisla *et al.*[26] and Entsch *et al.*[27] However, if extensive work in this area is planned, then it is worth investing in the procedure devised by Müller and van Berkel.[28] In this method, the enzyme is covalently adsorbed (presumably through Cys-116) to a column of AH-Sepharose substituted with a reactive thiol reagent. FAD is washed from the bound enzyme, and the apoprotein is eluted from the column with dithioerythritol. With this method, experience has shown that large amounts of apoprotein can be prepared with only modest loss of enzyme by denaturation (but the apoprotein loses activity with time[28]). The final dialysis step described in the method can be eliminated if the only purpose is to bind a new FAD analog. It is only necessary to incubate the apoprotein with a 2-fold excess of the new FAD for a few minutes, then precipitate the enzyme with solid ammonium sulfate to 70% saturation, followed by molecular sieve chromatography (Sephadex G-25) to remove excess flavin and other solutes as required.

Valuable information about the active site of the enzyme can be obtained from chemically modified FAD. For example, the chemical environment of the flavin can be tested by the incorporation of a reactive substituent, such as 7- and 8-halogen-substituted FAD[29] or 2-thio-FAD.[30] The reaction mechanism can be probed by appropriate changes to FAD. When 1-deaza-FAD was used as cofactor,[27] the enzyme carried out each step in catalysis except the transfer of oxygen to *p*-hydroxybenzoate. The derivative, 6-hydroxy-FAD, yielded a competent enzyme which had a lower turnover rate than the native enzyme.[31] Subtle changes in catalysis with 6-hydroxy-FAD established the physical orientation of substrate to flavin during catalysis.

Amino Acid Changes. This enzyme, perhaps as much as any other enzyme, catalyzes a reaction under the absolute control of the protein involved. Thus, with the knowledge of structure available, there is considerable interest in probing the role of amino acid residues, particularly in the active site. In Holland, the principal investigators, van Berkel and

[25] R. Spencer, J. Fisher, and C. Walsh, *Biochemistry* **15**, 1043 (1976).
[26] S. Ghisla, B. Entsch, V. Massey, and M. Husain, *Eur. J. Biochem.* **76**, 139 (1977).
[27] B. Entsch, M. Husain, D. P. Ballou, V. Massey, and C. Walsh, *J. Biol. Chem.* **255**, 1420 (1980).
[28] F. Müller and W. J. H. van Berkel, *Eur. J. Biochem.* **128**, 21 (1982).
[29] V. Massey, M. Husain, and P. Hemmerich, *J. Biol. Chem.* **255**, 1393 (1980).
[30] A. Claiborne, P. Hemmerich, V. Massey, and R. Lawton, *J. Biol. Chem.* **258**, 5433 (1983).
[31] B. Entsch, V. Massey, and A. Claiborne, *J. Biol. Chem.* **262**, 6060 (1987).

Müller, have made several studies of protein modification with reactive chemicals. They have, for example, attempted to target cysteine,[32] tyrosine,[33] and arginine residues.[34] The problem, as usual, is the inability to rationally target active site residues by this approach. Alternative approaches include natural variations and mutations. As with the chemical approach, there is little hope of predictive modifications, or low probability of successful results. The techniques of gene manipulation hold much greater hope for the future of this work.[10] Although these methods are labor intensive, they are becoming easier to implement all the time. In my laboratory, tyrosine residues 201, 222, and 385 (all in the active site of the enzyme) have been changed individually to phenylalanine. It was found that all three changes substantially modified the catalytic function of the enzyme in different aspects, without completely eliminating catalysis.[35] Thus, it should not be long before a picture of the protein function emerges.

Oxygen Reactions. The most comprehensive information on the oxygen reactions of flavoprotein oxygenases remains the 1976 paper by Entsch *et al.*[5] The great majority of the reaction model proposed at that time has been confirmed in other reports and investigations of other flavoproteins, mostly in association with Massey's laboratory. The enzyme studies have been supported by comprehensive chemical investigations of the initial reactions of appropriate reduced flavin models with oxygen.[36] One great puzzle left from the results reported in 1976 was the nature of a transient chemical species found associated with the transfer of oxygen from flavin to the substrate, the crucial step in the formation of an oxygenated product. It was the injection of a novel physical technique into the study of this enzyme which may have solved the problem. Anderson *et al.*[37] used pulse radiolysis to study the interactions of flavins with oxygen. They have chemical evidence from reactions of *p*-hydroxybenzoate with hydroxyl radicals that the unknown enzyme intermediate is a radical pair between flavin and oxygenated product, which has some kinetic stability under some conditions with the enzyme.

[32] W. J. H. van Berkel, W. J. Weijer, F. Müller, P. A. Jekel, and J. J. Beintema, *Eur. J. Biochem.* **145**, 245 (1984).
[33] W. J. H. van Berkel, F. Müller, P. A. Jekel, W. J. Weijer, H. A. Schreuder, and R. K. Wierenga, *Eur. J. Biochem.* **176**, 449 (1988).
[34] R. A. Wijnands, F. Müller, and A. J. W. G. Visser, *Eur. J. Biochem.* **163**, 535 (1987).
[35] B. Entsch, P. Bundock, and R. E. Wicks, *Proc. Australian Biochem. Soc.* **21**, P54 (1989).
[36] T. C. Bruice, *in* "Flavins and Flavoproteins" (R. C. Bray, P. C. Engel, and S. G. Mayhew, eds.), p. 45. de Gruyter, Berlin and New York, 1984.
[37] R. F. Anderson, K. B. Patel, and M. R. L. Stratford, *J. Biol. Chem.* **262**, 17475 (1987).

Genetics

Further advances in our understanding of protein structure and function in this enzyme will include an application of gene technology. It would thus be sensible to integrate this work if possible with genetics and bacterial gene control. The bacterium *P. fluorescens* has not been used as a model for bacterial genetics, but it does exist within a closely related group of bacteria known as the fluorescent pseudomonads, which includes favorite targets for genetic analysis, *P. aeruginosa* and *P. putida*.[38] It has been found from the accurate sequence of the gene for *p*-hydroxybenzoate hydroxylase *(pobA)* that the enzyme from *P. aeruginosa* is the same as that from *P. fluorescens* except for two residues, amino acids 228 and 249.[10] Both side chains are at the surface of the enzyme, and the changes have very little influence on catalysis.[13] Thus, the gene from *P. aeruginosa* is an ideal model to link the genetics of pseudomonads to detailed enzyme analysis.

Only one gene has been identified for the metabolism of *p*-hydroxybenzoate by microorganisms *(pobA)*. There must be a permease-type mechanism for the uptake of this compound into cells, but no information exists on this subject. The product of the enzyme reaction (protocatechuate, or 3,4-dihydroxybenzoate) is considered an integral member of the central aromatic degration pathways.[39] The gene in *P. aeruginosa* is located about the 30' position in the chromosome and may be close to, but not directly linked with, some genes involved in the degradation of protocatechuate via the β-ketoadipate pathway.[38] The gene is also located in the chromosome of *P. putida*. Earlier studies had shown that the enzyme is under tight control and is induced by its substrate, but only in the absence of some alternate carbon sources.[39] Recently, cloned *pobA* from *P. aeruginosa* has been identified, sequenced, and structurally analyzed.[10] It was found that the intact gene is not expressed in *Escherichia coli,* but expression is successful when the protein coding region is linked to an *E. coli* promoter. A successful strategy has now been developed[35] to engineer site-specific changes in the enzyme through manipulations of the gene with synthetic oligonucleotides, according to the method of Kunkel *et al.*[40] The same techniques can be used for gene manipulation to investigate gene control in *P. aeruginosa.*

It should be noted that similar work is now in progress in investigations of the *sal* gene for salicylate monooxygenase from *P. cepacia.*[41]

[38] B. W. Holloway and A. F. Morgan, *Annu. Rev. Microbiol.* **40**, 79 (1986).
[39] R. Y. Stanier and L. N. Ornston, *Adv. Microb. Physiol.* **9**, 89 (1973).
[40] T. A. Kunkel, J. D. Roberts, and R. A. Zakour, this series, Vol. 154, p. 367.
[41] Y. Kim and S.-C. Tu, *Arch. Biochem. Biophys.* **269**, 295 (1989).

[25] Polycyclic Aromatic Hydrocarbon Degradation by *Mycobacterium*

By Carl E. Cerniglia and Michael A. Heitkamp

In recent years, there has been a sharp increase in the number of publications concerned with the use of microorganisms to degrade hazardous wastes. We recently isolated a *Mycobacterium* species from oil-contaminated estuarine sediments that has the ability to degrade polycyclic aromatic hydrocarbons. This bacterium has been shown to mineralize naphthalene, phenanthrene, pyrene, fluoranthene, 1-nitropyrene, 6-nitrochrysene, and 3-methylcholanthrene and has potential applications for the bioremediation of polycyclic aromatic hydrocarbons in the environment.[1-3]

Isolation of Polycyclic Aromatic Hydrocarbon-Degrading *Mycobacterium*

The bacterium was isolated from a 500-ml microcosm containing 20 g of sediment, 180 ml of estuarine water and 100 μg of pyrene.[1] The sediment was obtained from a drainage pond chronically exposed to petrogenic chemicals. After incubation of the microcosm for 25 days under aerobic conditions, the sediment samples were serially diluted and screened for the presence of polycyclic aromatic hydrocarbon-degrading microorganisms by a method modified from that of Kiyohara *et al.*[4]

The screening medium consisted of a mineral salts medium[5] containing, per liter: NaCl, 0.3 g; $(NH_4)_2SO_4$, 0.6 g; KNO_3, 0.6 g; KH_2PO_4, 0.25 g; K_2HPO_4, 0.75 g; $MgSO_4 \cdot 7H_2O$, 0.15 g; LiCl, 20 μg; $CuSO_4 \cdot 5H_2O$, 80 μg; $ZnSO_4 \cdot 7H_2O$, 100 μg; $Al_2(SO_4)_3 \cdot 16H_2O$, 100 μg; $NiCl \cdot 6H_2O$, 100 μg; $CoSO_4 \cdot 7H_2O$, 100 μg; KBr, 30 μg; KI, 30 μg; $MnCl_2 \cdot 4H_2O$, 600 μg; $SnCl_2 \cdot 2H_2O$, 40 μg; $FeSO_4 \cdot 7H_2O$, 300 μg; agar, 20 g; and distilled water, 1000 ml. This medium is routinely mixed from stock solutions as described by Skerman.[5]

The surfaces of the agar plates were sprayed with a 2% (w/v) solution of

[1] M. A. Heitkamp and C. E. Cerniglia, *Appl. Environ. Microbiol.* **54,** 1612 (1988).
[2] M. A. Heitkamp, W. Franklin, and C. E. Cerniglia, *Appl. Environ. Microbiol.* **54,** 2549 (1988).
[3] M. A. Heitkamp and C. E. Cerniglia, *Toxic. Assess.* **1,** 103 (1986).
[4] H. Kiyohara, K. Nagao, and K. Yano, *Appl. Environ. Microbiol.* **43,** 454 (1982).
[5] V. B. D. Skerman, "A Guide to the Identification of the Genera of Bacteria," 2nd Ed. Williams & Wilkins, Baltimore, Maryland, 1967.

a polycyclic aromatic hydrocarbon dissolved in acetone–hexane (1 : 1, v/v) and dried overnight at 35° to volatilize the carrier solvents. This treatment results in a visible and uniform surface coat of the polycyclic aromatic hydrocarbon on the agar. Inocula (100 μl) from the 10^{-1}, 10^{-2}, 10^{-3}, and 10^{-4} dilutions of microcosm sediments were gently spread with sterile glass rods onto the agar surface; the plates were inverted and incubated for 3 weeks at 24° in sealed plastic bags to conserve moisture.

When colonies surrounded by clear zones arising from polycyclic aromatic hydrocarbon uptake and utilization were observed (usually after 2 to 3 weeks), they were subcultured into fresh mineral salts medium containing 250 μg/liter each of peptone, yeast extract, and soluble starch and 0.5 μg/ml of a polycyclic aromatic hydrocarbon dissolved in dimethylformamide. After three successive transfers, a bacterium was isolated which was able to degrade pyrene, a polycyclic aromatic hydrocarbon containing four aromatic rings. The bacterium was identified as a *Mycobacterium* species on the basis of its cellular and colony morphology, gram-positive and strong acid-fast reactions, diagnostic biochemical tests, G + C content of its DNA, and the presence of high molecular weight mycolic acids.[2]

Growth of Organism and Culture Conditions

The bacterium was grown in 125-ml Erlenmeyer flasks containing 30 ml of basal salts medium[5] supplemented with 250 μg/ml each of peptone, yeast extract, and soluble starch and 0.5 μg/ml of pyrene dissolved in dimethylformamide. The cultures were incubated in the dark at 24° for 72 hr on a rotary shaker operating at 150 rpm. The growth of cultures was measured by monitoring the increase in optical density at 500 nm (OD_{500}) on a standard spectrophotometer. Plating experiments, using known OD_{500} values, showed a relationship of 2.9×10^6 cells/ml at an OD_{500} of 0.10. Cells in the mid logarithmic phase of growth were harvested by centrifugation at 8000 g for 20 min at 4°. The harvested cells were resuspended in sterile 0.1 M tris(hydroxymethyl)aminomethane buffer (pH 7.5) at a concentration of 3×10^6 cells/ml and used as inoculum for studies of polycyclic aromatic hydrocarbon biodegradation.

Biodegradation Experiments

Biodegradation of polycyclic aromatic hydrocarbons by the *Mycobacterium* was monitored in a flow-through microcosm test system.[3] This system enables simultaneous monitoring of mineralization (complete degradation to CO_2) and the recovery of volatile metabolites, nonvolatile metabolites, and residual polycyclic aromatic hydrocarbon (Fig. 1). Micro-

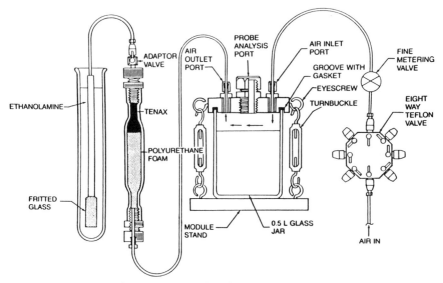

FIG. 1. Schematic representation of the flow-through microcosm test system.[3]

cosms in this test system consisted of 500-ml glass minitanks containing 100 ml of minimal basal salts medium[5] with 0.92 μCi of ^{14}C-labeled polycyclic aromatic hydrocarbon and 50 μg of nonlabeled polycyclic aromatic hydrocarbon. The polycyclic aromatic hydrocarbons used and their sources were as follows: [1,4,5,8-^{14}C]naphthalene (5.10 mCi/mmol), Amersham/Searle Corp., Arlington Heights, IL; [9-^{14}C]phenanthrene (19.3 mCi/mmol), Amersham/Searle; [3-^{14}C]fluoranthene (54.8 mCi/mmol), Chemsyn Science Laboratories, Lenexa, KS; [4-^{14}C]pyrene (30.0 mCi/mmol), Midwest Research Institute, Kansas City, MO; 3-[6-^{14}C]methylcholanthrene (13.4 mCi/mmol), New England Nuclear Corp., Boston, MA; and 6-nitro[5,6,11,12-^{14}C]chrysene (57.4 mCi/mmol), Chemsyn Science Laboratories.

Each microcosm was inoculated with 1.5×10^4 cells/ml, which were mixed twice weekly, incubated at 24° for 14 days, and continuously purged with compressed air. The gaseous effluent from each microcosm was directed through a volatile organic-trapping column containing 7 cm of polyurethane foam and 500 mg of Tenax GC (Alltech Associates, Inc., Deerfield, IL) and a [14]CO$_2$-trapping column (50 ml of monoethanolamine–ethylene glycol, 7:3, v/v). Mineralization was measured at various intervals by adding duplicate 1-ml aliquots from the [14]CO$_2$-trapping column to scintillation vials containing 15 ml of a 1:1 mixture of Fluoralloy and methanol (Beckman Instruments Co., Fullerton, CA). Autoclaved inoculated microcosms and microcosms lacking the *Mycobacterium* were included to detect abiotic polycyclic aromatic hydrocarbon degradation.

Some typical results of mineralization of polycyclic aromatic hydrocarbons by the *Mycobacterium* are illustrated in Fig. 2. The pyrene-induced cultures mineralized naphthalene (59.5%), phenanthrene (50.9%), fluoranthene (89.7%), pyrene (63.0%), 1-nitropyrene (12.3%), 3-methylcholanthrene (1.6%), and 6-nitrochrysene (2.0%) to carbon dioxide after 14 days of incubation.

Enzymatic Mechanism for Initial Ring Oxidation of Pyrene

Both *cis*- and *trans*-4,5-pyrene dihydrodiols occur as ring oxidation products during the degradation of pyrene by the polycyclic aromatic hydrocarbon-degrading *Mycobacterium*.[6] Studies with other aromatics have shown that ring oxidation reactions are usually dioxygenase-mediated in bacteria and monooxygenase-mediated in eukaryotic organisms. The enzymatic hydroxylation of aromatics by monooxygenase or dioxygenase enzymes involves insertion of either one or two atoms, respectively, from molecular oxygen into the aromatic nucleus. Therefore, mono- and dioxygenase-mediated reactions may be distinguished by incubating cultures in an atmosphere containing $^{18}O_2$ and measuring mass unit increases in molecular weights of metabolites that are characteristic of either single or double oxygen insertions into aromatic rings (Scheme 1). For example, a single oxygen insertion (monooxygenase) in the presence of $^{18}O_2$ instead of $^{16}O_2$ will produce a 2-unit mass increase in the molecular weight of the product, whereas a double oxygen insertion (dioxygenase) will result in a 4-unit mass increase. Experiments utilizing $^{18}O_2$ were conducted to determine the enzymatic mechanism for the initial oxidative attack of pyrene by this bacterium.

The *Mycobacterium* cultures for the $^{18}O_2$ experiments were grown in 125-ml Erlenmeyer flasks capped with rubber septa. The flasks, containing 30 ml of medium, as described above, were inoculated with 1.5×10^6 cells/ml. The air spaces above these cultures were evacuated and overlaid with argon 4 times prior to the addition of 1 atm of $^{18}O_2$ (99.8 atom%; Mound Facility, Miamisburg, OH) and 0.5 μg of pyrene per ml. The cultures were incubated for 48 hr in an $^{18}O_2$ atmosphere on a rotary shaker at 24° in the dark. The relative concentration of $^{18}O_2$ in the flasks was determined by direct-injection mass spectrometry analysis of the headspace and shown to be 96.8% $^{18}O_2$ at the beginning of the study and 95.5%

[6] M. A. Heitkamp, J. P. Freeman, D. W. Miller, and C. E. Cerniglia, *Appl. Environ. Microbiol.* **54**, 2556 (1988).

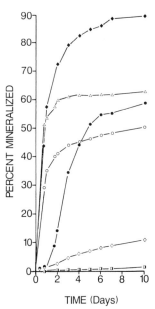

TIME (Days)

FIG. 2. Mineralization of naphthalene (●), phenanthrene (○), pyrene (△), fluoranthene (◆), 1-nitropyrene (◇), 6-nitrochrysene (□), and 3-methylcholanthrene (■) by a polycyclic aromatic hydrocarbon-degrading *Mycobacterium*. Each point is the mean of three replicate cultures.[1]

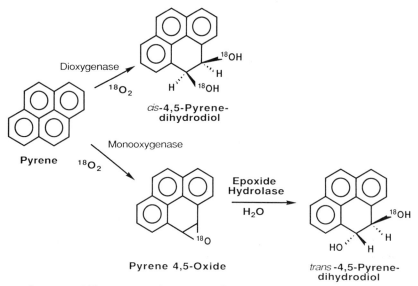

SCHEME 1. Different metabolic pathways of pyrene oxidation by *Mycobacterium*.

$^{18}O_2$ after 48 hr. The *Mycobacterium* cells and culture medium were extracted with six equal volumes of ethyl acetate dried with anhydrous Na_2SO_4, and evaporated *in vacuo* at 40° to 10 ml. The extracts were transferred to calibrated glass vials and evaporated to dryness under dry argon.

The residues were resuspended in methanol, and the pyrene dihydrodiols were purified by repeated injection and peak collection on a reversedphase, high-pressure liquid chromatography (HPLC) system. The instrumentation and chromatographic column for the HPLC analysis have been previously described in detail.[6] A methanol–water (55:45, v/v) isocratic solvent system at a flow rate of 1 ml/min was used to separate and purify the cis and trans isomers of the 4,5-pyrenedihydrodiols produced by the *Mycobacterium* species. The *trans*- and *cis*-4,5-pyrenedihydrodiols eluted at 19.2 and 21.8 min, respectively, from this HPLC system. The purified pyrene-4,5-dihydrodiol isomers were analyzed by mass spectrometry as previously described.[6]

A schematic diagram of the chemical structures from the results of these studies is presented in Fig. 3. The *trans*-4,5-pyrenedihydrodiol from cultures incubated in an $^{18}O_2$ atmosphere had a molecular weight of 238 $(C_{16}{}^{16}O_1{}^{18}O_1H_{10})$, which is a 2-mass unit increase (a single ^{18}O insertion) over the normal molecular weight of 236 $(C_{16}{}^{16}O_2H_{10})$ for pyrenedihydrodiol produced in a $^{16}O_2$ atmosphere. This 2-mass unit increase is characteristic of the monooxygenase-catalyzed hydroxylation of pyrene. The *cis*-4,5-pyrenedihydrodiol had a molecular weight of 240 $(C_{16}{}^{18}O_2H_{10})$ for the pyrene dihydrodiol produced in an $^{18}O_2$ atmosphere. This 4-mass unit increase is characteristic of the dioxygenase-catalyzed hydroxylation of pyrene. These results indicate that this *Mycobacterium* species oxidizes pyrene by both dioxygenase and monooxygenase enzymatic mechanisms (Scheme 1).

[26] Benzoate-CoA Ligase from *Rhodopseudomonas palustris*

By Jane Gibson, Johanna F. Geissler, and Caroline S. Harwood

$$\text{Benzoate} + \text{CoASH} + \text{MgATP} \longrightarrow \text{benzoyl-CoA} + \text{AMP} + \text{P} \sim \text{P}$$

Anaerobic metabolism of benzoate by phototrophic and denitrifying bacteria involves reductive saturation of the aromatic nucleus[1,2] and is initiated by formation of the coenzyme A (CoA) thioester.[3,4] *Rhodopseudomonas palustris* contains a soluble and specific ligase with a high affinity for benzoate; the purification scheme detailed below has been described in an earlier report.[5]

Assay Methods

Principle. Two assay methods have been used. Method 1 measures the rate of conversion of ^{14}C-labeled benzoate to benzoyl-CoA, which remains hydrophilic under acid conditions (pH 2). The method can be applied to crude extracts as well as to purified enzyme, and the sensitivity can be varied as required by changing the specific radioactivity of the aromatic substrate. Benzoate-CoA ligase (EC 6.2.1.25) can also be assayed (Method 2) by coupling the formation of AMP to the oxidation of NADH through adenylate kinase, pyruvate kinase, and lactate dehydrogenase in the presence of excess phosphoenolpyruvate. This method requires partial purification of the benzoate-CoA ligase to reduce ATPase activity, which gives a high background rate of NADH oxidation in the absence of aromatic substrate. Both reactions are absolutely dependent on added reduced coenzyme A (CoASH), even in crude extracts.

Reagents for Method 1 (Radioactive Assay). All stock reagents are prepared at 10-fold the concentration used in the actual assay.

Stock reagents

Triethanolamine-HCl buffer, pH 8.0, 0.2 M
ATP, 5 mM
CoASH (Li salt), 2.5 mM

[1] P. L. Dutton and W. C. Evans, *Biochem. J.* **113**, 525 (1969).
[2] R. J. Williams and W. C. Evans, *Biochem. J.* **148**, 1 (1975).
[3] G. N. Hutber and D. W. Ribbons, *J. Gen. Microbiol.* **129**, 2413 (1983).
[4] C. S. Harwood and J. Gibson, *J. Bacteriol.* **165**, 504 (1986).
[5] J. F. Geissler, C. S. Harwood, and J. Gibson, *J. Bacteriol.* **170**, 1709 (1988).

MgCl$_2$, 50 mM

[7-^{14}C]Benzoate, 40 μM, containing 1 – 2 μCi/ml

General reagents

Trichloroacetic acid, 50% (w/v)

HCl, 0.1 M

Ethyl ether or ethyl acetate

Reagents for Method 2 (Coupled Assay)

Stock reagents

Triethanolamine-HCl buffer, pH 8.0, 0.2 M

ATP, 5 mM

CoASH (Li salt), 2.5 mM

MgCl$_2$, 50 mM

Sodium benzoate, 2.5 mM

KCl, 100 mM

Phosphoenolypyruvate (tricyclohexylammonium salt), 100 mM

NADH, 3.5 mM

Pyruvate kinase; adenylate kinase; lactate dehydrogenase: purified commercial preparations

Procedure: Method 1. Each assay mixture contains 50 μl of each stock solution plus water to give a final volume of 0.5 ml. A volume of reaction mix appropriate to the number of assays to be carried out within the next 1 to 2 hr is prepared. Aliquots of 0.5 ml are prewarmed to 30°, and the reaction is initiated with 10 – 20 μl of enzyme preparation containing 10 – 100 microunits of activity. Standard reactions are stopped after 2 min by adding 50 μl of 50% (w/v) trichloroacetic acid. When using crude extracts, the precipitated protein is sedimented by centrifugation and the supernatant transferred quantitatively to a clean screw-capped vial; with purified preparations, centrifugation can be omitted. The reaction mixture is acidified further by the addition of 1 ml of 0.1 M HCl and then extracted 2 times by vigorous shaking with 2.5 ml ethyl ether or ethyl acetate. A sample (usually 0.2 ml) is removed for scintillation counting, and the amount of reaction product formed is obtained by comparing the water-soluble counts with total counts in the reaction mix, calculated by counting a sample either of the original reaction mix or of the acidified mixture after enzymatic reaction but before solvent extraction.

The single radioactive product formed in this assay comigrates with benzoyl-CoA in reversed-phase high-performance liquid chromatography (HPLC) on a C$_{18}$ ODS column, and is quantitatively converted to benzoate on incubation at 50° for 10 min at pH 12.

Procedure: Method 2. Reaction mixtures are prepared using 0.1 ml of each stock solution and the amount of coupling enzymes to provide a final concentration of approximately 2 units/ml, plus water to give a final volume of 1 ml. Cuvettes are equilibrated to 30°, and the reaction is initiated by adding $10-20 \mu l$ of enzyme preparation containing $1-10$ milliunits of benzoate-CoA ligase. Progress of the reaction is followed by decrease in absorbance at 340 nm, and the rate is calculated from the reaction stoichiometry of 2 mol of NADH oxidized per mole of benzoyl-CoA formed.

Units. One unit of benzoate-CoA ligase is defined as the amount of enzyme that catalyzes the formation of 1 μmol of benzoyl-CoA per minute at 30°, and specific activity as micromoles product formed per minute per milligram protein at that temperature.

Purification Procedure

The strain of *Rhodopseudomonas palustris* (CGA001) used most extensively was obtained from the collection of Dr. R. K. Clayton at Cornell University; this strain is physiologically very similar to *R. palustris* ATCC 11168, to other laboratory strains, and to new isolates of *R. palustris*.[6] Cultures are conveniently grown in 10-liter bottles that have been completely filled with JGR medium[5] containing 6 mM sodium benzoate and closed with butyl rubber stoppers while the medium was still hot from the autoclave. After cooling, the bottles are inoculated with $100-400$ ml of a benzoate-grown culture, gently stirred with a magnetic mixer, and illuminated with two 60- or 75-W incandescent light bulbs. Growth becomes light-limited after $5-6$ days, but continues more slowly thereafter. Approximately 15 g (wet weight) of cells are obtained on harvest after $8-12$ days. Cells collected by centrifugation are washed twice with 20 mM triethanolamine-HCl buffer, pH 7.5, and stored at $-20°$ if not used immediately. Neither centrifugation nor storage requires anoxic conditions.

Step 1: Preparation of Cell Extract. For enzyme purification, $10-20$ g (wet weight) of cells is suspended in $5-10$ volumes of 20 mM triethanolamine-HCl buffer, pH 7.5, 1 mM phenylmethylsulfonyl fluoride (PMSF) is added, and the suspension is passed 2 or 3 times through a chilled French pressure cell at $6-7$ MPa and collected on ice; all subsequent steps are carried out at $0-4°$. Unbroken cells are removed by low-speed centrifugation (10 min at 9,000 g), and 1.5% (w/v) streptomycin sulfate is added to the dark red supernatant. After 60 min on ice with occasional mixing, the preparation is centrifuged at high speed (1 hr at 200,000 g in a Beckman 70

6 C. S. Harwood and J. Gibson, *Appl. Environ. Microbiol.* **54**, 712 (1986).

Ti rotor), yielding a clear, pale brown supernatant. Enzyme activity in this extract is moderately stable, decreasing only by about 10% over 3–4 days at 0° on ice.

Step 2: First Hydroxylapatite Column Fractionation. The membrane-free extract is applied to a column (2.5 cm diameter; 100–150 ml) of Bio-Rad HTP hydroxylapatite, equilibrated with 20 mM potassium phosphate buffer, pH 7.5, at a flow rate of approximately 1 ml/min, and washed in with 300 ml of the same buffer. Colored material is retained by the column, but a pinkish band of cytochrome travels slowly through the column during this wash. The ligase is eluted with 300 ml of 80 mM potassium phosphate buffer.

Step 3: Phenyl-Sepharose–Sepharose 4B. The pale yellowish eluate is applied directly to a 2.5-cm-diameter column containing 50 ml of a 1 : 10 mixture of phenyl-Sepharose CL-4B and Sepharose 4B-200 (both from Sigma, St. Louis, MO) equilibrated in 20 mM triethanolamine-HCl, pH 7.5. The column is washed successively with 100 ml of the triethanolamine buffer and 300 ml of the same buffer containing 5 mM phenylalanine, which elutes a large quantity of inactive protein. Ligase is eluted in 300 ml of 0.1 mM ATP, 0.1 mM MgCl$_2$, plus 5 mM phenylalanine in 20 mM triethanolamine-HCl, pH 7.5.

Step 4: Second Hydroxylapatite Chromatography. Final purification and concentration are achieved by applying the effluent from Step 3 to a small hydroxylapatite column (1 cm diameter; 10 ml) equilibrated in 20 mM potassium phosphate buffer, pH 7.5, washing with 60 ml of 80 mM phosphate buffer, pH 7.5, and eluting the enzyme with 20 ml of 200 mM potassium phosphate, pH 7.5. The eluate is concentrated to 1–2 ml by ultrafiltration using a 30-kDa cutoff membrane, diluted to give a final buffer concentration of 20 mM, and reconcentrated in the same way, before storing in small aliquots in liquid N$_2$ or at −70°. Table I gives details of a typical preparation.

Properties

Activity and Stability. The best preparations, which are at least 90% pure as judged by sodium dodecyl sulfate–polyacrylamide gel electrophoresis (SDS–PAGE), have activities of 26–27 units/mg protein. Loss of activity (10–20% per day) occurs on storage of dilute solutions in ice and was not decreased by adding dithiothreitol, EDTA, ethylene glycol, or glycerol. In purer preparations, the decrease in activity could be correlated with loss of the 60-kDa protein band on SDS gels and the appearance of a new band at 45 kDa. This proteolysis is somewhat diminished, but not eliminated, by addition of PMSF (0.1 mM) to all solutions used during the

TABLE I
PURIFICATION OF BENZOATE-CoA LIGASE[a]

Step	Volume (ml)	Total protein (mg)	Specific activity (units/mg)	Total activity (units)	Recovery (%)	Purification (-fold)
Soluble protein	100	100	0.05	4.8	100	—
Hydroxylapatite (1st); 80 mM P$_i$	300	60	0.1	5.7	119	2
Phenyl-Sepharose; MgCl$_2$–ATP	300	0.26	18	3.6	74	286
Hydroxylapatite (2nd); 200 mM P$_i$	10	0.05	25	1.3	26	526

[a] Starting material 10 g (wet weight) of cells.

purification and by carefully maintaining the temperature at or below 4°. Storage in liquid N$_2$ for 3–6 months is associated with no greater losses of activity than could be attributed to freezing and thawing alone.

Molecular Size. Both denaturing electrophoresis and size-exclusion HPLC yield a molecular size of 60–62 kDa, indicating that the enzyme is monomeric. Isoelectric focusing shows a single component at pH 5.2.

Substrate Specificity. ATP, divalent metal ion, and reduced CoA are absolutely required for the reaction. GTP does not substitute for ATP, but Mn^{2+} (0.25 mM) replaces Mg^{2+} (2.5 mM) to give the same overall reaction rate. Of a total of about 25 aromatic, alicyclic, and fatty dicarboxylic acids tested, only benzoate and 2-fluorobenzoate are used at high rates and at concentrations of 50 μM or less; some activity could, however, be detected with 250 μM 4-fluorobenzoate, Δ^1-cyclohexene carboxylate, Δ^3-cyclohexene carboxylate, and picolinate. Cyclohexane carboxylate, 4-hydroxybenzoate, 1,4-dihydrobenzoate, benzoyl formate, *trans*-cinnamate, and coumarate, all of which support phototrophic growth, are not substrates for the purified enzyme.

pH Optimum. The ligase has a broad pH optimum between about pH 8.4 and 8.9. Activity at pH 8.0, the usual assay pH, is about 75% of that observed at the maximum.

Kinetic Parameters. The apparent K_m for benzoate under standard assay conditions is approximately 1 μM, supporting the suggestion that the enzyme plays an important role in establishing the high affinity of benzoate uptake by whole cells.[4] K_m values for CoASH and for ATP are approximately 100 and 2.5 μM, respectively.

Effect of Growth Conditions. The specific activities of extracts prepared from benzoate cultures at different stages of growth show little variation. However, virtually no activity is detected in succinate-grown cultures.

Activities in extracts of cells grown phototrophically with 4-hydroxyben-zoate, cyclohexane carboxylate, or protocatechuate are at least as high as those in benzoate-grown cultures, and some activity could be detected even in aerobic cultures using 4-hydroxybenzoate; *R. palustris* does not grow aerobically using benzoate. The presence of benzoate-CoA ligase in these extracts is confirmed in immunoblots using a highly specific rabbit antiserum prepared against pure enzyme.

[27] Lignin Peroxidase from Fungi: *Phanerochaete chrysosporium*

By T. Kent Kirk,* Ming Tien, Philip J. Kersten,*
B. Kalyanaraman, Kenneth E. Hammel, and Roberta L. Farrell

appropriately substituted aromatic → lignin peroxidase / H_2O_2 → aryl cation radical

Introduction

Lignin peroxidases are unusual oxidizing extracellular peroxidases produced by most of the ligninolytic fungi that cause white rot of wood.[1-4] In the presence of H_2O_2 they catalyze the one-electron oxidation of a wide variety of aromatics to yield, as initial products, aryl cation radicals that subsequently undergo substituent-dependent reactions of both radical and ionic nature.[1-3] These enzymes bring about the oxidative cleavage of lignin model compounds[1-3,5] and are thought to be of key importance in lignin

[1] T. K. Kirk and R. L. Farrell, *Annu. Rev. Microbiol.* **41**, 465 (1987).
[2] J. A. Buswell and E. Odier, *CRC Crit. Rev. Biotechnol.* **6**, 1 (1987).
[3] M. Tien, *CRC Crit. Rev. Microbiol.* **15**, 141 (1987).
[4] K. E. Hammel, B. Kalyanaraman, and T. K. Kirk, *J. Biol. Chem.* **261**, 16948 (1986).
[5] T. Umezawa and T. Higuchi, *FEBS Lett.* **218**, 255 (1987), and papers cited therein.

* This article was written and prepared by T. Kent Kirk and Philip J. Kersten as U.S. Government employees on official time.

biodegradation. They have also been shown to oxidize various polycyclic aromatic hydrocarbons and related structures, including pyrene, anthracene, benzo[a]pyrene, dibenzo[p]dioxin, and thianthrene.[4,6,7] Certain halogenated aromatics are also oxidized, including 2-chlorodibenzo[p]dioxin[4] and 2,4,6-trichlorophenol.[8] Lignin peroxidases, which are heme glycoproteins, were discovered in the extracellular broth of secondary metabolic cultures of the basidiomycete *Phanerochaete chrysosporium*,[9,10] from which various isoenzyme forms have been purified and studied.[11-14] The major studied lignin peroxidase has a p*I* of 3.5 and has been referred to as H8.[15] A nomenclature scheme has been proposed recently in which H8 is referred to as LiP1.[14] Lignin peroxidase has also been isolated from the lignin-degrading basidiomycetes *Phlebia radiata*[16] and *Trametes versicolor*.[17] The literature has been reviewed recently.[1-3] An earlier description of lignin peroxidase, by Tien and Kirk, is found elsewhere in this series.[18]

Assay Methods

Principles. Three simple and similar assay methods are described below. Although no assays based on oxidation of hydrocarbons have been used, a procedure for demonstrating the oxidation of pyrene is described below after the description of the assays.

The assay most commonly used is based on the oxidation of veratryl

[6] S. D. Haemmerli, M. S. A. Leisola, D. Sanglard and A. Fiechter, *J. Biol. Chem.* **261**, 6900 (1986).

[7] R. P. Schreiner, S. E. Stephens, Jr., and M. Tien, *Appl. Environ. Microbiol.* **54**, 1858 (1988).

[8] K. E. Hammel and P. J. Tardone, *Biochemistry* **27**, 6563 (1988).

[9] M. Tien and T. K. Kirk, *Science* **221**, 661 (1983).

[10] J. K. Glenn, M. A. Morgan, M. B. Mayfield, M. Kuwahara, and M. H. Gold, *Biochem. Biophys. Res. Commun.* **114**, 1077 (1983).

[11] M. Tien and T. K. Kirk, *Proc. Natl. Acad. Sci. U.S.A.* **81**, 2280 (1984).

[12] M. H. Gold, M. Kuwahara, A. A. Chiu, and J. K. Glenn, *Arch. Biochem. Biophys.* **234**, 353 (1984).

[13] A. Paszczyynski, V.-B. Huynh, and R. L. Crawford, *Arch. Biochem. Biophys.* **244**, 750 (1986).

[14] R. L. Farrell, K. E. Murtagh, M. Tien, M. D. Mozuch, and T. K. Kirk, *Enzyme Microb. Technol.* **11**, 322 (1989).

[15] T. K. Kirk, S. Croan, M. Tien, K. E. Murtagh, and R. L. Farrell, *Enzyme Microb. Technol.* **8**, 27 (1986).

[16] A. Kantelinen, R. Waldner, M.-L. Niku-Paavola, and M. S. A. Leisola, *Appl. Microbiol. Biotechnol.* **28**, 193 (1988).

[17] L. Jönsson, T. Johansson, K. Sjöström, and P. O. Nyman, *Acta Chem. Scand., Ser. B* **41**, 766 (1987).

[18] M. Tien and T. K. Kirk, this series, Vol. 161, p. 238.

alcohol (3,4-dimethoxybenzyl alcohol) to veratraldehyde in the presence of H_2O_2; the increase in absorbance at 310 nm is monitored.[11] The second method is based on the oxidation, in the presence of H_2O_2, of 1,4-dimethoxybenzene to 1,4-p-benzoquinone (monitored at 250 nm).[14,19] The third method is based on the oxidation, in the presence of H_2O_2, of 1,2,4,5-tetramethoxybenzene to the corresponding aryl cation radical, which is relatively stable under the conditions employed; the reaction is followed by monitoring the increase in absorbance at 450 nm.[19] [The course of oxidation of 1,4-dimethoxybenzene and 1,2,4,5-tetramethoxybenzene can also be followed readily by electron spin resonance (ESR) spectroscopy as described below. However, the radical from the 1,4-congener decomposes rapidly, which makes it more difficult to use in kinetic studies.] Final products of the oxidation of the 1,2,4,5-congener are 2,5-dimethoxy-p-benzoquinone and 4,5-dimethoxy-o-benzoquinone in a 4:1 ratio.[20]

The enzyme concentration for purified preparations may conveniently be determined from the absorbance at 409 nm, using an extinction coefficient of 168 mM^{-1} cm^{-1}.[21] The optimum temperature for the assays is approximately 37°.

Reagents for Method 1

10 mM veratryl alcohol (purified by distillation[11])
0.25 M sodium (+)-tartrate, pH 3.0
10 mM H_2O_2 (prepared fresh)

Reagents for Method 2

10 mM 1,4-dimethoxybenzene in water (gentle heat aids dissolution)
0.25 M sodium (+)-tartrate, pH 3.0
10 mM H_2O_2 (prepared fresh)

Reagents for Method 3

10 mM 1,2,4,5-tetramethoxybenzene[19] in 50% (v/v) ethanol in water
0.25 M sodium (+)-tartrate, pH 3.0
10 mM H_2O_2 (prepared fresh)

Procedure for Method 1. Reaction mixtures contain 2 mM veratryl alcohol, 0.4 mM H_2O_2, 50 mM tartrate, and enough lignin peroxidase to give an absorbance change of about 0.2 min^{-1} (1-cm cuvette). Reactions are started by adding the peroxide and are monitored at 310 nm; the molar extinction coefficient of the product, veratraldehyde, is 9300 M^{-1} cm^{-1}.[11]

[19] P. J. Kersten, M. Tien, B. Kalyanaraman, and T. K. Kirk, *J. Biol. Chem.* **260**, 2609 (1985).
[20] P. J. Kersten, B. Kalyanaraman, K. E. Hammel, B. Reinhammar, and T. K. Kirk, *Biochem. J.,* in press. (1990).
[21] M. Tien, T. K. Kirk, C. Bull, and J. A. Fee, *J. Biol. Chem.* **261**, 1687 (1986).

Procedure for Method 2. Reaction mixtures contain 0.15 mM 1,4-dimethoxybenzene, 0.4 mM H_2O_2, 100 mM tartrate, and enough lignin peroxidase to give an absorbance change of about 0.2 min^{-1} (1-cm cuvette). Reactions are started by adding the peroxide and are monitored by decrease in absorbance at 286 nm; the molar extinction of the substrate is 1260 M^{-1} cm^{-1}.

Procedure for Method 3. Reaction mixtures contain 0.2 mM 1,2,4,5-tetramethoxybenzene, 0.4 mM H_2O_2, 50 mM tartrate, and enough lignin peroxidase to give an absorbance change of 0.2 min^{-1}. Reactions are started by adding the peroxide and are monitored as increased absorbance at 450 nm; the molar extinction coefficient of the cation radical product is 9800 M^{-1} cm^{-1}.[19] Control reactions are necessary to verify that the rate of cation radical decay is less than the rate of its production.

Comments on Assays

Lignin peroxidase is most active below pH 3, but it is not very stable under these conditions; thus, reaction rates are linear only for about 2 min. For that reason, pH 3 is used here. The assays can be run at pH 4 also, but the reactions are slower. The kinetic data in Table I were determined at pH 4 for three of the substrates.

The enzyme is inactivated by H_2O_2 in the absence of a reducing substrate such as veratryl alcohol. Therefore, care should be taken to minimize the preincubation of enzyme with H_2O_2 in the absence of aromatic substrate. Lignin peroxidase activity shows a high temperature dependence, and therefore for reproducible results the temperature should be held constant; the rate approximately doubles with every 7° increase (M. Tien, unpublished).

We have described three assays. The first two are similar, both measuring two-electron oxidation products. Products in addition to veratraldehyde can be formed on the oxidation of veratryl alcohol, particularly at higher pH values,[22] although under the conditions described here such products are minor. Initial studies indicated that the oxidation of veratryl alcohol proceeds by a direct two-electron oxidation.[21] However, more recent studies by Marquez *et al.*[23] indicate that one-electron-oxidized intermediates are formed. Although the exact chemical mechanism of this oxidation is being investigated, this assay is a convenient and reliable method for measuring activity. In the second assay, *p*-benzoquinone and

[22] S. D. Haemmerli, H. E. Schoemaker, H. W. H. Schmidt, and M. S. A. Leisola, *FEBS Lett.* **220**, 149 (1987).

[23] L. Marquez, H. Wariishi, H. B. Dunford, and M. H. Gold, *J. Biol. Chem.* **263**, 10549 (1988).

methanol are the major products,[19] although other products can be detected. The third assay measures the cation radical directly and is perhaps most representative of the catalytic reaction per se.

Electron Spin Resonance Spectroscopy

Under conditions similar to those of assay Methods 2 and 3 above, the initial products of oxidation—the aryl cation radicals—are stable enough to be detected readily by ESR spectroscopy.[19] The radicals can be detected using a Varian E-9 and E-109 (or equivalent) spectrophotometer operating at 9.5 GHz and employing 100-kHz field modulation. Hyperfine splittings are measured from computer simulations of spectra.[24] The reaction mixtures for direct ESR measurements contain, in a volume of 2 ml, approximately 0.5 μM lignin peroxidase, 0.32 mM H_2O_2, 0.1 M sodium tartrate, pH 2.5, and 1–5 mM of the methoxybenzene substrate. The reactions, at room temperature, are initiated by addition of enzyme, and the mixtures are transferred immediately to an aqueous quartz flat cell positioned in the ESR spectrometer cavity. Figure 1 shows the ESR spectrum of the cation radical of 1,2,4,5-tetramethoxybenzene. Similar incubation conditions are used to detect phenoxy radicals formed from phenolic substrates.[25] The time course of cation radical formation is monitored at the maximum of the signal intensity with the magnetic field turned off. For certain substrates, the cation radicals or other radical intermediates are not readily detected, but they can be with the use of spin-trapping agents.[24]

Demonstration of Pyrene Oxidation[4]

[This is not described as an assay method.] A 1-ml reaction mixture is prepared in water–N,N-dimethylformamide (DMF), 8:2 by volume, that contains 20 mM sodium tartrate (final), pH 2.5 (solution A). To this solution at room temperature is added pyrene (as a 2 mM stock solution in DMF) to give 20 μM final concentration, and LiP1 (0.5 μM). The reaction is initiated with H_2O_2 (200 μM initially) and followed in repetitive scans between 600 and 220 nm in a double-beam UV–visible spectrophotometer. The reference cell contains solution A, plus DMF to replace the pyrene solution, and H_2O to replace the enzyme and H_2O_2 solutions.

The starting material exhibits peaks at 260, 272, 304, 318, and 334 nm, which decrease in intensity as the reaction progresses. The accumulation of products is accompanied by a general increase in absorbance from 350 to

[24] K. E. Hammel, M. Tien, B. Kalyanaraman, and T. K. Kirk, *J. Biol. Chem.* **260,** 8348 (1985).
[25] E. Odier, M. D. Mozuch, B. Kalyanaraman, and T. K. Kirk, *Biochimie* **70,** 847 (1988).

FIG. 1. Electron spin resonance spectrum of the cation radical of 1,2,4,5-tetramethoxy-benzene.

550 nm. Near isosbestic points occur at 246, 256, 276, 298, and 340 nm. The two major products (~85% of the total) are pyrene-1,6-quinone and pyrene-1,8-quinone.[4]

Enzyme Production

Methods for lignin peroxidase production in early studies used culture conditions optimized for lignin degradation in shallow stationary flasks.[9,15] Advantages with this type of experimental setup include the small scale and the ease with which the samples can be taken for analyses. These considerations remain important in studies of xenobiotic or radiolabeled substrate degradation studies. However, scale-up of enzyme production with shallow stationary cultures is labor intensive and becomes impractical. Consequently, a number of methods for lignin peroxidase production have been devised, including growing the fungus in rotating biological contactors,[15] in rotating drums,[13] and immobilized in gels.[26] For most studies, however, and for routine production of enzyme, the fungus is grown as pellets in agitated 2-liter flasks. Conditions required for lignin peroxidase production under nutrient nitrogen starvation (to achieve the necessary secondary metabolic state) are presented below with emphasis on key parameters required for stationary and agitated Erlenmeyer flask cultures. *Phanerochaete chrysosporium* (strain BKM-F-1767; ATCC 24725)

[26] Y.-Y. Linko, M. Leisola, N. Lindholm, J. Troller, P. Linko, and A. Fiechter, *J. Biotechnol.* **4**, 283 (1986).

has been investigated extensively under both conditions and produces the highest level of lignin peroxidase activity of the wild-type strains studied.[27]

The composition of agar slants for maintenance and spore production is 10 g glucose, 10 g malt extract, 2 g peptone, 2 g yeast extract, 1 g L-asparagine, 2 g KH_2PO_4, 1 g $MgSO_4 \cdot 7H_2O$, 1 mg thiamin-HCl, and 20 g agar per liter. After several days (preferably > 7) of incubation at 39°, dense conidiation is apparent over the agar surface; the slants are then stored in a refrigerator (5°) until needed but are used within 8 weeks. Spores are suspended in sterile water by gently rubbing the surface of the flooded agar with a sterile inoculation loop or glass rod. The spore suspension is passed through sterile glass wool to remove mycelia, and the spore concentration is determined at 650 nm (an absorbance of 1 cm^{-1} is approximately 5×10^6 spores/ml).

The culture media for both stationary and agitated cultures contain, per liter of culture: 2 g KH_2PO_4 (from a 40× stock solution), 0.5 g $MgSO_4 \cdot 7H_2O$ (40× stock), 0.1 g $CaCl_2 \cdot 2H_2O$ (40× stock), 1% glucose (10× stock), 10 mM pH 4.2 *trans*-aconitate or 2,2-dimethyl succinate (10× stock), 1 mg thiamin-HCl (1000× stock), 0.2 g ammonium tartrate (40× stock), 100 ml spore suspension inoculum (made fresh; OD$_{650nm}$ 0.5), and 10 ml of trace element solution (100× stock, see below). Sterile distilled water is added to make 1 liter. The thiamin and KH_2PO_4 stock solutions are filter sterilized; the others are autoclaved as separate solutions.

The 100× stock solution of trace elements is made by first dissolving 1.5 g of nitrilotriacetic acid in 800 ml water with adjustment of the pH to approximately 6.5 with KOH. The following are then added with dissolution: 0.5 g $MnSO_4 \cdot H_2O$, 1.0 g NaCl, 0.1 g $FeSO_4 \cdot 7H_2O$, 0.1 g $CoSO_4$, 0.1 g $ZnSO_4 \cdot 7H_2O$, 10 mg $CuSO_4 \cdot 5H_2O$, 10 mg $AlK(SO_4)_2 \cdot 12H_2O$, 10 mg H_3BO_3, and 10 mg $Na_2MoO_4 \cdot 2H_2O$. Water is added to 1 liter.

For shallow stationary cultures, 10 ml of culture fluid is used in rubber-stoppered 125-ml Erlenmeyer flasks at 30–39°. Cultures are flushed with 100% O_2 2 days after inoculation. Enzyme production can be stimulated by including 0.4 mM veratryl alcohol final concentration and trace elements (7 times the above concentration) at the time of inoculation.[15] A stock solution of veratryl alcohol is filter-sterilized through a 2.0-μm filter.

For agitated cultures, 750 ml of culture fluid is used in 2-liter flasks on a rotary shaker at 120 rpm (2.5-cm-diameter cycle) at 30–39°. In this case, 0.05% Tween 20 or Tween 80 (final concentration) is included in the medium at the time of inoculation. Cultures are flushed with 100% O_2 after

[27] T. K. Kirk, M. Tien, S. C. Johnsrud, and K.-E. Eriksson, *Enzyme Microb. Technol.* **8**, 75 (1986).

2 days and daily thereafter. Under both stationary and agitated conditions, maximal lignin peroxidase activity is observed between days 4 and 7, depending on temperature. Evidence suggests that enzyme production is highest near 30°, but that maximum titer takes longer to develop.[28] Enzyme production can be stimulated by increasing the concentration of trace elements 7-fold.

Enzyme Purification

The lignin peroxidase-containing extracellular fluid can be harvested by centrifugation (10,000 g for 10 min) at 4° or by passage through cheesecloth. The fluid is then concentrated by ultrafiltration in a Millipore Minitan unit with a 10-kDa cutoff membrane. Alternatively, the isoenzymes can be precipitated with 66% acetone at −10°, then dissolved in column buffer.[12] Four liters of culture supernatant are typically concentrated to about 50 ml. The concentrate is filtered (0.45-μm pore size) to remove precipitated mycelial polysaccharide and further concentrated (Amicon YM10, Danvers, MA, 10-kDa cutoff) to a final volume less than 15 ml. The preparation is then dialyzed overnight at 4° against 4 liters of either 10 mM sodium acetate, pH 6, for anion-exchange fast protein liquid chromatography (FPLC), or 5 mM sodium succinate, pH 5.5 for chromatography on DEAE-BioGel A. Approximately 90% recovery of enzymatic activity is typically obtained at this point.

The lignin peroxidase isoenzymes are purified by anion-exchange chromatography on DEAE-BioGel A, or on a larger scale by Q-Sepharose Fast Flow (Pharmacia, Piscataway, NJ) column chromatography, or by FPLC with a Mono Q column or Fast Flow Q column (Pharmacia).

For chromatography on DEAE-BioGel A,[18] all steps are performed at 4°. The column (1 × 16 cm) is equilibrated with 5 mM sodium succinate, pH 5.5. The sample is loaded and eluted from the column with a NaCl gradient (0.14 M, total volume of 600 ml). A typical elution profile is shown in Fig. 2. The predominant peak (fraction 240) is isoenzyme LiP1 (H8), and is over 95% pure as determined by sodium dodecyl sulfate–polyacrylamide gel electrophoresis (SDS–PAGE) or isoelectric focusing. The isolated lignin peroxidase isoenzymes are dialyzed against 5 mM potassium phosphate buffer, pH 6.5, for storage at −20°. Recovery of total lignin peroxidase activity in the culture supernatant in this purification is about 75%.

For purification by FPLC, the concentrated dialyzed extracellular fluid, free of polysaccharide, is used. The mobile phase consists of a gradient

---

[28] M. Asther, C. Capdevila, and G. Corrieu, *Appl. Environ. Microbiol.* **54**, 3194 (1988).

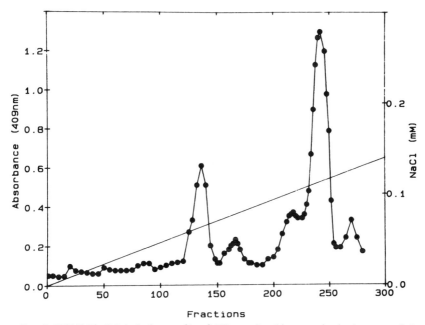

FᴵG. 2. DEAE-BioGel A elution profile of 409 nm-absorbing proteins in the extracellular fluid from 5-day flask cultures. The sloping line shows the NaCl gradient. Fractions of 2 ml were collected. The major peak is LiP1.

from 10 mM to 1 M sodium acetate, pH 6.0, applied over a 40-min period at a flow rate of 2 ml/min with monitoring at 409 nm (heme) and at 280 nm (protein). (Monitoring at 409 nm is far more sensitive than 280 nm.) A typical profile is shown in Fig. 3. The peaks designated MnP are associated with manganese peroxidase activity.[13,15,29] In most preparations the eluant containing each isoenzyme is about 85% pure; they can be purified further by repeated passages. The isoenzymes elute at the following sodium acetate molarities: LiP1, 0.43 M; LiP2, 0.18 M; LiP3, 0.16 M; LiP4, 0.34M; LiP5, 0.40M; and LiP6, 0.58 M. Total recovery of activity in this procedure is about 50% after purification of the isoenzymes to greater than 98% by repeated passages. Specific activities and other properties of the isoenzymes are given by Farrell et al.[14]

The extracellular culture fluid is prepared to Q-Sepharose Fast Flow column chromatography by the same procedure as for FPLC. The column is equilibrated with 10 mM sodium acetate buffer, pH 6.0 (NA buffer), at 4°. Concentrated culture fluid, dialyzed against NA buffer, is loaded onto the column, which is then washed with NA buffer. A gradient from 10 mM

---

[29] M. Kuwahara, J. K. Glenn, M. A. Morgan, and M. H. Gold, *FEBS Lett.* **169**, 247 (1984).

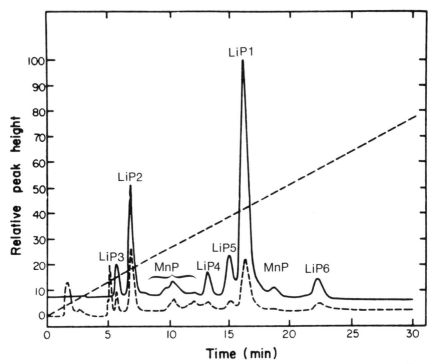

FIG. 3. Mono Q FPLC profile of a culture fluid sample. Full and dashed curves show absorbance at 409 and 280 nm, respectively. The lignin peroxidase isoenzymes are designated LiP1–6. The MnP designations are for peaks with manganese peroxidase activity. The sloping line shows the acetate gradient. Adapted from Kirk *et al.*[15]

to 1 *M* sodium acetate, pH 6.0, is then applied to the column run throughout at a flow rate of 8 ml/min. Fractions are collected and absorbance at 409 nm is monitored. Elution of the isoenzymes from the column is similar to that from Mono Q.

Purification of the isoenzymes by preparative isoelectric gel electrophoresis has also been described.[30]

Enzyme Properties

Up to 15 isoenzymes of the lignin peroxidase of *P. chrysosporium* BKM-F-1767 have been separated by isoelectric focusing from cultures grown under various conditions and for various culture times.[30] In our

[30] M. S. A. Leisola, B. Kozulic, F. Meussdoerffer, and A. Fiechter, *J. Biol. Chem.* **262**, 419 (1987).

work with this strain under nitrogen-limited growth, we have isolated and studied six isoenzymes using the fungal growth conditions and the FPLC purification scheme presented here.[14,15] These six isoenzymes have the following pI values and molecular weights (MW in parentheses): LiP1, 3.5 (42K); LiP2, 4.4 (38K); LiP3, 4.7 (38K); LiP4, 3.7 (40K, 42K); LiP5, approximately 3.6 (42K); and LiP6, 3.3 (43K). In earlier work, the designations were H1 (LiP3), H2 (LiP2), H6 (LiP4), H7 (LiP5), H8 (LiP1), and H10 (LiP6).[14] The relative abundances of these isoenzymes differ with culture time, but LiP1 is usually the major one in nitrogen-limited cultures.

At least four structural genes have been described which encode the lignin peroxidase isoenzymes in *P. chrysosporium*. Both cDNA and structural gene sequences have been published.[31-37]

Substrate Specificity. Although lignin peroxidase is readily oxidized by H_2O_2, it is not affected by other peroxides including *tert*-butyl hydroperoxide and cumene hydroperoxide. It is oxidized by *m*-chloroperbenzoic acid and *p*-(nitroperoxy)benzoic acid, however.[38] Specificity for the aromatic substrate, although it has not been studied per se, is apparently determined largely by the oxidation potential.[4,20] The enzyme oxidizes structurally diverse aromatic compounds; many different lignin-related structures have been studied.[1-3,5] Lignins are oxidized by lignin peroxidase, resulting in spectral changes and in both partial polymerization and partial depolymerization.[25,39,40]

Spectral Properties. Lignin peroxidase has absorbance maxima of 409 and 502 nm.[11,23,41] Spectral changes that accompany oxidation to Compound I, reduction of Compound I to Compound II, and reduction of Compound II to resting enzyme have been reported.[11,21,23] (Compound I is the two-electron oxidized state of peroxidases, and Compound II the one-

[31] M. Tien and C.-P. D. Tu, *Nature (London)* **328,** 742 (1987).

[32] H. A. deBoer, Y. Z. Zhang, C. Collins, and C. A. Reddy, *Gene* **60,** 93 (1987).

[33] R. L. Farrell, P. Gelep, A. Anilionis, K. Javahenian, and T. E. Maione, Eur. Patent Appl. 87,810,516.2 (1987).

[34] T. L. Smith, H. Schalch, J. Gaskell, S. Covert, and D. Cullen, *Nucleic Acids Res.* **16,** 1219 (1988).

[35] H. Schalch, J. Gaskell, T. L. Smith, and D. Cullen, *Mol. Cell. Biol.* **9,** 2743 (1989)

[36] Y. Asada, Y. Kimura, M. Kuwahara, A. Tsukamoto, K. Koide, A. Oka, and M. Takanami, *Appl. Microbiol. Biotechnol.* **29,** 469 (1988).

[37] I. Walther, M. Kalin, J. Reiser, F. Suter, B. Fritsche, M. Saloheimo, M. Leisola, T. Teeri, J. K. C. Knowles, and A. Fiechter, *Gene* **70,** 127 (1988).

[38] V. Renganathan and M. H. Gold, *Biochemistry* **25,** 1626 (1986).

[39] T. E. Maione, K. Javahenian, M. A. Belew, L. E. Gomez, and R. L. Farrell, *in* "Lignin Enzymic and Microbial Degradation" (E. Odier, ed.), pp. 177–183. INRA Publications, Paris, 1987

[40] S. D. Haemmerli, M. S. A. Leisola, and A. Fiechter, *FEMS Lett.* **35,** 33 (1986).

TABLE I
KINETIC PARAMETERS OF VARIOUS SUBSTRATES

TABLE I
KINETIC PARAMETERS OF VARIOUS SUBSTRATES

Kinetic parameter	Values for various substrates[a]				
	H_2O_2	Veratryl alcohol	1,4-Dimethoxy-benzene	1,2,4,5-Tetra-methoxybenzene	Model I
K_m (μM)	17	89	73	90	111
k_{cat} (sec^{-1})	—	1.3	8.6	11	0.5
k_{cat}/K_m (M^{-1} sec^{-1})	—	1.48×10^4	11.7×10^4	12×10^4	0.45×10^4

[a] Kinetic determinations for 1,2,4,5-tetramethoxybenzene were done in 50 mM sodium tartrate, pH 3.0;[20] those for the other substrates were determined in 100 mM sodium tartrate, pH 4.0.[14]

electron oxidized state.) Raman resonance spectroscopic properties have been described.[41]

pH Optimum. Lignin peroxidase exhibits a pH optimum of approximately pH 2.5–3.0.[11,21,23] Marquez et al.[23] have shown that conversion of Compound I to Compound II, and of the latter back to resting enzyme, in the presence of substrate, exhibits a pH optima below 3.0. As previously described, the enzyme is rapidly inactivated at pH values below about pH 3.0.[21]

Kinetic Properties of LiP1. Kinetic parameters for lignin peroxidase isoenzyme LiP1 are given in Table I for hydrogen peroxide and for the aromatic substrates veratryl alcohol, 1,4-dimethoxybenzene, 1,2,4,5-tetramethoxybenzene, and the lignin substructure model compound 1-(3,4-dimethoxyphenyl)-2-(2-methoxyphenoxy)propane-1,3-diol (Model I).

Summary

Lignin peroxidase of *Phanerochaete chrysosporium* can be produced and purified relatively easily. Described here is production and separation of six isoenzymes, using a commonly studied wild-type strain. The fungus is grown on a chemically defined nitrogen-limiting medium in stationary or agitated flasks, the extracellular broth containing the lignin peroxidases is concentrated, and the isoenzymes are separated and purified by anion-exchange chromatography. The isoenzymes have similar physical and catalytic properties but somewhat different kinetic properties. They are encoded at least in part by different structural genes. Lignin peroxidases, in

[41] L. A. Andersson, V. Renganathan, A. A. Chiu, T. M. Loehr, and M. H. Gold, *J. Biol. Chem.* **260**, 6080 (1985).

the presence of H_2O_2, oxidize their aromatic substrates by one electron to cation radicals, which undergo diverse reactions of ionic and radical nature. Purification and characterization of lignin peroxidases from other lignin-degrading fungi have also been reported.

Acknowledgments

The use of trade, firm, or corporation names in this publication is for the information and convenience of the reader. Such use does not constitute an official endorsement or approval by the U.S. Department of Agriculture of any product or service to the exclusion of others which may be suitable.

[28] Long-Chain Alcohol Dehydrogenase of *Candida* Yeast

By Mitsuyoshi Ueda and Atsuo Tanaka

$$CH_3(CH_2)_nCH_2OH + NAD^+ \rightarrow CH_3(CH_2)_nCHO + NADH + H^+$$

In the *n*-alkane-assimilating yeasts *Candida tropicalis* and *Candida lipolytica,* incorporated alkanes are hydroxylated by a cytochrome *P*-450 and NADPH–cytochrome-*P*-450 reductase system,[1] which is localized in the endoplasmic reticulum (microsomes),[2] to the corresponding long-chain alcohols. The long-chain alcohols so formed are not only degraded in the peroxisomal fatty acid β-oxidation system but are also utilized for lipid biosynthesis in mitochondria and the endoplasmic reticulum after dehydrogenation by long-chain alcohol dehydrogenase and aldehyde dehydrogenase to the corresponding fatty acids,[3-5] although the participation of alcohol oxidase is suggested.[6] Octanol dehydrogenase [octanol : NAD$^+$ oxidoreductase (EC 1.1.1.73)] has been found in *Saccharomyces cerevisiae* adapted to hydrocarbons[7] and hexadecanol dehydrogenase [hexade-

[1] W. Duppel, J. M. Lebeault, and M. J. Coon, *Eur. J. Biochem.* **36**, 583 (1973).

[2] S. Kawamoto, A. Tanaka, M. Yamamura, Y. Teranishi, S. Fukui, and M. Osumi, *Arch. Microbiol.* **112**, 1 (1977).

[3] M. Gallo, B. Roche, L. Aubert, and E. Azoulay, *Biochimie* **55**, 195 (1973).

[4] M. Gallo, B. Roche-Penverne, and E. Azoulay, *FEBS Lett.* **46**, 78 (1974).

[5] T. Yamada, H. Nawa, S. Kawamoto, A. Tanaka, and S. Fukui, *Arch. Microbiol.* **128**, 145 (1980).

[6] G. D. Kemp, F. M. Dickinson, and C. Ratledge, *Appl. Microbiol. Biotechnol.* **29**, 370 (1988).

[7] B. Roche and E. Azoulay, *Eur. J. Biochem.* **8**, 426 (1969).

canol:NAD^+ oxidoreductase (EC 1.1.1.164)] in rat liver[8] and *Euglena gracilis.*[9]

Assay Method

Principle. The reduction of NAD^+ is measured at 30° with a recording spectrophotometer at 340 nm by the method of Lebeault *et al.*[10] with slight modifications.

Reagents

Tris-HCl buffer, 50 mM, pH 8.5
Decanol, 12 mM dissolved in acetone
NAD^+, 50 mM dissolved in deionized water
NaN_3, 50 mM dissolved in deionized water
Procedures. To a quartz cuvette (10-mm light path) are added 1.1 ml of Tris-HCl buffer, 0.1 ml of decanol, 0.1 ml of NAD^+, 0.1 ml of NaN_3, and enzyme to a final volume 1.5 ml. The reaction is initiated by the addition of enzyme. The reading is corrected for the blank which is obtained with enzyme and all reagents mentioned except for decanol.

Growth of Microorganism

Candida tropicalis Berkhout pK 233 (ATCC 20336) is cultivated aerobically at 30° in a medium containing 5.0 g/liter of $NH_4H_2PO_4$, 2.5 g/liter of KH_2PO_4, 1.0 g/liter of $MgSO_4 \cdot 7H_2O$, 20 mg/liter of $FeCl_3 \cdot 6H_2O$, 1.0 ml/liter of corn steep liquor and a carbon source [16.5 g/liter of glucose, 10 g/liter of stearyl alcohol, or 10 ml/liter of *n*-alkane mixture (C_{10} 16.4, C_{11} 50.4, C_{12} 32.5, and C_{13} 0.7%, w/w)]. In the case of the alkane mixture and stearyl alcohol, 0.5 ml/liter of Tween 80 is added to the medium. The initial pH of the medium is adjusted to pH 5.2 with 1 M NaOH before sterilization. *Candida lipolytica* NRRL Y-6795 is also cultivated by the same method as that for *C. tropicalis.*

Preparation of Cell-Free Extracts

Cells harvested at the midexponential growth phase by centrifugation at 1,000 g for 5 min are washed twice with deionized water and suspended

[8] W. Stoffel, D. Lekim, and G. Heyn, *Hoppe-Seyler's Z. Physiol.* **351,** 875 (1970).
[9] P. E. Kolattukudy, *Biochemistry* **9,** 1095 (1970).
[10] J. M. Lebeault, B. Roche, Z. Duvnjak, and E. Azoulay, *Biochim. Biophys. Acta* **220,** 373 (1970).

into 50 mM potassium phosphate buffer (pH 7.2) to give a concentration of 2 g dry cells/liter. An aliquot of the cell suspension (10 ml) is subjected to ultrasonic disintegration (20 kHz, 5 min) at 0°. The cell homogenate is centrifuged at 10,000 g for 15 min, and the resulting supernatant is used as a cell-free extract.[11]

Subcellular Fractionation

Subcellular organelles [peroxisomes, mitochondria, and the endoplasmic reticulum (microsomes)] having long-chain alcohol dehydrogenase activity are isolated from the protoplast homogenate of alkane-grown yeast cells. Yeast cells (1.5 g dry weight) harvested at the exponential phase of growth are suspended in 200 ml of 40 mM potassium phosphate buffer (pH 7.2) containing 0.80 M sorbitol and 0.1 M 2-mercaptoethanol, and treated with 40 mg of Zymolyase 20T at 30° until the degree of lysis reaches about 70–80%. Thereafter all the procedures are carried out at 0–4°. The protoplasts so formed are collected by centrifugation at 3,000 g for 10 min, suspended in 30 ml of 50 mM potassium phosphate buffer (pH 7.2) containing 0.65 M sorbitol and 0.5 mM EDTA, and homogenized for 10 min in a Teflon homogenizer under cooling.

The homogenate is centrifuged at 3,000 g for 10 min to remove heavy particles (P_1). The supernatant (S_1) is again subjected to centrifugation at 20,000 g for 15 min to obtain the supernatant fraction (S_2) and the particulate fraction (P_2) containing peroxisomes and mitochondria. The P_2 fraction is suspended in 8.0 ml of 50 mM potassium phosphate buffer (pH 7.2) containing 20.0% (w/w) sucrose and 0.5 mM EDTA, and an aliquot (2.5 ml) is layered over a discontinuous sucrose density gradient composed of the same buffer containing 30.0, 40.0, 41.3, 42.5, and 50.0% sucrose (each 2.5 ml). Intact peroxisomes and mitochondria are isolated by centrifugation at 49,600 g for 2 hr. From top to bottom, 3.75, 2.5, 2.5, 2.5, 2.5, and 1.25 ml are respectively collected and named fractions 1 to 6 in that order. The fraction containing catalase is the peroxisomal fraction and that containing cytochrome oxidase is the mitochondrial fraction. The microsomal (P_3) and cytosolic fractions (S_3) are obtained by centrifugation of the S_2 fraction at 139,000 g for 2 hr.[12] To improve the separation of peroxisomes and mitochondria, slight modifications of the method have been developed.[13]

[11] M. Ueda, H. Okada, A. Tanaka, M. Osumi, and S. Fukui, *Arch. Microbiol.* **136**, 169 (1983).

[12] M. Yamamura, Y. Teranishi, A. Tanaka, and S. Fukui, *Agric. Biol. Chem.* **39**, 13 (1975).

[13] M. Ueda, K. Yamanoi, T. Morikawa, H. Okada, and A. Tanaka, *Agric. Biol. Chem.* **49**, 1821 (1985).

TABLE I
CELLULAR LEVELS OF LONG-CHAIN ALCOHOL
DEHYDROGENASE IN Candida tropicalis

Cells	Enzyme activity[a] (nmol/min/mg protein)
Glucose grown	5.85
Alkane grown	79.9
Stearyl alcohol grown	83.5

[a] Enzyme activities are measured with cell-free extracts.

Properties

Induction. The activity of long-chain alcohol dehydrogenase in alkane-grown cells of *C. tropicalis* is more than 10 times higher than that in glucose-grown cells. Stearyl alcohol-grown cells also have a high activity level of the enzyme (Table I).

Substrate Specificity. Alcohols having medium to long carbon chains (C_8–C_{16}) are dehydrogenated by the *C. tropicalis* enzyme. Tetradecanol is the substrate showing the maximum activity. Alcohols with much longer chains (C_{17}–C_{19}) do not serve as good substrates. This low activity may be in part due to the poor solubility of these longer-chain alcohols in water. The enzyme of *C. lipolytica* can dehydrogenate alcohols with C_6–C_{18} carbon chains, showing the maximum activity on decanol. Both enzymes are NAD-dependent, and activities of phenazine methosulfate-dependent dehydrogenase or those of alcohol oxidase cannot be detected under the experimental conditions employed.

Solubilization and Stability. This enzyme seems to be bound to membranes of the respective subcellular organelles. Solubilization with 1% Tween 80 or Triton X-100 at 4° stimulates the inactivation of the enzyme. Dithiothreitol (2 mM) or ethylene glycol (20%, w/w) does not serve as a stabilizer.

Subcellular Localization. The enzyme of *C. tropicalis* and *C. lipolytica* is distributed among microsomal, mitochondrial, and peroxisomal fractions. The long-chain alcohol dehydrogenase of these yeasts is quite different from ordinary (short-chain) alcohol dehydrogenase (EC 1.1.1.1) in terms of the subcellular localization (Table II).

Function. In peroxisomes, long-chain alcohol dehydrogenase is colocalized with long-chain aldehyde dehydrogenase, acyl-CoA synthetase (long-chain-fatty-acid–CoA ligase), and the fatty acid β-oxidation system, indicating that the enzyme is participating in fatty acid degradation. On the

TABLE II

Subcellular Distribution of Long-Chain Alcohol Dehydrogenase
in Alkane-Grown *Candida tropicalis* and *Candida lipolytica*[a]

Enzyme	Relative enzyme activity (%) in fractions								
	P_2	P_3	S_3	No. 1	2	3	4	5	6
C. tropicalis									
Catalase	83	2	15	9	10	10	15	50	6
Cytochrome oxidase	98	2	0	0	42	21	19	18	0
Long-chain alcohol dehydrogenase	75	21	4	5	23	13	15	38	6
Short-chain alcohol dehydrogenase	0	0	100	—	—	—	—	—	—
C. lipolytica									
Catalase	58	4	38	15	20	19	20	26	0
Cytochrome oxidase	90	10	0	2	68	11	7	9	3
Long-chain alcohol dehydrogenase	67	28	5	0	27	27	14	32	0
hort-chain alcohol dehydragenase	0	0	100	—	—	—	—	—	—

[a] The substrates of long-chain alcohol dehydrogenase and short-chain alcohol dehydrogenase are decanol and ethanol, respectively. Fractions, see text.

other hand, the enzyme localized in yeast mitochondria, in which fatty acid β-oxidation has not been detected, and in the endoplasmic reticulum is co-localized with long-chain aldehyde dehydrogenase, acyl-CoA synthetase, and glycerol-3-phosphate acyltransferase. The enzyme may have an indispensable role in lipid biosynthesis.[14]

[14] A. Tanaka, M. Osumi, and S. Fukui, *Ann. N.Y. Acad. Sci.* **386,** 183 (1982).

[29] Long-Chain Aldehyde Dehydrogenase of *Candida* Yeast

By Mitsuyoshi Ueda and Atsuo Tanaka

$$CH_3(CH_2)_nCHO + NAD^+ \rightarrow CH_3(CH_2)_nCOOH + NADH + H^+$$

In the *n*-alkane-assimilating yeasts *Candida tropicalis* and *Candida lipolytica,* long-chain alcohols formed by the hydroxylation system in the endoplasmic reticulum (microsomes) are transformed to the corresponding fatty acids by the actions of long-chain alcohol dehydrogenase and aldehyde dehydrogenase. The fatty acid products may then be utilized for energy production and the biosynthesis of cellular components via fatty acid β-oxidation in peroxisomes and for lipid biosynthesis in mitochondria and endoplasmic reticulum.[1-4]

Assay Method

Principle. The reduction of NAD^+ is measured at 30° with a recording spectrophotometer at 340 nm by the method of Lebeault *et al.*[5] with slight modifications.

Reagents

Tris-HCl buffer, 50 mM, pH 8.5
Decanal, 12 mM dissolved in acetone
NAD^+, 50 mM dissolved in deionized water
NaN_3, 50 mM dissolved in deionized water

Procedures. To a quartz cuvette (10-mm light path) are added 1.1 ml of Tris-HCl buffer (pH 8.5), 0.1 ml of decanal, 0.1 ml of NAD^+, 0.1 ml of NaN_3, and enzyme to a final volume of 1.5 ml. The reaction is initiated by the addition of enzyme. The reading is corrected for the blank which is obtained with enzyme and all reagents mentioned except decanal.

[1] S. Kawamoto, A. Tanaka, M. Yamamura, Y. Teranishi, S. Fukui, and M. Osumi, *Arch. Microbiol.* **112**, 1 (1977).

[2] M. Gallo, B. Roche, L. Aubert, and E. Azoulay, *Biochimie* **55**, 195 (1973).

[3] M. Gallo, B. Roche-Peneverne, and E. Azoulay, *FEBS Lett.* **46**, 78 (1974).

[4] T. Yamada, H. Nawa, S. Kawamoto, A. Tanaka, and S. Fukui, *Arch. Microbiol.* **128**, 145 (1980).

[5] J. M. Lebeault, B. Roche, Z. Duvnjak, and E. Azoulay, *Biochim. Biophys. Acta* **220**, 373 (1970).

TABLE I

CELLULAR LEVELS OF LONG-CHAIN ALDEHYDE DEHYDROGENASE
IN *Candida tropicalis*[a]

Cells	Long-chain aldehyde dehydrogenase (nmol/min/mg protein)
Glucose grown	8.79
Alkane grown	98.0
Stearyl alcohol grown	93.2

[a] Enzyme activities are measured with cell-free extracts.

Growth of Microorganism

Candida tropicalis Berkhout pK 233 (ATCC 20336) or *Candida lipolytica* NRRL Y-6795 is cultivated aerobically at 30° in a medium containing 16.5 g/liter of glucose, 10 g/liter of stearyl alcohol, or 10 ml/liter of *n*-alkane mixture (C_{10} 16.4, C_{11} 50.4, C_{12} 32.5, and C_{13} 0.7%, w/w).[6] Other medium components are as detailed in the preceding chapter.[7]

Preparation of Cell-Free Extracts and Subcellular Fractionation

Cell-free extracts and subcellular organelles [peroxisomes, mitochondria, and the endoplasmic reticulum (microsomes)] having long-chain aldehyde dehydrogenase activity are prepared from alkane-grown yeast cells as described in the preceding chapter.[7]

Properties

Induction. The activity of long-chain aldehyde dehydrogenase in alkane-grown cells of *C. tropicalis* is more than 10 times higher than that in glucose-grown cells, similar to the case of long-chain alcohol dehydrogenase. Stearyl alcohol-grown cells also exhibit a high activity level of the enzyme (Table I).

Substrate Specificity. Long-chain aldehyde dehydrogenase oxidizes primary aldehydes of C_7-C_{14}, the same as long-chain alcohol dehydrogenase.[7] The aldehyde dehydrogenase of *C. tropicalis* has little or no activity on substrates with a chain length of C_{15} or greater, although the yeast can assimilate $C_{15}-C_{19}$ alkanes. The reasons may be ascribed to the low solu-

[6] M. Ueda, H. Okada, A. Tanaka, M. Osumi, and S. Fukui, *Arch. Microbiol.* **136,** 169 (1983).
[7] M. Ueda and A. Tanaka, this volume [28].

TABLE II

SUBCELLULAR DISTRIBUTION OF LONG-CHAIN ALDEHYDE DEHYDROGENASE IN
ALKANE-GROWN *Candida tropicalis* AND *Candida lipolytica*[a]

Enzyme	Relative enzyme activity (%) in fractions								
	P_2	P_3	S_3	No. 1	2	3	4	5	6
C. tropicalis									
Catalase	83	2	15	9	10	10	15	50	6
Cytochrome oxidase	98	2	0	0	42	21	19	18	0
Long-chain aldehyde dehydrogenase	82	18	0	0	22	11	14	47	6
C. lipolytica									
Catalase	58	4	38	15	20	19	20	26	0
Cytochrome oxidase	90	10	0	2	68	11	7	9	3
Long-chain aldehyde dehydrogenase	85	13	2	2	46	15	18	18	1

[a] The substrate of long-chain aldehyde dehydrogenase is decanal. For designations of fractions, see the preceding chapter.[7]

bility in water of such longer-chain substrates or to the occurrence of other dehydrogenases depending on unknown hydrogen acceptors.

Solubilization and Stability. The enzyme bound to organelle membranes cannot be solubilized even with 1% Tween 80 or Triton X-100 at 4°. Dithiothreitol (2 mM) or ethylene glycol (20%, w/w) is not useful as a stabilizer.

Subcellular Localization. The enzyme of *C. tropicalis* and *C. lipolytica* is distributed among microsomal, mitochondrial, and peroxisomal fractions (Table II).

Function. In peroxisomes, this enzyme supplies fatty acids to be degraded *via* the β-oxidation pathway for production of energy and cellular components. On the other hand, in mitochondria and the endoplasmic reticulum, this enzyme supplies fatty acids for lipid biosynthesis.

Section II

Methylotrophy

A. Dissimilatory Enzymes of Methylotrophic Bacteria
Articles 30 through 54

B. Assimilation of Formaldehyde in Methylotrophic Bacteria
Articles 55 through 63

C. Methylotrophic Enzymes in Methanol-Utilizing Yeast
Articles 64 through 72

[30] Soluble Methane Monooxygenase from *Methylococcus capsulatus* Bath

By SIMON J. PILKINGTON and HOWARD DALTON

$$CH_4 + NADH + H^+ + O_2 \xrightarrow{\text{MMO}} CH_3OH + NAD^+ + H_2O$$

Methane monooxygenase (MMO) (EC 1.14.13.25) is the enzyme that catalyzes the initial step in the assimilation of methane in bacteria that grow with methane as the sole carbon and energy source. The enzyme has been shown to exist in two forms in *Methylococcus capsulatus* Bath. Cells grown under conditions of high copper availability produce a membrane-bound (particulate) methane monooxygenase, whereas cells grown under conditions of low copper availability produce a soluble methane monooxygenase. These two enzymes appear to be distinct in their properties and polypeptide composition.[1] It is the purification and characterization of the soluble methane monooxygenase that are considered here; however, the assay procedure may be applied to bacterial extracts containing either form of the enzyme.

Assay Methods

Soluble Methane Monooxygenase

Principle. In crude cell extracts addition of methane and NADH in air does not lead to the quantitative accumulation of methanol because methanol is further metabolized via methanol oxidase, methanol dehydrogenase, and even methane monooxygenase.[2] Reproducible assay methods have therefore relied on the accumulation from a methane monooxygenase-catalyzed reaction of products which are not substrates of other enzymes present in the extract. The substrate of choice is propylene because the product, propylene oxide, is not further metabolized and is produced at approximately the same rate that methanol is produced from methane. It is easily measured quantitatively at the nanomolar level using gas chromatography. Other methods, such as the disappearance of bromomethane,[3] have been used reliably, but assay methods based on the disappearance of methane in enclosed vials are notoriously variable and should be avoided.

[1] S. H. Stanley, S. D. Prior, D. J. Leak, and H. Dalton, *Biotechnol. Lett.* **5**, 487 (1983).

[2] J. Colby and H. Dalton, *Biochem. J.* **157**, 495 (1976).

[3] J. Colby, H. Dalton, and R. Whittenbury, *Biochem. J.* **151**, 459 (1975).

Reagents

20 mM Tris-HCl buffer, pH 7.0
100 mM NADH (ethanol-free)
Propylene gas

Procedure. Methylococcus capsulatus Bath grows at an optimal temperature of 45°, and so all enzyme assays are done at this temperature. Assays are carried out in conical flasks (7 ml internal volume) sealed with a rubber closure (Suba Seal No. 37, W. H. Freeman, Barnsley, Yorkshire, UK; or a crimped septum through which a hypodermic needle can be inserted). To the flask are added 0.95 ml of the reaction mixture, containing 20 mM Tris-HCl buffer, pH 7.0, plus extract, and the reaction flask is sealed with the closure. Three milliliters of the gas phase of the reaction flask is removed by syringe and replaced with 3 ml of propylene. The flask is then preincubated for 1–2 min in a gyratory water bath and the reaction initiated by the addition of 50 μl of 100 mM NADH (ethanol-free). Samples of 5 μl are removed at intervals from the liquid phase of the reaction flask and analyzed by gas chromatography for propylene oxide.

Gas chromatography is typically carried out using a 1 m × 4 mm glass column containing Porapak Q (Water Associates, Milford, MA), maintained isothermically at 190° with nitrogen as carrier gas at a flow rate of 30 ml/min. A flame ionization detector is used to detect the resolved products of the assay, the peak area being determined by an integrator. The gas chromatograph is calibrated by injection of 5 μl of a 2 mM solution of propylene oxide. Methane monooxygenase activities are therefore expressed as nanomoles of propylene oxide produced per minute per milligram of protein.

Commercially available NADH contains sufficiently high levels of ethanol (10–20 mol %) to interfere with the gas chromatographic analysis of propylene oxide in assays of methane monooxygenase. It is therefore necessary to remove ethanol from NADH by ether extraction of the NADH solution followed by evaporation under vacuum to remove the ether.

The soluble methane monooxygenase from *Methylococcus capsulatus* Bath is composed of three component proteins (see Purification Procedure), and when specific activity is plotted as a function of the concentration of extract in the assay a nonlinear relationship is obtained.[2] Maximum specific activity is obtained when a concentration of 5–7 mg/ml protein is present in assays of bacterial extracts. Assays of purified component proteins of the soluble methane monooxygenase are carried out in the presence of saturating amounts of the other two component proteins, which are present in equimolar ratios.

Protein C

Principle. Of the three component proteins of the soluble methane monooxygenase only protein C has independent catalytic activity. It can carry out the transfer of electrons from NAD(P)H to a variety of electron acceptors.[4]

Reagents

20 mM Tris-HCl buffer, pH 7.0
10 mM NAD(P)H
9 mM Potassium ferricyanide
    or 0.9 mM Dichlorophenolindophenol (DCPIP)
    or 0.5 mM Cytochrome *c*

Procedure. Two milliliters of 20 mM Tris-HCl buffer, pH 7.0, containing 0.19 mmol DCPIP or potassium ferricyanide or 50 μmol cytochrome *c* are placed in a 3-ml cuvette, sealed with a Suba Seal, and sparged with helium for 10 min to remove oxygen. After addition of the enzyme by injection through the Suba Seal, the reaction is initiated by the addition of 0.23 mmol of NAD(P)H and the reaction monitored spectrophotometrically at the appropriate wavelength for the electron acceptor used. All assays are carried out at 45°.

Purification Procedure

Growth of Organisms. Methylococcus capsulatus Bath is grown in a 100-liter batch fermentor at 45° sparged with air–methane (5:1, v/v) in a nitrate mineral salts media containing, per liter: 2 g KNO_3, 1 g $MgSO_4$, 0.2 g $CaCl_2$, 0.0012% FeEDTA, 10 ml 5% Na_2HPO_4/KH_2PO_4 buffer, pH 6.8, and 1 ml of stock trace element solution containing 200 mg $CuSO_4 \cdot 5H_2O$, 500 mg $FeSO_4 \cdot 7H_2O$, 400 mg $ZnSO_4 \cdot 7H_2O$, 15 mg H_3BO_4, 50 mg $CoCl \cdot 6H_2O$, 500 mg $Na_2MoO_4 \cdot 2H_2O$, 250 mg EDTA, 20 mg $MnCl_2 \cdot 4H_2O$, and 10 mg $NiCl_2 \cdot 6H_2O$ in 1 liter of water. The exponentially growing culture is harvested by continuous centrifugation and the cell paste resuspended in the minimum of 20 mM Tris-HCl buffer, pH 7.0, to effect easy drop freezing in liquid nitrogen. Cells in this form can be stored at −80° until required.

Step 1: Preparation of Cell-Free Extract. Cell paste is thawed and diluted with 20 mM Tris-HCl buffer, pH 7.0, containing 5 mM $MgCl_2$ plus 5 mM sodium thioglycolate. Cells are broken by two passages through a French pressure cell at 137 MPa (American Instrument Company, Silver

---

[4] J. Colby and H. Dalton, *Biochem. J.* **177**, 903 (1979).

TABLE I
PURIFICATION OF PROTEIN A[a]

Step	Protein (mg)	Specific activity (nmol/min/mg)	Total activity (nmol/min)	Purification factor	Yield (%)
Soluble extract	208	59.4	12,355	1.00	100
DEAE-cellulose	115	104.4	11,960	1.76	97
Mono Q	40.8	109.2	4455	1.83	36
		(227.7)	(9108)	(3.86)	(74)
TSK-G 3000 SWG	34.0	126.2	4289	2.12	35
		(264.8)	(9004)	(4.46)	(73)

[a] In practice much larger amounts of protein are prepared in a single DEAE-cellulose ion-exchange chromatography step. For ease of comparison figures are calculated from the results of a single TSK-G 3000 SWG preparation. Figures in parentheses represent activities after reconstitution of iron by incubation with sodium dithiothreitol and FeEDTA.

Spring, MD). Whole cells are removed by centrifugation for 10 min at 32,000 g. The extract can then be separated into soluble and particulate fractions by centrifugation at 80,000 g for 60 min.

Step 2: Resolution of Soluble Methane Monooxygenase into Component Proteins. Resolution of the soluble methane monooxygenase into its three component proteins is effected by ion-exchange chromatography.[5] Soluble extract from Step 1 is loaded onto a column of DEAE-cellulose (4.5 × 7.0 cm) equilibrated with 20 mM Tris-HCl buffer, pH 7.0, containing 5 mM sodium thioglycolate and 50 mM NaCl. Material not binding to the column is eluted with the above buffer, yielding a fraction containing protein A. The column is then eluted with successive batches of the buffer containing, respectively, 0.1 and 0.2 M NaCl, yielding protein B, and 0.5 M NaCl, yielding protein C. Having separated each protein of the soluble methane monooxygenase, their purification can be undertaken separately.

Purification of Protein A

Step 1: Mono Q Ion-Exchange Chromatography. Material containing protein A from the DEAE-cellulose step (up to 300 mg) is loaded onto a prepacked 8-ml Mono Q ion-exchange column (Pharmacia LKB Biotechnology, Uppsala, Sweden) and eluted with a 0–300 mM linear gradient of NaCl in 20 mM Tris-HCl buffer, pH 7.0, using a Pharmacia fast protein liquid chromatography (FPLC) system (Pharmacia LKB Biotechnology). Protein A elutes at approximately 150 mM NaCl as the second major peak to be eluted from the column.

[5] J. Colby and H. Dalton, *Biochem. J.* **171**, 461 (1978).

TABLE II
PURIFICATION OF PROTEIN B

Step	Protein (mg)	Specific activity (nmol/min/mg)	Total activity (nmol/min)	Purification factor	Yield (%)
Soluble extract	520	60	31,200	1.0	100
DEAE-cellulose	157	162.2	25,465	2.7	81.6
Sephadex G-100	15.6	1235	19,216	20.6	61.6
DEAE-Sepharose CL-6B	4.88	2759	13,464	50.00	43.2
Hydroxyapatite	1.56	7289	11,371	121.5	36.4

Step 2: TSK-G 3000 SWG Gel Filtration. Material from Mono Q ion-exchange chromatography containing protein A is concentrated over a PM30 ultrafiltration membrane (Amicon Corporation, Danvers, MA). Concentrated material (40–80 mg/ml) is applied in 2-ml aliquots to a TSK-G 3000 SWG (21.5 × 600 mm) gel permeation column preceded by a guard column, Utlra-Pac TSK-GSWPG (21.5 × 75 mm) (Tosoh Corporation, Tokyo, Japan). Buffer (20 mM Tris-HCl, pH 7.0) is supplied to the column at a flow rate of 3 ml/min.[6] Protein A elutes after approximately 100 ml, is concentrated as previously described, and is frozen dropwise in liquid nitrogen and stored at −80° until required. This method of storage can be applied to all intermediary steps of purification without loss of activity. A typical purification is summarized in Table I.

Purification of Protein B

Step 1: Sephadex G-100 Gel Filtration. Fractions containing protein B from the DEAE-cellulose separation are treated with 0.3 mM phenylmethylsulfonyl fluoride and concentrated by ultrafiltration over a YM5 membrane (Amicon). Concentrated fractions are applied to a Sephadex G-100 column (2.5 × 80 cm), equilibrated with 10 mM sodium phosphate buffer, pH 7.0, and eluted with the same buffer at a flow rate of 20 ml/hr.

Step 2: DEAE-Sepharose Ion-Exchange Chromatography. Pooled active fractions from Step 1 are applied to a DEAE-Sepharose CL-6B column (1.5 × 30 cm) equilibrated with 10 mM sodium phosphate buffer, pH 7.0. The column is eluted at a flow rate of 20 ml/hr with a linear gradient (200 ml) of 10–300 mM NaCl in the same buffer, after washing the column with 3 column volumes of buffer containing 10 mM NaCl.

Step 3: Hydroxyapatite Chromatography. Pooled active fractions from Step 2 are concentrated as in Step 1, diluted, and reconcentrated to reduce the ionic strength of the sample. The sample is then applied to a column of

[6] M. P. Woodland and H. Dalton, *Anal. Biochem.* **139,** 459 (1984).

TABLE III
PURIFICATION OF PROTEIN C

Step	Protein (mg)	Specific activity[a] (nmol/min/mg)	Total activity (nmol/min)	Purification factor	Yield (%)
Soluble extract	20,000	60	1200×10^3	1	100
DEAE-cellulose	271	2000	540×10^3	33	45
5'-AMP-Sepharose	68	6000	404×10^3	99	34

[a] Specific activities of purified protein C have been recorded as high as 38,000 nmol/min/ mg protein using very low quantities (10 μg) of protein C and 5 mg proteins A plus B in assays.

hydroxyapatite (2×7 cm) equilibrated with 10 mM sodium phosphate buffer, pH 7.0. The column is eluted at 20 ml/hr with a linear gradient (200 ml) of 10–150 mM sodium phosphate buffer, pH 7.0. Protein B is eluted at 45 mM phosphate. Protein B is then frozen dropwise in liquid nitrogen and stored at $-80°$ without loss of activity. A typical purification is summarized in Table II.

Purification of Protein C

Step 1: 5'-AMP-Sepharose Affinity Chromatography. Fractions containing protein C from the DEAE-cellulose ion-exchange step are pooled and concentrated by ultrafiltration over a PM30 membrane (Amicon). The sample is then applied to a column of 5'-AMP-Sepharose 4B (Pharmacia LKB Biotechnology) (1.5×11 cm) equilibrated with 20 mM Tris-HCl buffer, pH 7.0, containing 5 mM sodium thioglycolate. Nonspecifically bound protein is eluted at 60 ml/hr with 1 column volume of buffer containing 0.5 M NaCl followed by 1 column volume of buffer to remove excess salt. Protein C is finally eluted with buffer containing 1 mM NADH (ethanol-free) and can then be frozen dropwise in liquid nitrogen and stored at $-80°$ without loss of activity. A typical purification of protein C is summarized in Table III. *Note:* Some 5'-AMP Sepharose columns have not always proved to be reliable and appear to lose their ability to bind protein C when they have been used for a number of purifications. An alternative to 5'-AMP-Sepharose is Reactive Green Agarose (Sigma Chemical Company Ltd., Poole, Dorset), which has proved successful.

Properties

Mechanism. The three proteins of the soluble methane monooxygenase have each been assigned a role in the oxidation of methane. In summary, protein C acts as an NADH oxidoreductase, in which the FAD group

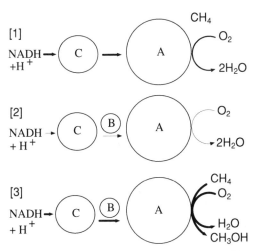

FIG. 1. Mechanism of the soluble methane monooxygenase from *Methylococcus capsulatus* Bath. [1] Proteins A and C catalyze the four-electron reduction of oxygen to water in the presence or absence of a hydroxylatable substrate. [2] The addition of protein B switches the enzyme from an oxidase to an oxygenase. In the absence of a hydroxylatable substrate electron flow between proteins A and C is shut down, preventing the reduction of oxygen to water. [3] The addition of a hydroxylatable substrate (methane) to the complete soluble MMO complex restores electron flow between proteins A and C and the oxygenase reaction is catalyzed to the complete exclusion of the oxidase reaction. The thickness of arrows indicates the relative flux through the system.

accepts two electrons from NADH and passes them one at a time to its Fe_2S_2 center.[7] From the Fe_2S_2 center electrons are passed to protein A, the site of substrate activation.[8] In the absence of protein B the hydrocarbon substrate is not oxidized and the electrons reduce oxygen exclusively to water such that the enzyme functions as an oxidase. In the presence of protein B electron flow to protein A only occurs in the presence of a suitable hydrocarbon substrate, and an oxygenase reaction is catalyzed. Protein B therefore acts to regulate the flow of electrons to the site of substrate activation, converting the enzyme from an oxidase to an oxygenase (Fig. 1).[9]

Steady-state kinetic analysis revealed a concerted substitution mechanism in which methane binds to the enzyme followed by NADH to give the first ternary complex which reacts to yield reduced enzyme and NAD^+.

[7] J. Lund and H. Dalton, *Eur. J. Biochem.* **147**, 291 (1985).
[8] J. Lund, M. P. Woodland, and H. Dalton, *Eur. J. Biochem.* **147**, 297 (1985).
[9] J. Green and H. Dalton, *J. Biol. Chem.* **260**, 15795 (1985).

TABLE IV
SUBSTRATES AND PRODUCTS OF SOLUBLE
METHANE MONOXYGENASE *Methylococcus
capsulatus* BATH

Substrate	Product(s)
Methane	Methanol
Ethane	Ethanol
Hexane	1-Hexanol, 2-hexanol
Ethylene	Epoxyethane
Propylene	Epoxypropane
trans-2-Butene	*trans*-2, 3-Epoxybutane, *trans*-2-buten-1-ol
Chloromethane	Formaldehyde
Dichloromethane	Carbon monoxide
Trichloromethane	Carbon dioxide
Dimethyl ether	Methanol, formaldehyde
Diethyl ether	Ethanol, acetaldehyde
Cyclohexane	Cyclohexanol
Benzene	Phenol, hydroquinone
Toluene	Benzyl alcohol, *p*-cresol
Styrene	Styrene epoxide
Pyridine	Pyridine *N*-oxide
Naphthalene	α- and β-Naphthol
Carbon monoxide	Carbon dioxide
Ammonia	Hydroxylamine

The reduced enzyme–methane complex binds O_2 to give a second ternary complex, which breaks down to release water and methanol. With methane as as substrate the K_m values for methane, NADH, and O_2 are 3, 55.8, and 16.8 μM, respectively. The K_m for NADH varies between 25 and 300 μM depending on the nature of the hydrocarbon substrate.[10]

Substrate Specificity. The soluble methane monooxygenase has been shown to insert an oxygen atom into a wide range of substrates.[11,12] To date over 150 compounds have been shown to act as oxidizable substrates; a representative list of such substrates and products can be found in Table IV.

Inhibitors. The soluble methane monooxygenase complex is strongly inhibited by 8-hydroxyquinoline and acetylene, with the latter thought to act as a suicide substrate, being oxidized to a highly reactive ketene species

[10] J. Green and H. Dalton, *Biochem. J.* **236,** 155 (1986).
[11] D. I. Stirling, J. Colby, and H. Dalton, *Biochem. J.* **177,** 361 (1979).
[12] D. I. Stirling and H. Dalton, *J. Gen. Microbiol.* **116,** 277 (1980).

TABLE V
PHYSICOCHEMICAL CHARACTERISTICS OF SOLUBLE
METHANE MONOOXYGENASE PROTEINS

Property	Protein A	Protein B	Protein C
Molecular weight[a]	220,000	17,000	38,000
Number of subunits	3	1	1
Subunit molecular weight[b]	α 54,000 (60,630) β 42,000 (44,720) γ 17,000 (19,840)	17,000	38,000 (38,550)
Subunit structure	$\alpha_2\beta_2\gamma_2$	—	—
Prosthetic group	μ-Hydroxy-bridged binuclear iron center (Fe–Fe distance 3.41 Å)	None	One FAD + one Fe_2S_2
Isoelectric point	5.2	4.2	4.8
Electronic absorption spectrum	Major peak 285 nm	Major peak 278 nm, shoulder 287 nm	Major peak 465 nm, shoulder 395 nm
ESR Spectrum g values	Oxidized 4.3, 2.01; half-reduced $g_z = 1.78$, $g_y = 1.88$, $g_x = 1.95$, 4.3, 2.01	Not done	Oxidized none; reduced 2.047, 1.960, 1.864
References	6, 15, 16, *c, d*	9	4, *c*

[a] Calculated from gel filtration data.

[b] Calculated from SDS—gel electrophoresis data. Figures in parentheses represent the values predicted from DNA sequence data.

[c] A. C. Stainthorpe, J. C. Murrell, G. P. S. Salmond, H. Dalton, and U. Lees, *Arch. Microbiol.* **152**, 154 (1989).

[d] M. P. Woodland, D. S. Patil, R. Cammack, and H. Dalton, *Biochim. Biophys. Acta* **873**, 237 (1986).

which then covalently binds to the active site of protein A.[13] Copper is a specific inhibitor of protein C which acts by interacting with the prosthetic groups FAD and Fe_2S_2.[14]

Physicochemical Properties. The physicochemical properties of the three component proteins that form the soluble methane monooxygenase complex are summarized in Table V.[15,16] The prosthetic groups of protein C (FAD/Fe_2S_2) and protein A (μ-hydroxo binuclear iron center) have been thoroughly described in Refs. 4, 5, and 15.

[13] S. D. Prior and H. Dalton, *FEMS Microbiol. Lett.* **29**, 105 (1985).

[14] J. Green, S. D. Prior, and H. Dalton, *Eur. J. Biochem.* **153**, 137 (1985).

[15] A. Ericson, B. Hedman, K. O. Hodgson, J. Green, H. Dalton, J. G. Bentsen, R. Beer, and S. J. Lippard, *J. Am. Chem. Soc.* **110**, 2330 (1988).

[16] J. Green and H. Dalton, *Biochem. J.* **259**, 167 (1989).

Kinetic Properties of Enzyme Complex. Electron transfer through the methane monooxygenase complex from NADH to methane/O_2 has been investigated and rate constants calculated.[16] The following rate constants have been determined: binding of NADH to protein C, 40×10^6 M^{-1} sec^{-1}; reduction of protein C by NADH, 2.9×10^6 M^{-1} sec^{-1}; and intramolecular transfer of electrons from the FAD to the Fe_2S_2 center of protein C, 3.6×10^6 M^{-1} sec^{-1}. The rate of electron transfer from protein C to protein A is altered by the presence of protein B and methane. In the absence of protein B and methane the rate constant is 2.5 sec^{-1}, with the addition of protein B it drops to 0.35 sec^{-1}, and with the addition of both protein B and methane is increased to 5.9 sec^{-1}. With saturating concentrations of methane and O_2 the observed rate constant for the conversion of methane to methanol is 0.26 sec^{-1}. All constants were calculated at 18°.

Stability. Protein A is unstable to successive freezing and thawing in its pure form. This is associated with loss of iron from the protein.[17] The addition of 10 mM sodium dithiothreitol and 0.5 mM FeEDTA to protein A restores iron to the protein and also activity, typically tripling the activity of pure protein A up to a maximum of 1400 nmol/min/mg (propylene oxide produced).[17] Protein B, in crude form, requires the addition of the protease inhibitor phenylmethylsulfonyl fluoride to maintain activity, suggesting that this protein may be particularly susceptible to proteases, a property possibly linked to its role as a regulatory protein. Protein C requires the presence of a thiol protective agent such as sodium thioglycolate throughout purification.

[17] J. Green and H. Dalton, *J. Biol. Chem.* **263**, 17561 (1988).

[31] Methane Monooxygenase from *Methylosinus trichosporium* OB3b

By BRIAN G. FOX, WAYNE A. FROLAND, DAVID R. JOLLIE, and
JOHN D. LIPSCOMB

$$\text{Methane} + \text{NAD(P)H} + \text{H}^+ + \text{O}_2 \rightarrow \text{methanol} + \text{NAD(P)}^+ + \text{H}_2\text{O}$$

Bacterial methane monooxygenase (EC 1.14.13.25) catalyzes the NAD(P)H-dependent activation of O_2 required for the biological oxidation of methane.[1] Higher alkanes, alkenes, and aromatic compounds are also oxidized.[2] The enzyme can be expressed in either a soluble or a membrane-bound form dependent on growth conditions.[3] Growth techniques that reproducibly promote expression of the more readily purified soluble enzyme are described here. Methods for the preparation of the soluble form of the methane monooxygenases from *Methylococcus capsulatus* Bath and *Methylobacterium* CRL-26 have been reported.[4-8] However, these purified enzymes exhibit less than 25% of the specific activity required to account for the *in vivo* growth rate of the organisms. By using stabilizing reagents, an improved purification protocol for the soluble methane monooxygenase from *Methylosinus trichosporium* OB3b has been developed, with the purified enzyme exhibiting the full activity observed *in vivo*. This preparation, which is described here, has been the subject of a previous report.[9] Methane monooxygenase consists of three protein components including an NAD(P)H oxidoreductase, a small protein termed B, and a hydroxylase. All three components are required for efficient substrate turnover coupled to NAD(P)H oxidation.

Assay Methods

Principle. Methane monooxygenase activity can be determined by the polarographic measurement of O_2 utilization or by the gas chromato-

[1] C. Anthony, "The Biochemistry of the Methylotrophs." Academic Press, London, 1982.
[2] I. J. Higgins, D. J. Best, and R. C. Hammond, *Nature (London)* **286**, 561 (1980).
[3] S. H. Stanley, S. D. Prior, D. J. Leak, and H. Dalton, *Biotechnol. Lett.* **5**, 487 (1983).
[4] M. P. Woodland and H. Dalton, *J. Biol. Chem.* **259**, 53 (1984); also see S. J. Pilkington and H. Dalton, this volume [30].
[5] J. Colby and H. Dalton, *Biochem. J.* **177**, 903 (1974).
[6] J. Green and H. Dalton, *J. Biol. Chem.* **260**, 15795 (1985).
[7] R. N. Patel and J. C. Savas, *J. Bacteriol.* **169**, 2313 (1987).
[8] R. N. Patel, *Arch. Biochem. Biophys.* **252**, 229 (1987).
[9] B. G. Fox and J. D. Lipscomb, *Biochem. Biophys. Res. Commun.* **154**, 165 (1988).

graphic detection of hydroxylated products. The components appear to form active complexes during catalysis. Consequently, the relative concentrations of the reductase, component B, and hydroxylase must be adjusted to give maximal activity. Although the relationship between the reductase concentration and the maximal activity of the complete methane monooxygenase system appears to exhibit saturation kinetic behavior, the concentration dependence of component B is more complex. High concentrations of component B have an inhibitory effect on the rate of both O_2 utilization and product formation in the *M. trichosporium* OB3b methane monooxygenase.

Procedure for Optimizing Concentrations of Protein Components

The optimal concentrations of the protein components required for maximal activity must be determined at each step of the purification. These concentrations are determined by varying the concentration of one component relative to fixed concentrations of the other two components. This determination can be most conveniently accomplished by continuous monitoring of the rate of O_2 uptake. For example, in the hydroxylase assay, the hydroxylase fraction and a catalytic amount of the purified component B are placed into the oxygen electrode chamber, and the purified reductase is then added until the maximal rate is observed. In a second assay, the hydroxylase fraction and the previously determined saturating concentration of the purified reductase are added to the oxygen electrode chamber. The purified component B is then added until the maximal rate is observed. After the suitable ratios of the protein preparations required for maximal activity have been determined by the polarographic method, the gas chromatographic measurement of product formation can be made. The specific activity values determined by these two techniques agree within 10% when measured at the same temperature.

Procedure for Polarographic Assays

Reagents

MOPS buffer (Sigma Chemical Co., St. Louis, MO, free acid, pH adjusted with NaOH), 25 mM, pH 7.5, 23°, fully equilibrated with air ($\sim 250 \mu M$ O_2)

Furan (Aldrich Chemical Co., Milwaukee, WI, 99+%): using a gastight syringe, add 150 μl of furan to 10 ml of MOPS buffer contained in a 30-ml vial (Pierce Chemical Co., Rockford, IL) sealed with a Teflon septum

NADH (Sigma, disodium salt, Grade III), 500 mM in 10 mM Tris, pH 8.0; residual ethanol is removed by lyophilization

Assay. The general procedures and instrumentation used for polarographic assays are described in another chapter in this volume.[10] The assay is performed in the following manner: (1) MOPS buffer is placed in the oxygen electrode chamber, and the chamber is closed with a capillary bored glass plug; (2) 100 μl of furan-saturated buffer is added to the oxygen electrode chamber using a gas-tight syringe; (3) reductase, hydroxylase, and component B are added to the oxygen electrode chamber; (4) a stable baseline is established; and (5) the hydroxylation reaction is initiated by the addition of 1 μl of NADH solution.

Definition of Unit of Activity. In the polarographic assay, a unit is defined as the amount of methane monooxygenase component required, in the presence of optimal concentrations of the other methane monooxygenase protein components and saturing concentrations of furan and NADH, to catalyze the consumption of 1 μmol of O_2 per minute at 23° in 25 mM MOPS buffer, pH 7.5, equilibrated with air.

Activity Calculations. The method for calculation of activity is described in another chapter in this volume.[10] Separate assays are performed to correct for the oxygen consumption rate observed in the absence of furan. Specific activity is defined as units per milligram of protein present in the assay. Specific activities are determined under conditions such that the component of interest is present in the assay in limiting amounts. Protein concentrations of fractions obtained during the purification procedures are determined colorimetrically. The concentrations of the purified components are determined colorimetrically using dialyzed and lyophilized standards of the protein components.

Procedure for Gas Chromatographic Assays

Reagents

MOPS buffer, 25 mM, pH 7.5, 23°, fully equilibrated with air (~250 μM O_2)

Propene (Aldrich, 99+%), stored in a 500-ml Erlenmeyer flask sealed by a septum and filled by repeated evacuation and flushing with propene

1,2-Epoxypropane (Aldrich, 99+%, Gold Label)

Chloroform [Fisher Chemical Co., Fairlawn, NJ, HPLC grade, stabilized with pentene (the pentene serves as a convenient standard for gas chromatography)]

NADH

Gas Chromatograph. A Hewlett Packard 5700A gas chromatograph and Hewlett Packard 3380A signal integrator are used with flame-ioniza-

---

[10] J. W. Whittaker, A. M. Orville, and J. D. Lipscomb, this volume [14].

tion detection on a Porapak Q column (6 ft. × ¼ in. I.D., nitrogen carrier gas flow rate 60 ml/min, injector 200°, column 205°, detector 250°).

Calibration of the Gas Chromatograph. Standards are made by successive dilution of 1,2-epoxypropane into chloroform. All standard solutions are made in Teflon-sealed reaction vials (Pierce), and transfers are made using gas-tight syringes. A plot of the concentration of 1,2-epoxypropane versus the ratio of integrated signal areas of 1,2-epoxypropane and the internal standard pentene is used to determine the instrument response. The instrumental conditions reported above for the gas chromatograph give baseline separation of propene, 1,2-epoxypropane, pentene, and chloroform.

Assay. The general procedures and instrumentation used in the gas chromatographic assays have been described elsewhere.[9,11] The assay is performed in the following manner: (1) hydroxylase, component B, and reductase are placed in a 5-ml reaction vial; (2) buffer is added to make a total volume of 500 μl; (3) the reaction vial is sealed using Teflon-backed silicon septa; (4) 3 ml of propene is added to the reaction vial using a syringe; and (5) the reaction is initiated by the addition of 1 μl of NADH solution. Reaction vials are shaken in a reciprocating water bath at 100 rpm and 30°. At appropriate time points, 1,2-epoxypropane is extracted from a 50-μl aliquot of the reaction mixture with an equal volume of chloroform. Typically, 1 μl of the chloroform layer is analyzed for the presence of 1,2-epoxypropane.

Definition of Unit of Activity. In the gas chromatographic assay, a unit is defined as the amount of methane monooxygenase component required, in the presence of optimal concentrations of the other methane monooxygenase components and saturating concentrations of propene and NADH, to catalyze the production of 1 μmol of 1,2-epoxypropane per minute at 30° in 25 mM MOPS buffer, pH 7.5 equilibrated with air.

Calculation of Activity. The concentration of 1,2-epoxypropane produced in the enzymatic reaction is determined from the standardization plots made for calibration of the gas chromatograph. The chloroform extraction procedure removes 80.0% of the 1,2-epoxypropane from the aqueous phase at 23°. The observed concentration of 1,2-epoxypropane is corrected to account for this extraction efficiency. Specific activity is defined as units per milligram of protein present in the assay. Protein concentrations are determined as described above.

---

[11] E. Bayer, "Gas Chromatography." Elsevier, New York, 1961.

Procedure for Assay of Reductase NADH Oxidoreductase Activity

Reagents

MOPS buffer, 50 mM, pH 7.0, 23°

2,6-Dichlorophenolindophenol (Sigma), 0.1 M, prepared fresh each day by dissolving 30 mg of 2,6-dichlorophenolindophenol in 1 ml of MOPS buffer

NADH, 0.2 M solution in 10 mM Tris, pH 8.0

Assay. This assay has been previously described.[12] Column fractions are screened for reductase activity by observing the catalyzed reduction of 2,6-dichlorophenolindophenol in the presence of NADH optically at 600 nm. The assay is performed in a 1.5-ml cuvette containing 10 μl of 2,6-dichlorophenolindophenol solution, 5 μl of NADH solution, 5 to 50 μl of reductase fraction, and buffer to a total volume of 1 ml. The assay is initiated by the addition of the reductase sample.

Definition of Unit of Activity. A unit is defined as the amount of reductase that catalyzes the NADH-dependent reduction of 1 μmol of 2,6-dichlorophenolindophenol per minute at 23° in the standard assay.

Calculation of Activity. The reduction of 2,6-dichlorophenolindophenol is indicated by the decrease in absorption at 600 nm. The observed rate of reduction of 2,6-dichlorophenolindophenol in the presence of the reductase and NADH is corrected by subtraction of the rate of the nonenzymatic reduction of 2,6-dichlorophenolindophenol in the presence of NADH alone. The observed ΔA_{600}/min is converted to $\Delta\mu$mol/min using an ϵ_{600} value of 13 mM^{-1} cm^{-1}. Specific activity is defined as units per milligram of reductase. The reductase concentration is determined as described above.

Growth of *Methylosinus trichosporium* OB3b

Reagents

Commercial grade methane (Air Products, Allentown, PA) is used as the sole carbon source; methane is analyzed by gas chromatography for the inhibitor acetylene by flame-ionization detection on a GS-Q column (Alltech Associates, Chicago, IL, 30 m × 0.53 μm I.D., hydrogen carrier gas flow rate 5.0 ml/min, injector 50°, column 50°, detector 50°)

Deionized and glass-distilled water is used in all purification procedures

---

[12] D. E. Hultquist, this series, Vol. 52, p. 463.

Growth of Organism. Since a mixture of methane in air greater than 5% (v/v) is explosive, extreme care should be taken in the growth of methanotrophic bacteria. The obligate methanotroph *M. trichosporium* OB3b can be obtained from the National Collection of Industrial and Marine Bacteria (Aberdeen, Scotland). The following concentrated salts solutions comprise the mineral salts medium used for culture of the organism: $100 \times$ salts solution (g per liter): $NaNO_3$, 85; K_2SO_4, 17; $MgSO_4 \cdot 7H_2O$, 3.7; $CaCl_2 \cdot 2H_2O$, 0.7; $100 \times$ phosphate solution (g per liter): KH_2PO_4, 53.0; Na_2HPO_4, 86.0; adjust to pH 7.0; $500 \times$ trace metals solution (mg per liter): $ZnSO_4 \cdot 7H_2O$, 288; $MnSO_4 \cdot 7H_2O$, 233; H_3BO_3, 62; $NaMoO_4 \cdot 2H_2O$, 48; $CoCl_2 \cdot 6H_2O$, 48; KI, 83; $CuSO_4 \cdot 5H_2O$, 125; add 1 ml of 0.25 M H_2SO_4 per liter of trace metals solution; $1000 \times$ iron solution: $FeSO_4 \cdot 7H_2O$, 1.12 g per 100 ml; add 5 ml of 0.25 M H_2SO_4 per 100 ml of iron solution. To prepare liquid medium, add 10 ml of $100 \times$ salts solution, 10 ml of $100 \times$ phosphate solution, and 2 ml of $500 \times$ trace metals solution per liter. After sterilization and cooling, add 1 ml of $1000 \times$ iron solution per liter by sterilization through a 0.2-μm filter (Gelman Sciences Inc., #4192, Ann Arbor, MI).

To prepare inocula for large-scale growth, single colonies from culture plates are transferred aseptically into 50 ml of liquid medium contained in each of four 250-ml Erlenmeyer flasks. Flask closures are made from rubber stoppers fitted with a 0.2 μm filter (Gelman #4225) and a three-way Luer-tipped stopcock (American Hospital Supply, McGaw Park, IL). The flasks are filled with a $1:4$ (v/v) mixture of methane in air. The cultures are incubated for a period of 1 to 3 days in a rotary shaker at 150 rpm and 30°. The culture flask headspace is replaced daily with a fresh methane–air mixture. When the optical density is 0.5 or greater at 540 nm, the contents of the four 250-ml Erlenmeyer flasks are transferred aseptically to 500 ml of serile medium contained in each of four 2-liter Fernbock flasks. The 2-liter flasks are filled with the methane–air mixture and incubated in a rotary shaker at 200 rpm. The methane–air mixture is replaced 3 times per day. Growth in the Fernbock flasks should be continued until the optical density at 540 nm is 2.5 or greater.

The cells used for enzyme purifications are grown at 30–34° in a 14-liter fermentor (New Brunswick Scientific, New Brunswick, NJ). The concentration of iron in the fermentor medium is doubled by the addition of 2 ml of the $1000 \times$ iron solution per liter. The fermentor vessel containing 10 liters of sterile medium is inoculated with 2 liters of bacterial culture grown in the Fernbock flasks as described above. The fermentor vessel is sparged at the following rates: methane, 600 to 1500 ml/min, 15 psig; air, 1800 to 4500 ml/min, 15 psig. The stirring and sparging rates are gradually increased from the minimum settings as the optical density of the culture increases. When the optical density at 540 nm is 8.0 or greater, 8 liters of the culture broth is harvested using an HPK 40 Pellicon cassette system equipped with 5 square feet of 0.45-μm HVLP membrane (Millipore Corp., Bedford, MA) and washed with cold 20 mM sodium phosphate buffer, pH 7. The harvested cells are centrifuged at 9000 g for 20 min. The cell paste is stored at $-80°$. Typically, 18 to 25 g of cell paste per liter of culture medium is obtained. The fermentor vessel is then refilled with fresh medium at a flow rate of approximately 600 ml/hr, and the cell harvesting procedure is repeated.

Purification of Methane Monooxygenase

Reagents

Fe(NH$_4$)$_2$(SO$_4$)$_2$·6H$_2$O (Sigma, ACS reagent grade)
Cysteine (Sigma, free base)
Dithiothreitol (Sigma, Sigma grade)
Ultrapure (NH$_4$)$_2$SO$_4$ (Schwarz-Mann Biotech, Cleveland, OH)
Sodium mercaptoacetate (Aldrich, 97%)
Buffer A, 25 mM MOPS, pH 7.0, containing 200 μM Fe(NH$_4$)$_2$(SO$_4$)$_2$ · 6H$_2$O and 2 mM cysteine

All purification procedures are performed at 4°. Buffers containing ferrous iron, cysteine, dithiothreitol, and sodium mercaptoacetate are prepared immediately before use. *It is essential to perform all steps of the purifications as rapidly as possible and to avoid unnecessary freeze–thaw cycles.* The results of the purification procedures for the three components of the soluble methane monooxygenase from *M. trichosporium* OB3b are summarized in Table I.

Step 1: Preparation of Cell-Free Extract. The cell paste (~ 200 g) is placed in a 600-ml stainless steel beaker and suspended in 200 ml of buffer A. The cell suspension is sonicated for 16 min (Branson Sonic Power Model 350, Danbury, CT, $\frac{3}{4}$ inch disruptor horn, 70% of maximum output). During the sonication, the temperature of the cell suspension is maintained at or below 4° by placing the stainless steel beaker in a saturated CaCl$_2$ ice bath. The sonicated suspension is diluted with an additional 200 ml of buffer A and centrifuged at 48,000 g for 90 min. The supernatant is carefully decanted, diluted with an additional 200 ml of buffer A, and adjusted to pH 7.0 by dropwise addition of 0.1 M NaOH.

Step 2: Separation of Methane Monooxygenase Components. The cell-free extract is loaded onto a Fast Flow DEAE-Sepharose CL-6B (Pharmacia LKB Biotechnology Inc., Piscataway, NJ) column (40 × 250 mm)

TABLE I
PURIFICATION OF SOLUBLE METHANE MONOOXYGENASE FROM *Methylosinus trichosporium* OB3b

Step	Volume (ml)	Total protein (mg)	Total activity[a] (mU)	Specific activity (mU/mg)	Yield (%)	Purification (−fold)
Hydroxylase						
Cell-free extract	630	11,150	2,230,000	200	100	1.0
DEAE-Sepharose CL-6B	177	1590	1,427,000	900	64	4.5
Sephacryl S-300	95	835	1,420,000	1700	63	8.5
Component B						
Cell-free extract	630	11,150	2,230,000	200	100	1.0
DEAE-Sepharose CL-6B	169	590	1,410,000	2400	63	12.0
Sephadex G-50	118	153	2,563,000	16,700	115	83.5
DEAE-Sepharose CL-6B	62	110	1,237,000	11,200	55	56.0
Reductase						
Cell-free extract	630	11,150	835,000	75	100	1.0
DEAE-Sepharose CL-6B	160	136	708,000	5200	85	69.3
DEAE-Sepharose CL-6B	68	68	651,000	9600	78	128.0
Ultrogel AcA 54	18	21	550,000	26,100	66	348.0

[a] A unit is defined as the amount of methane monooxygenase component required, in the presence of optimal concentrations of the other methane monooxygenase components and saturating concentrations of propene and NADH, to catalyze the production of 1 μmol of 1,2-epoxypropane per minute at 30° in 25 mM MOPS buffer, pH 7.5, equilibrated with air.

equilibrated in buffer A at a linear flow rate of 40 cm/hr. After loading, the column is washed with an additional 600 ml of buffer A at the same flow rate. All methane monooxygenase components are completely adsorbed under these conditions. Following the wash, a 2 liter gradient of 0.0 to 0.40 M NaCl in buffer A is used to elute the methane monooxygenase components at a linear flow rate of 15 cm/hr. Fractions containing the hydroxylase, component B, and the reductase elute at 0.08, 0.18, and 0.27 M NaCl, respectively. The relative positions for elution of the methane monooxygenase components as well as methanol dehydrogenase, cytochrome c-554, cytochrome c-550, and formate dehydrogenase are shown in Fig. 1. The complete purification of formate dehydrogenase is described elsewhere in this volume.[13]

Step 3: Further Purification of Hydroxylase. Fractions of the hydroxylase obtained from Step 2 exhibiting greater than 20% of the maximal activity are pooled and immediately concentrated by ultrafiltration on a YM100 membrane (Amicon). The concentrated protein is centrifuged at

[13] D. R. Jollie and J. D. Lipscomb, this volume [49].

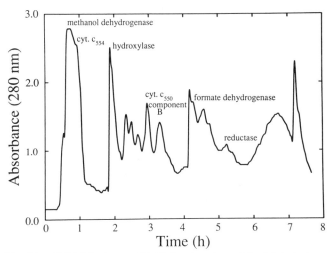

FIG. 1. Fast-Flow DEAE-Sepharose CL-6B fractionation of the *Methylosinus trichosporium* OB3b cell-free extract. The cell-free extract prepared as reported in Step 1 of "Purification of Methane Monooxygenase" was fractionated as described in Step 2.

10,800 *g* for 10 min and applied to a Sephacryl S-300 High Resolution (Pharmacia) column (40 × 500 mm) equilibrated in 25 m*M* MOPS, pH 7.0, containing 100 μM Fe(NH$_4$)$_2$(SO$_4$)$_2 \cdot$6H$_2$O, 2 m*M* cysteine, 1 m*M* dithiothreitol, and 5% (v/v) glycerol at a linear flow rate of 5 cm/hr. Hydroxylase fractions with activity greater than 20% of the maximum activity are pooled and concentrated via ultrafiltration on YM100. The concentrated hydroxylase is frozen in liquid nitrogen and stored at −80°.

Step 4: Further Purification of Component B. Fractions of the component B obtained from Step 2 exhibiting greater than 20% of the maximal activity are pooled, and solid (NH$_4$)$_2$SO$_4$ is added with slow stirring to make a 50% saturated solution at 4°. After 20 min, the solution is centrifuged at 30,000 *g* for 30 min. The pellet is suspended in 20 to 25 ml of 25 m*M* MOPS, pH 7.0, centrifuged at 30,000 *g* for 20 min to remove particulate material, and applied to a Sephadex G-50 (Pharmacia) column (40 × 950 mm) equilibrated in 25 m*M* MOPS, pH 7.0, at a linear flow rate of 3.5 cm/hr. Fractions of component B exhibiting greater than 20% of the maximal activity are pooled. The solution is adjusted to pH 6.5 by dropwise addition of 1 *M* HCl, and solid NaCl is added with slow stirring to make an 80 m*M* NaCl solution. This solution is applied to a Fast Flow DEAE-Sepharose CL-6B column (23 × 80 mm) equilibrated in 25 m*M* MOPS, pH 6.5, containing 80 m*M* NaCl. The component B is eluted with a 680 ml gradient from 80 m*M* to 0.25 *M* NaCl in 25 m*M* MOPS, pH 6.5, at a linear flow rate of 4 cm/hr. Fractions of the component B exhibiting greater than 20% of the maximal activity are pooled, concentrated via ultrafiltration on YM5, frozen in liquid nitrogen, and stored at −80°.

Step 5: Further Purification of Reductase. All buffers used to purify the reductase contain 5 mM sodium mercaptoacetate. Reductase fractions obtained from Step 2 exhibiting greater than 20% of the maximal activity in the 2,6-dichlorophenolindophenol reduction assay are pooled and diluted with an equal volume of 25 mM MOPS, pH 6.5. The protein solution is adjusted to pH 6.5 using a saturated solution of MOPS free acid and applied to a Fast Flow DEAE-Sepharose CL-6B column (23 × 80 mm) equilibrated in 25 mM MOPS, pH 6.5, at a linar flow rate of 30 cm/hr. After washing the column with 100 ml of equilibration buffer containing 0.15 M NaCl, the reductase is eluted in a 500 ml gradient from 0.15 to 0.32 M NaCl in 25 mM MOPS, pH 6.5, at a linear flow rate of 5 cm/hr. Fractions exhibiting an A_{458}/A_{340} ratio greater than 1.05 are pooled and diluted with an equal volume of 25 mM MOPS, pH 7.0. This solution is adjusted to pH 7.0 by dropwise addition of 0.1 M NaOH and applied to a Fast Flow DEAE-Sepharose CL-6B column (10 × 40 mm) equilibrated in 25 mM MOPS, pH 7.0. The reductase is eluted from this column using 25 mM MOPS, pH 7.0, containing 0.4 M NaCl and 20% (v/v) glycerol at a linear flow rate of 2 cm/hr. The concentrated reductase is applied to an Ultrogel AcA 54 (IBF Biotechnics, Savage, MD) column (25 × 450 mm) equilibrated in 25 mM MOPS, pH 7.0, and eluted at a linear flow rate of 3 cm/hr. Fractions exhibiting a constant A_{458}/A_{340} ratio of 1.3 and hydroxylation activity greater than 20% of the peak fraction are pooled, concentrated using the ion-exchange technique described above, frozen in liquid nitrogen, and stored at −80°.

Properties

The soluble methane monooxygenase can be reproducibly expressed in *M. trichosporium* OB3b cultured as described here when the optical density at 540 nm is greater than 8. The methane monooxygenase activity is stable in the cell paste at −80° for up to 1 year. The three components of the soluble methane monooxygenase are subsequently purified in high yield and with high specific activity from these cells. Table II contains a summary of the physical properties for each of the three purified methane monooxygenase components.[14]

Purity and Stability. All three components are pure as judged by denaturing or nondenaturing polyacrylamide gel electrophoresis, ultracentrifu-

---

[14] B. G. Fox, W. A. Froland, J. E. Dege, and J. D. Lipscomb, *J. Biol. Chem.* **264,** 10023 (1989).

TABLE II
PHYSICAL PROPERTIES OF SOLUBLE METHANE MONOOXYGENASE FROM *Methylosinus trichosporium* OB3b

Property	Hydroxylase	Component B	Reductase
Subunit structure (SDS–PAGE)	$(\alpha\beta\gamma)_2$	α_1	α_1
Subunit molecular masses (kDa)	54.4, 43.0, 22.7	—	—
Holoenzyme molecular mass (kDa)			
Sedimentation velocity	245	15.1	38.3
Gel filtration	245	15–31	38.4
Estimated from SDS–PAGE	241	15.8	39.7
Estimated from amino acid analysis	246	15	39.2
V_o estimate from amino acid analysis	0.73	0.73	0.75
$s_{20,w}^o$ (sec $\times 10^{13}$)	10.8	1.6	3.2
$D_{20,w}^o$ (cm^2/sec $\times 10^7$)	4.0	10.9	8.2
Stokes radius (Å)	50.3	19.6	26.2
Fe content (mol/mol)	4.3	None	2
Other metals	None	None	None
Inorganic S content (mol/mol)	None	None	2
FAD (mol/mol)	None	None	1
Specific activity (mU/mg)	1700	11,200	26,100
Percent recovery	63	55	66
Protein (mg obtained/200 g cell paste)	835	110	21

gation, and gel-filtration chromatography. Antibodies raised to each of the purified components do not cross-react with the other components. The purified components can be stored at $-80°$ for longer than 1 year without loss of catalytic activity.

Kinetic Properties. All three protein components are required for efficient substrate turnover activity coupled to NADH oxidation. For the complete methane monooxygenase, the following turnover numbers are observed: methane, 3.7 sec^{-1}; propene, 4.4 sec^{-1}; furan, 7.6 sec^{-1}. The combination of the reductase and hydroxylase yields product in the presence of NADH and O_2 at about one-tenth the rate observed when component B is present. The following values for $K_{m(app)}$ are observed; methane, 25 μM; furan, 35 μM; NADH, 50 μM. During the hydroxylation of propene and furan, optimal activity is observed at pH 7.5.

Cofactor of the Hydroxylase. The most active hydroxylase preparations contain up to about 4 mol of nonheme iron per mol protein. Mössbauer studies of the oxidized hydroxylase show that an even number of high-spin Fe(III) atoms are antiferromagnetically coupled ($\Delta E_Q = 1.07$ mm/sec, $\delta = 0.50$ mm/sec) to an $S = 0$ spin system.[15] No evidence for other types of

[15] B. G. Fox, K. K. Surerus, E. Münck, and J. D. Lipscomb, *J. Biol. Chem.* **263**, 10553 (1988).

catalytic iron sites are provided by Mössbauer studies. Electron paramagnetic resonance (EPR) signals observed at $g_{av} = 1.85$ upon partial reduction and at $g = 16$ upon complete reduction are consistent with the presence of a μ-oxo- or μ-hydroxo-bridged binuclear iron cluster in the hydroxylase.[4,15] The oxidized form of the hydroxylase exhibits an electronic absorption at 282 nm. No distinct optical features are observed above 300 nm. This is the first example of an oxygenase enzyme which appears to utilize an oxo-bridged binuclear iron cluster for oxygen activation chemistry.

Cofactors of Reductase. The oxidized reductase exhibits optical absorption maxima at 270, 340, 396, and 458 nm. Prominent shoulders are observed at 430 and 478 nm. The optical absorption smoothly decreases from 500 to 650 nm. In highly purified reductase preparations, the ratios of A_{270}/A_{458} and A_{458}/A_{340} are 2.8 and 1.3, respectively. The reductase contains 1 mol of acid-dissociable FAD, 2 mol of iron, and 2 mol of inorganic sulfide per mole of protein. The observed ratio of iron to inorganic sulfide and the optical spectral properties are consistent with the presence of a [2Fe–2S] cluster in the reductase.

Acknowledgments

This work was supported by grants from the National Institutes of Health (GM 24689 and GM 40466).

[32] Methanol Dehydrogenase from *Hyphomicrobium* X

By J. Frank and J. A. Duine

$$CH_3OH + dye \rightarrow CH_2O + dye \cdot H_2$$

Methanol dehydrogenase (EC 1.1.99.8) is a pyrroloquinoline quinone (PQQ)-containing oxidoreductase. The enzyme is involved in the oxidation of methanol by gram-negative methylotrophic bacteria.[1] It has a broad substrate specificity, and some facultative methylotrophs possibly use this enzyme for growth on higher alcohols.[2] Purification and (partial) characterization of the enzyme from *Hyphomicrobium* X has been reported previously.[3]

[1] C. Anthony, *Adv. Microb. Physiol.* **27**, 113 (1986).
[2] A. Groeneveld, M. Dijkstra, and J. A. Duine, *FEMS Microbiol. Lett.* **25**, 311 (1984).
[3] J. A. Duine, J. Frank, and J. Westerling, *Biochim. Biophys. Acta* **524**, 277 (1978).

Assay Methods

Principle. Activity measurements are based on the capability of certain cationic dyes to act as artificial electron acceptors for enzyme reduced by methanol in the presence of activator (NH_4Cl) at high pH. Different dyes and different detection methods can be used. In Method 1, activity is related to the amperometrically measured rate of oxygen consumption arising from reoxidation of reduced phenazine metho- or ethosulfate (PMS or PES). In Method 2, reduced PMS or PES is reoxidized by dichlorophenolindophenol (DCPIP), and the rate of reduction of the secondary electron acceptor is measured spectrophotometrically. Method 3 uses Wurster's blue, the perchlorate salt of the cationic free radical of N,N,N',N'-tetramethyl-*p*-phenylenediamine. PMS, PES and DCPIP are commercially available; Wurster's blue is not, but a convenient procedure for its synthesis has been described.[4]

Both PMS and PES are inherently unstable, especially at high pH, leading to aldehyde production owing to dealkylation. These aldehydes are substrates with high affinity for the enzyme so that fresh dye solutions should be used since otherwise the difference between the blank and the measured activity becomes too small. In this respect PES is to be preferred over PMS as it is the more stable of the two,[5] but its K_m value is 2 times higher. For the same reason, care should be exercised in preparing the solutions to avoid contamination with substrate (aldehydes are notorious contaminants in the laboratory atmosphere, and buffer salts are sometimes crystallized from alcoholic solutions). It should also be mentioned that enzyme preparations commonly contain varying amounts of an endogenous substrate of unknown nature.[3]

In principle, contaminating and endogenous substrate could be removed by preincubating the enzyme with electron acceptor, after which adequate measurements with the added substrate are possible. However, most methanol dehydrogenases are very labile under these conditions, that is, rapid inactivation occurs as soon as substrate is depleted.

Cyanide is used in cases where enzyme preparations are still contaminated with components of the respiratory chain, blocking the catalysis of the reoxidation of reduced electron acceptor by oxygen. Cyanide also prevents the aforementioned inactivation of enzyme by electron acceptor and masks the effects of contaminating and endogenous substrate. This property is related to the competitive binding of cyanide to PQQ. Therefore, kinetic parameters should be obtained from measurements in the absence of cyanide.

[4] L. Michaelis and S. Granick, *J. Am. Chem. Soc.* **65**, 1747 (1943).
[5] R. Gosh and J. R. Quayle, *Anal. Biochem.* **99**, 245 (1979).

Reagents

Sodium tetraborate buffer, 100 mM, pH 9.0, containing 2 mM methanol

Potassium cyanide, 10 mM, pH 9.0

Ammonium chloride buffer, 500 mM, pH 9.0

Phenazine methosulfate or phenazine ethosulfate, 10 mM

Dichlorophenolindophenol, 0.5 mM

Wurster's blue, 5.25 mM

Procedure for Method 1. The assay is begun by placing 0.65 ml sodium tetraborate buffer, 0.1 ml ammonium chloride buffer, 0.1 ml potassium cyanide, and 0.1 ml phenazine methosulphate in the thermostatted cell (20°) of a biological oxygen monitor provided with a Clark electrode (Yellow Springs Instruments, Yellow Springs, OH). To this mixture 50 μl of extract containing up to 50 milliunits of enzyme is added, and the rate of oxygen consumption is taken from the linear part of the curve. One unit is the amount of enzyme catalyzing the consumption of 1 μmol of oxygen per minute at 20° (assuming that H_2O_2 is formed and that no catalase is present).

Procedure for Method 2. To a cuvette (10-mm light path) are added the following (in the indicated sequence): 0.55 ml sodium tetraborate buffer, 0.1 ml DCPIP, 0.1 ml potassium cyanide, 0.1 ml ammonium chloride buffer, 50 μl of extract, and 0.1 ml PMS or PES. After mixing the decrease in absorbance is followed at 600 nm, and the rate is calculated using an ϵ_{600} value of 22,000 M^{-1} cm^{-1} for DCPIP.[6] One unit is the amount of enzyme catalyzing the reduction of 1 μmol of DCPIP per minute at 20°.

Procedure for Method 3. To a cuvette (10-mm light path) are added the following (in the indicated sequence): 0.65 ml sodium tetraborate buffer, 0.1 ml potassium cyanide, 0.1 ml ammonium chloride buffer, and 0.1 ml Wurster's blue. After 30–60 sec extract is added (50 μl), and the decrease in absorbance is followed at 640 nm. The rate is calculated using an ϵ_{640} value of 2143 M^{-1} cm^{-1} for Wurster's blue.[7] One unit is the amount of enzyme catalyzing the reduction of 2 μmol of Wurster's blue per minute at 20° (Wurster's blue is a one-electron acceptor).

Purification Procedure

Growth of Organism. Hyphomicrobium X, isolated by Attwood and Harder,[8] is grown at 30° on a mineral medium.[3] The basic medium

[6] J. McD. Armstrong, *Biochim. Biophys. Acta* **86**, 194 (1964).
[7] L. Michaelis, M. P. Schubert, and S. Granick, *J. Am. Chem. Soc.* **61**, 1981 (1939).
[8] M. M. Attwood and W. Harder, *J. Microbiol. Serol.* **38**, 369 (1972).

contains, per liter, 2.28 g $K_2HPO_4 \cdot 3H_2O$, 1.38 g $NaH_2PO_4 \cdot H_2O$, 0.5 g $(NH_4)_2SO_4$, and 0.2 g $MgSO_4 \cdot 7H_2O$ and is brought to pH 7.0 with 2 M NaOH. After autoclaving, the following solutions are added to 1 liter of this medium: 1 ml of spore solution (sterilized by autoclaving), containing 7.8 mg $CuSO_4 \cdot 5H_2O$, 10 mg H_3BO_3, 10 mg $MnSO_4 \cdot 4H_2O$, 70 mg $ZnSO_4$, and 10 mg MoO_3 per liter, 1 ml of calcium solution (sterilized by autoclaving) containing 25 mg $CaCl_2 \cdot 2H_2O$ per liter, 1 ml of iron solution (sterilized by filtration) containing 3.5 g $FeCl_3 \cdot 6H_2O$ and 24.2 g tricine per liter, and 5 ml of methanol (sterilized by filtration). Growth in a fermentor (20 liter) is performed with an aeration of 7 liter/min and a stirrer speed of 200 rpm. The pH is maintained at 7.0 ± 0.2 with 25% (v/v) ammonia solution. A second addition of methanol is made when the concentration of methanol falls below 0.1% (v/v) (as judged from gas chromatographic analysis[3] of the culture supernatant). The cells are harvested in the stationary phase by centrifugation at 48,000 g. Alternatively, growth is stopped by adjusting the pH of the culture to 4.0 (with concentrated HCl). The cells are allowed to settle by placing the fermentor in the cold room (4°) for 3 hr. After decanting most of the clear supernatant, the cells are collected from the remaining suspension by centrifugation at 5000 g for 15 min. The cell paste is stored at $-20°$.

All operations are carried out at room temperature, except dialysis and centrifugation which are performed at 4°.

Step 1: Preparation of Cell-Free Extract. Cell paste, 1000 g wet weight, is suspended in 1400 ml of 0.1 M Tris-HCl, 10 mM EDTA, pH 7.0. The suspension is mixed with 600 mg of lysozyme, allowed to stand for 10 min, and centrifuged at 27,300 g. The pellet is suspended in 1200 ml of 0.1 M Tris-HCl containing 0.5% Triton X-100, pH 9.0, and, after standing at room temperature for 15 min, centrifuged as above. The pellet is reextracted in the same way and the supernatants are combined to give the cell-free extract.

Step 2: Ammonium Sulfate Fractionation. Protein in the cell-free extract precipitating between 2.1 and 4.45 M $(NH_4)_2SO_4$ (at 4°) is collected by centrifugation at 13,000 g for 20 min; the pellet suspended in 200 ml of 20 mM potassium phosphate buffer, pH 7.0, and dialyzed overnight against 5 liters of the same buffer.

Step 3: Silica Gel Treatment. The dialyzed enzyme preparation from Step 2 is centrifuged at 48,900 g for 15 min and passed through a column of silica gel (4×20 cm). The silica gel (Merck, type Si 60, 0.063–0.2 mm) is pretreated by heating at 550° for 1 hr. Just before use, the material is suspended in 20 mM potassium phosphate buffer, pH 7.0, and the mixture poured into the column. The cytochromes c are retained by the column while methanol dehydrogenase passes through and is washed from the column with 20 mM potassium phosphate buffer, pH 7.0. [Cytochrome c_L and cytochrome c_H can be eluted with 20 mM potassium phosphate buffer, pH 7.0, containing 10% (w/v) polyethylene glycol 6000.]

Step 4: Treatment with DEAE-Sepharose. After centrifugation of the pooled active fractions of Step 3 at 48,900 g for 10 min, the solution is applied to a DEAE-Sepharose (Fast Flow) column (4 × 15 cm), equilibrated with 20 mM potassium phosphate buffer, pH 7.0. The column is washed with the same buffer until all activity is eluted.

Step 5: Hydroxyapatite Chromatography. The pooled fractions from Step 4 are applied to a hydroxyapatite column (4 × 10 cm), equilibrated with 20 mM potassium phosphate buffer, pH 7.0. After washing with 4 bed volumes of 20 mM potassium phosphate buffer, pH 7.0 methanol dehydrogenase is eluted with 0.2 M potassium phosphate buffer, pH 7.0 (greenish-yellow fractions).

Step 6: Phenyl-Sepharose Chromatography. Although the preparation obtained after Step 5 is usually pure, sometimes fluorescent contaminants are detected which can only be removed by including a chromatography step on Phenyl-Sepharose. For this, 200 mg of solid $(NH_4)_2SO_4$ is added to 1 ml of the enzyme preparation of Step 5, and this solution is applied to a column of Phenyl-Sepharose (1 × 3 cm), equlibrated with 20 mM potassium phosphate buffer, pH 7.0, containing 1.5 M $(NH_4)_2SO_4$. After washing the column with 3 bed volumes of the same buffer, containing 1.0 M $(NH_4)_2SO_4$, methanol dehydrogenase is eluted with 10 mM sodium phosphate buffer, pH 7.0.

The final methanol dehydrogenase preparation is sterilized by filtration (Acrodisc, Gelman, Ann Arbor, MI, 0.22 μm) and stored at +4°. The result of a typical purification is summarized in Table I.

Properties

Purity. Polyacrylamide gel electrophoresis of the native enzyme shows a major and two minor bands which all display enzyme activity.

Substrate Specificity. The enzyme oxidizes a large range of primary alcohols, as well as formaldehyde and acetaldehyde.[9]

Primary Electron Acceptors. Besides the phenazine derivatives and Wurster's blue, the free radical of 2,2'-azino-di-(3-ethylbenzthiazoline-6-sulfonic acid) also accepts electrons from methanol dehydrogenase. Typically negatively charged electron acceptors (ferricyanide, DCPIP) are not effective.

Activators. Activity depends on the presence of ammonium chloride. It

---

[9] J. A. Duine and J. Frank, *Biochem. J.* **187**, 213 (1980).

TABLE I
PURIFICATION OF METHANOL DEHYDROGENASE FROM *Hyphomicrobium* X

Fraction	Total volume (ml)	Total protein (mg)	Total activity[a] (units)	Specific activity (units/mg protein)	Yield (%)
Step 1. Cell-free extract	2353	24,941	7427	0.3	100
Step 2. (NH₄)₂SO₄ fraction (2.1–4.45 M)	327	6397	5073	0.8	68
Step 3. Silica gel percolate	398	5764	5073	0.9	68
Step 4. DEAE-Sepharose percolate	485	1411	3750	2.7	50
Step 5. Hydroxyapatite eluate	220	985	3088	3.1	42
Step 6. Phenyl-Sepharose eluate	294	926	2941	3.2	40

[a] Results obtained with Assay Method 3.

can be replaced by benzyl and ethyl esters of glycine, phenylpropylamine, and phenylbutyramine, substituted benzylamines, and 2-bromoethylamine.[10] Methylamine is a poor activator.

Inhibitors. Cyanide (k_i 1 mM) and hydroxylamine (k_i 12 μM) are competitive inhibitors with respect to methanol.[9] Cyclopropanol and cyclopropanone hydrate irreversibly and stoichiometrically (with respect to PQQ) inhibit the enzyme.[11] Irreversible inhibition has also been observed with methylhydrazine.[10]

Absorption Coefficients. Using the method of van Iersel *et al.,*[12] an $A_{280}^{0.1\%}$ of 2.04 has been determined. Molar absorption coefficients can be derived from this value for the three different redox forms of the enzyme: $\epsilon_{390} = 20,400\ M^{-1}\ cm^{-1}$ for the oxidized enzyme–HCN complex (MDH$_{ox}$·HCN), $\epsilon_{343} = 38,000\ M^{-1}\ cm^{-1}$ for the reduced enzyme (MDH$_{red}$), and $\epsilon_{348} = 29,000\ M^{-1}\ cm^{-1}$ is for the semiquinone form (MDH$_{sem}$).

Redox Forms. The purification scheme as presented here yields the MDH$_{sem}$ form as judged from the absorption spectrum (Fig. 1) and the free radical present in the preparation (Fig. 2). This form of the enzyme does not react with the alcohol substrate. It can be reduced photochemically with EDTA and deazalumiflavine[13] or with the methyl viologen cation radical to MDH$_{red}$[14] (Fig. 1). Oxidation with artificial electron acceptors

[10] J. A. Duine and J. Frank, unpublished results (1980).
[11] M. Dijkstra, J. Frank, J. A. Jongejan, and J. A. Duine, *Eur. J. Biochem.* **140,** 369 (1984).
[12] J. van Iersel, J. Frank, and J. A. Duine, *Anal. Biochem.* **151,** 196 (1985).
[13] J. Frank, M. Dijkstra, J. A. Duine, and C. Balny, *Eur. J. Biochem.* **174,** 331 (1988).
[14] R. de Beer, J. A. Duine, J. Frank, and J. Westerling, *Eur. J. Biochem.* **130,** 105 (1983).

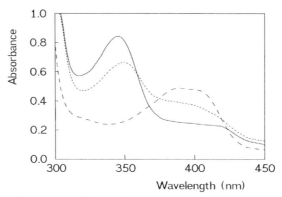

FIG. 1. Absorption spectra of the different redox forms of methanol dehydrogenase. MDH_{red} (——), MDH_{sem} (---), and $MDH_{ox} \cdot HCN$ (– – –), all at a concentration of 22 μM in 20 mM potassium phosphate buffer, pH 7.0.

results in the formation of MDH_{ox}which is only stable in the presence of a competitive inhibitor such as cyanide or hydroxylamine,[9] owing to adduct formation of these compounds with the cofactor. Upon oxidation in the presence of substrate (but without activator) a transient oxidized enzyme–substrate complex ($MDH_{ox} \cdot S$) can be observed.[13]

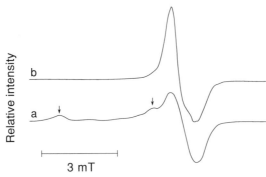

FIG. 2. X-band electron spin resonance spectra of methanol dehydrogenase. The spectrum is shown of MDH_{sem} in air-saturated buffer after cooling to liquid nitrogen temperature (a). The signals in spectrum a, indicated with arrows, do not appear when the solution is flushed with nitrogen prior to cooling (b).

Apoenzyme. In all attempts to reconstitute PQQ-depleted enzyme to holoenzyme, activity has not been observed.[10,15]

Molecular Weight, Subunit Structure, and Cofactor Content. A molecular weight of 120,000 is found upon gel-permeation chromatography on Sephadex G-200.[9] Sodium dodecyl sulfate–polyacrylamide gel electrophoresis shows two bands with molecular weights of 60,000 and 8,000,[10] indicating that the enzyme consists of four subunits. Two molecules of PQQ per enzyme molecule[13] can be extracted[16] from the enzyme.

Kinetic Parameters. With Wurster's blue as the electron acceptor (50 mM NH$_4$Cl as activator, Method 3) an apparent K_m of 0.3 mM is found for methanol. Methanol oxidation in the ES complex can be described by a single exponential and is fully rate limiting (k 0.06 sec^{-1}) in the absence of activator and only partly rate limiting (k 23 sec^{-1}) in its presence.[13] Oxidation of MDH$_{red}$ and MDH$_{sem}$ by Wurster's blue is rapid at pH 9.0 (k 2.22 × 10^5 and 0.75 × 10^5 M^{-1} sec^{-1}, respectively) and slow at pH 7.0 (k 3291 and 2280 M^{-1} sec^{-1}).[13]

Natural Electron Acceptor and Activator. In contrast to artificial electron acceptors, the natural one, cytochrome c_L, very actively oxidizes MDH$_{red}$ and MDH$_{sem}$ at pH 7.0 (k 1.9 × 10^5 and 2.1 × 10^5 M^{-1} sec^{-1}, respectively[17]), but not at pH 9.0. Autoreduction[18] of this cytochrome does not play a role in this process[19] (autoreduction is negligible when the cytochrome is purified on Phenyl-Sepharose[17]).

An oxygen-sensitive compound has been detected which is most probably the natural activator as it activates the substrate oxidation step at pH 7.0[20] (NH$_4$Cl has a very low activity at this pH).

Stability and pH Optimum. Storage of the enzyme preparations in frozen form induces changes in the absorption spectrum,[14] and significant loss of activity occurs. This might be related to the interaction of O$_2$ with PQQH$^{\cdot}$ at low temperatures as can be observed in the electron spin resonance (ESR) spectrum[14] (Fig. 2). For these reasons, sterile storage at 4° is recommended (loss of activity is not more than 30% per year in 0.2 M potassium phosphate buffer, pH 7.0).

[15] V. L. Davidson, J. W. Neher, and G. Cecchini, *J. Biol. Chem.* **17**, 9642 (1985).
[16] R. A. van der Meer, B. W. Groen, M. A. G. van Kleef, J. Frank, J. A. Jongejan, and J. A. Duine, this volume [41].
[17] M. Dijkstra, J. Frank, and J. A. Duine, *Biochem. J.* **257**, 87 (1989).
[18] D. T. O'Keefe and C. Anthony, *Biochem. J.* **190**, 481 (1980).
[19] M. Dijkstra, J. Frank, J. E. van Wielink, and J. A. Duine, *Biochem. J.* **251**, 467 (1988).
[20] M. Dijkstra, J. Frank, and J. A. Duine, *FEBS Lett.* **227**, 198 (1988).

[33] Methanol Dehydrogenase from *Methylobacterium extorquens* AM1

By DARREN J. DAY and CHRISTOPHER ANTHONY

Methanol dehydrogenase (MDH) (EC 1.1.99.8) is the periplasmic quinoprotein which catalyzes the oxidation of methanol to formaldehyde in all gram-negative bacteria during growth on methane or methanol.[1-3] Its immediate electron acceptor is cytochrome c_L, which is oxidized by way of a conventional small cytochrome c_L and a membrane oxidase. Although the methods presented here are for *Methylobacterium extorquens* AM1, most are suitable for MDHs from other gram-negative bacteria. *Methylobacterium extorquens* AM1 is a pink facultative methylotroph that grows on methanol and methylamine and multicarbon compounds but not on methane. It has previously been called *Pseudomonas* AM1 and *Methylobacterium* AM1.

Assay of Methanol Dehydrogenase

For most work with MDH a dye-linked assay is used with phenazine ethosulfate (PES) as primary electron acceptor; reduction of this is monitored by coupling to oxygen (measured in an oxygen electrode) or by coupling to a second dye, 2,6-dichlorophenolindophenol (DCPIP) (monitored spectrophotometrically). The oxygen electrode assay is best when longer periods of reaction are required as in studies of substrate specificity. The spectrophotometric assay is best when many assays are necessary as in enzyme purification; it can also be modified for very rapid analysis of column fractions.

An important and often confusing characteristic of most MDHs is their activity in the absence of added substrate. This activity requires the presence of activator (ammonia), it is not markedly diminished by dialysis or gel filtration, and its initial rate is the same as that with added substrate. It is not subtracted from rates measured in the presence of added substrate.

[1] C. Anthony, "The Biochemistry of Methylotrophs." Academic Press, New York, 1982.
[2] C. Anthony, *Adv. Microb. Physiol.* **27**, 113 (1986).
[3] C. Anthony, *in* "Bacterial Energy Transduction" (C. Anthony, ed.), p. 293. Academic Press, London, 1988.

Assay with Oxygen Electrode

Reagents

Tris-HCl buffer, 0.6 M, pH 9.0
Ammonium chloride, 0.45 M
Phenazine ethosulfate, 10 mM
Methanol, 0.2 M

Procedure. The reaction mixture contains, in an oxygen electrode vessel at 30°, 0.5 ml Tris buffer, 0.1 ml ammonium chloride, 0.1 ml phenazine ethosulfate, 0.1 ml methanol, and 60 units of MDH. The final volume is made to 2 ml with water. The reaction is started by addition of phenazine ethosulfate, and oxygen consumption is measured assuming that 450 nmol oxygen is dissolved in 2 ml of the buffer at 30°.

Units. One unit is defined as that amount which facilitates consumption of 1 μmol O_2 per minute. It should be noted that most preparations of MDH show endogenous dye reduction. This may be diminished, but not eradicated, by dialysis or gel filtration. The initial rate of this reaction is usually the same as that measured in the presence of added methanol, and activator (ammonia) is required. This endogenous dye reduction should be ignored (not subtracted) when calculating rates of enzyme reaction.

Spectrophotometric Assay

Reagents

Tris-HCl buffer, 0.6 M, pH 9.0
Ammonium chloride, 0.45 M
Phenazine ethosulfate, 33 mM
2,6-Dichlorophenolindophenol, 2.6 mM
Methanol, 0.3 M

Procedure. The reaction mixture contains, in a spectrophotometer cuvette at 30°, 0.5 ml Tris buffer, 0.1 ml ammonium chloride, 0.1 ml DCPIP, 0.1 ml methanol, and MDH. The final volume is made to 3 ml with water. The reaction is started by addition of 0.1 ml phenazine ethosulfate, and the decrease in absorption at 600 nm is measured. The reference cuvette contains water.

Units. One unit is defined as that amount which catalyzes the reduction of 1 μmol O_2 per minute between 15 and 30 sec after starting the reaction. It should be noted that most preparations of MDH show endogenous dye reduction. This may be diminished, but not eradicated, by dialysis or gel filtration. The initial rate of this reaction is usually the same as that measured in the presence of added methanol, and activator (ammonia) is required. This endogenous dye reduction should be ignored (not subtracted) when calculating rates of enzyme reaction.

Microtiter Assay

The microtiter assay is convenient for rapid location of MDH in fractions during purification. The reagents are as listed above. They are mixed as follows: 0.5 ml Tris buffer, 0.1 ml methanol, 0.1 ml ammonium chloride, 0.1 ml phenazine ethosulfate, 0.2 ml DCPIP, and 2 ml water. The mixture is stored in a lightproof container. Small volumes (5–50 μl) are placed in wells of a microtiter plate, and the reaction is started by addition of 200 μl of the assay mixture. When MDH is present the color in the wells changes from blue-green to yellow, the time taken (1–15 min) giving an indication of the activity of the MDH present.

Methanol:Cytochrome c Oxidoreductase Activity

Assay of methanol:cytochrome c oxidoreductase activity is the natural assay for MDH, although the rates determined are usually low.[3-5] Horse heart cytochrome c is able to react with the specific electron acceptor for MDH, cytochrome c_L, but is unable to react directly with MDH. The MDH may thus be assayed by coupling electron transfer from methanol, through MDH, to cytochrome c_L and thence to an excess of horse heart cytochrome c. Similar rates are obtained in such an assay if a large amount of cytochrome c_L is the sole cytochrome present, but more cytochrome c_L is then required so that it can be detected by the spectrophotometer. The pH optimum is usually pH 7.0, but activity should be tested at other pH values. Ammonia is not usually necessary, but it may activate some preparations and so is included in the assay mixture. This assay is not suitable for use with crude extracts which may contain compounds or enzymes able to oxidize or reduce cytochrome c. It should be noted that increases in ionic strength markedly decrease the rate of reaction.

Stopped-flow kinetic measurements of the separate components of the MDH/cytochrome c_L reaction have been described elsewhere.[5]

Reagents

MOPS buffer, 120 mM, pH 7.0
Methanol, 50 mM
Ammonium chloride, 50 mM
Cytochrome c (equine or bovine), 0.25 mM
Cytochrome c_L, 50 μM

[4] M. Beardmore-Gray, D. T. O'Keeffe, and C. Anthony, *J. Gen. Microbiol.* **132,** 1553 (1983).
[5] M. Dijkstra, J. Frank, and J. A. Duine, *Biochem. J.* **257,** 87 (1989).

The cytochrome is oxidized before the reaction by addition of a grain of potassium ferricyanide, the excess reagent being removed by gel filtration on a Pharmacia (Sweden) PD10 column.

Procedure. The reaction mixture contains, in a 1-ml reaction mixture in a spectrophotometer cuvette (1-cm light path), 0.1 ml MOPS buffer, 0.1 ml ammonium chloride, 0.1 ml methanol, and 0.2 ml of cytochrome c and cytochrome c_L. The reaction is started by addition of MDH solution in 20 mM MOPS buffer, pH 7.0, and the rate of reduction of cytochrome c is recorded at 550 nm. The rate of reaction is constant for at least 70% of the reaction. Any reaction occurring in the absence of methanol is not subtracted from the rate measured in its presence.

Units. One unit is defined as that amount of MDH catalyzing the reduction of 1 nmol of cytochrome c per minute. This is calculated from the extinction coefficient for the reduced minus oxidized difference spectrum for cytochrome c (19 mM^{-1} cm^{-1}). Thus, one unit will increase the measured absorption by 0.019/min. An increase in absorption of 1.0/min is due to 52.6 units of enzyme.

Purification Procedure

Growth of Organism. Methylobacterium extorquens AM1 (NCIB 9133) is maintained on nutrient agar plus 0.4% (w/v) methylamine hydrochloride or in glycerol as described elsewhere in this volume.[6] Bacteria are grown aerobically at 30° on 0.5% (v/v) methanol in batch culture in mineral salts medium containing the following (g/liter): $NaH_2PO_4 \cdot 2H_2O$, 0.57; K_2HPO_4, 1.53; NH_4Cl, 3.0; $MgSO_4 \cdot 7H_2O$, 0.2; and 5 ml/liter of trace elements solution containing (mg/liter) $CaCl_2 \cdot 2H_2O$, 530; $FeSO_4 \cdot 7H_2O$, 200; $MnSO_4 \cdot 4H_2O$, 20; $ZnSO_4 \cdot 7H_2O$, 20; $CuSO_4 \cdot 5H_2O$, 4; $CoCl_2 \cdot 6H_2O$, 4; Na_2MoO_4, 4; H_3BO_3, 3; and 1 M HCl, 1 ml.

Bacteria are harvested by centrifugation, washed in 20 mM Tris-HCl buffer, pH 8.0, and resuspended in the same buffer to a density of about 3 g wet weight/10 ml. They are broken by sonication in batches of 40 ml in 20 cycles of 30 sec followed by 30 sec of cooling. Whole cells and membranes are removed by centrifugation at 50,000 g for 1 hr at 4°. For breakage of cells the French press is equally suitable, and, for some bacteria, periplasmic material is readily obtained and makes a better starting point.[6] The MDH of some bacteria is more labile than that from *M. extorquens;* such enzymes may be stabilized during harvesting and purification by inclusion of 25 mM methanol or KCN.[4,6] In some bacteria a significant proportion of MDH may be associated with membrane fractions; there is no evidence

---

[6] A. R. Long and C. Anthony, this volume [34].

that this is more than an artifact of preparation, and this enzyme is usually readily released by washing with 0.5 M KCl; the salt must then be removed by dialysis prior to ion-exchange chromatography. Alternatively, detergent may be used to release the enzyme prior to purification.[7]

Purification. The high isoelectric point of MDH allows removal of unwanted protein by passage through ion-exchange media; cytochrome c_L binds to the anion-exchange material, and cytochrome c_H passes through together with the MDH. This purification procedure can therefore also be used for the first part of the purification of the soluble cytochromes of this organism.[8] The soluble fraction from about 15 g wet weight of bacteria is passed down a column (24 × 130 mm) of DEAE-Sepharose (Fast Flow, Pharmacia) equilibrated in 20 mM Tris-HCl, pH 8.0. MDH and cytochrome c_H are not adsorbed; they are dialyzed against 20 liters of 10 mM potassium phosphate buffer (pH 7.0) and applied to a hydroxyapatite column (16 × 100 mm) equilibrated in the same buffer. MDH is eluted (at about 90 mM phosphate) with a linear gradient of 10–250 mM potassium phosphate (pH 7.0) (in 200 ml). Active fractions are pooled, concentrated on a 50-kDa cutoff membrane in an Amicon (Danvers, MA) concentrating cell, desalted by gel filtration on a Pharmacia PH10 column equilibrated in 25 mM MES buffer (pH 5.5), and purified by cation-exchange chromatography on a Pharmacia S-Sepharose (Fast Flow) column (24 × 85 mm) equilibrated in the same buffer. MDH is eluted (at about 100 mM NaCl) using a gradient of 0–250 mM NaCl (200 ml total volume). This produces pure methanol dehydrogenase. For smaller amounts, the final step can be replaced by cation-exchange chromatography on a 1-ml Pharmacia Mono S column equilibrated in 20 mM MES buffer, pH 5.5, MDH being eluted with the same buffer containing 100 mM NaCl. Active fractions are stored at −20°. See Table I for a summary of purification.

The procedure described above is likely to be suitable for most other methylotrophic bacteria. Many equally suitable alternative procedures may be followed, the method of choice depending on the purpose of the work and the extent to which other proteins from the same extract are required (e.g., c-type cytochromes[8] or modifier protein[6]). Some procedures depend on the stability of MDH at pH 4.0; this provides a method for rapid removal of nucleic acids and unwanted protein[9,10] (not suitable if modifier protein is also required). MDH is very soluble and requires about 65% saturated ammonium sulfate before any precipitation occurs; this is proba-

[7] S. Ford, M. D. Page, and C. Anthony, *J. Gen. Microbiol.* **131**, 2173 (1985).

[8] D. Day and C. Anthony, this volume [44].

[9] D. T. O'Keeffe and C. Anthony, *Biochem. J.* **192**, 411 (1980).

[10] C. Anthony, this series, Vol. 18B, p. 818.

TABLE I

PURIFICATION OF METHANOL DEHYDROGENASE[a]

Purification stage	Volume (ml)	Total units	Specific activity	Purification factor	Yield (%)
Cell-free extract	75	25.0	34	1	100
DEAE-Sepharose	110	24.2	110	3.2	96
Hydroxyapatite	21.25	10.6	303	8.8	42
S-Sepharose	6.25	8.7	364	10.6	34

[a] Units are as defined for the spectrophotometric assay, and specific activity is in milliunits per milligrams of protein as assayed by the bicinchoninic acid assay [P. K. Smith, R.I. Krohn, G. T. Hermanson, A. K. Mallia, F. H. Gartner, M. D. Provenzano, E. K. Fujimoto, N. M. Goeke, B. J. Olson, and D. C. Klenk, *Anal. Biochem.* **150**, 76 (1985)].

bly best avoided if work on the activation of MDH by ammonia is envisaged. Gel filtration is suitable for removal of smaller proteins (e.g., the *c*-type cytochromes).

Preparation of Separate Subunits.[11] The subunits are difficult to dissociate; conditions leading to dissociation also dissociate the prosthetic group (PQQ) from the enzyme, and the process is irreversible. The following procedure separates the subunits and the PQQ from each other. Pure MDH is concentrated to about 200 μl containing about 4 mg protein. This is incubated at 80° for 20 min in 2% sodium dodecyl sulfate (SDS) in 50 mM Tris-HCl (pH 8.0) followed by gel filtration on a Pharmacia Superose 12 column equilibrated in 50 mM Tris-HCl (pH 8.0) containing 0.1 M NaCl and 0.2% SDS. An alternative procedure is to replace SDS with 6 M guanidium chloride (omitting the NaCl).

Purification of MDH from Other Methylotrophs. Methods are available for many different methylotrophic bacteria,[1] the following being especially important examples: *Methylophilus methylotrophus,*[4,12] *Hyphomicrobium* X,[5,13] *Paracoccus denitrificans,*[4,6] *Acetobacter methanolicus,*[14] and *Methylophilus glucoseoxidans.*[15]

Properties

Purity and Absorption Spectrum. The enzyme prepared as described here is pure. The faint band seen on SDS–polyacrylamide gel electrophoresis in addition to the main band is the β subunit. The absorption due to

[11] D. N. Nunn, D. Day, and C. Anthony, *Biochem. J.* **260**, 857 (1989).
[12] M. D. Page and C. Anthony, *J. Gen. Microbiol.* **132**, 1553 (1986).
[13] M. Dijkstra, J. Frank, J. E. van Wielink, and J. A. Duine, *Biochem. J.* **251**, 467 (1988).
[14] E. J. Elliott and C. Anthony, *J. Gen. Microbiol.* **134**, 369 (1988).
[15] A. P. Sokolov, N. I. Govorukhina, and Yu. A. Trotsenko, *Biokhimiya* **54**, 811 (1989).

the protein has a peak at 280 nm and a shoulder at 290 nm; that due to the prosthetic group has a peak at about 345 nm and a shoulder at about 400 nm. The green-fluorescent prosthetic group is released from the enzyme by boiling or by lowering the pH to a little below 4.

Substrate Specificity. Methanol dehydrogenase oxidizes a wide range of primary alcohols and also formaldehyde which exists as a *gem*-diol in solution.

Molecular Weight and Subunit Composition. The enzyme is a tetramer, consisting of two α subunits of about 60 kDa and two β subunits of 8.46 kDa. The primary sequence of the smaller subunit has been published as well as the N-terminal region of the α subunit.[11] Antibody raised to the whole enzyme reacts only very weakly with the small subunit.

pH Optimum, Activators, and Electron Acceptors. The pH optimum is about 9.0 when measured with artificial electron acceptors which include phenazine methosulfate, phenazine ethosulfate, or Wurster's blue. Ammonium chloride (or methylammonium chloride) is usually essential for activity with these acceptors. The physiological electron acceptor is cytochrome c_L, but the rate is usually much lower than when measured with artificial electron acceptors. With this cytochrome the pH optimum is usually 7.0 and activator is not usually required.

[34] Modifier Protein for Methanol Dehydrogenase of Methylotrophs

By Antony R. Long and Christopher Anthony

Modifier protein (M protein) modifies the substrate specificity of methanol dehydrogenase (MDH). This dehydrogenase has a high affinity for a wide range of primary alcohols, which are oxidized at rates very similar to those measured with methanol.[1] It is also able to catalyze the rapid oxidation of formaldehyde to formate; the function of the M protein is probably to prevent this oxidation by decreasing the affinity of MDH for formaldehyde.[2] Some substrates, including 1,2-propanediol, 1,3-propanediol, and 1,3-butanediol, have much lower affinities for MDH, and the M protein was first discovered because of its ability to facilitate the oxidation by pure MDH of 1,2-propanediol by increasing its affinity for the dehydrogenase.[3,4]

[1] C. Anthony, *Adv. Microb. Physiol.* **27,** 113 (1986).
[2] M. D. Page and C. Anthony, *J. Gen. Microbiol.* **132,** 1553 (1986).
[3] J. A. Bolbot and C. Anthony, *J. Gen. Microbiol.* **120,** 245 (1980).
[4] S. Ford, M. D. Page, and C. Anthony, *J. Gen. Microbiol.* **131,** 2173 (1985).

The assay systems for M protein thus depend on its ability to facilitate oxidation of a "poor" substrate or to prevent oxidation of formaldehyde.

Assay Methods

Principle. The main assay determines the effect of M protein on the oxidation of 1,3-butanediol by MDH in the dye-linked MDH assay system using phenazine ethosulfate in an oxygen electrode.

Reagents

Tris-HCl buffer, 0.5 M, pH 9.0
Ammonium chloride, 0.5 M
Phenazine ethosulfate, 10 mM
1,3-Butanediol, 0.2 M
Methanol dehydrogenase, 60 units

Procedure. The reaction mixture contains, in an oxygen electrode vessel at 30°, 0.5 ml Tris buffer, 0.1 ml ammonium chloride, 0.1 ml phenazine ethosulfate, 0.1 ml of 1,3-butanediol, and 60 units of MDH. The final volume is made to 2 ml with water. The reaction is started by addition of phenazine ethosulfate, and oxygen consumption is measured assuming that 450 nmol oxygen is dissolved in 2 ml of the buffer at 30°. To confirm that M protein is present in fractions during purification it is necessary to confirm that the oxidation of formaldehyde is diminished by addition of putative M protein. This is done by repeating the assay described here with 0.1 ml of 0.2 M formaldehyde. An amount of M protein that gives a good rate with butanediol will give about 50% inhibition of the rate of formaldehyde oxidation. It is essential to use this method of confirmation because some bacteria (e.g., *Paracoccus denitrificans*) contain other compounds that have been found to stimulate oxidation of 1,3-butanediol in crude extracts.

Units. One unit is defined as that amount which facilitates consumption of 1 nmol O_2 per minute. The unit for methanol dehydrogenase is the same when it is measured in this assay system with 0.1 ml of 0.2 M methanol instead of 1,3-butanediol. It should be noted that most preparations of MDH show endogenous dye reduction. This may be diminished, but not eradicated, by dialysis or gel filtration. The initial rate of this reaction is usually the same as that measured in the presence of added methanol, and activator (ammonia) is required. This endogenous dye reduction should be ignored (not subtracted) when calculating rates of enzyme reaction. It should also be noted that there is usually a rapid, transient oxygen consumption recorded with 1,3-butanediol and other low-affinity substrates. This is ignored when recording stimulations in activity on addition of M protein; the subsequent continuous (linear) oxygen consumption is used for all calculations.

Purification Procedures

Two procedures are given; one is for *Methylophilus methylotrophus,* an obligate methylotroph in which the properties of M protein were first described, and the other is for *Paracoccus denitrificans,* a facultative methylotroph (and autotroph) which has the merit of ease of formation of periplasmic fractions.

Growth of Methylophilus methylotrophus (NCIB 10515). The growth conditions do not markedly affect production of M protein. Stock cultures are maintained in 30% glycerol at $-15°$. Growth is on methanol either in batch culture or in oxygen-limited or carbon-limited continuous culture as previously described.[2,5,6] After harvesting by centrifugation, cells may be stored at $-20°$ after rapid freezing in liquid nitrogen.

Growth of Paracoccus denitrificans (NCIB 8944) and Preparation of Spheroplasts and Periplasmic Fractions. Growth conditions do not markedly affect production of M protein. Stock cultures are maintained on nutrient agar plus succinate (0.2%), and starter cultures are grown from these in nutrient broth (8 g/liter) plus succinate (0.2%). For production of M protein, growth is at $30°$ on methanol in aerated vessels. The culture medium contains the following (per liter): K_2HPO_4, 6 g; KH_2PO_4, 4 g; NH_4Cl, 1.6 g; $NaHCO_3$, 0.5 g; $MgSO_4 \cdot 7H_2O$, 0.2 g; $CaCl_2 \cdot 2H_2O$, 40 mg; ferrous EDTA, 0.12 g; oxoid yeast extract, 0.1 g; Hoagland's trace elements, 0.1 ml; and methanol, 0.5% (v/v). After growth to the end of the log phase, bacteria are harvested and washed in 25% of the growth volume of ice-cold 10 mM HEPES buffer (pH 7.3) containing 150 mM NaCl. The bacteria should be used immediately for preparation of periplasmic fractions.

The following method for preparation of spheroplasts is suitable for cells harvested from 20 liters of growth medium. The cells are suspended at room temperature in 800 ml of spheroplast buffer containing 200 mM HEPES (pH 7.3), 500 mM sucrose, 0.5 mM disodium EDTA, and 25 mM methanol (to stabilize the MDH). Lysozyme (1 g), dissolved in 15 ml cold water, is added, followed by 800 ml of cold methanol solution (25 mM). The suspension is divided into 4 volumes of 400 ml, incubated for 1 hr at $30°$ with gentle swirling, and chilled on ice-water for 5 min, and the spheroplasts are harvested by centrifugation at 5000 g for 20 min at $4°$.

[5] A. R. Cross and C. Anthony, *Biochem. J.* **192,** 421 (1980).
[6] A. R. Cross and C. Anthony, *Biochem. J.* **192,** 429 (1980).

The periplasmic fraction (supernatant) is decanted and protease inhibitors added (benzamidine hydrochloride, 5 mM; phenylmethylsulfonyl fluoride, 0.5 mM). This process usually yields about 1600 ml of periplasmic fraction containing 0.4 mg protein/ml.

Purification of M Protein from Methylophilus methylotrophus. The bacteria are suspended (0.25 g wet weight/ml) in 20 mM HEPES-HCl buffer (pH 7.5) containing 25 mM methanol and are disrupted in an ultrasonic disintegrator in batches of 50 ml in 10 cycles of 30 sec followed by 30 sec of cooling. Whole bacteria and debris are removed by centrifugation at 40,000 g for 20 min, and membranes are removed from the supernatant by centrifugation at 150,000 g for 1 hr at 4° The supernatant extract is applied to a column of DEAE-cellulose (8 × 6 cm) equilibrated with 12.5 mM HEPES-HCl (pH 7.5) at 4°. The column is then washed with 100 ml of the same buffer; the MDH, which does not adsorb to the column, is collected during this washing step.

The column is washed with 500 ml of the same buffer containing 50 mM KCl. The M protein is eluted with 200 ml of the same buffer containing 100 mM KCl. After concentration under pressure over an Amicon XM50 membrane, further purification is achieved by gel filtration on Sephadex G-150 (fine grade, 75 × 3.5 cm; upward flow, 16 ml/hr in 12.5 mM HEPES-HCl buffer, pH 7.5). Pooled active fractions may then be passed through the same column. The M protein is purified by anion-exchange chromatography on a Pharmacia Mono Q column or its equivalent equilibrated with HEPES buffer as before with a gradient of 0–0.5 M KCl (total volume, 26 ml). Further purification is achieved by desalting followed by a second run through Mono Q but using a shallower gradient (7 mM/ml). This separates the M protein from the bulk of the protein with 60 kDa subunits that was previously assumed to be the M protein itself. The purification process is summarized in Table I.

Methanol dehydrogenase from *M. methylotrophus* is prepared as described previously.[2]

Purification of M Protein from Paracoccus denitrificans. Bovine DNase I (10 mg) is added to the periplasmic fraction (5 liters, containing 2 g protein, derived from 60 liters of bacteria) which is then applied to a column of DEAE-Sepharose (12 × 5 cm) equilibrated in 20 mM HEPES buffer (pH 7.5). Proteins are eluted with the same buffer containing NaCl. The first fraction, eluting with 100 mM NaCl, just before cytochrome *c*-550, contains a nonprotein component that shows some activity in creasing the oxidation of 1,3-butanediol. This component also stimulates formaldehyde oxidation and so explains why formaldehyde is oxidized by crude extracts of *P. denitrificans* even in the presence of the inhibitory M protein. The nonprotein activator may be purified by passage through

TABLE I
PURIFICATION OF M PROTEIN FROM *Methylophilus methylotrophus*[a]

Purification stage	Volume (ml)	Total units	Specific activity	Purification factor	Yield (%)
Cell-free extract	200	107	4.4	1	100
DEAE-cellulose	220	49	16.4	3.8	50
Sephadex G-150	204	20	40.5	9.3	19
Mono Q (1)	202	18	77.3	17.7	17
Mono Q (2)	200	14	140.0	31.8	14

[a] In this example M protein was purified from 50 g bacteria grown in oxygen-limited continuous culture.[2] Units are those described in the text; specific activities are units per milligram protein.

an Amicon PM10 membrane (MW cutoff 10,000) which retains all protein. The activating component may then be purified on a Pharmacia Mono Q column equilibrated in 20 mM HEPES buffer (pH 7.5), elution occurring with 150 mM NaCl.

After complete elution of the cytochrome c-550 the M protein is eluted with buffer containing 200 mM NaCl, concentrated to 10 ml on an Amicon YM100 membrane, and purified further by gel filtration on a column of Sephacryl 200 (85 × 3.5 cm) equilibrated in 20 mM HEPES buffer (pH 7.5) containing 100 mM NaCl (upward flow, 10 ml/hr). Active fractions are purified by anion-exchange chromatography on a Pharmacia Mono Q column (1 ml) equilibrated in the same buffer and flowing at 1 ml/min. A gradient of 0–500 mM NaCl (total volume 26 ml) is used; M protein is eluted with 250 mM NaCl, giving a protein of about 85% purity. The purification process is summarized in Table II.

Purification of Methanol Dehydrogenase from Paracoccus denitrificans. The MDH from *P. denitrificans* has a low isoelectric point, in contrast to most MDHs which are basic proteins.[1] MDH is purified from the same periplasmic extract that is used for the purification of its M protein. It is eluted from the DEAE-Sepharose column after the M protein by increasing the NaCl concentration to 300 mM. Pooled active fractions are concentrated to 10 ml over an Amicon XM50 membrane and applied to a gel filtration column of Sephacryl S-200 (85 × 3.5 cm) equilibrated in 20 mM HEPES buffer (pH 7.5) containing 200 mM NaCl and 25 mM methanol at 4°. The purification is completed by anion-exchange chromatography on a column of Q Sepharose (10 × 2.5 cm) equilibrated in 20 mM HEPES buffer (pH 7.5) containing 25 mM methanol. After applying the protein, the column is washed in the same buffer containing 300 mM NaCl, then MDH is eluted in the same buffer containing 400

TABLE II
PURIFICATION OF M PROTEIN FROM *Paracoccus denitrificans*[a]

Purification stage	Volume (ml)	Total units	Specific activity	Purification factor	Yield (%)
DEAE-Sepharose	120	6000	37.5	1	100
Sephacryl-200	63	2016	75	2	34
Mono Q	32	576	88	2.35	10

[a] Because of the presence of a nonprotein component able to stimulate oxidation of 1,3-butanediol in the periplasmic fraction, it is not possible to express purification factors and yields in terms of the initial activity. It should be noted that the values cannot be compared directly (except for specific activities) with the values in Table I for *M. methylotrophus* because in the latter organism the starting material is a complete soluble fraction rather than the periplasmic fraction. Units are those described in the text; specific activities are units per milligram protein.

mM NaCl. For small amounts of MDH this anion-exchange step can be done on a Pharmacia Mono Q column. Salt is removed by concentration on an Amicon XM50 membrane followed by gel filtration on a PD10 column equilibrated with 20 mM HEPES buffer (pH 7.5) containing 25 mM methanol. The preparation is homogeneous on sodium dodecyl sulfate–polyacrylamide gel electrophoresis (SDS–PAGE). It is stored at $-20°$.

Properties

Stability of M Protein. The activity of M protein is very sensitive to extremes of pH and to the nature of the buffers used. All activity is lost when extracts are treated with protamine sulfate, ammonium sulfate, or low-pH buffers. MOPS buffer and phosphate buffer should not be used. HEPES buffer is the best for all purification steps. The M protein is stable when frozen at $-20°$.

Molecular Weight and Subunit Structure. The molecular weight of M proteins from *M. methylotrophus,*[2,4] *Methylobacterium extorquens* AM1,[2,4] and *P. denitrificans* are all similar in having a native molecular weight of about 130,000 as estimated by gel filtration. The subunit molecular weight was previously reported to be about 70,000,[2,4] but subsequent work has shown that the predominant band seen on SDS–PAGE is not due to M protein, but is a contaminant of it. The subunit molecular weight, based on SDS–PAGE, is about 45,000, indicating that the protein is a trimer or, more likely, a tetramer.

Isoelectric Point. The isoelectric point is acidic in all cases, the p*I* for the protein from *M. methylotrophus* being 5.6 as measured by isoelectric focusing.[2]

Effect on Formaldehyde Oxidation by Methanol Dehydrogenase. M proteins inhibit formaldehyde oxidation; in the case of *M. methylotrophus* and *M. extorquens* this is by decreasing the V_{max} and the affinity of MDH for this substrate;[2] the mechanism has not been investigated for *P. denitrificans.*

Effect on Alcohol Oxidation by Methanol Dehydrogenase. Alcohols that are poor substrates for MDH, because of their low affinity for the enzyme, are oxidized in the absence of M protein, but only in a transient manner because they fail to protect MDH from the inhibitory effects of the artificial electron acceptor phenazine ethosulfate.[2,4] In the presence of M protein they are oxidized because their affinity for MDH increases and the enzyme is thus protected. It should be noted that, because of this mode of action, in the presence of small amounts of M protein there is an initial rapid transient consumption followed by a slower linear progress curve.

Effect on Methanol Dehydrogenase in the Cytochrome-Linked System. The natural electron acceptor for MDH is a specific cytochrome *c* called cytochrome c_L.[1,7,8] The effect of M protein in this system mirrors that in the dye-linked system.[2] With pure proteins from *M. extorquens* and *M. methylotrophus* the rate of oxidation of all alcohols (methanol, ethanol, and 1,3-butanediol) is increased but the rate of oxidation of formaldehyde is halved. The assay system is described elsewhere in this volume.[9,10]

Periplasmic Location. Because of the difficulty of preparing stable spheroplasts of other methylotrophs, the location of the M protein has been determined only for *P. denitrificans.* No activity can be demonstrated in membrane or cytoplasmic fractions after disruption of spheroplasts; M protein is located exclusively, with MDH and soluble *c*-type cytochromes, in the periplasmic fraction of this organism.

[7] D. N. Nunn and C. Anthony, *Biochem. J.* **256,** 673 (1989).

[8] D. N. Nunn and M. E. Lidstrom, *J. Bacteriol.* **166,** 591 (1986).

[9] D. Day and C. Anthony, this volume [33].

[10] D. Day and C. Anthony, this volume [44].

[35] Methanol Dehydrogenase from Thermotolerant Methylotroph *Bacillus* C1

By N. ARFMAN and L. DIJKHUIZEN

$$\text{Methanol} + \text{NAD}^+ \rightleftharpoons \text{formaldehyde} + \text{NADH} + \text{H}^+$$

Thermotolerant methylotrophic *Bacillus* strains employ an NAD-dependent methanol dehydrogenase for the initial oxidation of methanol to formaldehyde.[1] The purification described below has been the subject of a previous report.[2]

Assay Methods

Principle. Two assay methods can be used, namely, measurement of the methanol-dependent NAD reduction (methanol dehydrogenase activity) and measurement of the formaldehyde-dependent NADH oxidation (formaldehyde reductase activity). Both reactions are followed spectrophotometrically, by measuring the change in absorbance at 340 nm. The methanol dehydrogenase activity in the methylotrophic bacilli, however, is very sensitive to dilution inactivation and is generally not proportional to the amount of enzyme added. These disadvantages are not observed with the formaldehyde reductase assay described below.

Reagents

Potassium phosphate buffer, pH 6.7, 100 mM
NADH, 3 mM
Formaldehyde, 100 mM, prepared from paraformaldehyde[3]
Dithiothreitol (DTT), 100 mM

Procedure. The assay mixture in a quartz cuvette (10-mm light path) contains 0.5 ml potassium phosphate buffer, 50 μl NADH, 10 μl DTT, formaldehyde reductase sample (1 – 10 μl), to give a change in absorbance of 0.1 – 0.2/min; and water to adjust the volume to 0.9 ml. This mixture is incubated at 50° for several minutes to record any endogenous NADH oxidation. The formaldehyde reductase reaction is then started by the addition of 0.1 ml of formaldehyde. The resulting rate of absorbance

---

[1] L. Dijkhuizen, N. Arfman, M. M. Attwood, A. G. Brooke, W. Harder, and E. M. Watling, *FEMS Microbiol. Lett.* **52**, 209 (1988).

[2] N. Arfman, E. M. Watling, W. Clement, R. J. van Oosterwijk, G. E. de Vries, W. Harder, M. M. Attwood, and L. Dijkhuizen, *Arch. Microbiol.*, **152**, 285 (1989).

[3] J. R. Quayle, this series, Vol. 90, p. 314.

decrease at 340 nm, minus the formaldehyde-independent rate, is taken as formaldehyde reductase activity.

Unit of Enzyme Activity. One unit is defined as the amount of enzyme catalyzing the oxidation of 1 μmol of NADH per minute at 50° under the reaction conditions described. Specific activity is expressed as units per milligram of protein. Protein is determined by the Bradford method,[4] with bovine serum albumin as a standard.

Purification Procedure

Growth of Organism. Bacillus C1 is grown at a relatively low dilution rate (0.1/hr) in a methanol-limited (50 mM methanol) continuous culture (working volume of 1 liter) on a mineral salts medium having the following composition (per liter):[5] KH_2PO_4, 1.0 g; $(NH_4)_2SO_4$, 1.5 g; $MgSO_4 \cdot 7H_2O$, 0.2 g; and trace element solution,[6] 0.2 ml. The temperature is controlled at 50° and the pH maintained at 7.3 by automatic adjustment with 1 N NaOH. Extracts of cells grown under these conditions display very high activities of formaldehyde reductase and possess a dominant 43,000-dalton protein band when subjected to sodium dodecyl sulfate–polyacrylamide gel electrophoresis (SDS–PAGE),[7] which suggests that the enzyme constitutes a high percentage of total cell protein. Steady-state cultures are harvested by centrifugation at 3,800 g for 10 min at 4°, washed twice with 50 mM potassium phosphate buffer, pH 7.5, containing 5 mM $MgSO_4$. The resulting pellet is resuspended in the same buffer and stored at $-20°$ until required for purification. Extracts and samples obtained in various purification steps (see below), are stored at $-80°$ in the presence of 5 mM DTT.

Step 1: Preparation of Cell-Free Extract. Cells are disrupted in the presence of 5 mM DTT by passage through a French pressure cell operating at 1.4×10^5 kN/m². Unbroken cells and debris are removed by centrifugation at 25,000 g for 20 min at 4°. The supernatants thus obtained contain approximately 10 mg protein/ml. The viscosity of the sample is lowered by adding protamine sulfate (10% solution) to a final concentration of 1 mg per 10 mg of protein. The resulting precipitate (nucleic acids) is removed by centrifugation at 25,000 g for 20 min at 4°. The supernatant is desalted in buffer A (20 mM Tris-HCl, pH 7.5, 5 mM $MgSO_4$, 5 mM 2-mercaptoethanol) by Sephadex G-25 gel filtration (PD10 column, Bio-

[4] M. M. Bradford, *Anal. Biochem.* **72**, 248 (1976).

[5] P. R. Levering, J. P. van Dijken, M. Veenhuis, and W. Harder, *Arch. Microbiol.* **129**, 72 (1981).

[6] W. Vishniac and M. Santer, *Bacteriol. Rev.* **21**, 195 (1957).

[7] U. K. Laemmli and K. Favre, *J. Mol. Biol.* **80**, 575 (1973).

Rad, Richmond, CA). The further purification steps are performed at room temperature with a Pharmacia fast protein liquid chromatography (FPLC) system.

Step 2: Anion-Exchange Chromatography. Protein (320 mg; see Table I) is applied to a Fast Flow Q-Sepharose (Pharmacia) column (25 ml gel; washed and equilibrated with buffer A) and eluted by applying a linear 0–1.0 M potassium chloride gradient (120 ml) in buffer A, at a flow rate of 1.0 ml/min. Fractions of 4 ml are collected and analyzed for enzyme activity. Formaldehyde reductase typically elutes from the column at KCl concentrations of approximately 300 mM. Formaldehyde reductase-containing fractions are pooled and prepared for hydrophobic interaction chromatography by adding ammonium sulfate to a final concentration of 1 M.

Step 3: Hydrophobic Interaction Chromatography. Samples (13–14 mg of protein) are applied to a FPLC Phenyl-Superose HR5/5 column (1 ml gel; Pharmacia), equilibrated with 50 mM Tris-HCl, pH 7.0, 5 mM MgSO$_4$, 5 mM 2-mercaptoethanol containing 1.0 M (NH$_4$)$_2$SO$_4$ (buffer B). Bound protein is eluted by applying a linear 1.0–0 M (NH$_4$)$_2$SO$_4$ gradient (20 ml) in buffer B, at a flow rate of 0.4 ml/min. Fractions of 0.5 ml are collected and analyzed for enzyme activity. Formaldehyde reductase elutes from the column at (NH$_4$)$_2$SO$_4$ concentrations of approximately 300 mM. The peak fractions are combined and used for further characterization of enzyme properties.

A typical purification is summarized in Table I.

Properties

Purity. The homogeneity of the enzyme preparation obtained by the procedures described above has been checked by several criteria: (1) in N-terminus analysis, only a single N-terminal amino acid was detectable, and (2) SDS–PAGE gave only a single band, even on overloaded gels. The 4.5-fold purification obtained indicates that the enzyme may constitute up to about 22% of total soluble protein at a dilution rate of 0.10/hr in methanol-limited chemostats, assuming that no inactive enzyme is present in the final preparation.

Molecular Weights. The molecular weight of the purified enzyme, determined by gel-filtration chromatography on a Pharmacia FPLC Superose 6 HR 10/30 column (equilibrated with 100 mM Tris-HCl, pH 7.5, 5 mM MgSO$_4$, 5 mM 2-mercaptoethanol),[8] is approximately 300,000.

---

[8] Samples of 0.2 ml (~200 μg of protein) are run together with gel filtration standards (1,35–670 kDa range; Bio-Rad, Richmond, CA).

TABLE I

PURIFICATION OF METHANOL DEHYDROGENASE FROM *Bacillus* C1[a]

Fraction	Protein (mg)	Total activity (units)[b]	Specific activity (units/mg)	Recovery (%)	Purification (-fold)
Step 1. Extract	320	1408	4.4	100	1.0
Step 2. Q-Sepharose pool	54.6	541	9.9	38	2.3
Step 3. Phenyl-Superose pool	24.6	482	19.6	34	4.5

[a] Reproduced, with permission, from N. Arfman, E. M. Watling, W. Clement, R. J. van Oosterwijk, G. E. de Vries, W. Harder, M. M. Attwood, and L. Dijkhuizen, *Arch. Microbiol.* **152**, 286 (1989).

[b] Formaldehyde reductase assay.

SDS–PAGE indicated that the enzyme is a hexamer or octamer composed of identical subunits with a molecular weight of 43,000.

Stability. The methanol dehydrogenase and formaldehyde reductase activities are stable in cell-free extracts, stored at −20° or −80°, provided DTT is present. During enzyme purification a remarkable difference is observed between the recoveries of methanol dehydrogenase and formaldehyde reductase activities. After the Q Sepharose and Phenyl-Superose column steps, 34% of the original formaldehyde reductase activity is recovered, whereas the preparation almost completely loses alcohol dehydrogenase activity (>95% loss). Addition of sucrose, glycerol, iron ions, zinc ions, or DTT does not prevent this specific loss of methanol dehydrogenase activity. Rather, the additional presence of a second protein (itself without alcohol dehydrogenase activity) is required to specifically restore alcohol dehydrogenase activity of the purified preparations. The properties and role of this second protein remain to be identified.

Specificity. The enzyme is NAD-specific, and partially purified preparations catalyze the oxidation of various primary alcohols at the following relative rates: methanol, 44%; ethanol, 100%; *n*-propanol, 71%; *n*-butanol, 87%.

Michaelis Constants. In extracts and partially purified preparations the enzyme displays biphasic kinetics toward methanol (apparent K_m values of 3.8 and 166 mM) and ethanol (apparent K_m values 7.5 and 70 mM). Biphasic kinetics is also observed toward NAD (apparent K_m values 15 and 190 μM), but not for formaldehyde (apparent K_m value 2.0 mM) in the reverse reaction.

[36] Methylamine Oxidase from *Arthrobacter* P1

By William S. McIntire

$$CH_3NH_3^+ + H_2O + O_2 \rightarrow \text{formaldehyde} + NH_4^+ + H_2O_2$$

Methylamine oxidase (EC 1.4.3.6) is the first enzyme responsible for the assimilation of carbon from the environment, in the form of methylamine or ethylamine,[1] by the facultative gram-positive methylotroph *Arthrobacter* P1. When grown on these compounds the organism also produces a large amount of catalase.[1,2] The formaldehyde produced from the oxidation of methylamine is assimilated via the Embden–Meyerhof fructose-bisphosphate aldolase/transaldolase variant of the ribulose monophosphate cycle[1,3,4] (also known as the hexulose phosphate cycle[5]). Curiously, both this and the serine pathway of formaldehyde assimilation are operative when *Arthrobacter* P1 is grown on choline.[6]

Assay Method

Principle. Two methods can be used to assay methylamine oxidase. Method 1 involves the direct polarographic measurement of O_2 concentrations using a Clark electrode as described by van Iersel *et al.*[2] Method 2 makes use of the absorbance change that occurs on enzymatic oxidation of benzylamine to benzaldehyde. Method 2 is more convenient since it requires a UV–visible spectrophotometer.

Reagents for Method 1

Potassium phosphate buffer, 50 mM, pH 8.0
Methylamine hydrochloride, 0.3 M

[1] P. R. Levering, J. P. van Dijken, M. Veenhuis, and W. Harder, *Arch. Microbiol.* **129**, 72 (1981).

[2] J. van Iersel, R. A. van der Meer, and J. A. Duine, *Eur. J. Biochem.* **161**, 415 (1986).

[3] C. Anthony, "The Biochemistry of Methylotrophs," p. 60. Academic Press, New York, 1982.

[4] P. R. Levering, L. Dijkhuizen, and W. Harder, *FEMS Microbiol. Lett.* **14**, 257 (1982); P. R. Levering, and L. Dijkhuizen, *Arch. Microbiol.* **144**, 116 (1986).

[5] P. J. Large, "Aspects in Microbiology, Volume 8: Methylotrophy and Methanogenesis," p. 25. Van Nostrand-Reinhold, Princeton, New Jersey, 1983.

[6] P. R. Levering, D. J. Binnema, J. P. van Dijken, and W. Harder, *FEMS Microbiol. Lett.* **12**, 19 (1981).

Reagents for Method 2

Potassium phosphate buffer, 100 mM, pH 7.2

Benzylamine hydrochloride, 100 mM

Procedure for Method 1. The buffer is thermostatted at 25° and saturated with air to ensure a constant $[O_2]$ for each assay. Buffer, 1.485 ml, is put into the Clark electrode cell (1.5 ml capacity) followed by 15 μl of methylamine hydrochloride. The cell is well stirred using a flea-type Teflon-coated magnetic stir bar. Oxygen consumption is monitored after syringe injection of enzyme (0.01–0.1 units) through the stopper of the cell.

During the purification, many samples will contain varying amounts of endogenous catalase. This will depress the magnitude of the measured rates from the true values by a factor between 1 and 2 depending on the amount of catalase present in each sample. To circumvent this problem, excess exogenous catalase is added and the observed rates multiplied by 2 to obtain the true values.

Procedure for Method 2. Buffer, 2.9 ml, saturated with air at the assay temperature of 30°, is pipetted into a 1-cm light path, 3-ml quartz cuvette. To this is added 100 μl of the benzylamine solution, also air-saturated at the assay temperature. After the addition of microliter quantities of crude or pure enzyme, the absorbance change associated with the formation of benzaldehyde is monitored at 250 nm. A $\Delta\epsilon_{250}$ of 12,500 is used to calculate the rates. Because of high protein concentration and turbidity, it is not possible to measure the activity of the crude bacterial extracts.

Definition of Unit and Specific Activity. For Method 1, one unit of activity is defined as the amount of methylamine oxidase needed to reduce 1 μmol of O_2 per minute at 25°, and for Method 2, one unit is defined as the amount of enzyme required to produce 1 μmol of benzaldehyde per minute at 30°. Specific activity is expressed as units per milligram of protein. Protein is determined by the Bradford method,[7] or from the ϵ_{280} value of 1.82 (mg/ml)$^{-1}$ for pure enzyme.[2]

Comments. A third method for assaying methylamine oxidase has been reported.[8] This procedure requires 2,2′-azinodi-3-ethylbenzthiazoline and peroxidase to measure the H_2O_2 formed when methylamine is oxidized.

Since catalase is the major contaminating protein in most steps of the

[7] M. M. Bradford, *Anal. Biochem.* **72**, 248 (1976).

[8] M. van Vleits-Smits, W. Harder, and J. P. van Dijken, *FEMS Microbiol. Lett.* **11**, 31 (1981).

[9] H. Lück, *in* "Methods of Enzymic Analysis" (H. U. Bergmeyer, ed.), 1st Ed., p. 885. Academic Press, New York, 1963; H. Aebi, *in* "Methods of Enzymic Analysis" (H. U. Bergmeyer, ed.) 2nd Ed., Vol. 2, p. 673. Verlag Chemie, Weinheim/Academic Press, New York, 1974.

methylamine oxidase purification, a general assay procedure[9] has been adapted to determine catalase activity in the various fractions.[2,8] The assay relies on the change at 240 nm resulting from the depletion of H_2O_2.

Purification Procedure

Growth of Organism. Arthrobacter P1, originally isolated from subtropical greenhouse soil by Levering *et al.*,[1] is obtained from the National Collection of Industrial Bacteria, Ltd., Aberdeen, Scotland (NCIB 11625). The organism is first grown at 30° on nutrient agar (oxoid 3CM, p. 99, 1986 NCIB catalog: Lab-Lemco beef extract, 1 g; yeast extract, 2 g; peptone, 5 g; NaCl, 5 g; agar, 15 g; distilled water, 1 liter, autoclaved). Once flourishing, the culture is grown on agar plates or in liquid medium containing, per liter of deionized water: $NaH_2PO_4 \cdot H_2O$, 0.5 g; K_2HPO_4, 1.55 g; $(NH_4)_2SO_4$, 1.0 g; $MgSO_4 \cdot 7H_2O$, 0.2 g;[1] 0.2 ml of trace metals solution[10] [$ZnSO_4 \cdot 7H_2O$, 22 g; $CaCl_2$, 5.54 g; $MnCl_2 \cdot 4H_2O$, 5.06 g; $FeSO_4 \cdot 7H_2O$, 4.99 g; $(NH_4)_6Mo_7O_{25} \cdot 4H_2O$, 1.1 g; $CuSO_4 \cdot 5H_2O$, 1.57 g; $CoCl_2 \cdot 6H_2O$, 1.61 g; add 50 ml of water and enough concentrated HCl to dissolve the salts, then add 63.68 g of disodium ethylenediaminetetraacetic acid; the volume is brought to ~900 ml, the pH is adjusted to ~6.0 with KOH, and the final volume is 1 liter]; and 0.3% (w/v) methylamine hydrochloride (added as a 0.2-μm filter-sterilized 30%, w/v, solution). One-milliliter aliquots of mid log phase liquid culture, containing 15% (v/v) glycerol, are stored at −70° in sterilized vials. These are used to initiate 50 ml liquid inoculums.

For large-scale preparation, the cells are grown in aerated medium containing 0.5% (w/v) methylamine hydrochloride in six 12-liter fermentor jars. Each jar is inoculated with 300 ml of hardy culture; after approximately 20 hr the A_{600} reaches a maximum of 2.0–2.2, and the cells are harvested in Sharples centrifuges. The yield is 220–260 g of cell paste, which is stored indefinitely at −70°.

Other C_1 compounds able to support growth of *Arthrobacter* P1 are[1] dimethylamine, trimethylamine, trimethylamine *N*-oxide, *N,N*-dimethylethylamine, and *N,N*-diethylmethylamine; methanol, formate, dimethyl sulfide, dimethyl sulfoxide, and tetramethylammonium chloride are not growth substrates. A variety of other compounds can be used as carbon sources for *Arthrobacter* P1 growth, including ethylamine, diethylamine, choline, acetylcholine, betaine, sarcosine, ethanol, yeast extract, nutrient broth, various amino acids, sugars, carboxylic acids, and citric acid cycle intermediates.

Purification. Methylamine oxidase was partially purified by van Vleits-

[10] W. Visniac and M. Santer, *Bacteriol. Rev.* **21**, 195 (1957).

Smits et al.[8] and first purified to homogeneity by van Iersel et al.,[2] however, the procedure described here represents an improvement over the latter method (see Table I). Steps 2, 3, and 5 are similar to steps described before.[2] All procedures are done at 4° unless otherwise noted.

Step 1: Crude Extract. Thawed cells (223 g) are suspended in 400 ml of 50 mM potassium phosphate buffer, pH 7.0, twice passed through a 50-ml capacity French pressure cell (American Instruments Co., Inc., Silver Spring, MD) at 7,500 psi, and centrifuged at 48,000 g for 15 min. The pellet, which contains a large volume of unbroken cells, is resuspended in 100 ml of buffer, passed through the pressure cell 5 times, and centrifuged as before. The combined supernatant is 595 ml.

Step 2: Polyethylene Glycol Treatment. The supernatant is brought to 10% (w/v) polyethylene glycol (PEG) (average molecular weight 8,000; Sigma, St. Louis, MO, #P-2139) by slow addition of the solid material to the stirred solution. This is stirred an additional 20 min at room temperature and centrifuged at 300,000 g (Beckman Ti60 rotor) for 20 min. The pellet is washed by resuspension in 10% PEG in buffer, recentrifuged, and the supernatants combined (775 ml). Suspended DEAE-Trisacryl M (IBF Biotechnics, Inc., Savage, MD, and Villeneuve-la-Garenne, France), 1200 ml (900 ml settled volume), is slowly stirred into the supernatant. Continue stirring for another 20 min. (A small sample is centrifuged, and if any activity is detected in the liquid add more DEAE-Trisacryl M.) The suspension is centrifuged in 450-ml bottles at 17,700 g by bringing the rotor up to speed and immediately back to 0 g. The Trisacryl M is suspended in 10 mM potassium phosphate buffer, pH 7.0, poured into a 10-cm-diameter column, and washed with the same buffer until the effluent is clear. The column is stripped with 0.5 M potassium phosphate buffer, pH 7.0. The effluent is dialyzed for 18 hr against 14 liters of H$_2$O, then 4 hr against 14 liters of 10 mM potassium phosphate buffer, pH 7.0, using Spectrapor 12,000–14,000 molecular weight cutoff dialysis tubing, 0.0009 inch thickness, 2 ml/cm (Spectrum Medical Industries, Inc., Los Angeles, CA).

Step 3: DEAE-Trisacryl M Chromatography. The dialyzed solution is applied to a 5 × 55 cm DEAE-Trisacryl M column previously equilibrated with 10 mM potassium phosphate buffer, pH 7.0, and eluted with a 4 liter linear gradient from 10 to 400 mM potassium phosphate buffer, pH 7.0. The purest methylamine oxidase fraction (DE-I, 595 units) elutes between 1760 and 1850 ml as determined by UV–visible spectral and activity measurements. A fraction (DE-II) eluting from 1850 to 2280 ml, and containing about two-thirds of the methylamine oxidase activity (1097 units), also contains a large amount of catalase. These fractions are processed separately.

TABLE I
PURIFICATION OF METHYLAMINE OXIDASE FROM *Arthrobacter* P1

Fraction	Total volume (ml)	Total activity[a] (units)	Yield (%)
Step 1. Crude extract[b]	775	—	—
Step 2. PEG treatment	1180	1732	100
Step 3. DEAE-Trisacryl chromatography			
a. DE-I	111	595	35
b. DE-II	426	1097	63
Step 4. Ammonium sulfate fractionations			
a. Final 35% fraction	91.3	1370	79
b. Final 40% fraction	31.5	168	10
Step 5. Ultrogel AcA 34 chromatography			
a. 35% $(NH_4)_2SO_4$ fraction			
1. Main fraction	201	933 ⎤	
2. Side fraction	185	274 ⎬	76[c]
b. 40% $(NH_4)_2SO_4$ fraction, main fraction	45	115 ⎦	

[a] Assay Method 2.
[b] The activity could not be determined.
[c] The yield of pure enzyme from these 3 fractions is 1.62 g.

Step 4: Ammonium Sulfate Fractionations. Fraction DE-II is made 40% saturated $(NH_4)_2SO_4$ by the addition of the solid salt, then stirred for 2 hr. After centrifugation (48,000 g, 15 min), the pellet is redissolved in a minimum volume of 10 mM potassium phosphate buffer, pH 7.0, the supernatant is brought to 45% saturated $(NH_4)_2SO_4$, and the mixture is stirred for 30 min. This is centrifuged, and the process repeated to produce 45–50%, 50–60%, and 60–80% saturated $(NH_4)_2SO_4$ fractions. The redissolved pellets from each fraction are put into separate dialysis bags and dialyzed against 6.5 liters of 10 mM potassium phosphate buffer, pH 7.0. In a similar fashion, the resulting 40–45%, 45–50%, and 50–60% fractions are refractionated with $(NH_4)_2SO_4$. Some fractions from these, which contained substantial amounts of methylamine oxidase and contaminated with small amounts of catalase, as determined by UV–visible spectroscopy, are also refractionated. After dialysis, the purest samples from this procedure (920 units total) are combined with the DE-I fraction, which is of comparable purity. Final fractionations of 0–35%, 35–40%, 40–45%, 45–50%, and 50–60% saturated $(NH_4)_2SO_4$ result in the purest methylamine oxidase in the redissolved pellets of the 0–35% and 35–40% cuts.

During the repeated $(NH_4)_2SO_4$ fractionations, the major contaminant is catalase. The visible spectrum of catalase is dominated by a sharp peak at 405 nm, which has a molar extinction over 30 times that of methylamine oxidase at this wavelength; thus, the visible spectrum offers a sensitive test for the level of purity in each methylamine oxidase sample.

Step 5: Ultrogel AcA 34 Gel-Filtration Chromatography. The final 0–35% $(NH_4)_2SO_4$ fraction is concentrated to 35 ml in an ultrafiltration cell fitted with a Diaflo YM30 membrane filter (Amicon Corp., Lexington, MA) and dialyzed against 1 liter of 50 mM potassium phosphate buffer, pH 7.0, containing 150 mM KCl, for 16 hr. The sample is applied to a 5×100 cm column containing Ultrogel AcA 34 (IBF Biotechnics) (20,000–350,000 molecular weight range) equilibrated with the same buffer. The purest methylamine oxidase elutes from 1300 to 1520 ml. This fraction is concentrated to 23 ml (YM30 membrane filter) and stored at $-70°$ as 0.25-ml aliquots. The redissolved pellet from the final 35–40% $(NH_4)_2SO_4$ step is similarly processed.

Properties

Purity. The enzyme is pure as judged by sodium dodecyl sulfate–polyacrylamide electrophoresis and by the value of the A_{278}/A_{486} ratio, which is between 63 and 65 (the literature value[2] is 72.9). The visible spectrum (Fig. 1) indicates that the enzyme is devoid of catalase contamination.

The purity was also checked by high-performance liquid chromatography (HPLC) on a Pharmacia Mono Q HR5/5 ion-exchange column. A linear gradient from 0 to 1 M KCl in 10 mM Bis-Tris Propane (1,3-bis[tris(hydroxymethyl)amino]propane) buffer, pH 7.0, was employed. When the gradient is run in 30 min, a single peak (280 nm detection) with a retention time of 15.8 min is observed, but when the gradient time is 150 min, several overlapping peaks are seen (retention times 50 to 53 min). Samples from each peak give identical UV–visible spectra, and there is no interconversion of the methylamine oxidase forms in these samples. This phenomenon is not an artifact of the Mono Q column, because similar results are obtained when either 0.75×7.5 cm TSK DEAE-5PW or 0.46×25 cm TSK DEAE-2SW HPLC columns are employed (Phenomenex, Rancho Palos Verdes, CA). It is not known whether these peaks are a result of methylamine oxidase isozyme or modified forms generated during purification. Although a detailed analysis has not been conducted, each form seems to have the same properties.

Substrate Specificity. The enzyme can oxidase the primary amines methyl-, ethyl-, propyl-, butyl-, ethanol-, and benzylamine; however, tyramine, spermine, putracine, trimethylamine, and dimethylamine are not substrates.[2,8] Only O_2 is known to act as a reoxidizing substrate for methylamine oxidase.

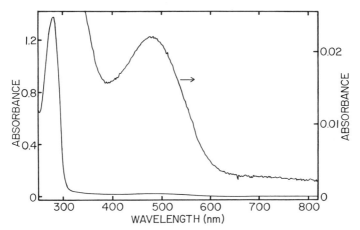

Fig. 1. Absorbance spectrum of methylamine oxidase (4.2 μM) taken at 25° in 50 mM potassium phosphate buffer, pH 7.5, containing 0.5 M KCl. The arrow points to axis for this curve.

Inhibitors. Methylamine oxidase is reversibly inhibited by ammonium ions and the anions chloride and acetate. It is irreversibly inhibited by hydrazine, methylhydrazine, and phenylhydrazine, which form covalent adducts with the enzyme-bound redox cofactor, and diethyl dithiocarbamate, which removes the essential copper from the enzyme. Other inhibitors are α,α'-bipyridyl, CN^-, p-chloromercuriphenylsulfonic acid, and iodoacetic acid.[2,8]

Stability. The enzyme is stable in the pH range from 6.3 to 8.0 at 25°. When stored at $-70°$ it appears to be stable indefinitely; however, several freezing and thawing cycles cause the enzyme to denature. During the purification, samples not being processed are stored on ice with a few drops of toluene added, rather than freezing.

Steady-State Kinetic Parameters. The true maximal velocity is not known, because the K_m value for O_2 has not been determined. In air-saturated buffer, the apparent value with methylamine as the substrate is 0.87 units/mg at 25° and pH 8.0 (assay Method 1),[2] whereas the apparent K_m for methylamine is 0.20 mM.[2] With benzylamine as the substrate, at 30° and pH 7.2 (assay Method 2), the apparent V_{max} is 0.83 units/mg.

Molecular Weight and Subunit Composition. Methylamine oxidase is a homodimer of molecular weight 170,000 as determined by gel-filtration chromatography using Sephadex G-200[2] or a TSK-3000SW HPLC column (Phenomenex). Gel-filtration chromatography on a Pharmacia Superose 6 HR10/30 HPLC column offers a value of 155,000. The subunit molecular weight has been estimated to be 82,000 by chromatography on a polyol Si-300, 3 μm, HPLC column (Serva Fine Biochemicals, Inc., Westbury,

NY) in the presence of sodium dodecyl sulfate,[2] and 80,000 using a TSK-4000SW HPLC column (Phenomenex) with guanidine hydrochloride as the denaturant. Each subunit contains one molecule of a quinone-type cofactor covalently attached to the polypeptide chain and one atom of Cu(II).[2]

Spectral and Oxidation–Reduction Properties. The enzyme absorbs maximally in the UV–visible spectral region at 278 nm (ϵ 328 mM^{-1} cm^{-1}) and at 486 nm (ϵ 5,240 M^{-1} cm^{-1}) (Fig. 1). The values for the extinction coefficients are obtained from the anaerobic titration of the enzyme with a standardized solution of methylamine hydrochloride. Each molecule of enzyme-bound cofactor consumes two electrons, without the occurrence of quinone radical or reduced copper. A stoichiometric concentration of methylamine causes nearly complete bleaching from 350 to 820 nm. An attempted titration of the enzyme with sodium dithionite resulted in minor spectral changes attributed to a reaction at the copper site. 5-Deazariboflavin radical[11] could not reduce the enzyme.

The visible spectrum is insensitive to pH (at pH 6.30, $\epsilon_{486} = 5,040$ M^{-1} cm^{-1}, and at pH 8.80, $\epsilon_{486} = 5,340$ M^{-1} cm^{-1}), whereas the λ_{max} undergoes a hypsochromic shift on addition of NH$_4^+$ (at pH 6.30, $\lambda_{max} = 458$ nm, $\epsilon = 4,900$ M^{-1} cm^{-1}, and at pH 8.80, $\lambda_{max} = 446$ nm, $\epsilon = 5,900$ M^{-1} cm^{-1}).

Methylamine oxidase from *Arthrobacter* P1 is a prototype of various eukaryotic amine oxidases, for example, bovine plasma amine oxidase, porcine kidney diamine oxidase, and plant diamine oxidase. These eukaroytic enzymes, like methylamine oxidase, are homodimers of approximately molecular weight 170,000 and contain 1 mol of quinone cofactor and 1 g atom of Cu(II) per mole per subunit.[2,12,13] The copper site in all are identical as judged by extended X-ray absorption fine structure, spin echo envelope modulation, and other spectral techniques.[14] As with methylamine oxidase, exposure of plasma amine oxidase and porcine diamine oxidase to substrate results in the two-electron reduction of the quinone moiety only. In the presence of CN$^-$, which reacts with the Cu(II), sub-

[11] V. Massey and P. Hemmerich, *Biochemistry* **17**, 9 (1978).

[12] W. S. McIntire, *Pharmacol. Res. Commun.* **20**, Suppl. 4, 155 (1988).

[13] R. A. van der Meer, J. A. Jongejan, J. Frank, and J. A. Duine, *FEBS Lett.* **206**, 111 (1986); C. L. Lobenstein-Verbeek, J. A. Jongejan, J. Frank, and J. A. Duine, *FEBS Lett.* **170**, 305 (1984); R. S. Moog, M. A. McGuirl, C. E. Coté, and D. M. Dooley, *Proc. Natl. Acad. Sci. U.S.A.* **83**, 8435 (1986); P. F. Knowles, K. B. Pandeya, F. X. Ruis, C. M. Spenser, R. S. Moog, M. A. McGuirl, and D. M. Dooley, *Biochem. J.* **241**, 603 (1987); U. Bachrach, *in* "Structure and Function of Amine Oxidases" (B. Mondavi, ed.), p. 5. CRC Press, Boca Raton, Florida, 1985; A. Rinaldi, G. Floris, and A. Giatosio, in "Structure and Function of Amine Oxidases" (B. Mondavi, ed.), p. 51. CRC Press, Boca Raton, Florida, 1985.

[14] D. M. Dooley, personal communication (1989).

strate treatment results in the formation of reduced copper and semiquinone radical for methylamine oxidase and the mammalian and plant diamine oxidases.[12,15]

Other Properties. When bovine plasma amine oxidase is treated with NaB^3H_3CN in the presence of [^{14}C]benzylamine, 1 mol of ^{14}C is incorporated into the enzyme; however, no ^3H is incorporated. These findings are consistent with a quinone-type redox cofactor for plasma amine oxidase, but not pyridoxal phosphate as suggested by earlier research on this enzyme.[16] Repeating this trapping experiment with methylamine oxidase produces virtually the same results,[17] which provides further evidence for the striking similarity of this bacterial copper-containing amine oxidase and the plasma amine oxidase. Unlike the eukaryotic oxidases, methylamine oxidase is not a glycoprotein. Lastly, the isoelectric point for methylamine oxidase is 4.6.[2]

Acknowledgments

The research on this enzyme by the author is supported by a Department of Veterans Affairs grant (Dr. Thomas P. Singer, principal investigator), Program Project Grant HL16251-16 from The Heart, Lung, Blood Institute of The National Institutes of Health, and an Academic Senate grant from the University of California, San Francisco.

[15] D. M. Dooley, M. A. McGuirl, J. Peisech, and J. McCracken, *FEBS Lett.* **214**, 274 (1987); A. Finazzi-Argo, A. Rinaldi, G. Floris, G. Rotilio, *FEBS Lett.* **176**, 378 (1984).
[16] C. Hartmann and J. P. Klinman, *Biofactors* **1**, 41 (1988).
[17] C. Hartmann and J. P. Klinman, *FEBS Lett.* **261**, 441 (1990).

[37] Methylamine Dehydrogenase from *Thiobacillus versutus*

By JOHN E. VAN WIELINK, JOHANNES FRANK, and JOHANNIS A. DUINE

$$CH_3NH_3^+ + H_2O + dye \rightarrow CH_2O + NH_4^+ + dye \cdot H_2$$

Thiobacillus versutus is one of the methylotrophic bacteria which use methylamine dehydrogenase for growth on methylamine. The enzyme from this organism has been purified[1] and its quinoprotein nature established.[2] Small-scale (Method 1) and large-scale (Method 2) purification procedures are presented here. How the natural electron acceptors (amicyanin, cytochrome c-550, and cytochrome c-552) can be obtained is indicated in the large-scale procedure.

[1] F. M. D. Vellieux, J. Frank Jzn, M. B. A. Swarte, H. Groendijk, J. A. Duine, J. Drenth, and W. G. J. Hol, *Eur. J. Biochem.* **154**, 383 (1986).
[2] R. A. van der Meer, J. A. Jongejan, and J. A. Duine, *FEBS Lett.* **221**, 299 (1987).

Assay Method

Principle. The measurement of enzyme activity is routinely performed using the perchlorate salt of the cationic free radical of N,N,N',N'-tetramethyl-p-phenylenediamine (Wurster's blue) as an artificial electron acceptor. This compound is not commercially available, but a convenient synthesis has been described.[3]

Procedure. To 1 ml of 0.1 M potassium phosphate buffer, pH 7.0, containing 10 mM methylamine hydrochloride and 120 μM Wurster's blue (prepared daily) in a cuvette, 5–50 μl of sample is added. After mixing, the decrease in absorbance at 600 nm is followed in a spectrophotometer for 1 min. The rate can be calculated using an ϵ_{600} value 9000 M^{-1} cm^{-1}. Because the enzyme is assayed under nonsaturating conditions (with respect to Wurster's blue), the absorbance change should not exceed 0.2. One unit is the amount of enzyme catalyzing the conversion of 1 μmol of methylamine per minute, which is equivalent to 2 μmol of Wurster's blue per minute.

Purification Procedure

Growth of Organism. Thiobacillus versutus[4] (ATCC 25364), formerly[5] called *Thiobacillus* A2, is grown aerobically at 30° on a mineral medium. The basic medium (sterilized by autoclaving) contains, per liter: 1.45 g K_2HPO_4, 0.65 g $NaH_2PO_4 \cdot 2H_2O$, and 12.5 g methylamine hydrochloride. To this medium is added 10 ml of a solution (sterilized by autoclaving) containing 30 g $MgSO_4 \cdot 7H_2O$ per liter and 1 ml of a trace element solution (sterilized by autoclaving) prepared according to Vishniac and Santer[6] but with a 5 times higher concentration of copper salt.

During the entire period of growth the culture is maintained at 90% oxygen saturation or higher (regulated by aeration and agitation rate of the fermentor), and the pH is maintained at 7.3 (with NaOH). Antifoam (Serva No. 35109) is added when necessary.

Cells are harvested in the stationary growth phase by centrifugation at 48,000 g for 20 min at 4°. Cell paste (obtained with a typical yield of 6 g wet weight per liter) is stored at $-20°$.

[3] L. Michaelis and S. Granick, *J. Am. Chem. Soc.* **65**, 1747 (1943).
[4] A. P. Harrison, Jr., *Int. J. Syst. Bacteriol.* **33**, 211 (1983).
[5] B. F. Taylor and D. S. Hoare, *J. Bacteriol.* **100**, 487 (1969).
[6] W. Vishniac and M. Santer, *Bacteriol. Rev.* **21**, 195 (1957).

Purification Method 1

All operations are carried out at room temperature unless indicated otherwise.

Step 1: Preparation of Cell-Free Extract. Frozen cell paste (40 g) is suspended in an equal volume of 10 mM potassium phosphate buffer, pH 7.0. Cells are disrupted in a precooled (4°) French pressure cell at 125 MPa. After centrifugation at 48,000 g for 20 min at 4°, the pellet is resuspended in an equal volume of 10 mM potassium phosphate buffer, pH 7.0, and the procedure repeated. The two supernatants are combined (80 ml) and dialyzed for 16 hr against 10 liters of 10 mM potassium phosphate buffer, pH 7.0, at 4°, yielding the cell-free extract.

Step 2: DEAE-Sepharose Chromatography. The dialyzed extract of Step 1 is applied to a DEAE-Sepharose CL-6B column (2 × 15 cm) equilibrated with 10 mM potassium phosphate buffer, pH 7.0. After washing with 3 bed volumes of a 0.2 M Tris-HCl buffer, pH 8.0, methylamine dehydrogenase is eluted with a linear gradient (0.2–1.0 M Tris/HCl, pH 8.0, 10 bed volumes).

Step 3: Dialysis. The combined active fractions (yellow-green) from Step 2 are dialyzed against 0.1 M sodium acetate, pH 5.0. This leads to the precipitation of a large quantity of contaminating protein, which is removed by centrifugation at 48,000 g for 10 min. The supernatant is concentrated to a protein concentration of at least 10 mg/ml and stored at −20°.

Heat Treatment. If there is no interest in obtaining other components of the respiratory chain, a heat treatment step[7] can be performed after Step 1. Cell-free extract is heated for 20 min at 70°, after which it is centrifuged at 48,000 g for 20 min. The pellet is washed with an equal volume of 10 mM potassium phosphate buffer, pH 7.0, and the suspension centrifuged as above. The supernatants are combined. The heat treatment leads to a significant increase of specific activity.

Table I summarizes the purification steps of Method 1.

Purification Method 2

To obtain methylamine dehydrogenase and its electron acceptors (amicyanin, cytochrome c-550, and cytochrome c-552) in larger quantities, a procedure has been developed that can be easily scaled up to the required amounts using radial flow columns.[8]

[7] G. W. Haywood, N. S. Janschke, P. J. Large, and J. M. Wallis, *FEMS Microbiol. Lett.* **15,** 79 (1982).

[8] V. Saxena, A. E. Weil, R. T. Kawahata, W. C. McGregor, and M. Chandler, *Int. Lab.* **18,** 50 (1988).

TABLE I
PURIFICATION OF METHYLAMINE DEHYDROGENASE: METHOD 1

Fraction	Total protein (mg)	Total activity (U)	Yield (%)	Specific activity (U/mg)
Step 1. Cell-free extract[a]	6030	550	100	0.1
Step 2. DEAE-Sepharose	1440	570	104	0.4
Step 3. Dialysis residue	260	520	95	2.0

[a] Prepared from 40 g cell paste.

Step 1: Preparation of Cell-Free Extract. Cell paste (130 g) is suspended in an equal volume of 10 mM potassium phosphate buffer, pH 7.0, to which 4 ml of a 0.1 M solution of phenylmethylsulfonyl fluoride in ethanol is added and 10 mg each of DNase and RNase. The cells (the suspension is at 4°) are disrupted by passage through a Gaulin homogenizer (Manton-Gaulin S.A.) at 70 MPa. After centrifugation at 48,000 g for 20 min at 4°, the pellet is suspended in an equal volume of buffer and passed again through the homogenizer. After centrifugation at 48,000 g for 20 min at 4°, the supernatants are combined (~250 ml), giving the cell-free extract.

Step 2: Hydrophobic Interaction Chromatography. To the cell-free extract (at room temperature), $(NH_4)_2SO_4$ is added to a concentration of 1.0 M. The mixture is centrifuged at 48,000 g for 10 min, and the resulting supernatant is applied to a 500-ml radial flow column (Sepragen) containing Phenyl-Sepharose CL-4B (Pharmacia) equilibrated with 10 mM potassium phosphate buffer, pH 7.0, containing 1.0 M $(NH_4)_2SO_4$. The column is washed with 3 bed volumes of the same buffer (cytochrome c-550 passes through and is collected). Amicyanin, cytochrome c-552, and methylamine dehydrogenase are eluted (in that order) with a linear gradient of $(NH_4)_2SO_4$ in 10 mM potassium phosphate, pH 7.0 (1.0–0 M, 18 bed volumes), at a flow rate of 30 ml/min. The enzyme and the electron acceptors can be easily detected in the fractions (250 ml) according to their color: methylamine dehydrogenase, yellow-green; amicyanin, blue; cytochrome c, red-yellow.

Step 3: Anion-Exchange Chromatography. Pooled fractions of methylamine dehydrogenase are concentrated by means of cross-flow filtration over 30,000 NMWL polysulfone filters (with a total filter area of 240 cm², Minitan, Millipore) at 4°. The concentrate (80 ml) is diluted to 500 ml with 10 mM potassium phosphate buffer, pH 7.0, and concentrated again to 80 ml. The concentrate is applied to a 500-ml radial flow column (Sepragen) packed with DEAE-Sepharose CL-6B Fast Flow (Pharmacia,

TABLE II
PURIFICATION OF METHYLAMINE DEHYDROGENASE: METHOD 2

Fraction	Total protein (mg)	Total activity (U)	Yield (%)	Specific activity (U/mg)
Step 1. Cell-free extract[a]	10,600	1060	100	0.1
Step 2. Phenyl-Sepharose	1980	1200	113	0.6
Step 3. DEAE-Sepharose	660	1050	99	1.6
Step 4. Fractogel	420	1050	99	2.5

[a] Prepared from 130 g cell paste.

Uppsala, Sweden), equilibrated with 10 mM potassium phosphate, pH 7.0. The column is washed with 3 bed volumes of 10 mM potassium phosphate, pH 7.0. Subsequently, methylamine dehydrogenase is eluted with a linear salt gradient (0–0.5 M NaCl in 10 mM potassium phosphate buffer, pH 7.0, 18 bed volumes) at a flow rate of 100 ml/min. The methylamine dehydrogenase-containing fractions are pooled and centrifuged at 48,000 g for 10 min.

Step 4: Gel Filtration. The pooled fractions from Step 3 are first concentrated to 80 ml by cross-flow filtration, as described under Step 3, and subsequently, by means of pressure filtration (on a Pellicon type PTHK membrane, cutoff 10^5 Da., Millipore) to 6 ml. Samples of 3 ml are applied to a Fractogel (Merck, Darmstadt, FRG) 50S column (2 × 35 cm), equilibrated with 10 mM potassium phosphate buffer, pH 7.0, and eluted with the same buffer at a flow rate of 15 ml/hr. Enzyme-containing fractions are pooled and concentrated by pressure filtration until a protein concentration of 10 mg/ml is attained. The final preparation is stored at −20°.

Table II summarizes the results of a typical purification with Method 2.

Properties

Purity. Both procedures give final preparations that are homogeneous, as revealed by electrophoresis and by a peak purity check method.[9]

Structure of Enzyme and Nature of Cofactor. The enzyme has a molecular weight of 123,500[1] and contains two types of subunits (molecular weights of 47,500 and 12,900[1]), forming an $\alpha_2\beta_2$ tetramer. After treatment of the oxidized form of the enzyme with phenylhydrazine, a product can be detached from the protein which is identical to the model hydrazone prepared from authentic pyrroloquinoline quinone (PQQ). Based on this observation, it has been proposed[2] that each small subunit contains one

[9] J. Frank Jzn, A. Braat, and J. A. Duine, *Anal. Biochem.* **162,** 65 (1987).

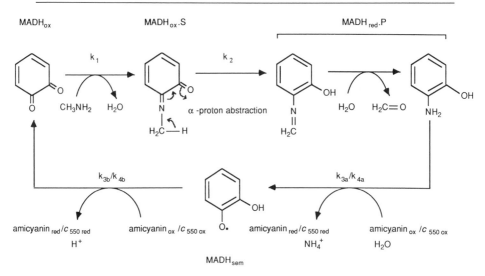

SCHEME 1. The redox cycle of methylamine dehydrogenase. Only the quinone moiety, participating in the redox reactions, is shown.

covalently bound PQQ (the quite different hexanol extraction procedure[10] gives identical results). However, studies on the three-dimensional structure of the semiquinone form of the enzyme have shown that the structure of the cofactor *in situ* is pro-PQQ,[11] which has been tentatively assigned a structure consisting of the indole moiety of PQQ to which a glutamate residue is attached (the latter being linked to the protein chain of the small subunit at two positions).

Spectral Properties. The $A_{280}^{0.1\%}$ value of the enzyme is 1.14, as determined by a spectrophotometric method.[12] Three redox forms of methylamine dehydrogenase are known. The oxidized enzyme form shows a characteristic absorption band at 440 nm (21,000 M^{-1} cm^{-1}); the semiquinone form (formed by titration of the oxidized enzyme form with sodium dithionite in an anaerobic cuvette) has an absorption maximum at 426 nm (33,000 M^{-1} cm^{-1}). Reaction of oxidized enzyme with methylamine leads to the reduced form with an absorption band at 328 nm (40,000 M^{-1} cm^{-1}).

[10] R. A. van der Meer, B. W. Groen, M. A. G. van Kleef, J. Frank, J. A. Jongejan, and J. A. Duine, this volume [41].

[11] F. M. D. Vellieux, F. Huitema, H. Groendijk, K. H. Kalk, J. Frank Jzn, J. A. Jongejan, J. A. Duine, K. Petratos, J. Drenth, and W. G. J. Hol, *EMBO J.* **8**, 2171 (1989).

[12] J. van Iersel, J. Frank Jzn, and J. A. Duine, *Anal. Biochem.* **151**, 196 (1985).

Enzyme Cycle and Reaction with Natural Electron Acceptors. Kinetic data[13] support an enzyme cycle as depicted in Scheme 1. Using the *in vitro* kinetic constants ($k_1 = 3.80 \times 10^5 \ M^{-1} \ \text{sec}^{-1}$, $k_2 = 150 \ \text{sec}^{-1}$, $k_3 = 2.0 \times 10^4 \ M^{-1} \ \text{sec}^{-1}$, and $k_4 = 1 \times 10^3 \ M^{-1} \ \text{sec}^{-1}$) and estimated concentrations of the components in the periplasm, an overall rate of methylamine oxidation can be calculated which has the same magnitude as the measured rate of methylamine-dependent respiration *in vivo*.[13]

In *Thiobacillus versutus* both amicyanin and cytochrome c-550 are able to accept electrons from methylamine dehydrogenase. From the kinetic model it can be concluded that 40% of the reducing equivalents flow directly from methylamine dehydrogenase to cytochrome c-550 in spite of the fact that $k_3 = 20k_4$ (the rate of electron transfer between amicyanin and cytochrome c-550 is $1.5 \times 10^3 \ M^{-1} \ \text{sec}^{-1}$).

[13] J. E. van Wielink, J. Frank Jzn, and J. A. Duine, *in* "PQQ and Quinoproteins" (J. A. Jongejan and J. A. Duine, eds.), p. 269. Kluwer Academic Publishers, Dordrecht, 1989.

[38] Methylamine Dehydrogenases from Methylotrophic Bacteria

By Victor L. Davidson

$$CH_3NH_2 + H_2O + 2 \ A \rightarrow CH_2O + NH_3 + 2 \ A^- + 2 \ H^+$$

Methylamine dehydrogenase (MADH) catalyzes the oxidation of methylamine to formaldehyde and ammonia, and in the process transfers two electrons from the substrate to some electron acceptor. MADHs have been isolated from a variety of methylotrophic and autotrophic bacteria.[1-7] Each enzyme possesses, as a prosthetic group, a covalently bound form of pyrroloquinoline quinone (PQQ).[8] The MADHs that have been isolated thus far exhibit similar kinetic and spectral properties, molecular weights, and amino acid compositions, but a wide range of p*I* values. The enzymes

[1] M. Husain and V. L. Davidson, *J. Bacteriol.* **169**, 1712 (1987).
[2] G. W. Haywood, N. S. Jansche, P. J. Large, and J. M. Wallis, *FEMS Microbiol. Lett.* **15**, 79 (1982).
[3] W. C. Kenney and W. McIntire, *Biochemistry* **22**, 3858 (1983).
[4] S. A. Lawton and C. Anthony, *Biochem. J.* **228**, 719 (1985).
[5] T. Matsumoto, *Biochem. Biophys. Acta* **522**, 291 (1978).
[6] S. Shirai, T. Matsumoto, and J. Tobari, *J. Biochem.* **83**, 1599 (1978).
[7] F. M. D. Vellieux, J. Frank, M. B. A. Swarte, H. Groendijk, J. A. Duine, J. Drenth, and W. G. J. Hol, *Eur. J. Biochem.* **154**, 383 (1986).
[8] J. A. Duine, J. Frank, and J. A. Jongejan, *Adv. Enzymol.* **59**, 165 (1987).

isolated from autotrophic bacteria, such as *Paracoccus denitrificans,* exhibit acidic pI values, whereas the enzymes isolated from obligate and restricted methylotrophs, such as bacterium W3A1, exhibit basic pI values. As such, different purification schemes are required for acidic and basic MADHs.

Assay Method

Principle. MADH activity is assayed spectrophotometrically with a dye-linked assay which is a modification of that first described by Eady and Large.[9] Oxidized phenazine ethosulfate (PES) is used as an electron acceptor. It is then reoxidized by 2,6-dichlorophenolindophenol (DCPIP), the reduction of which can be monitored by a decrease in absorbance at 600 nm.

Reagents

PES, 60 mM (freshly prepared)
DCPIP, 1.7 mM
Potassium phosphate buffer, 0.2 M, pH 7.5
Methylamine-HCl, 10 mM

Procedure. The standard assay mixture contains, in a volume of 1 ml, 0.5 ml of potassium phosphate buffer, 80 μl of PES, 0.1 ml of DCPIP, 10 μl of methylamine-HCl, and an appropriate amount of the enzyme or extract. The reaction is initiated by the addition of methylamine. Assays are performed at 30°. Initial velocities are determined from the rate of reduction of DCPIP, which is monitored at 600 nm ($\epsilon = 21,500 \ M^{-1} \ \text{cm}^{-1}$).

Definition of Unit and Specific Activity. One unit of enzyme activity is defined as the amount of enzyme required to reduce 1 μmol of DCPIP per minute at 30°. Specific activity is defined as units per milligram of protein as determined by the method of Bradford[10] with Pierce Protein Reagent (Pierce, Rockford, IL).

Purification Procedures

Paracoccus denitrificans Methylamine Dehydrogenase[1]

Growth of Organism. Paracoccus denitrificans (ATCC 13543) is grown aerobically at 30° in a medium based on that of Kornberg and Morris[11] and containing the following: K_2HPO_4 (0.6%); KH_2PO_4 (0.4%);

[9] R. R. Eady and P. J. Large, *Biochem. J.* **106,** 245 (1968).
[10] M. M. Bradford, *Anal. Biochem.* **72,** 248 (1976).
[11] H. L. Kornberg and J. G. Morris, *Biochem. J.* **95,** 577 (1968).

NH_4Cl (0.32%); $MgSO_4 \cdot 7H_2O$ (0.02%); $Na_2MoO_4 \cdot 2H_2O$ (0.0015%); $MnSO_4 \cdot H_2O$ (0.0005%); yeast extract (0.01%); $CaCl_2$ (0.004%); $NaHCO_3$ (0.05%); iron–citrate, prepared by combining 0.11% $FeSO_4 \cdot 7H_2O$ and 0.105% citric acid (5 ml/liter); and as a carbon source $CH_3NH_2 \cdot HCl$ (0.3%). The latter four components are separately sterilized by filtration and added to the remainder of the medium after it has been autoclaved and cooled. Cells are harvested in late log phase.

Preparation of Periplasmic Extract. MADH is purified from the periplasmic fraction of cells, which is prepared essentially as described by Alefounder and Ferguson.[12] Cells are suspended and homogenized in 20 ml per g wet weight of cells of 0.2 M Tris-HCl, pH 7.3, 0.5 M sucrose, and 0.5 mM EDTA. After warming the suspension to 30°, 5 mg per g wet weight of cells of lysozyme, which has been dissolved in water to 10 mg/ml, is added. After swirling briefly, an equal volume of water warmed to 30° is added to the suspension to induce a mild osmotic shock. The suspension is incubated for 20 min at 30° while being stirred with a glass rod. The suspension is centrifuged for 20 min at 20,000 g to remove spheroplasts and any unbroken cells from the periplasmic fraction.

DEAE-Cellulose Chromatography. The periplasmic fraction is concentrated by ultrafiltration and dialyzed against 10 mM potassium phosphate, pH 7.2. This sample is applied to a Whatman DE-52 cellulose column which has been equilibrated in the dialysis buffer. The column is washed with approximately 3 bed volumes of the starting buffer, then eluted with a linear gradient of 0–400 mM NaCl in 10 mM potassium phosphate, pH 7.2. Fractions containing methylamine dehydrogenase are pooled, concentrated by ultrafiltration, and dialyzed against 10 mM potassium phosphate, pH 7.2.

DEAE-Trisacryl Chromatography. The MADH-containing fractions from the DE-52 column are applied to a DEAE-Trisacryl (IBF Biotechnics, Savage, MD) column which has been equilibrated in 10 mM potassium phosphate, pH 7.2. After washing with 1 bed volume of the starting buffer, the column is eluted with a linear gradient of 100–300 mM NaCl in the equilibration buffer. Yellowish green fractions exhibiting MADH activity eluted at approximately 200 mM NaCl. These fractions are pooled, concentrated, and stored frozen in 10% ethylene glycol for future use.

Bacterium W3A1 Methylamine Dehydrogenase

Growth of Organism. Bacterium W3A1 (NCIB 11348) is grown aerobically at 30° in the mineral base E medium of Owens and Keddie,[13] with

---

[12] P. R. Alefounder and S. J. Ferguson, *Biochem. Biophys. Res. Commun.* **98**, 778 (1981).
[13] J. D. Owens and R. M. Keddie, *J. Appl. Bacteriol.* **32**, 338 (1969).

$CH_3NH_2 \cdot HCl$ (0.3%) present as the carbon source. The latter is separately sterilized by filtration and added to the remainder of the medium after it has been autoclaved and cooled. Cells are harvested in late log phase.

Preparation of Cell-Free Extract. Cells are frozen and thawed, then suspended and homogenized in 0.1 M potassium phosphate, pH 7.2, which includes 1% aprotinin, 1 mM $MgCl_2$, and DNase. This suspension is sonicated on ice for 5 min with a Branson Sonifier 250 at power setting 6 and 50% duty cycle. After centrifugation at 20,000 g for 20 min, the supernatant is saved and the pellet is again homogenized and sonicated as described above. The supernatant from the second treatment is combined with the first, concentrated by ultrafiltration, and dialyzed against 10 mM potassium phosphate, pH 7.2.

DEAE-Cellulose Chromatography. The dialyzed supernatant is applied to a Whatman DE-52 cellulose column which has been equilibrated in 10 mM potassium phosphate, pH 7.2. The column is washed with approximately 2 bed volumes of the equilibration buffer. MADH is not retained on this column.

CM-Cellulose Chromatography. The eluant from the DE-52 column is then directly applied to a Whatman CM-52 cellulose column which has been equilibrated in 10 mM potassium phosphate, pH 7.2. The column is washed with approximately 1 bed volume of the equilibration buffer and then eluted with a linear gradient of 0–150 mM NaCl in 10 mM potassium phosphate, pH 7.2. MADH elutes as a yellowish greenish band at approximately 75 mM NaCl.

Comments on Purification Procedures

Either of these two protocols may be applied to purification of MADH from total cell extracts or periplasmic cell fractions. Each of the MADHs for which a subcellular localization is thus far known functions in the periplasmic space.[1,14] As such, it is desirable to start from a periplasmic fraction, as this starting material is greatly enriched for MADH. Although the fractionation procedure works quite well for *P. denitrificans,* we have experienced difficulty in cleanly fractionating the periplasm from bacterium W3A1 and other obligate methylotrophs. For this reason the latter protocol uses the total cell extract as a starting material.

In *P. denitrificans* and bacterium W3A1, and perhaps all bacteria, MADH is an inducible enzyme which is present only during growth on methylamines as a sole source of carbon.[1,15] Thus, the choice of carbon

[14] A. A. Kasprzak and D. J. Steenkamp, *J. Bacteriol.* **156,** 348 (1983).
[15] V. L. Davidson, *J. Bacteriol.* **164,** 941 (1985).

source is critical if one is to obtain a reasonable yield of the enzyme. Yields of MADH are variable owing to variability in the levels of synthesis of the enzyme and in the efficiency of the fractionation procedures used. Average yields are approximately 2 to 3 mg/g wet weight of cells for bacterium W3A1 and 1 to 2 mg/g wet weight of cells for *P. denitrificans*. MADH is stable for at least 2 months when stored frozen in 10% ethylene glycol at −20°. For longer storage it is recommended that samples be kept at −80°.

Resolution and Reconstitution of Subunits[1,16]

The large and small subunits of MADH can be prepared by incubation of the holoenzyme overnight at 25° in 10 mM potassium phosphate, pH 7.0, which contains 6 M guanidine-HCl, followed by gel filtration with Sephadex G-100 which has been equilibrated with the incubation buffer. The larger subunit is monitored by absorbance at 280 nm, and the smaller subunit, which retains the covalently bound PQQ, may be monitored by absorbance at 440 nm. Following dialysis in 10 mM potassium phosphate, pH 7.0, to remove guanidine-HCl, reconstitution of the *P. denitrificans* and bacterium W3A1 MADHs from their respective subunits is achieved by incubating equimolar concentrations of the isolated large and small subunits for 24 hr at 4° in 10 mM potassium phosphate, pH 7.0. For each enzyme, reconstitution of approximately 70% of the original activity of the native enzyme is obtained.

Properties

Physical Properties.[1,3,17] Each MADH which has been isolated thus far exists as an $\alpha_2\beta_2$ complex of subunits, with PQQ covalently bound to each smaller subunit. Several physical properties of the enzymes from *P. denitrificans* and bacterium W3A1 are summarized in Table I. All MADHs have characteristic absorbance spectra. Extinction coefficients at various wavelengths for the different redox forms of *P. denitrificans* MADH have been obtained from redox titrations.[17] The oxidized enzyme exhibited absorbance maxima at 280 nm ($\epsilon = 178$ mM^{-1} cm^{-1}) and 440 nm ($\epsilon = 26.2$ mM^{-1} cm^{-1}) with a shoulder at 330 nm ($\epsilon = 20.6$ mM^{-1} cm^{-1}). Reduction by methylamine caused a loss of most of the absorbance at 440 nm ($\epsilon = 1.2$ mM^{-1} cm^{-1}), with a residual peak centered at 416 nm, and an increase in absorbance at 330 nm ($\epsilon = 56.4$ mM^{-1} cm^{-1}).

Stability.[1] MADHs are very resistant to denaturation at extremes of

[16] V. L. Davidson and J. W. Neher, *FEMS Microbiol. Lett.* **44**, 121 (1987).
[17] M. Husain, V. L. Davidson, K. A. Gray, and D. B. Knaff, *Biochemistry* **26**, 4139 (1987).

TABLE I
PROPERTIES OF METHYLAMINE DEHYDROGENASES

Properties	Bacterium W3A1	P. denitrificans
Molecular weight	120,000	124,000
Subunit molecular weights	42,000, 15,500	46,700, 15,500
pI	>8.0	4.3
Oxidation–reduction midpoint potential	N.D.[a]	+100 mV
pH optimum	7.5	7.5
A_{440}/A(oxidized)	7.2	6.8
Heat stable at 70°	Yes	Yes

[a] Not determined.

temperature and pH. *Paracoccus denitrificans* MADH retained 100% of its activity after a 30-min incubation at 70° and approximately 65% of its activity after incubation at 80°. At temperatures above 80° denaturation occurred more readily. When incubated for 48 hr at 30° in buffers with pH values ranging from 3.0 to 10.5, no decrease in activity is observed at any of the pH values.

Kinetic Properties.[18] The data obtained from steady-state kinetic analysis of *P. denitrificans* MADH with methylamine and PES as substrates was indicative of a Ping-Pong type of mechanism in which the aldehyde is released from the enzyme prior to the interaction with the reoxidant. *Paracoccus denitrificans* MADH exhibited K_m values of 10 μM for methylamine and 300 μM for PES and a V_{max} value of 15.3 μmol/min/mg of protein, which corresponds to a k_{cat} value of 1900 min^{-1}. The enzyme reacted with a variety of primary aliphatic amines, but not secondary, tertiary, and aromatic amines and amino acids. The substrate specificity of MADH was determined primarily by the K_m value for that amine. The enzyme exhibited K_m values of 170 μM for propylamine, 4.5 mM for butylamine, 3.5 mM for 1,3-diaminopropane, 630 μM for ethanolamine, 1.5 mM for histamine, and 45 mM for spermidine. Very similar results have been obtained for bacterium W3A1 MADH. For *P. denitrificans* MADH, a deuterium kinetic isotope effect with a V_{max} of 3.0 was also observed with CD_3NH_2, suggesting a mechanism in which removal of a methyl proton contributes significantly to the rate of methylamine oxidation.

Acknowledgment

This work was supported in part by Grant GM-41574 from the National Institutes of Health.

[18] V. L. Davidson, *Biochem. J.* **261**, 107 (1989).

[39] Methylamine Dehydrogenase from *Methylobacillus flagellatum*

By MICHAEL Y. KIRIUKHIN, ANDREY Y. CHISTOSERDOV, and
YURI D. TSYGANKOV

Methylamine + pyrroloquinoline quinone + $H_2O \rightarrow$
formaldehyde + reduced pyrroloquinoline quinone

Methylamine dehydrogenase catalyzes the oxidation of methylamine to formaldehyde and ammonia in the presence of an artificial electron acceptor. The enzyme is induced during growth of the bacterium on methylamine. Methylamine dehydrogenases have been purified from various methylotrophic bacteria.[1-5]

Assay Method

Principle. The assay method involves the continuous measurement of the reduction of the artificial electron acceptor phenazine methosulfate (PMS).[6] 2,6-Dichlorophenolindophenol (DCPIP) is used as a redox indicator. The molar extinction coefficient for DCPIP is taken from the work of Armstrong.[7]

Reagents

Potassium phosphate buffer, 200 mM, pH 7.5
Phenazine methosulfate, 10 mM
2,6-Dichlorophenolindophenol, 1 mM
Methylamine hydrochloride, 200 mM

Procedure. The assay mixture contains, in a final volume of 1 ml, potassium phosphate buffer, 0.5 ml; water, 0.19 ml; DCPIP, 0.1 ml; PMS, 0.1 ml; cell extract, 10 μl. The mixture is incubated for 5 min at 30°. After preincubation 0.1 ml of methylamine hydrochloride is added, and the rate of decrease in absorbance is followed at 600 nm.

[1] C. Anthony, "The Biochemistry of Methylotrophs." Academic Press, London, 1982.
[2] M. Husain and V. Davidson, *J. Bacteriol.* **169,** 1712 (1987).
[3] W. C. Kenny and W. McIntire, *Biochemistry* **22,** 3858 (1983).
[4] G. W. Heywood, N. S. Janschke, P. J. Large, and J. M. Wallis, *FEMS Microbiol. Lett.* **15,** 79 (1982).
[5] F. M. D. Vellieux, M. D. J. Frank, M. B. A. Swarte, H. Groendijk, J. A. Duine, J. Drenth, and W. G. J. Hol, *Eur. J. Biochem.* **154,** 383 (1986).
[6] R. R. Eady and P. J. Large, *Biochem. J.* **106,** 245 (1968).
[7] J. Armstrong, *Biochim. Biophys. Acta* **86,** 194 (1964).

Units. One unit of methylamine dehydrogenase is defined as that amount of enzyme catalyzing reduction of 1 μmol of DCPIP per minute at 30°. Specific activity is expressed as units per milligram of protein.

Purification Procedure

Growth of Organism. The strain of *Methylobacillus flagellatum*[8] is grown at 42° in a fermentor in a mineral salts medium containing, per liter: Na_2HPO_4, 6 g; KH_2PO_4, 3 g; NaCl, 0.5 g; NH_4Cl, 1 g; $CaCl_2$, 0.11 g; $MgSO_4 \cdot 7H_2O$, 0.5 g; and methylamine, 3 g. After 72 hr of growth, the culture is harvested by centrifugation at 3500 *g* for 60 min, and the pellet is rinsed with cooled 100 mM potassium phosphate buffer, pH 7.0. The cell paste is stored at $-20°$.

Step 1: Preparation of Cell-Free Extract. Cell paste (30 g wet weight) is suspended in 150 ml of 100 mM Tris-HCl buffer, pH 8.0. The suspension is disrupted by ultrasonication for 5 min (10 times, 30 sec each) at 0° in an MSE-150 ultrasonic disintegrator and centrifuged at 25,000 *g* for 60 min, yielding the cell-free extract as supernatant.

Step 2: Heat Treatment. The crude extract is heated at 75° for 20 min with slow mixing. This suspension is cooled to 5° and centrifuged at 25,000 *g* for 20 min. The supernatant is used further.

Step 3: Ammonium Sulfate Fractionation. To the supernatant, solid $(NH_4)_2SO_4$ is added with stirring up to 70% of saturation and kept at 4° for 1 hr. The pellet is collected by centrifugation at 25,000 *g* for 20 min and dissolved in a minimal volume of 100 mM Tris-HCl buffer, pH 7.9.

Step 4: Gel Filtration. The protein solution from Step 3 is applied to a column (2.6 × 35 cm) of Sephadex G-75 previously equilibrated with 20 mM Tris-HCl buffer, pH 7.9. Methylamine dehydrogenase is eluted with equilibration buffer. Fractions containing activity are pooled.

Step 5: DEAE-Toyopearl Chromatography. The enzyme solution from Step 4 is applied to a column (1.6 × 17 cm) of DEAE-Toyopearl, equilibrated with 20 mM Tris-HCl buffer, pH 7.9. The column is washed with 2 volumes of the same buffer (rate 0.8 ml/min). A portion of the methylamine dehydrogenase activity is eluted during washing. This fraction (Fraction 1) does not bind to the ion exchanger but is slightly retained on the column. Another portion of the activity is eluted with a linear gradient (200 ml) of 0 to 500 mM KCl in starting buffer. Methylamine dehydrogenase is eluted in a single peak in the interval 25 to 50 mM of KCl. Fractions containing activity are pooled and adjusted with glycerol up to a

[8] N. I. Govorukhina, L. V. Kletsova, Y. D. Tsygankov, Y. A. Trotsenko, and A. I. Netrusov, *Mikrobiologija* **56**, 849 (1988).

concentration of 50% (Fraction 2). The glycerol solution of methylamine dehydrogenase is stored at $-20°$.

A typical purification scheme is shown in Table I.

Properties

Purity. The enzyme preparation of Fraction 2 is homogeneous according to gel filtration and electrophoresis data. When subjected to sodium dodecyl sulfate–polyacrylamide gel electrophoresis (SDS–PAGE),[9] the preparation exhibited two bands with M_r values of 43,000 and 16,000 after staining with Coomassie blue. Gel filtration on Sephacryl G-200 and fast protein liquid chromatography (FPLC) chromatography on a Superose 12 column (Pharmacia LKB) revealed one symmetrical peak with an M_r value 60,000 ± 5,000. In both cases elution was carried out with 0.2 M NaCl in 0.3 M Tris-HCl buffer, pH 7.5.

The enzyme from Fraction 1 is not homogeneous: analysis of 80 μg by SDS–PAGE showed a set of protein bands. Two of these bands correspond to bands of Fraction 2. They are major. Gel filtration under the same conditions reveals a peak with an M_r value 55,000 ± 5,000.

Isoelectric Focusing. Isoelectric focusing of methylamine dehydrogenase Fraction 2 showed two bands of pI 7.40 and 7.80 after Coomassie blue staining. Isoelectric focusing of Fraction 1 showed one major band of pI 8.80.

Substrate Specificity. Neither fraction oxidizes alanine, glycine, 1,4-diaminobutane, lysine, 1-phenylethylamine, or dimethylamine. Activities with ethylamine comprise 20% for the Fraction 2 preparation and 43% for Fraction 1 in comparison with methylamine. Allylamine irreversibly inhibits both enzymes.

Stability. Both enzyme fractions are stable at room temperature. Solutions in 100 mM Tris-HCl buffer, pH 8.0, are stable for 1 month at 4°; in the same buffer with 50% glycerol they are stable for at least 4 months at $-20°$. Enzyme activities, under the assay conditions, increase linearly up to 75° (6-fold increase compare to 30°). Purified methylamine dehydrogenase does not change its activity after a 30-min incubation at 70° and loses 95% of activity after a 15-min incubation at 80°.

pH Optimum. The enzyme has optimum activity in both fractions at pH 7.5–8.0.

Michaelis Constants. The apparent K_m for the Fraction 2 enzyme is 1.4×10^{-5} M for methylamine (determined at 1.5 mM PMS concentration); for PMS it is 1.45×10^{-4} M (determined at 200 μM methylamine).

[9] U. K. Laemmli, *Nature (London)* **227**, 680 (1970).

TABLE I

PURIFICATION OF METHYLAMINE DEHYDROGENASE FROM *Methylobacillus flagellatum*

Fraction	Total volume (ml)	Total protein (mg)	Total activity (units)	Specific activity (units/mg protein)	Yield (%)
Step 1. Cell-free extract	131	2096	114	0.054	100
Step 2. Heat treatment, 75°, 20 min	125	1125	107	0.095	94
Step 3, 4. $(NH_4)_2SO_4$ fraction (0–70%) after Sephadex G-75 gel filtration	21	63	43	0.68	38
Step 5. Pooled fraction 1 after washing DEAE-Toyopearl	22	6.2	9	1.45	8
Step 5. Pooled fraction 2 after elution with linear gradient KCl (0–0.5 M)	12	11	30	2.72	26

For the Fraction 1 enzyme the apparent K_m for methylamine is $2 \times 10^{-5} M$ (determined at 2.0 mM PMS); for PMS it is $2.10 \times 10^{-4} M$ (determined at 200 μM methylamine).

Molecular Weight and Subunit Structure. Methylamine dehydrogenase from *Methylobacillus flagellatum* is heterodimer ($\alpha\beta$) (see above). Methylamine dehydrogenases from other bacteria have four subunits ($\alpha_2\beta_2$).[1-5]

[40] Trimethylamine Dehydrogenase from Bacterium W3A1

By WILLIAM S. McINTIRE

$$(CH_3)_3NH^+ + H_2O + \text{acceptor} \rightarrow (CH_3)_2NH_2^+ + \text{formaldehyde} + \text{reduced acceptor}$$

$$(CH_3)_2NH_2^+ + H_2O + \text{acceptor} \rightarrow CH_3NH_3^+ + \text{formaldehyde} + \text{reduced acceptor}$$

Trimethylamine dehydrogenase is the first enzyme involved in the assimilation of trimethylamine and dimethylamine by the gram-negative restricted facultative methylotrophs bacterium W3A1[1,2] and *Hyphomicrobium* X.[3] Trimethylamine dehydrogenase (EC 1.5.99.7) has been purified

[1] J. Colby and L. J. Zatman, *Biochem. J.* **148**, 505 (1975).
[2] O. Jenkins and D. Jones, *J. Gen. Microbiol.* **133**, 453 (1987).
[3] T. Urakami and K. Komagata, *J. Gen. Appl. Microbiol.* **33**, 521 (1987); W. Harder and M. M. Atwood, *Adv. Microb. Physiol.* **17**, 303 (1978).

from these organisms[4-6] and other methylotrophs.[7,8] *Hyphomicrobium* X also produces a distinct dimethylamine dehydrogenase.[5,6] The physical, chemical, spectral, and kinetic properties of the trimethylamine dehydrogenases from W3A1 and *Hyphomicrobium* X and those of dimethylamine dehydrogenase are very similar.[6,9-11] The formaldehyde produced by W3A1 is assimilated via the Entner–Doudoroff, 2-keto-3-deoxy-6-phosphogluconate aldolase/sedoheptulose bisphosphate variant of the ribulose monophosphate cycle[12] (also known as the hexulose phosphate cycle[13]).

An electron-transferring flavoprotein (ETF), the *in vivo* electron acceptor for reduced trimethylamine dehydrogenase (TMADH), has been purified from bacterium W3A1. The FAD bound to ETF is converted to the anionic (red) semiquinone radical:[14]

$$TMADH_{red} + 2\ ETF \rightarrow TMADH_{ox} + 2\ ETF^{\overline{\cdot}}$$

ETF is often purified as a radical, which can be reoxidized with $K_3Fe(CN)_6$; however, reoxidized ETF is unstable, in that it loses its electron acceptor activity and tends to precipitate.[14] Since only limited amounts of stable, oxidized ETF can be prepared, routine assays are carried out with phenazine methosulfate (PMS) as the reoxidizing substrate.

$$TMADH_{red} + PMS \rightarrow TMADH_{ox} + PMSH_2$$

Assay Method

PMS, originally used as a reoxidant for mitochondrial oxidoreductases,[15,16] was first employed to assay trimethylamine dehydrogenase by

[4] D. J. Steenkamp and J. Mallinson, *Biochim. Biophys. Acta* **429**, 705 (1976).

[5] J. B. M. Meiberg and W. Harder, *J. Gen. Microbiol.* **106**, 265 (1978).

[6] D. J. Steenkamp, *Biochem. Biophys. Res. Commun.* **88**, 244 (1979).

[7] J. Colby and L. J. Zatman, *Biochem. J.* **143**, 555 (1974).

[8] V. L. Davidson, M. Husain, and J. W. Neher, *J. Bacteriol.* **166**, 812 (1986).

[9] D. J. Steenkamp and H. Beinert, *Biochem. J.* **207**, 241 (1982).

[10] A. A. Kasprazak, E. J. Papas, and D. J. Steenkamp, *Biochem. J.* **211**, 535 (1983).

[11] R. C. Stevens, W. R. Dunham, R. H. Sands, T. P. Singer, and H. Beinert, *Biochim. Biophys. Acta* **869**, 81 (1986).

[12] J. Colby and L. J. Zatman, *Biochem. J.* **148**, 513 (1975); C. Anthony, "The Biochemistry of Methylotrophs," p. 60. Academic Press, New York, 1982.

[13] P. Large, "Aspects of Microbiology, Volume 8: Methylotrophy and Methanogenesis," p. 25. Van Nostrand-Rienhart, Princeton, New Jersey, 1983.

[14] D. J. Steenkamp and M. Gallup, *J. Biol. Chem.* **253**, 4086 (1978).

[15] T. P. Singer and E. B. Kearney, *in* "Methods of Biochemical Analysis" (D. Glick, ed.), Vol. 4, p. 307. Wiley, New York, 1957.

[16] T. P. Singer, *in* "Methods of Biochemical Analysis" (D. Glick, ed.), Vol. 22, p. 123. Wiley, New York, 1974.

Colby and Zatman.[17] The enzymatic reduction of PMS is monitored by measuring the rapid reoxidation of $PMSH_2$ by disodium 2,6-dichlorophenolindophenol (DCPIP).

Reagents

Sodium pyrophosphate buffer, 0.1 M, pH 7.7
PMS, 20 mM in H_2O
DCPIP, 0.05% (w/v in H_2O)
Trimethylamine hydrochloride, 0.1 M in H_2O

Procedure. The assay is started by mixing 2.46 ml of buffer, 0.2 ml of DCPIP ($\sim$90 μM final), 0.3 ml PMS, and 40 μl of trimethylamine hydrochloride in a 3-ml cuvette (1-cm light path) positioned in the cell holder of a spectrophotometer. A blank rate is recorded before addition of 2–50 μl of crude or pure enzyme. The rate of DCPIP bleaching is measured at 600 nm. The ϵ_{600} value of 21.7 mM^{-1} cm^{-1} for DCPIP is employed to calculate the rate, since reduced DCPIP does not absorb at 600 nm. Inclusion of 0.1 to 0.5 mM KCN is advisable in crude fractions to prevent reoxidation of reduced DCPIP by cytochrome oxidase.[16]

Definition of Unit and Specific Activity. One unit of activity is defined as the amount of trimethylamine dehydrogenase required to reduce 1 μmol of DCPIP per minute at 30°. The specific activity is defined as units per milligram of protein. Protein is determined by the procedure of Lowry *et al.*[18]

Precautions. The recommended source for PMS is Sigma Chemical Co. (St. Louis, MO). Most batches are light brown to golden yellow. If the solid is dark brown, it has been improperly purified or has decomposed. Stock solutions should appear bright yellow; brownish or greenish solutions should not be used, although the author has found that some enzymes are insensitive to PMS contaminants. Do not prepare stock solutions of PMS in buffer. A stable acidic solution results when PMS is dissolved in water. PMS is extremely photosensitive and must be protected from light.[16] Stock solutions of 20 mM PMS can be kept for as long as 2 months at 4° in the dark.

Commercial preparations of disodium DCPIP are also of variable quality. As some batches of DCPIP dissolve slowly, it is recommended that 50 mg be stirred in approximately 80 ml of water for 1–2 hr, the resulting solution passed through a 0.45-μm filter, and the A_{600} measured after dilution of 0.2 ml with 2.8 ml of buffer. The A_{600} is adjusted to about 2.0 for the diluted sample, by the appropriate addition of water to the stock

[17] J. Colby and L. J. Zatman, *Biochem. J.* **132**, 101 (1973).
[18] O. H. Lowry, N. J. Rosenbrough, A. L. Farr, and R. L. Randall, *J. Biol. Chem.* **193**, 265 (1951).

solution. This solution is usable for several weeks if stored at 4° and protected from light. Stock solutions need to be warmed to the assay temperature, and possibly sonicated, before use to redissolve solid DCPIP which may have formed during storage.

The ϵ_{600} value of 21.7 mM^{-1} cm^{-1} presented earlier can only be used at pH 7.7, because DCPIP undergoes dramatic spectral changes as the pH is altered. The literature value of the pK_a determined from these changes is 5.90 ± 0.02 at 26°. The value of ϵ_{600}, for any pH, can be calculated from the value at pH 7.0 (20.60 mM^{-1} cm^{-1}).[19]

The measured rate from this assay is a hyperbolic function of [DCPIP].[7,20] In the author's hands, the above defined assay provides the true rate of PMS reduction using DCPIP from General Biochemical, Inc. (Chagrin Falls, OH) (DCPIP from another manufacturer and used by others[7] required a higher [DCPIP] to provide the true rate). This effect causes a greater problem at higher pH; for example, at pH 9.1, twice the [DCPIP] is needed ($\sim$ 180 μM), which results in an A_{600} of approximately 4.0 in a 1-cm light path cuvette. This necessitates using a 0.5-cm light path cell.[20] (A spectrophotometer with a linear response at $A \approx 2.0$ is an essential component of the assay.) For routine work, such as assaying fractions during purification, no problem arises, as long as the same concentrations of PMS and DCPIP are used; however, to obtain accurate and true kinetic measurements from this coupled dye assay, proper tests and controls must first be performed.

Under no circumstances should PMS and DCPIP be premixed. These dyes interact to produce inhibitory compounds.

Finally, PMS is attacked by HO$^-$, producing 2-hydroxy-5-methyldihydrophenazine,[21] which is oxidized to 2-hydroxy-5-methylphenazine by DCPIP. This phenomenon accounts for a blank rate, which increases as the pH is raised. A correction for this blank should be made for all assays. It may be desirable to substitute PMS with phenazine ethosulfate, since the latter provides a smaller blank rate.

Purification

Growth of Organism. Bacterium W3A1,[22] originally isolated by Colby and Zatman from soil,[1] is obtained from the National Collection of Indus-

[19] J. M. Armstrong, *Biochim. Biophys. Acta* **86**, 194 (1964).

[20] W. S. McIntire, Ph.D. Thesis, University of California, Berkeley, 1983.

[21] G. A. Swan and D. G. Felton, "The Chemistry of Heterocyclic Compounds: Phenazines," p. 14. Interscience, New York, 1957.

[22] Recently this organism was classified in the taxon *Methylophilus*[2] which is not recognized as a legitimate genus. We will continue to refer to this organism as bacterium W3A1 until it is classified in an accepted genus.

trial Bacteria, Ltd., Aberdeen, Scotland (NCIB 11348). The bacteria are grown at pH 6.7 in liquid cultures or on agar plates at 30° using mineral base E of Owens and Keddie,[23] which contains, per liter of deionized water: $(NH_4)_2SO_4$, 0.5 g; K_2HPO_4, 1.04 g; KH_2PO_4, 0.75 g; $CaCl_2 \cdot 2H_2O$, 25 mg; $MgSO_4 \cdot 7H_2O$, 0.2 g; NaCl, 0.1 g; 3 ml of a trace metals solution ($ZnSO_4 \cdot 7H_2O$, 0.55 g; $MnSO_4 \cdot H_2O$, 0.11 g; $FeSO_4 \cdot 7H_2O$, 0.13 g; $CoCl_2 \cdot 6H_2O$, 40 mg; $CuSO_4 \cdot 5H_2O$, 40 mg; $Na_2MoO_4 \cdot 2H_2O$, 40 mg; these are put in a 250-ml graduated cylinder with 10–15 ml water and 0.5 ml concentrated HCl and stirred until the salts dissolve; to this is added 1.25 g of disodium ethylenediaminetetraacetic acid and 90 ml of water, with the pH being adjusted to 4–5 with KOH, and then brought to 250 ml). The trace metals solution is stable for several months. Add the appropriate amount of 30% (w/v) filter-sterilized (0.2 μm) solution of the carbon source to the cooled autoclaved medium.

As received from the Culture Collection, W3A1 grew slowly and unpredictably on agar plates containing 0.3% (w/v) methylamine hydrochloride.[24] Some colonies, isolated from such plates, grew on methylamine, trimethylamine, and dimethylamine, and methanol, whereas others grew well on these C_1 compounds and various other carbon sources (e.g., citrate, glutamate, glucose, nutrient broth, tetramethylammonium chloride). Being a restricted facultative methylotroph, W3A1 grows rapidly only on methylamine, dimethylamine, trimethylamine, and methanol, but poorly on glucose; it does not grow on any other carbon source.[1,2] The correct strain was purified by alternately growing in small liquid culture and on agar plates containing 0.3% methylamine hydrochloride through several cycles, and testing for growth on other carbon sources. The pure W3A1 culture was slimy and had the defined growth requirements.

After subculturing W3A1 15 to 20 times in 0.3% (w/v) methylamine liquid medium, two forms of the bacterium were observed on agar plates (0.3% methylamine). Like the original pure culture, one form appeared raised and opaque, whereas the other was flat and transparent. In liquid culture, the opaque variant grew in clumps, and after centrifugation, the pellet was flocculant and pale yellow, as expected for "wild-type" W3A1. The other variant grew as a well-dispersed culture in liquid medium and produced a tight pink pellet when centrifuged. The former variant, which derives its appearance from a polysaccharide capsule engulfing each bacterium, was designated W3A1-S (slimy). The latter variant, W3A1-NS (nonslimy), probably retains a much reduced polysaccharide capsule. Apparently, W3A1-S can convert to W3A1-NS, but the converse has not been

[23] J. D. Owens and R. M. Keddie, *J. Appl. Microbiol.* **32**, 338 (1969).
[24] W. S. McIntire and W. Weyler, *Appl. Environ. Microbiol.* **53**, 2183 (1987).

observed. Except for the carbohydrate content, the variants are nearly identical.[24] Currently, only W3A1-NS is used as a source of the enzymes of interest to the author, as the slime from the "wild-type" has been troublesome in the past.[4] One-milliliter aliquots from mid-log phase liquid cultures, containing 15% (v/v) glycerol, are stored at $-70°$ in sterilized vials. These are used to inoculate liquid cultures (50 ml). Since the glycerol adversely affects growth, subculturing is required before inoculating larger culture flasks.

For large quantities of cells for enzyme isolation, W3A1-NS is grown in 12-liter fermentor jars with aeration. From a 50–100 ml inoculum, the culture will grow to a maximal optical density at 600 nm of approximately 0.9, in 16–20 hr. Six 12-liter fermentors provide 130 g of cell paste when harvested in a Sharples centrifuge. The paste can be stored at $-70°$ indefinitely.

It has been found that the pH of the medium drops from an initial value of 6.7 to 5.3 in stationary phase, when the bacteria are grown on trimethylamine, and from 6.7 to 5.6, when methylamine is used. Growth of W3A1-NS initiated in medium containing 0.3% (w/v) methylamine hydrochloride and adjusted to pH 7.2 with methylamine free base, gives a slightly higher yield of harvested cells. In contrast, when the culture is brought back to pH 7.2, with methylamine free base, and back to 0.3% methylamine hydrochloride just as it is entering stationary phase, the recovery of cell paste increases by 60–70%. Although it has not been tested, a higher yield may result when trimethylamine free base and trimethylamine hydrochloride are used in a similar fashion.

A caution is offered concerning the commercial source of C_1 compounds needed for bacterial growth. Using dimethylamine free base (BDH Chemicals, Poole, UK) to grow W3A1, trimethylamine dehydrogenase with low specific activity and modified spectral properties was obtained. It was found that a mechanism-based inhibitor contaminated the dimethylamine.[25] Problems have never been encountered by the author or others[25] when methylamine hydrochloride, trimethylamine hydrochloride, or dimethylamine hydrochloride supplied by Sigma Chemical Co. were used as the carbon source for cell growth.

Purification Procedure. The purification procedure has changed little since its development by Steenkamp and Mallinson. Table I, reproduced from their paper,[4] presents the pertinent information regarding the purification of trimethylamine dehydrogenase from "wild-type" W3A1 harvested from 12 liters of medium. (The presence of slime precluded the determination of the true wet weight of the cells.) The following is a

[25] D. J. Steenkamp, *Biochem. Biophys. Res. Commun.* **132**, 352 (1985).

TABLE I
PURIFICATION OF TRIMETHYLAMINE DEHYDROGENASE FROM BACTERIUM W3A1[a]

Fraction	Total volume (ml)	Total protein (mg)	Total activity (units)	Specific activity (units/mg protein)	Yield (%)
Step 1. Crude extract	75	660	37.3	0.057	100
Step 2. Ammonium sulfate fractionation	35	188	41.3	0.22	111
Step 3. DEAE-cellulose chromatography	16	27.8	26.6	0.96	71
Step 4. Sephadex G-200 chromatography	30	20.4	21.8	1.07	58

[a] From Steenkamp and Mallinson[4]

description of the procedure currently used by the author; differences from the original method are pointed out. All steps are done at 4° unless otherwise noted.

Step 1: Crude Extract. Frozen W3A1-NS cell paste (130 g) is thawed, suspended in approximately 450 ml of 50 mM potassium phosphate buffer, pH 7.2, at 4°, and passed through a 50-ml french pressure cell (American Instruments Co., Inc., Silver Spring, MD) at 5000–6000 psi to homogenize the suspension and decrease the viscosity caused by the small amount of slime still associated with this organism. (For W3A1-S a much larger amount of slime is present, which can be made more manageable by several passes through the pressure cell or homogenization in a Waring blendor.) The cells are disrupted by ultrasonication (Heat Systems-Ultrasonics, Inc., Cell Disruptor Model W-225R, Plainview, NY) in 110-ml batches at 100 W using a $\frac{3}{4}$-inch tip. By sonicating for 0.5–1 min/cycle, the temperature is kept between 0 and 7°; total sonication time is 3 min. The resulting suspension is centrifuged at 48,000 g for 15 min to remove unbroken cells and debris.

Step 2: Ammonium Sulfate Fractionation. The centrifuged crude extract is brought to 50% saturated $(NH_4)_2SO_4$ by the addition of the solid salt. The mixture is stirred for 1.5 hr and centrifuged at 48,000 g for 10 min. The supernatant is similarly brought to 80% saturated $(NH_4)_2SO_4$ and centrifuged. The pellet is dissolved in 69 ml of 50 mM potassium phosphate buffer, pH 7.2, and dialyzed against 6 liters of the same buffer for 16 hr.

Step 3: DEAE-Cellulose Chromatography. The dialyzed sample is applied to a 5×55 cm column containing Whatman DE-52, previously equilibrated with 50 mM potassium phosphate buffer, pH 7.2, and eluted with a 3-liter linear gradient from 0 to 0.5 M NaCl in this buffer. Trimethylamine dehydrogenase elutes at 0.18 M NaCl. Another enzyme of interest, methylamine dehydrogenase, does not bind to the DE-52, and ETF elutes at 0.25 M NaCl.

Step 4: Sephadex G-200 Chromatography. The tubes from the DE-52 column containing the purest trimethylamine dehydrogenase are combined, diluted to 350 ml with 50 mM potassium phosphate buffer, pH 7.0, and concentrated to about 10 ml in an ultrafiltration cell fitted with a Diaflow YM30 membrane filter (Amicon Corp., Danvers, MA). The solution is applied to a 1.5 × 115 cm Sephadex G-200 column previously equilibrated with the 50 mM elution buffer. Measurements of A_{278}/A_{443} ratios indicated that the purest enzyme (146 mg) elutes between 350 and 385 ml.

Step 5: Purification of ETF.[14] The ETF fraction from the DE-52 column is concentrated to a few milliliters (Diaflo YM30 membrane) and chromatographed on a 2.5 × 115 cm Sephadex G-100 column with 50 mM potassium phosphate buffer. This step yields another 11.4 mg of partially pure trimethylamine dehydrogenase (eluted between 270 and 300 ml) and 22.3 mg of pure ETF (eluted between 345 and 380 ml). Cold ethylene glycol (20%, v/v) is added to all fractions, which are stored at $-20°$.

Properties

Purity. Samples of trimethylamine dehydrogenase prepared by the method of Weber and Osborne[26] were analyzed by sodium dodecyl sulfate–polyacrylamide electrophoresis.[27] When properly denatured, a single band was observed.[10] By high-performance liquid chromatographic analysis on a Mono Q HR5/5 column (Pharmacia LKB Biotechnology, Piscataway, NJ), pure trimethylamine dehydrogenase eluted as a single peak (a 100-min linear gradient from 0 to 1.0 M KCl in 10 mM potassium phosphate buffer, pH 7.0; retention time 30.3 min).

Substrate Specificity. Trimethylamine dehydrogenase from bacterium W3A1 is known to oxidize trimethylamine, dimethylamine, triethylamine, and diethylamine, whereas only PMS and ETF have been used as the reoxidizing substrate for this enzyme.[4] A wider range of substrates have been tested for the analogous enzyme from bacterium 4B6[7] (best to worst for each group based on specific activity): (A) tertiary amines: trimethylamine, 2-hydroxyethyldiethylamine, 2-chloroethyldimethylamine, ethyldimethylamine, diethylmethylamine, 2-aminoethyldimethylamine, 2-dimethylamino-2-methylpropanol, dimethylaminoacetonitrile, triethy-

---

[26] K. Weber and M. Osborne, *J. Biol. Chem.* **244,** 4406 (1969).
[27] U. K. Laemmli, *Nature (London)* **227,** 680 (1970).

lamine; (B) secondary amines: Diethylamine, dimethylamine, ethylmethylamine, O,N-dimethylhydroxylamine (primary, quaternary, poly-, and diamines are not substrates); (C) electron-accepting substrates: PMS, methylene blue, brilliant cresyl blue, pyocyanine; DCPIP alone is not an electron acceptor.

Inhibitors. Trimethylamine dehydrogenase from bacterium W3A1 (1) is inhibited by the substrates diethylamine and PMS, (2) is subject to inhibition of amine oxidation by acetaldehyde, ethylamine, dimethylamine, and tetramethylammonium chloride, (3) is subject to inhibition of PMS reduction by ethylamine and acetaldehyde,[4,28] (4) is subject to mechanism-based inhibition by phenylhydrazine,[29] and (5) is inhibited by tris(hydroxymethyl)aminomethane buffer. Inhibitors for the enzyme from bacterium 4B6 include Cu^{2+}, Co^{2+}, Ni^{2+}, Hg^{2+}, Ag^+, iodoacetamide, N-ethylmaleimide, p-chloromercuribenzoate, trimethylsulfonium chloride (reversible), tetramethylammonium chloride (reversible), other quaternary amines, choline, betaine, and various mechanism-based hydrazine- and hydrazide-type monoamine oxidase B inhibitors.[7]

Stability. The enzyme from bacterium W3A1 is stable for several years if stored at $-20°$ in 20% (v/v) ethylene glycol or as a concentrated solution in buffer (~ 100 mg/ml).

Steady-State Kinetic Parameters. The maximal velocity is in the range of 1.1–1.4 unit/mg, for trimethylamine dehydrogenase from bacterium W3A1.[6] The computer-estimated values for the Michaelis constants, derived from the graphical data presented by Steenkamp and Mallinson,[4] are as follows: $K_{\text{trimethylamine}} = 2.0$ μM, $K_{\text{PMS}} = 0.42$ mM; $K_{\text{diethylamine}} = 1.4$ mM, and $K_{\text{PMS}} = 0.56$ mM (Ping-Pong type mechanism). The inhibition constants are as follows: for dimethylamine, $K_{\text{I}} = 0.59$ mM (mixed-type noncompetitive[30] with trimethylamine); for ethylamine, $K_{\text{I}} = 9.8$ and 7.4 mM (mixed-type noncompetitive with diethylamine and PMS, respectively); for acetaldehyde, $K_{\text{I}} = 5.6$ and 4.2 mM (mixed-type noncompetitive with dimethylamine and uncompetitive with PMS, respectively).

Molecular Weight and Subunit Structure. Trimethylamine dehydrogenase is a homodimer of molecular weight 166,000, as measured by denaturing electrophoresis. The amino acid composition of the enzyme is known.[10] Each subunit contains one molecule of FMN covalently bound via a cysteinyl thioether at the 6-position of the flavin,[31] one Fe_4S_4 cluster

[28] D. J. Steenkamp and H. Beinert, *Biochem. J.* **207**, 233 (1982).

[29] J. Nagy, W. C. Kenney, and T. P. Singer, *J. Biol. Chem.* **254**, 2684 (1979).

[30] I. H. Segal, "Enzyme Kinetics," p. 170. Wiley, New York, 1975.

[31] D. J. Steenkamp, W. C. Kenney, and T. P. Singer, *J. Biol. Chem.* **253**, 2812 (1978); D. J. Steenkamp, W. McIntire, and W. C. Kenney, *J. Biol. Chem.* **253**, 2818 (1978).

of the ferredoxin type,[32] and one molecule of ADP.[33] A 2.4-Å X-ray structure and "electron density 'X-ray' sequence" have been deduced.[33] The 8α-methyl group of the flavin is about 4 Å from and within van der Waals contact of a cysteinyl sulfur of the Fe_4S_4 cluster.

Trimethylamine dehydrogenase and glutathione reductase have been found to be structural homologs. The ADP-binding domain of trimethylamine dehydrogenase is very much like that surrounding the ADP moiety of FAD in the reductase, whereas the second domain of trimethylamine dehydrogenase is homologous with the NADPH-binding domain of the latter enzyme.[34] On the other hand, the structure of the FMN-binding domain is very similar to the β_8/α_8 barrel motif found in both glycolate oxidase and the flavin-binding domain of bakers' yeast flavocytochrome b_2 (also known as lactate dehydrogenase).[35]

Spectral and Oxidation–Reduction Properties. Trimethylamine dehydrogenase absorbs in the visible region with a λ_{max} of 443 nm ($\epsilon = 54.6$ mM^{-1} cm^{-1}).[10] The A_{278}/A_{443} ratio of 7.30 for pure enzyme can be used as a criterion of purity.

An anaerobic titration, monitored in the UV–visible spectral region, indicates that the enzyme consumes 3 electron equivalents/mole of subunit in the form of sodium dithionite, as expected for the presence of FMN (2 electrons) and Fe_4S_4 (1 electron). When reduced by trimethylamine at pH 7.7, only 2 electron equivalents/mole of subunit can be accommodated by trimethylamine dehydrogenase; the 2-electron reduced enzyme contains an anionic (red) flavin radical and reduced Fe–S cluster. As the titration with trimethylamine proceeds, an intense spin-coupled species forms, which is characterized by unique UV-visible and electron paramagnetic resonance (EPR) spectral properties (signals at $g \approx 2$ and $g \approx 4$). The amount of this species increases in a sigmodal fashion from zero at approximately 0.3 mol of trimethylamine/mole of subunit, to a maximum level (1.4 electrons/subunit by double integration of the EPR signals), when a stoichiometry of 7 mol of trimethylamine/mole of subunit has been achieved. Similar results are obtained at different pH values or when other substrates are used. The enzyme, again, consumes 2 electrons/subunit when titrated with dithionite in the presence of tetramethylammonium chloride, and the UV-visible and EPR spectra are attributed to the spin–

[32] C. L. Hill, D. J. Steenkamp, R. H. Holm, and T. P. Singer, *Proc. Natl. Acad. Sci. U.S.A.* **74**, 547 (1977).

[33] L. W. Lim, F. S. Mathews, and D. J. Steenkamp, *J. Biol. Chem.* **263**, 3075 (1988).

[34] L. W. Lim, F. S. Mathews, D. J. Steenkamp, R. Hamlin, and N. h. Xuong, *J. Biol. Chem.* **261**, 15140 (1986).

[35] Z.-X. Xia, N. Shamala, P. H. Bethge, L. W. Lim, H. D. Bellamy, N. h. Xuong, F. Lederer, and F. S. Mathews, *Proc. Natl. Acad. Sci. U.S.A.* **84**, 2629 (1987).

spin interaction species, which reaches its maximal concentration at the equivalence point. These results provide strong evidence for a site (possibly the active site) where substrate, or its analog, binds to modulate the interaction between the redox centers.[11,28,36]

Oxidation–reduction potentials for trimethylamine dehydrogenase from W3A1 are known: $+44$ and $+36$ mV for the FMN/FMN$^{\overline{\cdot}}$ and FMN$^{\overline{\cdot}}$/FMNH$^-$ couples, respectively, and $+102$ mV for the Fe$_4$S$_4$ cluster (microcoulometry, pH 7.0); $+55$ mV for the FMN/FMNH$^-$ 2-electron couple, and $+90$ mV for the Fe–S cluster (potentiometric/spectrophotometric titration, pH 7.0); -20 mV for the FMN/FMNH$^-$ 2-electron couple and $+72$ mV for the Fe–S center (potentiometric/spectrophotometric titration, pH 9.1).[37]

Acknowledgments

The research on trimethylamine dehydrogenase by the author is supported by a Department of Veterans Affairs grant (Dr. Thomas P. Singer, principal investigator), and Program Project Grant HL16251-16 from The Heart, Lung, and Blood Institute of The National Institutes of Health.

[36] D. J. Steenkamp, H. Beinert, W. McIntire, and T. P. Singer, in "Mechanisms of Oxidizing Enzymes" (T. P. Singer and R. N. Ondarza, eds.), p. 127. Elsevier, Amsterdam, 1978.
[37] M. J. Barber, V. Pollock, and J. T. Spence, Biochem. J. 256, 657 (1988); M. J. Barber, personal communication (1989).

[41] Isolation, Preparation, and Assay of Pyrroloquinoline Quinone

By R. A. van der Meer, B. W. Groen, M. A. G. van Kleef, J. Frank, J. A. Jongejan, and J. A. Duine

Pyrroloquinoline quinone (PQQ, 2,4,7-tricarboxy-1H-pyrrolo[2,3-f] quinoline-4,5-dione) is the cofactor of quinoproteins, enzymes occurring in prokaryotes as well as eukaryotes, and is distributed over several classes of enzymes.[1-3] Quinoproteins relevant to this volume are those catalyzing the oxidation of alcohols (methanol dehydrogenase, quinoprotein alcohol dehydrogenase, and quinohemoprotein alcohol dehydrogenase) and methylamine (methylamine dehydrogenase and methylamine oxidase). Since interest in the role and distribution of PQQ is rapidly expanding, procedures are given for its preparation and determination.

[1] J. A. Duine, J. Frank, and J. A. Jongejan, FEMS Microbiol. Rev. 32, 165 (1986).
[2] J. A. Duine, J. Frank, and J. A. Jongejan, Adv. Enzymol. 42, 169 (1987).
[3] J. A. Duine and J. A. Jongejan, Annu. Rev. Biochem. 58, 403 (1989).

Isolation from Biological Materials

Principle. Substantial amounts of free and directly extractable PQQ are produced by methylotrophic bacteria, certain alcohol-grown *Pseudomonas* species, and acetic acid bacteria.[4] The compound can be obtained from the spent culture medium or by extraction from whole cells. A convenient procedure exists[5] for purification that consists of adsorption to an anion exchanger, elution with a methanolic salt solution, and a crystallization step.

Pretreatment of Anion Exchanger. The weakly basic anion exchanger Amberlyst A21 is suspended in 5 times its own volume of water, and, after settling, the fines are removed by decanting. This process is repeated until the supernatant remains clear. After decanting, the ion exchanger is suspended in 3 times its own volume of 90% (v/v) methanol and poured into a chromatographic column. The column is washed with 2 bed volumes each of 90% (v/v) methanol and 50% (v/v) methanol – 0.5 M NaCl, followed by 3 bed volumes of 50% (v/v) methanol – 1 M NaCl and 3 bed volumes of water. The column can be used as such or the material removed for batchwise adsorption.

Isolation from Culture Medium

After removal of bacteria and debris, pretreated Amberlyst A21 ion exchanger (a minimum of 1 g per 2 mg of PQQ is required) is added to the culture medium. The mixture is slowly stirred for 3 hr to keep the beads in suspension. After settling of the ion exchanger, the overlayer is decanted. The Amberlyst suspension is poured into a chromatographic column and washed with water (2 bed volumes) and 90% (v/v) methanol (2 bed volumes). Impurities (e.g., flavin derivatives) are removed with 50% (v/v) methanol – 0.5 M NaCl. This is continued until no more UV-absorbing material ($\lambda = 254$ nm) is detected in the eluate. PQQ is eluted with 50% (v/v) methanol – 1 M NaCl. The methanol is removed in a rotatory evaporator. PQQ crystallizes in tiny brick red needles when the remaining aqueous solution is cooled to 4° or when the water is allowed to evaporate slowly at room temperature. The crystals are filtered off, washed with a minimum of cold water, and dried over P_2O_5. At this stage the preparation is more than 98% pure, as judged by high-performance liquid chromatography (HPLC) and ^1H nuclear magnetic resonance (NMR).

[4] M. A. G. van Kleef and J. A. Duine, *Appl. Environ. Microbiol.* **55**, 1209 (1989).
[5] J. Frank and J. A. Duine, *Biochem. J.* **187**, 221 (1980).

Isolation from Whole Bacteria

Cell paste is suspended in an equal volume of 20 mM potassium phosphate buffer, pH 7.0, and 9 volumes of methanol is slowly added under stirring. After centrifugation (23,000 g, 10 min) the clear supernatant is passed through a pretreated Amberlyst A21 column (at least 1 g of ion exchanger per 2 mg of PQQ is required) and the procedure continued as indicated above.

Chemical Synthesis

The preparation of PQQ has been described by Corey and Tramontano,[6] Gainor and Weinreb,[7,8] Hendrickson and de Vries,[9,10] MacKenzie et al.,[11,12] and Buechi et al.[13] Preliminary evaluations of individual methods have appeared.[2,14] The following procedure is adapted and modified[15] from the method of Corey and Tramontano.[6]

Procedure for Synthesis of PQQ: Scheme 1

2-Methoxy-5-nitroformanilide (2). A mixture of 1 mol of 2-methoxy-5-nitroaniline (1) (Fluka, Buchs, Switzerland) and 200 ml of formic–acetic anhydride is stirred at 50° for 30 min. The cream-colored microcrystalline precipitate that is deposited from the transient dark yellow solution is filtered after cooling to ambient temperature and washed with copious amounts of methanol. Air-dried material (0.9 mol, 90%, mp 200–205°) can be used in the next step without further purification. Analytical samples (mp 203–205°) can be obtained by recrystallization from boiling methanol.

2-Methoxy-5-aminoformanilide (3). Hydrogenation of the anilide is conducted at 30° and atmospheric pressure in a well-stirred vessel containing 80 mmol 2 and 0.5 g of prereduced platinum oxide in 350 ml of methanol (separate reduction of platinum oxide is essential as activation of

[6] E. J. Corey and A. Tramontano, *J. Am. Chem. Soc.* **103**, 5599 (1981).

[7] J. A. Gainor and S. M. Weinreb, *J. Org. Chem.* **46**, 4317 (1981).

[8] J. A. Gainor and S. M. Weinreb, *J. Org. Chem.* **47**, 2833 (1982).

[9] J. B. Hendricksen and J. G. de Vries, *J. Org. Chem.* **47**, 1148 (1982).

[10] J. B. Hendricksen and J. G. de Vries, *J. Org. Chem.* **50**, 1688 (1985).

[11] A. R. MacKenzie, C. J. Moody, and C. W. Rees, *J. Chem. Soc., Chem. Commun.,* 1372 (1983).

[12] A. R. MacKenzie, C. J. Moody, and C. W. Rees, *Tetrahedron* **42**, 3259 (1986).

[13] G. Buechi, J. H. Botkin, G. C. M. Lee, and K. Yakushijin, *J. Am. Chem. Soc.* **107**, 5555 (1985).

[14] Y. Naruta and K. Maruyama, *in* "The Chemistry of Quinonoid Compounds" (S. Patai and Z. Rappoport, eds.), Vol. 2, Part 1, p. 374. Wiley, Chichester, New York, 1987.

[15] J. A. Jongejan and J. A. Duine, *Tetrahedron* submitted (1990).

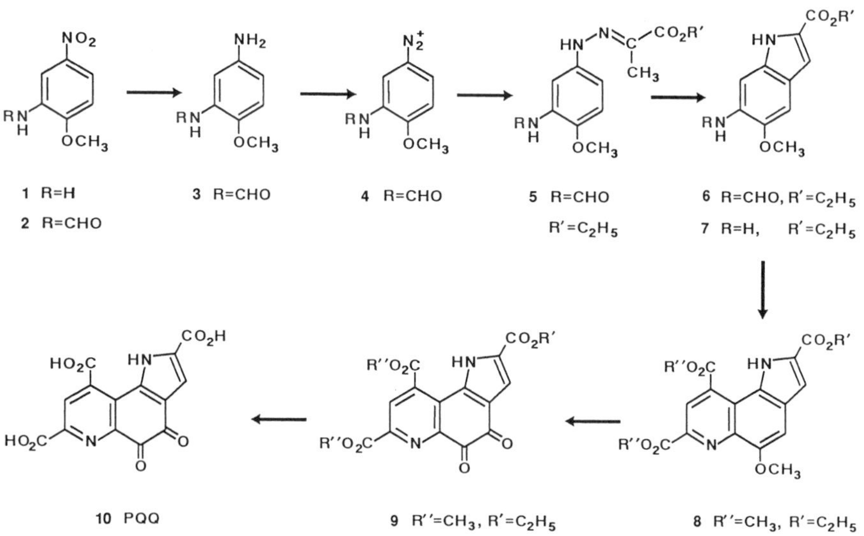

SCHEME 1

the catalyst in the presence of the anilide leads to sluggish and incomplete hydrogen uptake). A limiting amount of hydrogen (97% of theoretical value) is consumed in 2 hr. The catalyst is removed by filtration. Provided that exposure to oxygen of the highly pyrophoric material is avoided, the catalyst can be reused for up to 100 runs. The clear filtrate is concentrated to a small volume under reduced pressure. Aminoformanilide (3) is deposited as colorless crystals (76 mmol, 95%, mp 147.5–148.5°). Remaining traces of platinum may cause severe darkening of the preparation. ^1H NMR (deuterated dimethyl sulfoxide, DMSO-d_6), δ (from tetramethylsilane, TMS): 3.70 (s, 3H); 4.66 (br, 2H) 6.26 (dd, 1H); 6.73 (d, 1H); 7.57 (d, 1H); 8.27 (d, 1H); 9.38 (br, 1H).

Ethyl Pyruvate 3-Formylamino-4-methoxyphenylhydrazone (5). A solution of the diazonium salt 4, obtained by treatment of 3 (100 mmol) in 0.3 M HCL (670 ml) with sodium nitrite (100 mmol) at 0–5° and stirring for 20 min at 5°, is added over a 30-min period to a mixture of ethyl α-methylacetoacetate (110 mmol) and potassium hydroxide (110 mmol) in 2:1 methanol–water (1 liter) containing sodium acetate (25–30 g) at 0–2°. The resulting yellow solution is stirred for 4 hr at 2°, during which time a semicrystalline yellow solid is formed. The precipitate is collected by filtration and dried in air (27.5 g, 95% on 3). Thin-layer chromatography (TLC) (Bakerflex IB-2F), R_f 0.45 (dichloromethane–ethyl acetate, 4:1), R_f 0.83 (ethyl acetate; dark spot at 254 nm detection). Freshly crystallized material (from boiling ethanol) is virtually colorless; a red color develops upon prolonged exposure to air. ^{1}H NMR (DMSO-d_6), δ (from TMS): 1.24 (t, 3H); 2.02 (s, 3H, β-CH$_3$); 3.76 (s, 3H, OCH$_3$); 4.18 (q, 2H); 6.96 (sm. d, 1H + s, 1H, ØH's); 8.18 (br.s, 1H, NCH=O); 8.29 (sm.d. 1H, m-ØH); 9.5 (br.s, NHC=O); 9.72 (s, 1H, N$_\alpha$-H).

Slightly alkaline conditions are required for the formation of the hydrazone. Conditions reported by Corey and Tramontano[6] afford ethyl α-(3-formylamino-4-methoxyphenylazo)-α-methylacetoacetate as a major product. Application of the azo compound in the subsequent Fischer indolization reaction leads to considerable loss of material.

Ethyl 6-Formylamino-5-methoxyindole-2-carboxylate (**6**). Air-dried hydrazone (**5**) (100–200 mmol) is added to once-distilled formic acid (200–400 ml). The resulting dark brown solution is stirred at 80° for 10–12 hr. Cooling to ambient temperature affords **6** as virtually colorless crystals (92–96%). The product is crystallized from boiling ethanol (mp 233.5–234.5°; mp 215.4–217° is reported[6] for the corresponding methyl ester). ^{1}H NMR (DMSO-d_6), δ (from TMS): 1.37 (t, 3H); 3.87 (s, 3H); 4.32 (q, 2H); 7.00 (d, 1H); 7.17 (s, 1H); 8.38 (d, 1H, formyl-H); 8.45 (s, 1H); 9.72 (br, 1H, N^1-H); 11.72 (br, 1H, amide N-H). TLC, R_f 0.24 (dichloromethane–ethyl acetate, 4:1; blue fluorescence).

Ethyl 6-Amino-5-methoxyindole-2-carboxylate Hydrochloride (7 · HCl). Refluxing of **7** (100 mmol) for 2 hr in a mixture of concentrated HCl–acetone, 4:g6 (800 ml; v/v), gives **7** · HCl. The white precipitate is collected by filtration and washed with acetone (90–95%, after drying over P$_2$O$_5$, *in vacuo*, room temperature).

Ethyl 6-Amino-5-methoxyindole-2-carboxylate (**7**). A suspension of **7** · HCl (100 mmol) in 0.1 M aqueous sodium bicarbonate (~1 liter), containing a slight excess of base, is stirred for 15 min at 30–50° during which time the free amine (**7**) precipitates. Initially colorless crystals turn brown upon prolonged exposure to air [95%, material once crystallized from small volumes of chloroform, mp 166.6–167.0° *(in vacuo)*. ^{1}H NMR (CDCl$_3$), δ (from TMS): 1.35 (t, 3H); 3.83 (s, 3H); 3.93 (br, 2H, N^6-H); 4.33 (q, 2H); 6.55 (s, 1H, H-7); 6.87 (s, 1H, H-4); 7.03 (d, 1H, H-3); 8.85 (br, 1H, N^1-H). TLC, R_f 0.33 (dichloromethane–ethyl acetate, 4:1; blue fluorescence).

7,9-Dimethoxycarbonyl-2-ethoxycarbonyl-5-methoxy-1H-pyrrolo[2,3-f]quinoline (**8**). Amine **7** (100 mmol) is dissolved in dichloromethane (200 ml) containing dimethyl 2-oxoglutaconate (105 mmol; Fluka). After stirring for 12 hr at ambient temperature, dry hydrogen chloride (2–5 mmol) is bubbled into the dark brown solution. The vessel containing the solution is attached to an all-glass manometric apparatus filled with O$_2$. Variation of the volume resulting from changes of temperature and barometric pressure can be corrected for by recording the volume of an identical manometric apparatus to which an identical vessel, containing an equivalent volume of dichloromethane, is attached. During the first 2 hr, a rapid uptake of O_2 can be noted. A limiting amount, corresponding to approximately 50 mmol O_2, is consumed in 5–6 hr (volumes may be biased by the extremely high vapor pressure of the solvent). Evaporation of the solvent affords **8** in virtually quantitative yield. Note that similar incubations conducted in a (loosely) stoppered vessel result in the formation of an equivalent amount of material. However, in addition to the fully aromatic species, approximately 10–15% of a compound tentatively designated as the dihydroquinoline is present: TLC, R_f 0.40–0.45 (dichloromethane–ethyl acetate, 4:1; yellow). Subjecting such mixtures to oxidation with ceric ammonium nitrate (CAN) in the next step results in lower (60%, reported by Corey and Tramontano[6]) yields of PQQ triester. Analytical samples of **8** are obtained by crystallization from dichloromethane (mp 217.0–217.7°; mp 224–225° is reported for the trimethyl ester[6]). TLC, R_f 0.45 (dichloromethane–ethyl acetate, 4:1; yellow). ^{1}H NMR (CDCl$_3$), δ (from TMS): 1.43 (t, 3H); 4.07 (s, 6H); 4.13 (s, 3H); 4,47 (q, 2H); 7.23 (d, 1H, H-3); 7.28 (s, 1H, H-4); 8.93 (s, 1H, H-8); 12.15 (br, 1H).

7,9-Dimethoxycarbonyl-2-ethoxycarbonyl-1H-pyrrolo[2,3-f]quinoline-4,5-dione (**9**). Powdered CAN (105 mmol, Merck, Darmstadt, FRG) is added gradually to a stirred solution of **8** (20 mmol) in acetonitrile–water, 4:1 (200 ml, v/v), at 0–5° over a 30-min period. After further reaction for 15 min, water (500 ml) is added, and the resulting yellow-orange mixture is extracted with ethyl acetate–dichloromethane, 4:1 (4 times 150 ml each). Evaporation of the solute and crystallization of the orange residue from acetonitrile affords PQQ dimethyl ethyl triester (**9**) as orange-red crystals (90–95%, mp 226.4–227°, dec. TLC, R_f 0.31 (for PQQ trimethyl triester, 260–263° and R_f 0.14 are reported[6]). ^{1}H NMR (CDCl$_3$), δ (from TMS): 1.42 (t, 3H); 4.03 (s, 3H); 4.17 (s, 3H); 440 (q, 2H); 7.37 (d, 1H, H-3); 8.87 (s, 1H, H-8); 12.8 (br, 1H).

2,7,9-Tricarboxy-1H-pyrrolo[2,3-f]quinoline-4,5-dione (PPQ) (**10**). Hydrolysis of PQQ triester (**9**) is performed by dissolving the ester in excess 0.3 M potassium hydroxide at 50° for 30 min. The clarified solution is acidified (pH 1.5) using 3 MHCl. The copious precipitate is dissolved in hot water (90°) and adjusted to pH 1.5 (3 M HCl). From the deep-red solution (1–2 g of PQQ in 1000 ml water), dark red crystals are deposited over the course of several days at 80°. Air-dried crystals analyze correctly for PQQ·H_2O. $C_{14}H_8N_2O_9$ (348.2) requires C 48.29; H 2.60; N 8.04%. Found C 48.35, 48.42; H 2.55, 2.53; N 8.10, 8.12%. Extensive drying (100°, P_2O_5, > 6 hr, *in vacuo*) results in a loss of weight of 5.12%, in agreement with the presence of 1 mol of water of hydration. Samples obtained by a single precipitation step at ambient temperature are likely to contain PQQ, its covalent hydrate, as well as its (partial) potassium salt.[16]

Procedures for Preparation of 3-²H- and 2-¹³C-Labeled PQQ

[3-¹³C]PQQ can be prepared[17] according to the method of Corey and Tramontano.[6] Application of [α-¹³C]methylacetoacetate ester, obtained from [¹³C]methyl iodide (Sigma) and acetoacetic ester by the method of Folkerts and Adkins,[18] affords [3-¹³C]PQQ in an overall yield of 20% (from [¹³C]methyl iodide). [3-¹³C]PQQ dimethyl ethyl triester, ¹H NMR (CDCl₃), δ (from TMS): 7.37 (dd, $J_{H-3-N^1H} = 2.8$; $J_{^{13}C-3-H-3} = 184.5$ Hz).

Preparation of [3-²H]PQQ requires the use of deuterated acid during the Fischer indolization step[17] (Scheme 1). [3-²H]PQQ of high isotopic purity (> 95% ²H) is obtained in moderate yield by conducting the Fischer indolization of hydrazone 5, from [α-²H₃]methylacetoacetate ester, in [²H]trifluoroacetic acid (99% ²H, Aldrich) for 12 hr at 23°.

Procedure for Preparation of [8-²H]PQQ: Scheme 2

Preparation of [8-²H]PQQ from deuterated 2-oxoglutaconic acid ester has not been attempted. Instead, Pfitzinger quinoline synthesis employing isatin (12) and pyruvic acid in deuterated alkali affords [8-²H]PQQ in reasonable yields.[17]

N-(2-Carboethoxy-5-methoxyindole-6-yl)-2-(hydroxylimino)acetamide (11). A solution of the hydrochloride 7 · HCl (5 mmol) in water (100 ml) is added gradually to a mixture of chloral hydrate (6 mmol), hydroxylamine hydrochloride (15 mmol), and (anhydrous) disodium sulfate (20 g, partially dissolved) that has been stirred for 30 min at 35–40°. The resulting light brown solution is stirred at 35–40° for another 24 hr. During this time, an almost colorless precipitate is formed. After cooling to ambient temperature, the precipitate is collected by filtration, washed with water, and dried (P₂O₅, *in vacuo,* room temperature). The crude isonitroso compound (1.43 g, 4.35 mmol, 87%) is obtained as an off-white fluffy powder. Attempts to crystallize the product result in the formation of dark brown tarry material, similar in appearance to the tarry deposit observed when the Sandmeyer procedure is conducted at higher temperatures for shorter periods.

[16] H. van Koningsveld, J. A. Jongejan, and J. A. Duine, *in* "Proceedings of the First International Symposium on PQQ and Quinoproteins" (J. A. Jongejan and J. A. Duine, eds.), p. 243. Kluwer Academic Publishers, Dordrecht, 1989.

[17] J. A. Jongejan, R. P. Bezemer, and J. A. Duine, *Tetrahedron Lett.* **29,** 3709 (1988).

[18] K. Folkerts and H. Adkins, *J. Am. Chem. Soc.* **53,** 1416 (1931).

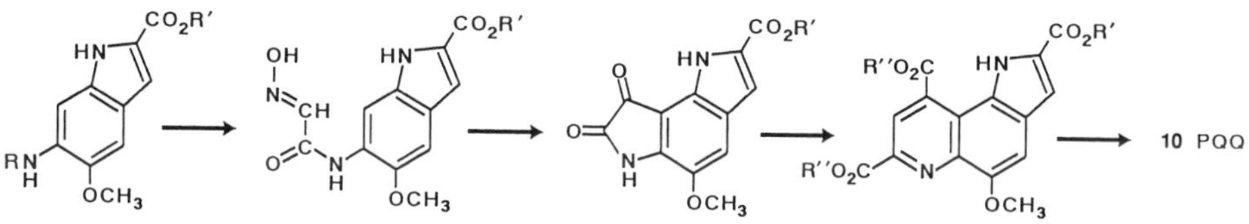

7 R=H
   R'=C₂H₅

11 R=H
    R'=C₂H₅

12 R=H
    R'=C₂H₅

8 R''=R'=H

SCHEME 2

2-Ethoxycarbonyl-5-methoxy-1H-pyrrolo[2,3-g]indole-7,8-dione (**12**).
Dried **11** (4 mmol) is added to a stirred syrup of polyphosphoric acid,
prepared according to Fieser and Fieser[19] at 100°, and stirred for another
hour at 140–150°. During this time, the color changes from brown to deep
violet. The mixture is cooled and poured onto crushed ice (200 g). The
resulting violet solution is (continuously) extracted with chloroform for
extended periods of time. Evaporation of the solvent affords a small
amount of crude dark amorphous material (95 mg). Crystallization from
acetic acid (1.5 ml) gives violet needles (45 mg) of the isatin derivative (**12**).
^{1}H NMR (DMSO-d_6), δ (from TMS): 1.32 (t, $3H$); 3.85 (s, $3H$); 4.23 (q,
$2H$); 7.08 (s, $1H$); 7.63 (s, $1H$); 11.3 (br, $2H$).

[8-^{2}H]-2,7,9-Tricarboxy-5-methoxy-1H-pyrrolo[2,3-f]quinoline by Pfit-
zinger Quinoline Synthesis. Compound **12** (45 mg) is suspended in (30%,
w/w) aqueous deuterated sodium hydroxide (5 ml; >95% ^{2}H) at 95°. After
solution is complete, 33% (w/v) aqueous pyruvic acid (0.5 ml) is added
dropwise. The solution is stirred for several hours at 95°. The extent of
conversion can be estimated from the increase of absorption at 310 nm of
neutralized (0.1 M HCl) samples diluted with water. Complete conversion
requires at least 4 hr.

When no further change is noted, the alkaline mixture is poured onto
crushed ice and adjusted to pH 2 with concentrated HCl. The yellow
precipitate is collected and dried (P_2O_5, *in vacuo,* room temperature).
Dried material (40 mg) is suspended in diethyl ether and treated overnight
with an etheral diazomethane solution, prepared according to Arndt,[20]
containing an approximately 5-fold excess of diazomethane. The residue,
obtained after evaporation of the solvent, is crystallized from a small
volume of acetonitrile. The final product contains less than 10% ^{1}H at the
8-position of PQQ as judged by ^{1}H NMR. Conversion of **8** [8-^{2}H]trimethyl
ester to [8-^{2}H]PQQ is performed by routine procedures.

[19] L. F. Fieser and M. Fieser, "Reagents for Organic Synthesis," Vol. 1, p. 894. Wiley, New
York, 1967.

Determination of Free PQQ

Identification and quantification of free PQQ can be carried out by a number of chromatographic methods and bioassays. In all these procedures it should be realized that the capability of PQQ to form adducts with nucleophiles at the C-5 position is a disturbing factor. It is known, for instance, that the adducts formed with certain amino acids are easily converted to condensation products, the so-called oxazoles, which are biologically inactive and show quite different chromatographic and spectral behavior.[21] Therefore, if the assay cannot be performed immediately, a clean-up step should be performed before storage of the sample. This treatment is also applied when a preconcentration step is necessary or when contaminants, interfering with the assay, should be removed.

Sample Preparation

Noncovalently bound PQQ can be detached from proteins and extracted from whole cells in several ways (Methods 1–3). Acid extraction (Method 2) is recommended if the amount of reduced cofactor ($PQQH_2$) should be determined (significant autoxidation of $PQQH_2$ occurs at pH values above 4). If information is required on the (denatured) protein or the presence of other cofactors, sodium dodecyl sulfate (SDS) extraction (Method 3) is the method of choice.

Method 1. To 1 volume of sample 9 volumes of methanol is added. Precipitated material is removed by centrifugation (5 min at 12,000 g). Removal of methanol by evaporation or dilution, eventually in combination with the Sep-Pak procedure (see below), is required prior to the application of HPLC methods or bioassays.

Method 2. To 1 volume of sample, 1 volume of 1.0 M NaH_2PO_4 (brought to pH 1.0 with concentrated HCl) is added, followed by the addition of 4 volumes of methanol. Further processing is as described for Method 1. Neutralization of the sample is required when a bioassay is used.

Method 3. The sample is brought to 0.2% (w/v) SDS and heated at 60° for 2 min. After centrifugation the supernatant can be directly analyzed (see Example 3). However, if reversed-phase HPLC or a bioassay is preferred, the detergent has to be removed by adding KCl to 0.1 M and, after cooling to 4°, centrifugation (5 min at 12,000 g).

[20] "Organic Syntheses, Collective Volume 2," 6th Ed., p. 165. Wiley, New York, 1950.
[21] M. A. G. van Kleef, J. A. Jongejan, and J. A. Duine, *Eur. J. Biochem.* **183,** 41 (1989).

Prepurification and Concentration

If chromatographic methods are used for the assay, prepurification of the sample is sometimes necessary (e.g., certain *Pseudomonas* species produce pigments that interfere with detection, or protein in the sample precipitates on the reversed-phase column material, compromising its functioning). As indicated in the section on Isolation from Biological Materials, the ion-exchange step is able to remove contaminants and to concentrate the PQQ. However, if small amounts of PQQ are present, this step leads to unacceptable losses. In that case the so-called Sep-Pak procedure is an alternative. The Sep-Pak C_{18} cartridge (Waters, Etten-Leur, The Netherlands) is washed with 10 ml of methanol and subsequently with 10 ml of water. The sample is acidified with HCl to pH 2.0 and applied to the cartridge (if SDS is present, this should be removed prior to this step). After washing with 10 ml of 2 mM HCl, PQQ is eluted with 1 ml of 70% (v/v) methanol. If necessary, methanol is removed as indicated above. Recovery is excellent as long as more than 80 nmol (25 μg) of PQQ is present.

Chromatographic Methods

PQQ has been chromatographed on reversed-phase HPLC columns by applying either ion-suppression or ion-pairing conditions.[22] Ion suppression (Example 1), achieved by using a methanolic solvent acidified to pH 2.0, gives rise to a rather broad peak of PQQ (most probably because of adduct formation of PQQ with methanol and water). Although a sharper peak is obtained with ion pairing (Example 2), this system is not attractive for (semi)preparative purposes since the quaternary ammonium salt used is difficult to remove. Gel permeation chromatography in the presence of SDS, combined with photodiode array detection and multicomponent analysis,[23] can be used to check the identity of the protein part and the presence and quantity of other cofactor(s) in a (denatured) homogeneous quinoprotein preparation.

Monitoring of the eluates normally occurs with a fixed-wavelength UV detector at 254 nm (0.3 μg of PQQ can be detected at a signal-to-noise ratio of 10). Owing to the fluorescing properties of the hydrate of PQQ, fluorescence detection can also be applied (excitation at 360 nm and detection of the light emitted at wavelengths above 418 nm). Fluorescence detection is more selective, and a higher sensitivity can be achieved.[23] To identify a

---

[22] J. A. Duine, J. Frank, and J. A. Jongejan, *Anal. Biochem.* **133**, 239 (1983).
[23] M. Dijkstra, J. Frank, J. A. Jongejan, and J. A. Duine, *Eur. J. Biochem.* **140**, 369 (1984).

peak in a chromatogram as belonging to PQQ, spiking or prederivatization with an aldehyde or ketone is helpful (Example 1).

Example 1. Isocratic HPLC on μBondapak C_{18} or Novapak C_{18} columns (Waters) is performed with 27% (v/v) methanol, 0.4% (v/v) H_3PO_4 at a flow rate of 1.5 ml/min. To identify the peak of PQQ, 200 μl of sample is mixed with 100 μl of 0.2 M $Na_2B_4O_7$ buffer, adjusted to pH 8.0 with concentrated HCl, and 90 μl of a 0.5% (v/v) solution of acetone or butyraldehyde. After 30 min at room temperature, conversion is complete and the PQQ peak should have moved to the position expected for the adduct.

At 20° the retention time of PQQ on the μBondapak C_{18} column is 12 min, whereas the acetone, propionaldehyde, and butyraldehyde adducts elute at 7.2, 13.3, and 26 min, respectively. Note that the retention times are temperature-dependent.

Gradient elution reduces retention times of hydrophobic derivatives of PQQ. PQQ elutes at 14.4 min in a 30 min linear gradient of 70–50% solvent A in solvent B [solvent A is 10% (v/v) methanol, 0.4% (v/v) H_3PO_4; solvent B is 80% (v/v) methanol, 0.4% (v/v) H_3PO_4], at a flow rate 1.5 ml/min.

Example 2. Adequate retention of PQQ on a Novapak C_{18} column is achieved with 5 mM tetrabutylammonium hydrogen sulfate in 48% (v/v) methanol at pH 3–7. In this system PQQ has a retention time of 7 min, which shifts to 14 min when the concentration of methanol is lowered to 40%.

Example 3. HPLC gel permeation of quinoproteins denatured with SDS is performed on a Polyol Si-300 column (4.1 × 250 mm, Serva) coupled to a photodiode array detector (HP1040, Hewlett-Packard). The eluant is 0.1 M sodium phosphate buffer, pH 6.5, containing 0.1% (w/v) SDS, at a flow rate of 0.4 ml/min. Protein and PQQ elute at 5.2 and 8.8 min, respectively. The eluate is monitored at 280 and 350 nm, and spectra are taken in the apex and the slopes of the peaks to check homogeneity and identity.

Biological Assays

The assays are based on the ability of quinoprotein dehydrogenase apoenzymes to reconstitute with PQQ to active holoenzymes. A wide variety of naturally occurring and self-prepared apoenzymes is available.[24] They can be used in unpurified form — present in whole cells (Example 1),

---

[24] J. A. Duine and J. A. Jongejan, *Vitamins Hormones* **45**, 223 (1989).

attached to membrane particles,[25,26] or cell-free extract (Example 2)—or in purified form (Examples 3 and 4).

Important factors to be considered in making a choice are the availability, the ease of preparation, and the stability of the apoenzyme; absence of background activity; a high turnover number; rapid reconstitution; and no interference of other enzymes converting the substrate or the product. No stringent requirements are set for the samples, except that they should not contain substances harmful to the enzyme or for the reconstitution process (chelators may inhibit the reconstitution since certain divalent metal ions are required). As is inherent to most biological assays, the high sensitivity requires that precautions are taken against contamination of buffers, columns, and laboratory glassware with the cofactor. For this reason, frequent checks should be made. Handling instructions in critical cases have been described.[27]

Example 1

Principle. PQQ is assayed with an *Acinetobacter calcoaceticus* PQQ⁻ mutant by reconstituting the activity of glucose dehydrogenase apoenzyme.[27] The extent of reconstitution is determined by incubating the cells with glucose and recording the rate of disappearance of glucose and/or the formation of gluconate (the strain of *A. calcoaceticus* employed is unable to convert glucose other than via glucose dehydrogenase and does not grow on gluconate).

Reagents

Spore solution: the spore solution is prepared according to Vishniac and Santer[28] and contains, per liter, 50 g EDTA, 2.2 g $ZnSO_4 \cdot 7H_2O$, 5.54 g $CaCl_2$, 5.06 g $MnCl_2 \cdot 4H_2O$, 5.0 g $FeSO_4 \cdot 7H_2O$, 1.1 g $(NH_4)_6Mo_7O_{24} \cdot 4H_2O$, 1.57 g $CuSO_4 \cdot 5H_2O$, and 1.61 g $CoCl_2 \cdot 6H_2O$; After dissolving, the pH is adjusted to pH 7.0 with 6 M NaOH

PGD solution: the solution contains (per 100 ml) 1.32 g glycylglycine plus 142 mg $MgCl_2$ (adjusted to pH 8.0 with 2 M NaOH) and 10 μl 6-phosphogluconate dehydrogenase (ammonium sulfate suspension, Boehringer Mannheim, FRG)

[25] M. Ameyama, M. Nonobe, E. Shinagawa, K. Matsushita, and O. Adachi, *Anal. Biochem.* **151**, 263 (1985).

[26] O. Geiger and H. Gorisch, *Anal. Biochem.* **164**, 418 (1987).

[27] M. A. G. van Kleef, P. Dokter, A. C. Mulder, and J. A. Duine, *Anal. Biochem.* **162**, 143 (1987).

[28] W. Vishniac and M. Santer, *Bacteriol. Rev.* **21**, 95 (1957).

NADP solution: the solution contains (per 25 ml) 60 mg NADP and 150 mg ATP

Growth of Organism. Acinetobacter calcoaceticus PQQ⁻ mutant LMD 82.43, derived from strain LMD 79.41,[29] is grown at 30° on a mineral medium. The basic medium contains, per liter, 4.6 g KH_2PO_4, 1.15 g K_2HPO_4, 2.5 g $(NH_4)_2SO_4$, and 0.2 g $MgSO_4 \cdot 7H_2O$. After autoclaving, 1 ml spore solution and 25 ml of 8% (w/v) sodium succinate solution (sterilized by autoclaving) are added to 1 liter of the basic medium. Growth is performed with 100 ml medium in a 500-ml flask on a reciprocal shaker (200 rpm), until the stationary growth phase is achieved (~ 17 hr).

Procedure. Ten-milliliter portions of the culture are centrifuged, then washed with and suspended in 10 ml sterile mineral salts medium. After addition of the sample (1 ml, containing at least 0.3 ng but maximally 30 ng PQQ) and of glucose solution (10 mM final concentration), incubation occurs at 37° in 100-ml flasks with shaking on a reciprocal shaker (200 rpm). The disappearance of glucose and/or the production of gluconic acid are measured in the following way. After centrifugation of the samples, glucose can be determined with the GOD–PAP method[30] (Boehringer, Mannheim). Gluconic acid is assayed as follows: to the cuvette (1-cm optical pathway), 1.70 ml of the PGD solution, 1.00 ml NADP, and 0.1 ml sample solution are added. After mixing, the blank absorbance at 340 nm is measured. The reaction is started by adding 10 μl of a gluconate kinase suspension (Boehringer Mannheim), and the absorbance is measured again after standing 30 min at room temperature. The amount of gluconic acid is calculated from the NADPH formed by taking the absorbance difference at 340 nm (before and after addition of gluconate kinase), and using a molar absorption coefficient of 6.3×10^3 M^{-1} cm^{-1} at this wavelength. The PQQ concentration of the sample is calculated from a calibration curve.

Example 2

Principle. Growth of *Pseudomonas aeruginosa* on gluconate induces the production of a membrane-bound quinoprotein glucose dehydrogenase[22] (GDH). PQQ is easily removed from this type of GDH by dialysis of the cell-free extract against an EDTA-containing buffer. After removal of EDTA, the extract is used for determination of PQQ by measuring the extent of reconstitution. GDH activities are measured spectrophotometrically using Wurster's blue as the electron acceptor.

[29] N. Goosen, D. A. M. Vermaas, and P. van der Putte, *J. Bacteriol.* **169,** 303 (1987).
[30] H. Möllering and H. U. Bergmeyer, *in* "Methods of Enzymatic Analysis" (H. U. Bergmeyer, ed.), 3rd Ed., Vol. 6, p. 220. Verlag Chemie, Weinheim, Deerfield Beach, Florida, Basel, 1984.

Reagents

KCN solution, 0.15 M

Wurster's blue,[31] 0.2 mM

Glucose solution, 40 mM

Potassium phosphate buffer, 20 mM, adjusted to pH 8.0 with concentrated HCl

Growth of Organism. Pseudomonas aeruginosa strain LMD 76.39 (ATCC 10145) is cultured aerobically at 30° in a mineral medium containing 30 mM sodium gluconate.[32] Cells are harvested in the stationary phase and stored at −20°.

GDH Apoenzyme Preparation. Cell paste (10 g wet weight) is suspended in 10 ml of 20 mM potassium phosphate buffer, pH 7.2. The cells are disrupted in a French pressure cell at 110 MPa. A few milligrams of DNase is added, and, after incubation at room temperature for 5 min, the suspension is centrifuged (48,000 g) at 4° for 20 min. The supernatant (17 ml) is dialyzed at 4° against 2 liters of 0.20 M Tris-HCl, pH 8.0, containing 10 mM EDTA and a few drops of toluene, for 2 days with one change of buffer. The preparation is further dialyzed for 1 day against 2 liters of 0.2 M Tris-HCl, pH 8.0, containing 10 mM MgCl$_2$, yielding the apoenzyme preparation, which is stored at −80°.

Procedure. Apoenzyme preparation (20 μl) is incubated with 30 μl sample at room temperature for 30 min in a cuvette (1-cm optical pathway). Subsequently, 30 μl KCN solution in potassium phosphate buffer and 500 μl glucose solution in 0.2 M Tris-HCl, pH 8.0, are added. The reaction is started by the addition of 500 μl Wurster's blue solution. The rate of discoloration is measured at 610 nm, at room temperature. The amount of PQQ present in the sample is calculated from a calibration curve.

Example 3

Principle. Reconstitution is measured of a homogeneous preparation of quinohemoprotein alcohol dehydrogenase apoenzyme from *Comamonas testosteroni.*[32] The amount of PQQ is determined by the extent of reconstitution, measured by the butanol-dependent reduction of Wurster's blue or potassium ferricyanide.

---

[31] L. Michaelis and S. Granick, *J. Am. Chem. Soc.* **65,** 1747 (1943).

[32] B. W. Groen and J. A. Duine, this volume [7].

Reagents

Apoenzyme from *Comamonas testosteroni* is prepared as described;[32] the enzyme is diluted with Tris–Ca buffer to a concentration of 1 μM (A_{410} 0.081)

Tris–Ca buffer: 0.1 M Tris-HCl buffer, pH 7.5, containing 3 mM CaCl$_2$

1-Butanol, 5 mM

Wurster's blue, 2 mM

Potassium ferricyanide, 24 mM

Procedure. Two-tenths milliliter of apoenzyme solution is mixed with 10–500 μl sample (containing at least 5 nM of PQQ) in a cuvette (1-cm optical pathway). After incubating for 10 min at 20°, 50 μl of Wurster's blue (or 50 μl of ferricyanide solution) and Tris–Ca buffer to a final volume of 975 μl are added. The reaction is initiated by adding 25 μl of the butanol solution. The reduction rate of Wurster's blue is measured spectrophotometrically at 610 nm, that of ferricyanide at 420 nm. The amount of PQQ is determined from a calibration curve.

Example 4

Principle. PQQ is assayed by measuring the extent of reconstitution of the soluble form of glucose dehydrogenase (GDH, EC 1.1.99.17) apoenzyme originating from *A. calcoaceticus* LMD 79.41. Since the levels of this enzyme in the parent strain are low and attempts to purify the apoenzyme from a PQQ⁻ mutant failed, apoenzyme is isolated from an *Escherichia coli* strain which contains a plasmid with the gene for this enzyme.[33] Since *E. coli* strains do not produce free PQQ,[27] high levels of GDH apoenzyme can be obtained in this way. Holoenzyme activity is measured with phenazine methosulfate (PMS) as the primary dichlorophenolindophenol (DCPIP) as the secondary electron acceptor.

Reagents

Tris–Ca buffer: 0.1 M Tris-HCl, pH 7.5, containing 3 mM CaCl$_2$ and 0.01% bovine serum albumin

DCPIP, 2.5 mM

PMS, 15 mM, stored in the dark

Glucose, 0.8 M

Apoenzyme, 0.2 μM ($\epsilon_{280\,nm}$ 109,000 M^{-1} cm^{-1}) in Tris–Ca buffer

Growth of Organism. Escherichia coli PGP 492 contains a plasmid with a gene from *A. calcoaceticus* LMD 79.41, coding for the protein part of

---

[33] A.-M. Cleton-Jansen, N. Goovsen, K. Vink, and P. van der Putte, *in* "Proceedings of the First International Symposium on PQQ and Quinoproteins" (J. A. Jongejan and J. A. Duine, eds.), p. 79. Kluwer Academic Publishers, Dordrecht, 1989.

soluble GDH.[33] The strain is grown on Luria broth containing (per liter): 10 g bacto tryptone, 5 g yeast extract, and 5 g NaCl, supplemented with 0.5% casein hydrolysate. After sterilization of this medium by autoclaving, 1 ml of a solution (sterilized by filtration) containing 0.5 mg thiamin and 40 mg ampicillin is added. A culture (40 ml complete medium in a 100-ml flask) is incubated for 15 hr at 30° on a reciprocal shaker (250 rpm). This culture serves as an inoculum for the second culture consisting of 1 liter complete medium in a 3-liter flask. After 15 hr of shaking (250 rpm) at 30°, the cells are harvested in the stationary growth phase by centrifugation and stored at $-40°$.

Step 1: Preparation of Cell-Free Extract. Cell paste (8.5 g) is suspended in 8.5 ml of 10 mM Tris-HCl buffer, pH 7.5, containing 0.3 mM CaCl$_2$, and passed twice through a French pressure cell at 110 MPa. Cells and cell debris are removed by centrifugation at 5000 g for 15 min.

Step 2: Detachment of the Apoenzyme. To the cell-free extract of Step 1 (15 ml), 25 ml of 10 mM Tris-HCl, pH 7.5, containing 0.3 mM CaCl$_2$, 1 ml of 0.1 M MgCl$_2$, and 3 mg of DNase (grade I, Boehringer Mannheim) are added. After incubating for 17 hr at 20°, the mixture is centrifuged at 48,000 g for 20 min.

Step 3: CM-Sepharose Chromatography. The supernatant from Step 2 (40 ml) is adjusted to pH 6.5 with dilute HCl and applied to a CM-Sepharose Fast Flow column (1.5 × 10 cm), previously equilibrated with 10 mM potassium phosphate, pH 6.5. After washing the cation-exchange column with the same buffer, elution of the enzyme is performed using a linear gradient of 0–0.6 M KCl (300 ml) in 10 mM potassium phosphate, pH 6.5. The active fractions are pooled and concentrated by ultrafiltration to 2 ml.

Step 4: Mono S Chromatography. The concentrated enzyme fraction from Step 3 is applied in 100-μl portions to a Mono S (0.5 × 5.0 cm, Pharmacia) cation-exchange column, equilibrated with 10 mM potassium phosphate, pH 6.5. Elution of the apoenzyme is performed using a linear gradient of 0–0.6 M KCl (15 ml) in 10 mM potassium phosphate, pH 6.5, at a flow rate of 0.5 ml/min. The active fractions are pooled, and a fully reconstituted preparation has a specific activity of 750 U/mg protein in the indicated assay. A typical example of the sensitivity and linearity of the assay is shown in Fig. 1.

Assay Procedure. To a cuvette (1-cm light path) are added 200 μl Tris–Ca buffer, 20 μl apoenzyme, and 10–500 μl sample solution. After an incubation time of 20 min at 25°, 25 μl DCPIP solution, 25 μl PMS solution, and Tris–Ca buffer to a final volume of 950 μl are added. The reaction is initiated by adding 30 μl glucose solution, and the discoloration of DCPIP is measured spectrophotometrically at 600 nm (ϵ_{600} 22,000 M^{-1} cm^{-1}). The PQQ concentration is determined from a calibra-

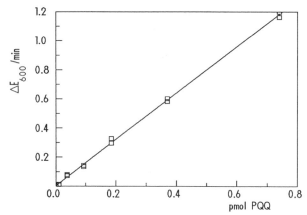

Fig. 1. Assay of PQQ with purified glucose dehydrogenase apoenzyme.

tion curve. One unit of holoenzyme is the amount catalyzing the conversion of 1 μmol of DCPIP per minute.

Evaluation

The assay developed with whole cells of PQQ$^-$ mutants of *A. calcoaceticus* is extremely sensitive. A concentration of 1 nM PQQ can be detected easily in a sample, and the response is linear over several decades. Interference may occur with samples containing glucose/gluconic acid or compounds inhibiting the GDH activity or the respiratory chain of the cells. The major drawbacks of the method, however, are its cumbersomeness and the relatively long time of the assay (up to 1 day) at low PQQ concentrations. In case no PQQ$^-$ mutants of *A. calcoaceticus* are available, a strain of *Acinetobacter lwoffii,* a naturally occurring PQQ$^-$ mutant, may be used.

Pseudomonas aeruginosa strains are readily available, and the procedure for obtaining apoenzyme from this type of GDH is very easy. The lower detection level with the indicated assay is 2 nM, and the response is linear over one decade. The practicability of the method may suffer from varying amounts of residual holoenzyme activity in the preparation. Although suitable for routine analysis, the method is therefore not recommended for samples containing very low amounts of PQQ. Moreover, reconstitution times are rather long (30 min).

In principle, apoquinohemoprotein alcohol dehydrogenase from *C. testosteroni* has no background activity. Reconstitution times are very short (less than 10 min), and the response is linear over several decades. In addition, the sensitivity is high; 0.5 pmol PQQ can be determined when a 500-μl sample is used (this corresponds to 1 nM PQQ in the sample).

Since soluble GDH has a high turnover number, PQQ in concentrations down to 0.02 nM can be determined accurately under the conditions described in Method 4.

Redox Cycling Assay

Recently, the so-called redox cycling assay was propagated for the determination of PQQ.[34] This is based on the property of PQQ to decarboxylate certain amino acids while it becomes concomitantly reduced (however, a noncyclic process owing to condensation of PQQ with amino acids to form oxazoles also occurs[21]). Reoxidation is achieved with a tetrazolium dye, a process which can be followed colorimetrically.[35] Using this assay, the presence of large amounts of free PQQ was claimed in certain foods and other materials originating from eukaryotic organisms.[36] However, it appears that the redox-cycling assay is not reliable for biological materials since several biochemical compounds (e.g., ascorbic acid, riboflavin) mimic the behavior of PQQ in the assay.[37]

Determination of Covalently Bound PQQ

Hydrazine Method

Enzymes are treated with a hydrazine inhibitor to convert the cofactor to a stable product which is detached from the protein by pronase, purified, and compared with a model compound prepared from authentic PQQ. Depending on the conditions and on the enzyme used, the product is either the hydrazone or the so-called azo compound.[38] Since the hydrazones are better defined [the structure of the hydrazone of PQQ and 2,4-dinitrophenylhydrazine (DNPH) has been solved by X-ray diffraction[39]] and are more stable, examples are presented where PQQ is determined by quantitative hydrazone formation. In the case of the copper-containing amine oxidases, this requires high O_2 concentrations, as illustrated with bovine serum amine oxidase treated with DNPH (Example 1). In other cases, this

[34] P. M. Gallop, M. A. Paz, R. Flückiger, and H. M. Kagan, *Trends Biochem. Sci.* **14**, 343 (1989).

[35] M. A. Paz, P. M. Gallop, B. M. Torrelio, and R. Flückiger, *Biochem. Biophys. Res. Commun.* **154**, 1330 (1988).

[36] J. Killgore, C. Smidt, L. Duich, M. Romero-Chapman, D. Tinker, K. Reiser, M. Melko, D. Hyde, and R. B. Rucker, *Science* **245**, 850 (1989).

[37] R. A. van der Meer, B. W. Groen, J. A. Jongejan, and J. A. Duine, *FEBS Lett.* **261**, 131 (1990).

[38] R. A. van der Meer, J. A. Jongejan, and J. A. Duine, *FEBS Lett.* **221**, 299 (1987).

[39] H. van Koningsveld, J. C. Jansen, J. A. Jongejan, J. Frank, and J. A. Duine, *Acta Crystalogr.* **C41**, 89 (1985).

is not critical, but long incubation times are required, as illustrated with dopamine β-monooxygenase derivatization with phenylhydrazine (PH) (Example 2). As a further check, a procedure is given (Example 2) to convert the hydrazone to PQQ.

Instrumentation. Chromatography is performed with a Waters HPLC system consisting of a U6K injector, a M6000A and M45 solvent delivery system with a M720 controller, and a Hewlett-Packard HP 1040 A photodiode array detector. The eluate is monitored at either 450 nm (for the DNPH hydrazone) or 335 nm (for the PH hydrazone), while taking absorption spectra (190–600 nm) throughout the chromatographic peaks. Reversed-phase chromatography is performed on a combination of a RP C_{18} guard column (4.0 × 25 mm) and a Waters 5 μm RCM cartridge (8.0 × 100 mm) in a RCM-100 module (Waters). The entire system is equilibrated with 10 mM sodium phosphate plus 10 mM NH_4Cl (pH 7.0)–methanol (93:7, v/v). After injecting the sample, a 20-min linear gradient (flow rate 1.5 ml/min) from 10 mM sodium phosphate plus 10 mM NH_4Cl (pH 7.0)–methanol (93:7, v/v) to 10 mM sodium phosphate (pH 7.0)–methanol (37:63, v/v) is applied, followed by holding the final condition for 5 min. In order to obtain reproducible results, it is necessary to reequilibrate the column with the starting buffer for 10 min between each run.

All solvents are prepared from Milli-Q water (Waters), filtered through a 0.22-μm filter, and degassed before use. Methanol (HPLC grade) was obtained from Rathburn Chemicals Ltd, Walkerburn, Scotland.

Reagents

DNPH solution: the DNPH solution is prepared according to Fieser and Fieser[40] in the following way: 2.0 g of DNPH is dissolved in 50 ml of 85% phosphoric acid by heating; the solution is cooled, and 50 ml 95% (v/v) ethanol is added; after cooling, the resulting 0.1 M DNPH solution is clarified by suction filtration from a trace of solid material

PH solution: freshly made 0.1 M phenylhydrazine-HCl

Pronase solution: 10 mg/ml pronase E (Merck), dissolved in water

Model Compounds. The C-5 hydrazone of PQQ and DNPH is prepared by adding a slight molar excess of the DNPH solution to a solution of PQQ in methanol (5 g/liter) at 40–50°. The suspension is stirred for 10 min at 50°. The precipitate formed is collected by suction filtration and washed with methanol. The orange-red solid is dissolved in large amounts of

---

[40] L. F. Fieser and M. Fieser, "Reagents for Organic Synthesis," Vol. 1, p. 167. Wiley, New York, 1967.

boiling methanol, and a microcrystalline orange solid is obtained upon cooling. This compound is the hydrazone[39] and has a retention time in the HPLC system of 14.6 ± 0.5 min (absorption maxima at 269,354, and 446 nm).

The C-5 hydrazone of PQQ and PH is prepared by applying conditions developed for reaction of hydrazines with ketones.[41] A hot aqueous solution (100 ml) containing PH·HCl (0.15 mmol), sodium acetate (4.5 mmol), and acetic acid (0.9 ml) is added to a hot solution of PQQ (0.1 mmol) in water (100 ml). The oily precipitate formed solidifies on cooling, after which it is removed by suction filtration. The precipitate is washed with 95% ethanol and dried over P_2O_5 under reduced pressure. The solid mainly consists of a compound considered to be the hydrazone,[38] having a retention time of 12.9 ± 0.5 min in the HPLC system and absorption maxima at 280 and 335 nm.

Example 1. A solution of 20 ml bovine serum amine oxidase[42] (0.05 to 1 mg enzyme) in 0.2 M sodium phosphate buffer, pH 7.0, is saturated with O_2 (monitored with a Clark electrode) by blowing O_2 over the surface at 40°. DNPH solution (also saturated with O_2 at 40°) is added to a 2-fold molar excess (with respect to the concentration of enzyme subunits). The mixture is kept at 40° for 16 hr while blowing O_2 over the solution. Excess reagent is removed by gel filtration on a Sephadex G-25 (Pharmacia) column (1.6 × 30 cm) in 20 mM sodium phosphate (pH 7.5), and fractions containing derivatized enzyme are pooled (the derivatized enzyme is retarded on the column material with a K_{av} of 2). Proteolysis is performed by incubating this solution (30 ml) with 1 ml pronase solution at 40° for 4 hr. After proteolysis, the solution is brought to pH 2.0 with 1 M HCl, and the so-called Sep-Pak procedure is applied: the solution is passed through a Sep-Pak C_{18} cartridge (Waters) equilibrated with 10 mM HCl. After washing successively with 10 mM HCl (100 ml), 10% methanol (10 ml), and water (100 ml), the red product is eluted with methanol (1 ml).

Identity (comparison with the model compound) as well as purity of the product are established by HPLC and monitoring by the photodiode-array detector. In case the Sep-Pak eluate appears to be pure, the amount of hydrazone can be calculated from the absorbance at 446 nm, using the molar absorption coefficient of PQQ-DNPH hydrazone in methanol $(31,400 \ M^{-1} \ cm^{-1})$.[43]

[41] F. J. Stevens and D. H. Higginbotham, *J. Am. Chem. Soc.* **76**, 2206 (1953).
[42] C. L. Lobenstein-Verbeek, J. A. Jongejan, J. Frank Jzn, and J. A. Duine, *FEBS Lett.* **170**, 305 (1984).
[43] R. A. van der Meer, J. A. Jongejan, J. Frank Jzn, and J. A. Duine, *FEBS Lett.* **206**, 111 (1986).

Example 2. To 20 ml of dopamine β-monooxygenase[44] (0.05 to 1 mg enzyme) in 0.2 M sodium phosphate buffer, pH 7.0, PH solution is added to a 2-fold molar excess (with respect to the concentration of enzyme subunits). The mixture is incubated for 16 hr at 40°. Removal of excess reagent, proteolysis with pronase, and isolation of the product is performed as indicated for Example 1.

Identity and purity of the product are checked by comparison with the model hydrazone of PQQ and PH, using the reversed-phase HPLC column in combination with photodiode array detection. In case the hydrazone in the Sep-Pak eluate is homogeneous, the amount can be calculated from the absorbance of the eluate at 335 nm, using the estimated molar absorption coefficient of 19,000 M^{-1} cm^{-1}. A more accurate determination can be performed after quantitative conversion of the hydrazone to PQQ. For that purpose, the methanol in the Sep-Pak eluate is removed in a rotary evaporator, and the residue is dried over P_2O_5 *in vacuo.* The solids are dissolved in approximately 2 ml dimethyl sulfoxide (Merck, pro analyse Art. 2931) and incubated at room temperature for 4 hr. The incubate is mixed with 10 ml of 10 mM HCl, and the Sep-Pak procedure is applied. Purity and identity of the resulting PQQ can be checked by reversed-phase HPLC (see section on Chromatographic Methods). Quantification is achieved either with a biological assay or by measuring the absorbance at 249 nm and using the molar absorption coefficient for PQQ at 249 nm (25,400 M^{-1} cm^{-1}).

Evaluation

A large number of quinoprotein oxidoreductases have been investigated with the hydrazine method. As suggested by the data,[45] quantitative extraction of hydrazone was achieved in all cases. The choice of the hydrazine is determined by the specificity of the enzyme (e.g., methylhydrazine effectively reacts with methylamine oxidase but the aromatic hydrazines do not;[46] treatment of dopa decarboxylase with PH scarcely leads to hydrazone formation[47]). Quantitative conversion requires high O_2 tensions and long incubation times (in the case of the amine oxidases, the so-called azo compound is formed in the first instance[38]). For unknown reasons, certain proteases are detrimental for the yield of hydrazone which is obtained.[42,45] In our hands, pronase E is able to detach the product from the protein in a quantitative way. Although HPLC combined with photo-

[44] R. A. van der Meer, J. A. Jongejan, and J. A. Duine, *FEBS Lett.* **231**, 303 (1988).
[45] R. A. van der Meer, J. A. Jongejan, and J. A. Duine, *in* "Proceedings of the First International Symposium on PQQ and Quinoproteins" (J. A. Jongejan and J. A. Duine eds.), p. 111. Kluwer Academic Publishers, Dordrecht, 1989.
[46] J. van Iersel, R. A. van der Meer, and J. A. Duine, *Eur. J. Biochem.* **161**, 415 (1986).
[47] B. W. Groen, R. A. van der Meer, and J. A. Duine, *FEBS Lett.* **237**, 98 (1988).

diode array detection provides an adequate tool to compare product with model compound, if any doubt exists or if the equipment is not available, the hydrazone can be converted to PQQ (as indicated in Example 2).

Hexanol Extraction Procedure[48]

Refluxing of PQQ with hexanol in the presence of HCl gives the 5,5-dihexyl ketal compound of PQQ. It appears that under these conditions, in certain enzymes PQQ is detached from the protein chain and ketal formation prevents reaction with amino acids, as exemplified by galactose oxidase (Example 1). For unknown reasons, in some cases esterification of the carboxylic acid group(s) with hexanol also occurs (this is apparent from the chromatogram described for dopa decarboxylase[47]). As is illustrated by glutamate decarboxylase (Example 2), however, ester formation can be prevented by changing the conditions, and formation of the stable compound 4-hydroxy-5-hexoxypyrroloquinoline (4-hydroxy-5-hexoxy-PQ) can be achieved. Also with the hexanol extraction procedure, the products obtained from the enzyme are compared with model compounds prepared from authentic PQQ.

Instrumentation. The results presented here were obtained with the instrumentation described for the hydrazine method. The eluates are monitored at either 357 nm (for the PQQ-5,5-dihexyl ketal) or 318 nm (for 4-hydroxy-5-hexoxy-PQ) while taking absorption spectra throughout the peaks. Reversed-phase chromatography is performed on a combination of a RP C_{18} guard column (4.0 × 25 mm) and a 5 μm C_{18} RCM cartridge (8.0 × 100 mm) in a RCM-100 module (Waters). The entire system is equilibrated with 10 mM sodium acetate (pH 4.5). After injection of the sample, a 20-min linear gradient (flow rate 1.5 ml/min) from the equilibration buffer to 10 mM sodium acetate (pH 4.5)–acetonitrile (40 : 60, v/v) is applied, holding the final condition for 5 min, and subsequently a 5-min linear gradient from the 40 : 60 ratio eluent to the equilibration buffer. In order to obtain reproducible results, it is necessary to reequilibrate the column between each run with starting buffer for 10 min.

Model Compounds. PQQ-5,5-dihexyl ketal is prepared from PQQ in the following way: 100 mg PQQ, dissolved in 150 ml of 3 M HCl, is mixed with 50 ml hexanol. This mixture is refluxed for 2 hr (157°). After cooling to room temperature, the hexanol layer is removed and washed with 200-ml portions of water until most of the acid is removed (pH 7.0). The hexanol is evaporated under reduced pressure, and any traces of volatile material are removed by keeping the residue overnight *in vacuo* over P_2O_5 (yield 85 mg). The ketal compound is homogeneous, as shown by the

---

[48] R. A. van der Meer, A. C. Mulder, J. A. Jongejan, and J. A. Duine, *FEBS Lett.* **254,** 99 (1989).

reversed-phase HPLC system (retention time 24.3 ± 0.5 min, absorption maxima at 239, 313, and 357 nm).

4-Hydroxy-5-hexoxy-PQ is prepared from PQQ-5,5-dihexyl ketal by dissolving 40 mg of the latter compound in 50 ml water and flushing the solution with argon for 15 min, after which 40 mg NaBH$_4$ is added. Reduction is complete after approximately 2 hr, as can be checked by reversed-phase HPLC (retention time of the product is 12.5 ± 0.5 min). The product is recovered using the Sep-Pak procedure, after which methanol is evaporated under reduced pressure. Conversion to 4-hydroxy-5-hexoxy-PQ is performed by dissolving the dried preparation (36 mg) in 10 ml concentrated H$_2$SO$_4$. After 20 min at 50°, the reaction is complete (as shown by reversed-phase HPLC), and the mixture is diluted with water. Recovery of the product is achieved by applying the Sep-Pak procedure (yield 29 mg). Using the indicated HPLC system, it appears that 4-hydroxy-5-hexoxy-PQ has a retention time of 14.5 ± 0.5 min (absorption maxima at 270, 318, and 375 nm).

Reagents. All solvents are prepared from Milli-Q water (Waters), filtered through a 0.22-μm filter, and degassed before use. Methanol and acetonitrile (both HPLC grade) are obtained from Rathburn Chemicals Ltd. Hexanol is from Merck (more than 98% pure, Art. 804393). Hydrochloric acid is of pro-analyse quality (Merck).

Example 1. 7.5 ml of a galactose oxidase preparation from *Dactylium dendroides*[49] (containing 0.1 to 10 mg enzyme) in 0.1 M sodium phosphate, pH 7.5, is mixed with 7.5 ml of 6 M HCl and 15 ml hexanol. The mixture is boiled under reflux conditions for 2 hr (157°). After cooling to room temperature, the organic layer is separated and washed with 200-ml portions of water until neutral. The hexanol is removed in a rotary evaporator under reduced pressure at elevated temperature. Traces of hexanol are removed by keeping the residue overnight *in vacuo* over P$_2$O$_5$. The solids are dissolved in a small amount of methanol, and the solution is diluted with water to 10% methanol. Subsequently, the Sep-Pak procedure is applied. Checking of homogeneity and identity of the extracted product occurs by comparison with the model compound, using reversed-phase HPLC and photodiode-array detection. If the product appears to be homogeneous, quantification is possible by measuring the absorbance at 357 nm in the Sep-Pak eluate and using the molar absorption coefficient for PQQ-5,5-dihexyl ketal at this wavelength ($8700\ M^{-1}\ cm^{-1}$, measured in methanol[48]). It is also possible to convert the ketal to PQQ itself by performing a hydrolysis step at high pH (bringing the solution to ~2 M sodium carbonate) at 90° for 2 hr. The resulting free PQQ can be quantified (after neutralization) by, for example, a biological assay.

[49] R. A. van der Meer, J. A. Jongejan, and J. A. Duine, *J. Biol. Chem.* **264**, 7792 (1989).

Example 2. To a 7.5-ml sample of glutamate decarboxylase from *E. coli*[50] (containing 0.1 to 10 mg of enzyme/ml) in 0.2 *M* sodium phosphate, pH 7.0, 7.5 ml of 6 *M* HCl and 10 ml hexanol are added. After refluxing for 30 min (157°), the water phase is removed by evaporation. Another 10 ml of hexanol is added and refluxing restarted for an additional 1 hr. The hexanol is then removed by evaporation under reduced pressure and elevated temperature. Traces of hexanol are removed by keeping the residue overnight under vacuum over P_2O_5. The dried residue is dissolved in methanol, and the solution is diluted with water to a methanol concentration of not more than 10% so that the Sep-Pak procedure can be applied. After checking homogeneity and identity of the product with reverse-phase HPLC, quantification is performed by measuring the absorbance at 318 nm in the Sep-Pak eluate, using the molar absorption coefficient for 4-hydroxy-5-hexoxy-PQ at this wavelength (39,000 M^{-1} cm^{-1}, measured in methanol). Attempts to oxidize the compound to PQQ with one-electron acceptors were successful,[48] but no information on quantitative aspects is available.

Evaluation and Remarks

Compared with the hydrazine method, the hexanol extraction procedure gives the same results.[48] In principle, the latter has a broader applicability since derivatization is not related to enzymatic activity. It has been applied to several enzymes and appears to be reproducible (it should be noted, however, that although the procedure involving ketal product formation works well for certain enzymes, the procedure developed to produce the 4-hydroxy-5-hexoxy-PQ has a more general applicability since, in contrast to the first method, so far no quinoproteins have been found where it failed). The mechanism involved in formation of the reduced product is unknown. Thus, it is unclear at the moment whether impurities in, for example, the hexanol used are responsible for the reduction step required for 4-hydroxy-5-hexoxy-PQ formation. With respect to this uncertainty, it is advised to use the same reagents as indicated here and to apply the method to homogeneous enzyme preparations. Finally, it should be noted that both the hydrazine method as well as the hexanol extraction procedure do not discriminate between pro-PQQ[51] and covalently bound PQQ. Also from other recent examples (unpublished results) it appears that the term "covalently bound PQQ" should be interpreted to include precursor-like structures being transformed into PQQ by the procedures discussed here.

[50] R. A. van der Meer, B. W. Groen, and J. A. Duine, *FEBS Lett.* **246**, 109 (1989).
[51] F. M. D. Vellieux, F. Huitema, H. Groendijk, K. H. Kalk, J. Frank, J. A. Jongejan, J. A. Duine, K. Petratos, J. Drenth, and W. G. J. Hol, *EMBO J,* **8**, 2171 (1989).

[42] Blue Copper Proteins Involved in Methanol and Methylamine Oxidation

By CHRISTOPHER ANTHONY

Introduction

The majority of methylotrophic bacteria, when grown on methylamine as the sole source of carbon and energy, synthesize a specific quinoprotein, methylamine dehydrogenase, which is located in the periplasm, together with its specific electron acceptor, a blue copper protein or cupredoxin, called amicyanin.[1-3] Although amicyanin is the best (usually only) electron acceptor for methylamine dehydrogenase, it is not detectable in all methylotrophs growing on methylamine by way of this enzyme. The difficulty of determining whether amicyanin is essential for methylamine oxidation is due to large variations in amounts of amicyanin found in those bacteria that do produce it, variations in amounts of blue copper proteins brought about by varying copper or iron concentrations during growth, and variations in the ease of dissociation of copper from the proteins when studied *in vitro*. In some methylotrophs it is possible that cytochrome c is able to replace amicyanin, but the extent to which this occurs *in vivo* is uncertain.[4-9]

Provided sufficient copper and iron are included in the growth medium, a second blue copper protein (or cupredoxin) is also synthesized. This is usually called azurin, or pseudoazurin because it has no function in reduction of nitrite and does not necessarily have a primary sequence similar to that of other azurins. It has no known function, but it is probably able to replace cytochrome c in mediating the oxidation of amicyanin by oxidase, thus leading to a complete methylamine oxidase electron-transport chain in which soluble cytochromes are completely replaced by blue copper proteins.[8,9] Although azurin cannot replace cytochrome c_L as elec-

[1] J. Tobari and Y. Harada, *Biochim. Biophys. Acta* **101**, 502 (1981).
[2] C. Anthony, *in* "Bacterial Energy Transduction" (C. Anthony, ed.), p. 293. Academic Press, London, 1988.
[3] C. Anthony, *Antonie van Leeuwenhoek* **56**, 13 (1989).
[4] R. Chandrasekar and M. H. Klapper, *J. Biol. Chem.* **261**, 3616 (1986).
[5] Y. Fukumori and T. Yamanaka, *J. Biochem. (Tokyo)* **101**, 441 (1987).
[6] S. A. Lawton and C. Anthony, *J. Gen. Microbiol.* **131**, 2165 (1985).
[7] S. A. Lawton and C. Anthony, *Biochem. J.* **228**, 719 (1985).
[8] K. A. Auton and C. Anthony, *Biochem. J.* **260**, 75 (1989).
[9] K. A. Auton and C. Anthony, *J. Gen. Microbiol.* **135**, 1923 (1989).

tron acceptor for methanol dehydrogenase, during growth of some organisms on methanol azurin may be present at a sufficiently high level to mediate the oxidation of cytochrome c_L by the oxidase.[8,9]

Assay of Blue Copper Proteins from Absorption Spectra

Concentrations of amicyanin and azurin are determined from the visible absorption spectra of protein oxidized with a few grains of potassium ferricyanide; the reference cuvette contains water. The oxidized proteins have characteristic broad absorption bands in the visible region, with peaks between 590 and 630 nm. Concentrations can be calculated from the appropriate extinction coefficients. If these are not known it is convenient to assume a value of 4 mM^{-1} cm^{-1} (1 μmol in 1 ml has an absorption of 4.0; an absorption of 1.0 is due to 250 nmol). After the blue copper proteins have been purified and extinction coefficients and relative proportions of each protein determined, then the concentrations of each blue copper protein can be determined in crude extracts (see Table I).

Assay of Blue Copper Proteins as Electron Acceptor for Methylamine Dehydrogenase

Reagents

MOPS–KOH buffer, 0.1 M, pH 7.0
Methylamine hydrochloride, 0.1 M, adjusted to pH 7.0
Methylamine dehydrogenase, 150 nM

Procedure and Units. The reaction is measured in a 1-ml spectrophotometer cuvette at 20°. The 1-ml reaction mixture contains 0.25 ml MOPS buffer, 0.1 ml methylamine dehydrogenase, and electron acceptor (blue copper protein) sufficient to give an absorption in the oxidized form of at least 0.1. Methylamine (0.1 ml) is added to start the reaction. Absorption is measured at the wavelength maxima of the protein being studied. One unit of methylamine dehydrogenase reduces 1 μmol of amicyanin per minute (see above for calculation). In this assay, 15 nM methylamine dehydrogenase (15 pmol in 1 ml) catalyzes the reduction of about 30 nmol of amicyanin per minute or about 1 nmol of horse heart cytochrome c per minute, or about 1.5 nmol oxygen in the dye-linked assay.

Purification of Blue Copper Proteins from Organism 4025

The purification presented below is for the blue copper proteins from an obligate methylotroph, similar to *Methylophilus methylotrophus,* called

TABLE I
PURIFICATION OF AMICYANIN AND AZURIN FROM ORGANISM 4025

Purification step	Volume (ml)	Copper protein (nmol/mg protein)	Copper protein (total, μmol)	Yield (%)	Purification (-fold)
Amicyanin					
Crude extract	80	1.5	5.47	100	1
DEAE-cellulose	83	6.0	4.44	69	3.9
Sephadex G-150	135	33.9	3.07	56	22.2
Mono S	77	76.5	2.44	45	50.0
Mono Q	83	96.8	2.33	43	63.3
Axurin					
Crude extract	150	0.083	0.75	100	1
DEAE-cellulose	100	0.91	0.50	67	10.9
Sephadex G-150	75	2.37	0.40	53	28.5
CM-cellulose	84	17.26	0.31	41	208.0
Mono S	92	81.3	0.29	32	978.6

organism 4025. It is unusual in requiring relatively high copper concentrations for growth on methylamine, during which exceptionally large amounts of both amicyanin and azurin are synthesized. Iron must also be added to growth medium because blue copper proteins are not synthesized in its absence.[9] The methods presented here are probably suitable for other methylotrophs, but, for convenience, references to alternative methods for use with other important methylotrophs are given: *Methylobacterium extorquens* AM1, the methylotroph from which amicyanin was first isolated,[10] and *Paracoccus denitrificans,* in which a convenient starting point can be the periplasmic fraction.[11-13] It should be noted that some blue copper proteins lose copper rather easily; this can be prevented by including 1 μM CuSO$_4$ in purification buffers.

Growth of Organism 4025. Bacteria are grown aerobically in 18-liter batch cultures at 30° on methylamine (0.5%, w/v) for preparation of amicyanin or on methanol (1%, v/v) for preparation of azurin. Growth on methanol leads to production of azurin only and so avoids the problem of extremely high concentrations of amicyanin. The purification methods can be used, however, for both proteins from methylamine-grown bacteria.

The growth medium is that previously described,[14] with the following trace element solution added to give a final concentration in the growth

[10] R. P. Ambler and J. Tobari, *Biochem. J.* **232,** 451 (1985).

[11] M. Husain and V. L. Davidson, *J. Biol. Chem.* **260,** 14626 (1985).

[12] K. Martinkus, P. J. Kennelly, T. Rea, and R. Timkovitch, *Arch. Biochem. Biophys.* **199,** 465 (1980).

[13] A. R. Long and C. Anthony, this volume [34].

[14] Y. Amano, H. Sawada, N. Takada, and G. Terui, *J. Ferment. Technol.* **53,** 315 (1975).

medium of 1 mg/liter. The trace element solution contains the following (g/liter): $FeSO_4 \cdot 2H_2O$, 0.2; H_3BO_3, 0.003; $MnSO_4 \cdot 4H_2O$, 0.02; $ZnSO_4 \cdot 7H_2O$, 0.02; Na_2MoO_4, 0.004; $CaCl_2 \cdot 2H_2O$, 0.53; $CoCl_2 \cdot 6H_2O$, 0.004. $CuSO_4 \cdot 5H_2O$ (5 mg per liter) is added separately. The growth medium described here is that used for the purifications described below. Alternative methods for growth of the organism in continuous culture have been described elsewhere.[9]

After harvesting by centrifugation, bacteria are washed twice at 4° in 20 mM Tris-HCl buffer (pH 8.0), suspended in the same buffer (100 g wet weight in 270 ml), and disrupted by sonication for 10 min in 1-min periods. Whole bacteria and debris are removed by centrifugation at 6000 g for 10 min, and membranes are removed by centrifugation at 138,000 g for 1 hr.

Purification of Amicyanin, Azurin, Methylamine Dehydrogenase, and c-Type Cytochromes from Methylamine-Grown Bacteria. The crude extract is passed down a column of DEAE-cellulose (5 × 23 cm) equilibrated with 20 mM Tris-HCl buffer (pH 8.0) containing 1.2 μM $CuSO_4$. Methylamine dehydrogenase, amicyanin, azurin, and cytochrome c_H are not adsorbed. The fractions containing these proteins are pooled and concentrated under nitrogen on an Amicon (Danvers, MA) YM2 filter and subjected to gel filtration on a column (4 × 84 cm) of Sephadex G-150 equilibrated with 20 mM Tris-HCl (pH 8.0) which separates the dehydrogenase from the smaller electron-transport proteins. The fractions containing these proteins are pooled and dialyzed against 10 mM MES–KOH buffer (pH 5.5), concentrated, and subjected to cation-exchange chromatography on a Pharmacia (Piscataway, NJ) fast liquid protein chromatography (FPLC) Mono S column (1 ml) equilibrated with the same buffer.

Amicyanin is not adsorbed to the Mono S column. It is dialyzed against 10 mM MOPS–KOH and subjected to anion-exchange chromatography on a Pharmacia Mono Q column equilibrated with the same buffer. Amicyanin is eluted at 300 mM NaCl in a gradient of 0–1.0 M NaCl in the same buffer. The pure amicyanin is dialyzed against the same MOPS buffer and stored at −20°.

Methylamine dehydrogenase from the initial gel filtration column is dialyzed against 10 mM potassium phosphate buffer (pH 7.0), applied to a column (5 × 14 cm) of CM-cellulose equilibrated with the same buffer, and eluted (at 55–65 mM phosphate) by using a linear gradient of 10–100 mM potassium phosphate buffer (pH 7.0). The pure dehydrogenase is dialyzed against 10 mM MOPS–KOH buffer (pH 7.0) and stored at −20°. The specific activity of the pure protein is 0.67 μmol of oxygen consumed per minute per milligram protein in the published dye-linked assay system;[7] the yield is about 50%.

Cytochrome c_L is eluted from the first DEAE-cellulose column with a linear gradient of 20–300 mM Tris-HCl buffer (pH 8.0). It is concentrated under nitrogen on an Amicon YM2 filter and partially purified by gel filtration on columns (2.5 × 76 cm) of Sephadex G-50, followed by Sephadex G-150, both equilibrated with 20 mM MOPS–KOH buffer (pH 7.0).

Azurin (containing cytochrome c_H) is eluted from the Mono S column at 310 mM NaCl by using a linear gradient of 0–1.0 M NaCl in 10 mM MES–KOH buffer (pH 5.5). The azurin produced by this method is 98% pure, the sole contaminant being cytochrome c_H, which can be removed by further runs through the Mono S column.

Purification of Azurin from Organism 4025 Grown on Methanol. The purification of azurin from methanol-grown organism 4025 has the advantage that there is no possibility of contamination with amicyanin, which is not induced during growth on methanol. Crude extract, prepared as described above, is subjected to anion-exchange chromatography on a column (6 × 15 cm) of DEAE-cellulose and gel filtration on Sephadex G-150 (4 × 84 cm) as described for amicyanin purification (above). Fractions containing azurin are dialyzed against 10 mM potassium phosphate buffer (pH 6.0) followed by cation-exchange chromatography on a column (6 × 6 cm) of CM-cellulose equilibrated in the same buffer. Azurin is eluted (at 40–50 mM) by using a linear gradient of 10–100 mM potassium phosphate buffer (pH 6.0). It is dialyzed against 10 mM MES–KOH buffer (pH 5.5) and purified to homogeneity on a Pharmacia Mono S column as described above. It is dialyzed against 10 mM MOPS–KOH buffer (pH 7.0) and stored at −17°.

Properties of Blue Copper Proteins from Organism 4025[6–8]

Amicyanin. Amicyanin has a molecular weight of 11,500, as measured by sodium dodecyl sulfate–polyacrylamide gel electrophoresis (SDS–PAGE). It has an isoelectric point of 5.3 (isoelectric focusing), which is markedly lower than those of amicyanins from *Methylobacterium extorquens* AM1 (pI 9.3).[1] It has a midpoint redox potential at pH 7.0 of +294 mV. In the oxidized state it has an absorption maximum at 620 nm with an extinction coefficient of 3.8 mM^{-1} cm^{-1}. It reacts as electron acceptor with methylamine dehydrogenase and with both cytochrome c and azurin but not with the cytochrome oxidase of organism 4025 (a cytochrome *co*).[8]

Azurin. Azurin has a molecular weight of 12,500 (measured by SDS–PAGE) and an isoelectric point of 9.4 (by isoelectric focusing), which is identical to that of azurin from *Methylobacterium extorquens* AM1.[1] It has a midpoint redox potential at pH 7.0 of +323 mV. In the oxidized state it has an absorption maximum at 620 nm with an extinction coefficient of 16 mM^{-1} cm^{-1}, which is about 4 times higher than that of most other blue copper proteins. Azurin is not an electron acceptor for methylamine dehydrogenase, but it is able to react with amicyanin, cytochrome *c*, and the oxidase (cytochrome *co*).

[43] Soluble Cytochromes c from *Methylomonas* A4

By ALAN A. DISPIRITO

Introduction

Methanotrophs oxidize methane to carbon dioxide by a series of two-electron steps with methanol, formaldehyde, and formate as intermediates.[1] Soluble *c*-type cytochromes have been shown or proposed to be directly involved as either an electron acceptor or electron donor in each step of the methane oxidation pathway except the last (formate oxidation). Soluble *c*-type cytochromes have been isolated and characterized in several different methanol-oxidizing methylotrophic bacteria.[2-4] However, in methanotrophs much less is known. This chapter describes the soluble cytochrome composition in the marine methanotroph *Methylomonas* strain A4 and examines the soluble cytochrome composition in several different freshwater methanotrophs.

Growth and Harvesting of Cells

Methylomonas A4[5] is grown in NMS mineral salts medium[6] plus 1.5% NaCl and a vitamin mixture[5] at 37° under an atmosphere of 20% methane and 80% air (v/v). Cells are grown in a 10-liter fermentor and harvested by centrifugation at 13,200 *g* for 15 min at 4°. Cells are resuspended in 50 m*M* Tris-HCl, 250 m*M* NaCl, pH 8.0, buffer and centrifuged at 13,200 *g* for 15 min at 4°. Approximately 2.5 g cell protein (10–15 g wet weight) is obtained per 10 liters, and at least 25 g cell protein should be

---

[1] C. Anthony, "The Biochemistry of Methylotrophs." Academic Press, New York, 1982.
[2] C. Anthony, *Adv. Microb. Physiol.* **27,** 113 (1986).
[3] D. Day and C. Anthony, this volume [44].
[4] J. Frank and J. A. Duine, this volume [45].
[5] M. E. Lidstrom, *Antonie van Leeuwenhoek* **54,** 189 (1988).
[6] R. Whittenbury and H. Dalton, *in* "The Prokayotes" (M. P. Starr, H. G. Stolp, A. Truper, A. Balows, and H. G. Schlegel, eds.), Vol. 1, p. 894. Springer-Verlag, Berlin and New York, 1981.

obtained before purification procedures are initiated. Cells can be stored at -70 or $-20°$.

Assay Methods

Spectroscopy

The absorption spectrum (250 to 750 nm) of either oxidized or reduced (following addition of a few crystals of sodium dithionite) cytochrome c is the usual assay. Rough quantitation during purification can be obtained using the extinction coefficient for the α absorption maximum of the reduced form. The heme content of a sample is determined by the pyridine ferrohemochromogen method.[7] The pyridine ferrohemochromogen spectrum does not change with NaOH concentrations varying from 0.02 to 0.5 M NaOH with pyridine concentrations varying from 1.2 to 6.5 M, but concentrations of 25% pyridine and 0.075 N NaOH are suggested. For example, the sample (1 – 100 μl) is mixed with 0.1 N NaOH to a total volume of 800 μl (for 1-ml cuvette), a few crystals of sodium dithionite are added with mixing, followed by the addition of 200 μl of 70% (v/v) pyridine (in H_2O), and the absorption spectrum (350 to 600 nm) is read immediately. The pyridine ferrohemochromogen absorption spectrum of c-type cytochromes shows absorption maxima at 414, 520, and 550 $\pm$ 1 nm with a $\Delta\epsilon$ (550 nm) of 29.1 cm^{-1} nM^{-1}.

Protein concentrations can be determined by the method of Lowry et al.[8] Bovine serum albumin can be used as a standard during the initial purification steps, but equine cytochrome c (Sigma, St. Louis, MO, type IV) should be used on the final sample since the absorbance from c-type cytochromes is approximately 1.2 times higher than bovine serum albumin in the Lowry assay.

Heme Stain

c-Type cytochromes can also be detected in polyacrylamide gels by labeling these proteins with 3,3'-diaminobenzidine (DAB).[9] Samples (whole cells or purified samples) are solubilized for 5 min or more at room temperature in 65 mM Tris-HCl, 2% sodium dodecyl sulfate, and 5% 2-mercaptoethanol and electrophoresed in a 10 to 18% polyacrylamide gel

---

[7] J.-H. Fuhrop, "Laboratory Methods in Porphyrin and Metalloporphyin Research." Elsevier, New York, 1975.

[8] O. H. Lowry, N. J. Rosebrough, A. L. Farr, and R. J. Randall, *J. Biol. Chem.* **193**, 265 (1951).

[9] A. McDonnel and L. A. Staehelin, *Anal. Biochem.* **117**, 40 (1981).

according to the method of Laemmli.[10] Following electrophoresis, the gel is fixed in 7% acetic acid for 15 to 20 min. The gel is then incubated with shaking in 0.5 M Tris-HCl, pH 7.0, buffer for 15 min. The buffer is decanted and this step repeated several (2–5) more times until the pH of the wash solution remains at pH 7.0. The Tris buffer is decanted and the gel incubated in 0.5 M Tris HCl, 1.4 mM DAB (Sigma), pH 7.0, buffer for 30 min at room temperature. The DAB–Tris-HCl solution is decanted and the gel transferred to a solution of 50 mM citrate, 2.8 mM DAB, pH 4.0, buffer. The reaction is started by adding 40 μl of a 30% (v/v) H_2O_2 solution per milliliter of citrate–DAB solution. The gel is then incubated for 8–12 hr in the dark at 0°. Labeled proteins will appear as reddish brown bands.

Purification Procedures

All isolation steps should be performed at 0 to 4°.

Cell Lysis. Methylomonas A4 can be lysed by a number of different methods. Since *Methylomonas* A4 is a marine methanotroph, cells can be lysed by centrifuging at 13,200 g for 15 min and resuspending in dilute buffer (less than 10 mM) or distilled water. Cells can also be lysed by passage through a French pressure cell 3 times at 20,000 psi. Cell suspensions can also be lysed by sonication for 10 min in 2- to 3-min sets. After lysis, deoxyribonuclease I is added to the cell suspension to a final concentration of 1–10 μg/ml.

Isolation of Soluble Fraction and Ammonium Sulfate Fractionation. Following lysis the cell slurry is centrifuged at 13,200 g for 30 min to remove unlysed cells and cell debris. The pellet is discarded and the supernatant is centrifuged at 150,000 g for 90 min to pellet the membrane fraction. The supernatant is then brought to 30% saturation with solid ammonium sulfate, stirred for 1.5–3 hr and centrifuged for 15 min. The pellet is discarded, and the concentration of ammonium sulfate in the supernate is raised to 60% saturation. The solution is stirred for 1.5–3 hr, centrifuged at 13,200 g for 15 min, and the pellet resuspended in a minimal volume of 10 mM Tris-HCl, pH 8.5, buffer (buffer B) (30–60% ammonium sulfate fraction). The concentration of ammonium sulfate in the supernatant is raised to 80% of saturation and the solution stirred and centrifuged as before (60–80% ammonium sulfate fraction). The resulting precipitate from the 60–80% ammonium sulfate fractionation step is resuspended in a minimal volume of buffer B. The proteins precipitating in 30–60% and 60–80% ammonium sulfate and the supernatant obtained

---

[10] U. K. Laemmli, *Nature (London)* **227**, 680 (1970).

from the above centrifugations (soluble 80% ammonium sulfate fraction) are dialyzed for 10–12 hr against 3–4 changes of buffer B. The three ammonium sulfate fractions are concentrated with a stirred cell (Diaflow, Amicon Corp., Danvers, MA) using a YM10 filter or concentrated in dialysis bags using polyethylene glycol.

Purification of Cytochrome c-554. Cytochrome c-554 is observed in all three ammonium sulfate fractions. The major portion (76%) of cytochrome c-554 is observed in the soluble 80% ammonium sulfate fraction. The cytochrome in this fraction is purified in one step by separation on a 10×20 cm preparative isoelectric focusing bed containing 4% Ultrodex, 1% ampholine, pH 4.0–6.0, and 1% ampholine, pH 3.5–10.0 (Fig. 1A). Cytochrome c-554 migrated as two major bands focusing at pH 6.4 and 5.65 (bands A and B; Fig. 1A) called cytochrome c_H-554 and cytochrome c_L-554, respectively, for the purpose of comparison, although the amino acid compositions and the N-terminal amino acid sequences of both are identical.[11] Both bands are eluted from the Ultrodex with 50 mM Tris-HCl buffer, pH 7.5 (buffer C). The two cytochromes are dialyzed against 3 changes of buffer C and concentrated with an Amicon Centricon with a YM10 filter.

Approximately 12% of cytochrome c-554 is observed in the 60–80% ammonium sulfate fraction. This fraction is separated on a preparative isoelectric focusing bed as described above. The major red band (Fig. 1B; band A) focusing at pH 6.4 is eluted with buffer A and concentrated with a stirred cell (YM10 filter). The sample is then loaded on a Sephadex G-50 column (2.5×40 cm) equilibrated with buffer A and the eluting colored fraction concentrated with a Centricon (YM10 filter).

Cytochrome c-554 is also observed in the 30–60% ammonium sulfate fraction. This fraction is loaded on a DEAE-cellulose column (2.5×35 cm) equilibrated with buffer B. The column is washed with approximately 2 bed volumes of column buffer, and a red band slowly elutes with 50 mM Tris-HCl, pH 8.0 buffer. The red band is concentrated with a stirred cell (YM10 filter) and applied to a preparative isoelectric focusing bed containing 4% Ultrodex, 1% ampholine, pH 5.0–8.0, and 1% ampholine (Fig. 1C). Cytochrome c-554 migrates as two bands at pH 5.6 and 6.4 (bands A and B; Fig. 1C). The samples are eluted from the Ultrodex and concentrated as described above.

Purification of Cytochrome c-551. The third major red band observed in the isoelectric focusing bed of the DEAE-cellulose column fraction (Fig. 1C; band E) is cytochrome c-551. This band is eluted from the Ultrodex and concentrated with a Centricon (YM10 filter) as described above. The

---

[11] A. DiSpirito and M. Lidstrom, unpublished data.

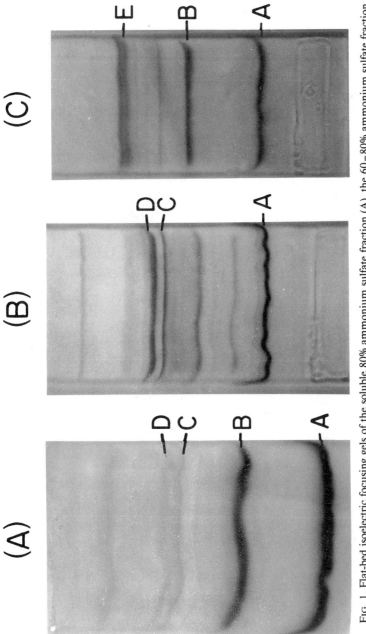

FIG. 1. Flat-bed isoelectric focusing gels of the soluble 80% ammonium sulfate fraction (A), the 60–80% ammonium sulfate fraction (B), and the DEAE-cellulose column fraction (C). Samples are loaded and focused for 14–17 hr at 500 V and 4°. Cytochromes *c*-554 (oxidized; band A), *c*-554 (reduced; band B), *c*-553 (band C), *c*-552 (band D), and *c*-551 (band E) separating at pH 6.4, 5.65, 4.85, 4.7, and 4.75, respectively, are shown.

TABLE I
PURIFICATION OF CYTOCHROMES c-554, c-553, AND c-552 FROM *Methylomonas* A4[a]

		Protein		Heme c		
Cytochrome	Purification step	Total (mg)	%	Total (mol)	%	nmol/mg protein
	Sonicated cells	25,220	100.0	42.9	100.0	1.7
	Soluble S144	15,700	62.2	37.2	86.7	2.4
	Soluble 80% ammonium sulfate	1,775	7.0	20.8	48.5	11.7
	60–80% ammonium sulfate	2,200	8.7	10.3	24.0	4.7
	30–60% ammonium sulfate	6,725	26.7	4.6	10.7	0.7
c-554, I	Soluble 80% ammonium sulfate	—	—	—	—	—
	Isoelectric focusing	73.50	0.29	9.7	22.6	132
c-554, II	60–80% ammonium sulfate	—	—	—	—	—
	Isoelectric focusing	14.0	0.06	1.8	4.2	129
	Sephadex G-50	11.9	0.05	1.6	3.7	134
c-554, III	30–60% ammonium sulfate	—	—	—	—	—
	DEAE-cellulose (CP3)	470.0	1.9	3.4	7.9	7.2
	Isoelectric focusing	12.6	0.05	1.7	4.0	135
c-554, I + II + III		98.0	0.39	13.0	30.3	133
c-553	60–80% ammonium sulfate	—	—	—	—	—
	Isoelectric focusing	57.5	0.23	2.0	4.7	3.48
c-552	60–80% ammonium sulfate	—	—	—	—	—
	Isoelectric focusing	103.3	0.41	6.5	15.2	62.9
	DEAE-sepharose CL-6B	73.5	0.29	6.0	14.0	81.6
c-551	30–60% ammonium sulfate	—	—	—	—	—
	isoelectric focusing	23.6	0.09	2.1	4.9	89
	Sephadex G-75	20.4	0.08	2.0	4.7	98

[a] The heme c concentration was measured using the pyridine ferrohemochrome method[10] using an ϵ (550 nm) of 29.1 cm^{-1} mM^{-1}.

sample is loaded on a Sephadex G-75 column (2.5 × 45 cm) equilibrated with buffer A. The red fraction that elutes from this column is then concentrated with a Centricon.

Purification of Cytochromes c-552 and c-553. The two red bands migrating at pH 4.85 and pH 4.7 in the isoelectric focusing bed of the 60–80% ammonium sulfate fraction are cytochrome c-553 (band C; Fig. 1B) and cytochrome c-552 (band D; Fig. 1B), respectively. The two cytochromes are eluted from the Ultrodex, dialyzed against 3 changes of buffer B, and concentrated with a stirred cell (YM10 filter). Cytochrome

TABLE II
PROPERTIES OF SOLUBLE CYTOCHROMES FROM *Methylomonas* A4

	Cytochrome			
Property	c-554	c-553	c-552	c-551
Percent soluble c-heme	57.3	7.5	26.4	8.8
nmol heme c/nmol cytochrome	1.1	1.6	0.9	1.6
Molecular weight[a]	8,500	34,000	11,500	16,500
Isoelectric point	5.65[b], 6.4[c]	4.85	4.7	4.75
Extinction coefficient (α band; cm^{-1} mM^{-1}	21.3	25.0	24.0	28.7
Absorption maxima (nm)				
Oxidized	413	406	408	410
Reduced				
γ band	418	419	417	416
β band	524	524	522	551
α band	554	553	552	551
Binding of CO	−	+	−	−

[a] The molecular weight of cytochrome c-554 was determined by gel fitration, and the molecular weights of cytochromes c-553, c-552, and c-551 were determined on denaturing gels.
[b] Oxidized form.
[c] Reduced form.

c-553 is then loaded on a DEAE-Sepharose CL-6B column (1.25 × 25 cm) (Pharmacia P-L Biochemicals) equilibrated with the dialysis buffer. A red band slowly elutes from the column with the dialysis buffer, and the sample is concentrated with a Centricon (YM10 filter). Cytochrome c-552 is also loaded on a DEAE-Sepharose CL-6B column (1.25 × 30 cm) equilibrated with buffer B. The column is washed with approximately 2 bed volumes of column buffer and with 25 mM Tris-HCl buffer, pH 8.0, and the sample is eluted with 50 mM Tris-HCl buffer, pH 8.0.

The purification of the four soluble cytochromes from *Methylomonas* A4 is summarized in Table I.

Properties of Soluble Cytochromes from *Methylomonas* A4

The properties of the four soluble cytochromes from *Methylomonas* A4 are listed in Table II.

TABLE III
HEME-STAINING PROTEINS AND IMMUNOBLOTTING OF METHANOTROPH SOLUBLE
PROTEINS USING ANTISERA AGAINST CYTOCHROMES c-554 AND c-552 FROM
Methylomonas A4[a]

Methanotroph	Heme-staining Protein	Relative intensity of cross-reaction	
		With Cyt c-554	With Cyt c-552
Methylomonas A4	37,500	−	−
	35,500	−	−
	34,000	−	+
	16,500	−	−
	11,500	−	+++
	8,500	+++	−
Methylomonas A1	35,500	−	−
	34,000	−	+
	16,500	−	−
	11,500	−	+++
	8,500	+++	−
Methlomonas C1	11,500	−	+++
	8,500	+++	−
Methylomonas MM2	12,000	−	−
	8,500	−	−
Methylomonas MN	37,000	−	+++
	13,000	−	+++
	8,500	+++	−
Methylomonas albus BG8	60,500	−	−
	11,500	−	+++
	8,500	−	−
Methylosinus trichosporium OB3b	12,000	−	−
Methylocystis parvus OBBP	12,000	−	−
Methylococcus capsulatus (Bath)	11,000	−	−
	10,500	−	−

[a] Symbols: +++, very strong cross-reaction; ++, moderate; +, weak but reproducible; −, no cross-reaction.

Soluble Cytochromes in Other Methanotrophs

The soluble *c*-type cytochromes have been examined in only one other methane utilizer, *Methylomonas capsulatus* (Bath).[12] Four soluble *c*-type cytochromes were also reported in *M. capsulatus,* two major and two minor. One of the major cytochromes, cytochrome *c*-555, has been purified and showed a molecular weight of 11,000. The complete amino acid sequence of this cytochrome was determined by Ambler *et al.,*[12] but other properties were not determined.

We have recently looked at the heme-staining proteins in 8 other methanotrophs and compared them to the two major monoheme cytochromes (cytochromes *c*-554 and *c*-552) from *Methylomonas* A4. Identification of the heme-staining proteins in polyacrylamide gels is determined as described above. The identification of heme-staining proteins which cross-react with antisera from cytochromes *c*-554 and *c*-552 is carried out by denaturing (sodium dodecyl sulfate-containing) polyacrylamide gel electrophoresis followed by Western blotting and cross-reaction of transferred polypeptide bands with polyclonal antibodies raised against the two cytochromes from *Methylomonas* A4. Proteins from the stained gels are blotted onto nitrocellulose membranes using a Bio-Rad Trans-Blot Cell (Bio-Rad Laboratories, Richmond, CA) according to the manufacturer's instructions. The nitrocellulose containing transferred polypeptides is incubated in Tris-HCl-buffered blocking solution, then with antibodies against cytochrome *c*-554 or cytochrome *c*-552, and the antibodies are detected with the alkaline phosphatase assay according to the protocol of the manufacturer (Bio-Rad).

Table III lists the heme-staining proteins from eight different methanotrophs and shows that the number of detectable heme-staining proteins varied from 1 to 6. All the methanotrophs tested showed at least one major heme-staining protein with a molecular weight of 11,000–13,000. Table III also shows the heme-staining proteins which cross-reacted with antisera from cytochromes *c*-554 and *c*-552. Only the heme-staining proteins from other *Methylomonas* isolates cross-reacted with antisera from the two cytochromes from *Methylomonas* A4.

[12] R. P. Ambler, H. Dalton, T. B. Meyer, R. G. Bartsch, and M. D. Kamen, *Biochem. J.* **233,** 333 (1986).

[44] Soluble Cytochromes c of Methanol-Utilizing Bacteria

By Darren J. Day and Christopher Anthony

Introduction

Methanol-utilizing bacteria typically contain relatively large amounts of at least two soluble, periplasmic c-type cytochromes.[1,2] These are functionally distinct but spectrally almost indistinguishable; they were therefore named according to their isoelectric points. Cytochrome c_H has a higher isoelectric point (it is usually basic) and is now known to correspond to the typical soluble c-type cytochromes mediating oxidation by membrane oxidases in mitochondria and many bacteria. Cytochrome c_L occurs only in methylotrophs; it is always acidic (pI 3.0–5.0) and its mass (M_r 17,000–22,000) is about twice that of typical soluble c-type cytochromes. It is the specific electron acceptor for methanol dehydrogenase[1-3] and is sometimes induced to higher levels during growth on methanol, when it constitutes 30–70% of the total soluble c-type cytochrome.

Although these are the main soluble c-type cytochromes in methylotrophic bacteria, many bacteria contain others that appear very similar but which show slight differences of isoelectric point or molecular weight, and some methylotrophs also contain additional c-type cytochromes, some of which have unusual features but no known function.

It should be noted that multiple c-type cytochromes are often observed in membrane fractions having properties identical to those isolated from soluble fractions. It is also important to remember that ion-exchange chromatography is able to separate the oxidized and reduced forms from each other; it is sometimes convenient therefore to include 1 mM sodium ascorbate in buffers to keep all the cytochromes in the reduced state. Deamidation (conversion of amides to acids) of cytochromes may lead to the appearance of additional cytochromes, and in the case of at least one methylotroph *(Methylophilus methylotrophus)* a 4K fragment appears to be readily lost from cytochrome c_L, thus leading to the appearance of two forms of this cytochrome.

The purification methods given below are for the two main soluble c-type cytochromes of *Methylobacterium extorquens* AM1. These are likely

[1] C. Anthony, *Adv. Microb. Physiol.* **27**, 113 (1986).
[2] C. Anthony, *in* "Bacterial Energy Transduction" (C. Anthony, ed.), p. 293. Academic Press, London, 1988.
[3] M. Beardmore-Gray, D. T. O'Keeffe, and C. Anthony, *J. Gen. Microbiol.* **132**, 1553 (1983).

to be readily adaptable to most other organisms, but for convenience references to alternative purification methods for the following important methylotrophs are given: *M. extorquens*,[4] *Methylophilus methylotrophus*,[3,5,6] *Acetobacter methanolicus*,[7] *Hyphomicrobium* X,[8] *Paracoccus denitrificans,* [9-11] *Methylomonas* J,[12] and organism 4025.[13]

Assay of c-Type Cytochromes

The assay to be used depends on the purpose and the state of purification of the cytochromes. The assay depends on the characteristic absorption spectra of the cytochromes. These are so similar that different *c*-type cytochromes cannot be distinguished in crude extracts and a measure of total cytochrome *c* must be used. The main components are usually separated early on in any procedure, either by ion-exchange chromatography or by gel filtration, and the proportions of each cytochrome can then be determined.

Units. The amount of cytochrome is recorded in micromoles, nanomoles, etc. This is calculated from the visible absorption spectrum of the reduced form at the absorption maximum (about 550 nm). For a rough estimate during purification the extinction coefficient for the reduced form may be used. In crude extracts it should be noted that the baseline in the visible region is likely to be sloping, and it is often convenient to use an artificial base of a line between about 535 and 575 nm for calculating peak heights. For more accurate estimates the reduced minus oxidized difference spectrum is used, with the appropriate extinction coefficient. A few grains of dithionite is used to reduce cytochromes, and a few grains of potassium ferricyanide or ammonium persulfate is used for oxidation. When either the proportions of the various cytochromes in crude extracts or the extinction coefficients are not known, a reasonable estimate of the amount of cytochromes can be made by assuming extinction coefficients from the literature for other cytochromes. Useful average values are 29 mM^{-1} cm^{-1} for cytochrome c_H, and 25 mM^{-1} cm^{-1} for cytochrome

[4] D. T. O'Keeffe and C. Anthony, *Biochem. J.* **192,** 411 (1980).

[5] A. R. Cross and C. Anthony, *Biochem. J.* **192,** 421 (1980).

[6] S. J. Froud and C. Anthony, *J. Gen. Microbiol.* **130,** 3319 (1984).

[7] E. J. Elliott and C. Anthony, *J. Gen. Microbiol.* **134,** 369 (1988).

[8] M. Dijkstra, J. Frank, and J. A. Duine, *Biochem. J.* **257,** 87 (1989).

[9] M. Husain and V. L. Davidson, *J. Biol. Chem.* **261,** 8577 (1986).

[10] K. A. Gray, D. B. Knaff, M. Husain, and V. L. Davidson, *FEBS Lett.* **207,** 239 (1986).

[11] G. Bosma, M. Braster, A. H. Stouthamer, and H. W. van Verseveld, *Eur. J. Biochem.* **165,** 665 (1987).

[12] S. Ohta and J. Tobari, *J. Biochem. (Tokyo)* **90,** 215 (1981).

[13] S. A. Lawton and C. Anthony, *Biochem. J.* **228,** 719 (1985).

c_L. For a pure cytochrome the extinction coefficient is calculated from the measured molecular weight and the heme content as determined from the pyridine hemochrome spectrum.

Molecular Weights by Denaturing Electrophoresis. Although not strictly an assay for cytochrome *c*, an important additional analysis is the determination of molecular weights by polyacrylamide gel electrophoresis in the presence of sodium dodecyl sulfate (SDS). This is important for confirming the presence of the expected cytochrome in fractions. Electrophoresis is done at pH 8.3 using SDS – Tris – glycine with final acrylamide concentrations of 10 – 15%, and gels are stained with Coomassie Brilliant Blue as previously described;[14] suitable molecular weights standards are used [BDH 44262 2I (Poole, UK), Sigma Dalton Mark VII (St. Louis, MO), and Pharmacia PMW kits (Piscataway, NJ)]. To confirm the position of cytochromes, they are visualized prior to protein staining by their activity in the peroxidase assay; if this is to be used, mercaptoethanol is not included in the electrophoresis system.

Reagents for Peroxidase Assay

TMBZ (3,3',5,5'-tetramethylbenzidine), 136 mg in 90 ml methanol (prepare fresh)

Sodium acetate buffer, 0.25 *M* (pH 5.0) (14.3 ml glacial acetic acid plus 900 ml water, adjusted to pH 5.5 with 1 *M* NaOH and made up to 1 liter)

TMBZ – acetate, 90 ml TMBZ solution plus 210 ml sodium acetate buffer

Hydrogen peroxide, 30% (w/v)

2-Propanol – acetate buffer, 3 parts 2-propanol plus 7 parts acetate buffer

Procedure. The gel is immersed in acetate buffer for 15 min to fix the protein; it is then immersed in the TMBZ – acetate mixture for 30 – 60 min in the dark. Hydrogen peroxide is added (1 ml/300 ml of solution). The green-staining cytochromes become visible within 15 min, and staining intensifies over 45 min. After sufficient staining is observed the gel is immersed in 2-propanol – acetate buffer mixture to remove excess stain and to clear the gel; the mixture can be replaced a few times to improve clarity. The gels can be photographed or scanned at this stage prior to staining for total protein in Coomassie blue. It is recommended that a range of protein concentrations be used for each fraction to ensure observation of minor components.

[14] K. Weber and M. Osborn, "The Proteins," 3rd ed. (H. Neurath and R. L. Hill, eds.). Vol. 1, p. 179. Academic Press, London, 1975.

Purification of Cytochromes from *Methylobacterium extorquens*
   AM1

Growth of *Methylobacterium extorquens* AM1 and preparation of extracts are described in the chapter on the methanol dehydrogenase of this organism.[15]

Cytochrome c_L. The soluble fraction from about 15 g wet weight of bacteria is passed down a column (24 × 130 mm) of DEAE-Sepharose (Pharmacia, Fast Flow) equilibrated in 20 mM Tris-HCl (pH 8.0). Methanol dehydrogenase and cytochrome c_H are not adsorbed; cytochrome c_L is eluted (at about 100 mM NaCl) with a gradient of 0–250 mM NaCl (total, 200 ml) in the same buffer. Without dialysis, solid ammonium sulfate is added to give 40% saturation at 4°, and the precipitate formed after centrifugation for 10 min at 10,000 g is discarded. The supernatant is applied to a Pharmacia Phenyl-Superose column (1 ml) equilibrated in 20 mM MOPS containing 2.0 M ammonium sulfate (pH 7.0) and cytochrome eluted (at about 1.6 M ammonium sulfate) with a decreasing gradient of 2–1 M ammonium sulfate (total volume about 20 ml). The cytochrome is desalted on a PD10 column equilibrated in 20 mM Tris-HCl (pH 8.0) and applied to a Pharmacia Mono Q column equilibrated in the same buffer. Pure cytochrome c_L is eluted (at about 100 mM) with a gradient of 0–200 mM NaCl in the same buffer (total volume about 10 ml). It is desalted on a PD10 column and stored at −20° (See Table I for purification steps.)

Cytochrome c_H. After passage through DEAE-Sepharose (see above) the extract is dialyzed overnight at 4° against 20 liters of 10 mM potassium phosphate buffer (pH 7.0) and applied to a hydroxylapatite column (16 × 100 mm) equilibrated in the same phosphate buffer. This is eluted with a linear gradient of 10–250 mM potassium phosphate (pH 7.0), the cytochrome c being eluted just prior to the dehydrogenase (at about 80 mM phosphate). It is dialyzed against 20 mM sodium acetate buffer (pH 5.0), applied to a column (24 × 85 mm) of S-Sepharose (Pharmacia, Fast Flow) equilibrated with 25 mM MES buffer (pH 5.5), and eluted (at about 90 mM NaCl) with a linear gradient of NaCl in the same buffer (0–250 mM, 200 ml total volume). For smaller amounts S-Sepharose can be replaced by a Mono S column, equilibrated in 25 mM MES buffer (pH 5.5), and the cytochrome c_H eluted as a very sharp peak using a linear gradient of 0–150 mM NaCl in the same buffer (total volume, 15 ml). The pure protein is passed through PD10 and stored at −20°.

---

[15] D. J. Day and C. Anthony, this volume [33].

TABLE I

PURIFICATION OF SOLUBLE CYTOCHROMES FROM *Methylobacterium extorquens* AMI[a]

Purification stage	Volume (ml)	Total cytochrome (nmol)	Specific content (nmol/mg)	Purification (-fold)	Yield (%)
Crude extract	75	1200	1.63	1	100
Cytochrome c_H					
DEAE-Sepharose	110	406	1.84	1.1	34
Hydroxylapatite	19	348	10.0	6.5	29
S-Sepharose	4.3	240	160.0	98	20
Cytochrome c_L					
DEAE-Sepharose	19	298	19.6	12.0	25
Phenyl-Superose	11	240	39.5	24.2	20
Mono Q	8	216	45.0	27.6	18

[a] The cytochrome recorded in the crude extract is total c-type cytochrome. The percentage values for each separate cytochrome relate to the initial total cytochrome content.

Properties of Cytochrome c_L of *Methylobacterium extorquens* AM1[4,16-18]

Cytochrome c_L has a molecular weight of 18,735 (from the gene sequence), a pI of 4.2, a midpoint redox potential at pH 7.0 of $+256$ mV (the apparent standard midpoint potential at pH 0 is $+345$ mV). Absorption maxima in the reduced form are at 416 and 549 nm, with extinction coefficients of 163 and 26 mM^{-1} cm^{-1}. In the oxidized form the absorption maxima are at 410 and 695 nm, with an extinction coefficient at 695 nm of 0.35 mM^{-1} cm^{-1}. The extinction coefficient at 549 nm measured from the reduced minus oxidized difference spectrum is 21.8 mM^{-1} cm^{-1}.

Cytochrome c_L has a single low-spin heme prosthetic group which has histidine and methionine axial ligands. It reacts with CO slowly and incompletely (72% binding in buffer saturated with CO at room temperature). It is rapidly autoreducible at high pH and it reacts with methanol dehydrogenase. Four ionizations affect the redox potential between pH 4.0 and 9.5, two in the oxidized form (pK values about 3.6 and 5.6) and two in the reduced form (pK values about 4.4 and 6.4), suggesting that the ionizing groups involved may be the two propionate side chains of the heme. The amino acid sequence (derived from the gene sequence) shows that cytochrome c_L constitutes a novel class of c-type cytochromes.

[16] M. Beardmore-Gray, D. T. O'Keefe, and C. Anthony, *Biochem. J.* **207**, 161 (1982).
[17] D. T. O'Keefe and C. Anthony, *Biochem. J.* **190**, 481 (1980).
[18] D. N. Nunn and C. Anthony, *Biochem. J.* **256**, 673 (1988).

Properties of Cytochrome c_H *Methylobacterium extorquens*
AM1[4,16,17]

Cytochrome c_H has a molecular weight of about 11,000, a pI of 8.8, and a midpoint redox potential at pH 7.0 of $+294$ mV (the apparent standard midpoint potential at pH 0 is $+404$ mV). Absorption maxima in the reduced form are at 416.5 and 550.5 nm, with extinction coefficients of 162 and 31 mM^{-1} cm^{-1}. In the oxidized form the absorption maxima are at 410 and 695 nm, with an extinction coefficient at 695 nm of 0.5 mM^{-1} cm^{-1}. The extinction coefficient at 550 nm measured from the reduced minus oxidized difference spectrum is 22.5 mM^{-1} cm^{-1}.

Cytochrome c_H has a single low-spin heme prosthetic group which has histidine and methionine axial ligands. It reacts with CO slowly and incompletely (36% binding in buffer saturated with CO at room temperature). It is rapidly autoreducible at high pH but does not react with methanol dehydrogenase. Four ionizations affect the redox potential between pH 4.0 and 9.5, two in the oxidized form (pK values about 3.5 and 5.5) and two in the reduced form (pK values about 4.5 and 6.5), suggesting that the ionizing groups involved may be the two propionate side chains of the heme. The amino acid sequence shows that cytochrome c_H is a typical class 1 c-type cytochrome.

[45] Cytochrome c_L and Cytochrome c_H from *Hyphomicrobium* X

By J. FRANK and J. A. DUINE

All gram-negative methylotrophic bacteria investigated so far contain at least two soluble cytochromes c, indicated as cytochrome c_L and cytochrome c_H, the subscripts referring to a low and a high isoelectric point, respectively.[1-3] It is generally accepted that cytochrome c_L is the electron acceptor of methanol dehydrogenase.[4,5] Different views exist, however, on the mechanism involved in this electron-transfer process,[5,6] the discrepancy being related to the autoreduction behavior of the isolated cytochrome c_L. As reported already, the procedure developed for the isola-

[1] D. T. O'Keeffe and C. Anthony, *Biochem. J.* **190**, 481 (1980).
[2] D. T. O'Keeffe and C. Anthony, *Biochem. J.* **192**, 411 (1980).
[3] M. Dijkstra, J. Frank, J. E. van Wielink, and J. A. Duine, *Biochem. J.* **251**, 467 (1988).
[4] C. Anthony, *Adv. Microb. Physiol.* **27**, 162 (1986).
[5] M. Dijkstra, J. Frank, and J. A. Duine, *Biochem. J.* **257**, 87 (1989).
[6] M. Beardmore-Gray, D. T. O'Keeffe, and C. Anthony, *J. Gen. Microbiol.* **115**, 523 (1983).

tion of cytochrome c_L from *Hyphomicrobium* X provides a preparation with a very low autoreduction rate[3] but with an excellent electron acceptor capacity for methanol dehydrogenase.[5] It has been shown that cytochrome c_H isolated from *Methylophilus methylotrophus* is oxidized 50 times faster by the cytochrome-*c* oxidase from this organism than cytochrome c_L,[7] suggesting that cytochrome c_H might be a mediator between cytochrome c_L and the oxidase.

Assay

To verify the cytochrome *c* character and to determine the heme *c* content in the proteins, the pyridine hemochrome method[8,9] is applied. The two cytochromes can be distinguished by isoelectric focusing or by chromatography on DEAE ion exchangers,[10] the latter procedure being used for the determination of the relative amounts of the cytochromes in cell-free extracts. The suitability of the cytochromes *c* to function as an electron acceptor for methanol dehydrogenase (MDH) can be established by following the electron-transfer reaction in a stopped-flow spectrophotometer.

Stopped-Flow Spectrophotometric Assay with Methanol Dehydrogenase

Principle. Oxidized cytochrome c_L accepts electrons from fully reduced (MDH_{red}) or pyrroloquinoline quinone (PQQ) semiquinone-containing (MDH_{sem}) methanol dehydrogenase, and the progress of the reaction can be followed spectrophotometrically. The reaction is so fast that a stopped-flow spectrophotometer is required. Excess methanol dehydrogenase over cytochrome *c* is required to assure that the reaction takes place preferentially with only one of the forms of methanol dehydrogenase. The preparation of MDH_{sem} is described elsewhere in this volume.[11] MDH_{red} can be obtained from this preparation by deazalumiflavin-mediated photoreduction.[12]

Procedure. Cytochrome c_L ($\sim 4 \mu M$, $\epsilon_{280} = 3.96 \times 10^4 \ M^{-1} \ cm^{-1}$) is oxidized with a slight excess of potassium ferricyanide and dialyzed against 10 mM MOPS buffer, pH 7.0. One of the syringes of the stopped-flow

---

[7] C. Anthony, *Adv. Microb. Physiol.* **27**, 196 (1986).

[8] J. H. Fuhrhop and K. M. Smith, *in* "Porphyrins and Metalloporphyrins" (K. M. Smith, ed.), p. 804. Elsevier, Amsterdam, 1975.

[9] J. S. Rieske, this series, Vol. 10, p. 488.

[10] S. J. Froud and C. Anthony, *J. Gen. Microbiol.* **130**, 3319 (1984).

[11] J. Frank and J. A. Duine, this volume [32].

[12] J. Frank, M. Dijkstra, J. A. Duine, and C. Balny, *Eur. J. Biochem.* **174**, 331 (1988).

spectrophotometer is filled with the dialyzed ferricytochrome c_L solution, the other with a solution of methanol dehydrogenase in the same buffer (at least 6 μM, $\epsilon_{280} = 2.45 \times 10^5 \ M^{-1} \ cm^{-1}$). After mixing, monoexponential reaction traces are observed from which k_{obs} can be obtained by nonlinear regression. The reduction of ferricytochrome c_L can be followed specifically at 550 nm, whereas the oxidation of methanol dehydrogenase can be observed at 337 nm (an isosbestic point of the cytochrome c_L spectra). The highest sensitivity is, however, obtained at 418 nm, where the reaction of cytochrome c_L with methanol dehydrogenase displays maximal absorption increases. The second-order rate constant for the reaction is obtained from a plot of k_{obs} versus the concentration of methanol dehydrogenase.

Purification Procedure

Growth of Organism. *Hyphomicrobium* X is grown aerobically at 30° on the mineral medium described elsewhere in this volume.[11] All operations during purification are performed at room temperature, except dialysis and centrifugation, which are carried out at 4°.

Separation of Cytochrome c_L, Cytochrome c_H, and Methanol Dehydrogenase

Step 1: Preparation of Cell-Free Extract. Cell paste (88 g wet weight) is suspended in 88 ml of 36 mM Tris–39 mM glycine buffer, pH 9.0. After addition of DNase, the cell suspension is passed twice through a French pressure cell at 110 MPa, and the mixture is centrifuged for 20 min at 48,000 g, giving the cell-free extract.

Step 2: DEAE-Sepharose Chromatography. The cell-free extract is applied to a DEAE-Sepharose column (4.5×30 cm), equilibrated with 36 mM Tris–39 mM glycine buffer, pH 9.0. Cytochrome c_H is eluted by washing the column with the same buffer and collected by pooling the red fractions. Prior to the elution of cytochrome c_L, methanol dehydrogenase is removed from the column with 36 mM Tris–21 mM H_3PO_4 buffer, pH 6.5. (Further purification of methanol dehydrogenase can be carried out as described elsewhere in this volume,[11] starting at Step 3). Cytochrome c_L is eluted with 36 mM Tris–21 mM H_3PO_4–100 mM NaCl buffer, pH 6.5, and collected by pooling the red fractions.

Purification of Cytochrome c_L

Step 3a: Gel Filtration Chromatography. The pooled fractions from Step 2 are concentrated by ultrafiltration (Millipore, Bedford, MA, Type PTGC membrane, cut-off 10^4 daltons) to less than 5 ml and applied to a

Fractogel TSK HW-50S gel-filtration column (1.6 × 56.6 cm) in 50 mM potassium phosphate–0.1 M NaCl buffer, pH 7.0. The red-colored fractions (fraction size 5 ml) are pooled.

Step 4a: DEAE-Sepharose Chromatography. The pooled fractions from Step 3a are dialyzed against 10 mM Tris-HCl, pH 8.0, containing 5 mM sodium ascorbate and applied to a DEAE-Sepharose column (1.0 × 9.5 cm) equilibrated with 10 mM Tris-HCl, pH 8.0. The cytochrome is eluted with a concave Tris-HCl gradient (10–125 mM, pH 8.0, 5 bed volumes) and concentrated by ultrafiltration as described in Step 3a.

Step 5a: Phenyl-Sepharose Chromatography. To the preparation obtained in Step 4a, 200 mg (NH$_4$)$_2$SO$_4$ per milliliter is added, and the solution is applied to a phenyl-Sepharose column (1.6 × 5 cm) equilibrated with 1.5 M (NH$_4$)$_2$SO$_4$–20 mM potassium phosphate buffer, pH 7.0. After washing the column with 2 bed volumes of the same buffer, cytochrome c_L is eluted with 0.5 M (NH$_4$)$_2$SO$_4$–20 mM potassium phosphate buffer, pH 7.0. The pooled red fractions are dialyzed against 10 mM MOPS buffer, pH 7.0, and stored at −20°.

Purification of Cytochrome c_H

Step 3b: Adsorption to Silica Gel. The pooled cytochrome c_H containing fractions from Step 2 are applied to a column (2.4 × 10 cm) of silica gel in 20 mM potassium phosphate buffer, pH 7.0 [the silica gel, type Si 60, 0.063–0.2 mm, Merck (Darmstadt, FRG) is pretreated for 1 hr at 550° and (after cooling) suspended in potassium phosphate buffer just before use]. The column is washed with 2 bed volumes of the same buffer, and cytochrome c_H is eluted with 0.2 M potassium phosphate buffer, pH 7.0, containing 10% (w/v) polyethylene glycol 6000. The red fractions are pooled.

Step 4b: CM-Sepharose Chromatography. The pooled fractions from Step 3b are dialyzed against 5 mM sodium acetate buffer, pH 5.0, containing 5 mM sodium ascorbate and applied to a CM-Sepharose column (1 × 10.5 cm), equilibrated with 5 mM sodium acetate buffer, pH 5.0. Cytochrome c_H is eluted with a linear NaCl gradient (0–0.2 M in 5 mM sodium phosphate buffer, pH 5.0, 4 bed volumes). The red cytochrome c_H-containing fractions are pooled.

Step 5b: Phenyl-Sepharose Chromatography. To the preparation from Step 4b, 300 mg (NH$_4$)$_2$SO$_4$ per milliliter is added, and the solution is applied to a Phenyl-Sepharose column (1 × 5 cm), equilibrated with 2.3 M (NH$_4$)$_2$SO$_4$–20 mM potassium phosphate buffer, pH 7.0. After washing the column with two bed volumes of the same buffer, cytochrome c_H is eluted with 1.5 M (NH$_4$)$_2$SO$_4$–20 mM potassium phosphate buffer, pH 7.0. The pooled red fractions are dialyzed against 10 mM MOPS buffer, pH 7.0, and stored at −20°.

TABLE I

PURIFICATION OF CYTOCHROME c_L AND CYTOCHROME c_H FROM *Hyphomicrobium* X

Fraction	Protein (mg)	Cytochrome (nmol)	nmol cytochrome c/ mg protein	Yield (%)
Cytochrome c_L				
Step 2. DEAE-Sepharose	154	754	4.9	100
Step 3a. Fractogel	18.2	544	29.9	72
Step 4a. DEAE-Sepharose gradient	8.7	463	53.1	61
Step 5a. Phenyl-Sepharose	6.2	330	53.3	44
Cytochrome c_H				
Step 2. DEAE-Sepharose	232	2097	9.1	100
Step 3b. Silica gel	69.9	2068	29.6	98
Step 4b. CM-Sepharose	18.7	1500	80.0	71
Step 5b. Phenyl-Sepharose	15.9	1274	80.1	61

The purification of cytochrome c_L and cytochrome c_H is summarized in Table I.

Properties

A number of properties are shown in Table II.[3,5]

Purity. A single band is found for both cytochrome c_L and cytochrome c_H on gradient polyacrylamide gel electrophoresis in the presence of sodium dodecyl sulfate (SDS) and staining with Coomassie blue.

Autoreduction and Reaction with Methanol Dehydrogenase. Cytochrome c_L and cytochrome c_H preparations obtained after Step 4 display an autoreduction rate (at pH 9.0) of 1.2×10^{-5} and 3.5×10^{-5} sec^{-1}, respectively. After hydrophobic interaction chromatography (Step 5), these values decrease to less than 1×10^{-6} sec^{-1}. Much higher values were reported[6] by Anthony and co-workers for the cytochromes c of *Methylophilus methylotrophus* and *Methylobacterium* AM1.

Very efficient electron transfer occurs between the reduced methanol dehydrogenase forms and ferricytochrome c_L at pH 7.0 (Table II). Nearly complete inhibition is observed in the presence of EDTA (0.1 mM), NaCl (0.2 M), or phosphate (0.2 M). At pH 9.0, a much lower reaction rate is found (Table II). Evidence has been presented[5] indicating that the rate-limiting step in the reaction at pH 7.0 is complex formation by the components; at pH 9.0, it is the electron-transfer step itself. Biphasic reaction traces are obtained at pH 9.0, probably as a consequence of the coexistence of two conformations of cytochrome c_L at that pH.

Quite different conclusions have been put forward by Anthony and co-workers[1,6] with respect to optimal pH, reaction rates, and the mecha-

TABLE II
PROPERTIES OF CYTOCHROME c_L AND CYTOCHROME c_H

Property	Cytochrome c_L	Cytochrome c_H
M_r native	19,500	15,500
M_r denatured	19,600	14,000
Isoelectric point	4.3 (4.2, 4.5)[a]	7.4[b], 7.5[b]
Spectroscopic data		
$A_{280}^{0.1\%}$	2.02	2.48
Reduced form		
Absorption maxima (nm)	414, 520, 550	414, 520, 550.6
ϵ (M^{-1} cm^{-1})	21,600 (550 nm)	23,700 (550.6 nm)
Ratio α/β	1.5	1.7
Oxidized form		
Absorption maxima (nm)	408	408
ϵ (M^{-1} cm^{-1})	6,700 (550 nm)	6,600 (550.6 nm)
Midpoint potential, pH 7.0 (mV)	+270	+292
Reaction rate with		
MDH$_{red}$, pH 7.0 (M^{-1} sec^{-1})	1.9×10^5	1.6×10^3
MDH$_{sem}$, pH 7.0 (M^{-1} sec^{-1})	2.1×10^5	—
MDH$_{red}$, pH 9.0 (sec^{-1})	0.23	—
K_{asso} (M^{-1})	8.2×10^3	—
MDH$_{sem}$, pH 9.0 (sec^{-1})	0.23	—
K_{asso} (M^{-1})	8.0×10^3	—
Heme content (mol/mol protein)	1.2	1.3

[a] Minor components.
[b] Equal amounts.

nism involved in electron transfer. It should be realized, however, that these authors did not measure the electron-transfer step, as such, but the whole reaction cycle with horse heart cytochrome c as the final electron acceptor. There are good reasons to assume that, depending on the conditions of the assay, a different step is rate-limiting in the redox cycle.

[46] Electron-Transfer Flavoproteins from *Methylophilus methylotrophus* and Bacterium W3A1

By MAZHAR HUSAIN

Electron-transfer flavoproteins (ETFs) are FAD-containing redox proteins that play a crucial role as biological electron-transfer links between enzymes of mitochondrial and bacterial degradation pathways and their respective electron-transport chains. The methylotrophs *Methylophilus methylotrophus* and bacterium W3A1 each synthesize an inducible ETF when grown on trimethylamine as the sole source of carbon.[1] In each of these methylotrophs, ETF serves as a physiological electron acceptor for trimethylamine dehydrogenase (TMADH) which is also induced by growth on trimethylamine.[1] TMADH, an iron-sulfur flavoprotein, which has recently been shown to contain a tightly bound ADP of unknown catalytic function, catalyzes the oxidative N-demethylation of trimethylamine to dimethylamine and formaldehyde.[2,3] The natural electron acceptors for methylotrophic ETFs have not yet been identified but are likely to be cytochromes of unusually high redox potentials.[4,5]

Purification Procedure

The purification of ETF from *M. methylotrophus* and bacterium W3A1 is performed essentially as described by Steenkamp and Gallup.[6] The buffer used throughout the purification is 20 mM potassium phosphate, pH 7.2, unless specified otherwise. All steps are carried out at $0-4°$.

Growth of Bacteria. *Methylophilus methylotrophus* (NCIB 10515) and bacterium W3A1 (NCIB 11348) are grown aerobically at 40 and 30°, respectively, in the mineral base E of Owens and Keddie,[7] supplemented with 0.3% trimethylamine as a carbon source. This liquid medium contains the following (per liter): 0.75 g KH_2PO_4, 1.04 g K_2HPO_4, 0.5 g $(NH_4)_2SO_4$, 0.1 g NaCl, 25 mg $CaCl_2$, 0.2 g $MgSO_4 \cdot 7H_2O$, and 3 ml of a trace metal mixture. The composition of the trace metal mixture is as

[1] V. L. Davidson, M. Husain, and J. W. Neher, *J. Bacteriol.* **166**, 812 (1986).

[2] D. J. Steenkamp and J. Mallinson, *Biochim. Biophys. Acta* **429**, 705 (1976).

[3] L. W. Lim, F. S. Mathews, and D. J. Steenkamp, *J. Biol. Chem.* **263**, 3075 (1988).

[4] C. M. Byron, M. T. Stankovich, M. Husain, and V. L. Davidson, *Biochemistry* **28**, 8582 (1989).

[5] A. R. Cross and C. Anthony, *Biochem. J.* **192**, 429 (1980).

[6] D. J. Steenkamp and M. Gallup, *J. Biol. Chem.* **253**, 4086 (1978).

[7] J. D. Owens and R. M. Keddie, *J. Appl. Bacteriol.* **32**, 338 (1969).

follows (per liter): 5.0 g EDTA, 2.2 g $ZnSO_4 \cdot 7 H_2O$, 0.57 g $MnSO_4 \cdot 4H_2O$, 0.5 g $FeSO_4 \cdot 7H_2O$, 0.16 g $CoCl_2 \cdot 6H_2O$, 0.16 g $CuSO_4 \cdot 5H_2O$, 0.15 g $Na_2MoO_4 \cdot 2H_2O$ (adjusted to pH6 with 40%, w/v, KOH solution). A stock solution of trimethylamine hydrochloride (10%, w/v) is filter-sterilized and added to the medium after it has been autoclaved and cooled.

Stock cultures of *M. methylotrophus* and bacterium W3A1 grown in the liquid medium are maintained at $-70°$ as small aliquots after the addition of 10% glycerol. Large batches of bacteria are grown in 15-liter fermentors and harvested in late log phase. The cell paste is stored frozen at $-20°$ until used. The yield of cells is about 1.5 g per liter of medium.

Step 1: Preparation of Crude Extract. Frozen cells (40 g) are softened overnight in the cold room and homogenized with 3 volumes (w/v) of the phosphate buffer. The cells are disrupted by two passages through a French pressure cell at about 20,000 psi followed by centrifugation at 18,000 g for 30 min. The supernatant is decanted and stored at 4°, and the pellet is resuspended in about 30 ml of the phosphate buffer. The resuspended pellet is then rehomogenized, sonicated for 10 min using a Branson Sonifier 250 at power setting 7 and 50% duty cycle, and centrifuged for 30 min at 18,000 g. The supernatant is combined with that of the first centrifugation, and the pellet is discarded. The combined supernatants (crude extract) are dialyzed for 16 hr against 4 liters of the buffer with 2 changes of dialysate.

Step 2: DEAE-Cellulose Chromatography. A glass column (4.5 × 40 cm) is packed with DEAE-cellulose (Whatman DE-52) and equilibrated with the phosphate buffer. The dialyzed crude extract from Step 1 is applied to the column, which is then washed with the buffer until the effluent is nearly colorless. The column is then eluted with a linear gradient formed with 600 ml of the phosphate buffer and 600 ml of the buffer containing 500 mM NaCl. Fractions of 10 ml are collected and tested for TMADH activity according to the procedure of Colby and Zatman.[8] TMADH is eluted at a NaCl concentration of approximately 150 mM. The yellow fractions containing ETF, eluted immediately after TMADH, are evaluated by sodium dodecyl sulfate–polyacrylamide gel electrophoresis (SDS–PAGE) performed according to the method of Laemmli and Favre,[9] except for the inclusion of 0.5 M urea in the resolving and stacking gels and 4 M urea and 4% SDS (w/v) in the final sample buffer. ETF peak fractions, which show the characteristic 42- and 38-kDa subunit bands and only small amounts of other contaminants, are pooled, concentrated to approx-

[8] J. Colby and L. J. Zatman, *Biochem. J.* **132,** 101 (1973).
[9] U. K. Laemmli and M. Favre, *J. Mol. Biol.* **80,** 575 (1973).

imately 5 ml by ultrafiltration using a PM30 membrane (Amicon Corp., Danvers, MA), and dialyzed for 16 hr against 500 ml of the phosphate buffer with 2 changes of dialysate.

Step 3: Gel Filtration on Sephadex G-100. The dialyzed ETF solution from Step 2 is applied to a column of Sephadex G-100 (3.5 × 100 cm) equilibrated with the phosphate buffer. ETF is eluted with the same buffer at a flow rate of about 60 ml/hr, and fractions of 4–5 ml are collected. The peak fractions are pooled, concentrated by ultrafiltration using a PM30 Amicon membrane, and dialyzed overnight against the phosphate buffer containing 10% (v/v) ethylene glycol with 2 changes of dialysate. Approximately 40 mg of purified ETF is obtained from 40 g of wet cell paste, although considerable variation in yield has been observed.

Properties

Based on several of their properties (summarized in Table I), ETFs from *M. methylotrophus* and bacterium W3A1 are very similar to each other and clearly different from the mitochondrial ETF, which has been studied in greater detail.[10]

Purity and Stability. As judged by polyacrylamide gel electrophoresis and isoelectric focusing, the purified preparations of ETFs from *M. methylotrophus* and bacterium W3A1 are homogeneous.[1,6] Low levels of TMADH activity can sometimes be detected in the purified preparations of methylotrophic ETFs. ETF preparations with undetectable amounts of TMADH can be obtained, when desired, by performing an additional gel-filtration step.

Purified ETF can be stored for a few months at −20° or longer at −80° without any effect on its spectral or catalytic properties. ETF from *M. methylotrophus* has been shown to undergo a dramatic spectral change when stored at −20° for an extended period of time (>4 months).[4] The reasons for this storage-related spectral change have not been fully investigated.

Physical and Chemical Properties. ETFs from *M. methylotrophus* and bacterium W3A1 are dimers of two nonidentical subunits (M_r 42,000 and 38,000) containing one molecule of noncovalently bound FAD per dimer.[1,6] The purified ETF from *M. methylotrophus* is nonfluorescent and exhibits an isoelectric point of 4.1.[1]

Spectral and Catalytic Properties. ETFs from *M. methylotrophus* and bacterium W3A1 as isolated are partially reduced and consist of a mixture

[10] C. Thorpe, *in* "Chemistry and Biochemistry of Flavoenzymes" (F. Müller, ed.), Vol. 3. CRC Press, Boca Raton, Florida, in press.

TABLE I
PROPERTIES OF ELECTRON-TRANSFER FLAVOPROTEINS

Property	*Methylophilus methylotrophus*	Bacterium W3A1[a]	Pig liver
Molecular weight			
Native	ND	77,000	68,000
Subunits	42,000, 38,000	42,000, 38,000	38,000, 32,000
FAD content (mol/mol)	1	1	1
Isoelectric point	4.1	ND	6.75
Fluorescence (relative to free FAD)	0	ND	2.1
λ_{max}	380, 438, 460	378, 438, 458	375, 436, 460
$E^{\circ\prime}$ (mV)			
$ETF_{ox}/ETF^{\circ-}$	+196	ND	+4
$ETF^{\circ-}/ETFH_2$	−197	ND	−50

[a] ND, Not determined.

of fully oxidized (ETF_{ox}) and anionic semiquinone ($ETF^{\circ-}$) forms of the flavoprotein.[1,6] Fully functional ETF_{ox} is, however, readily obtained by addition of a slight excess of potassium ferricyanide, followed by dialysis or by chromatography on Sephadex G-25. The ferricyanide-treated, fully oxidized ETF from bacterium W3A1, but not that from *M. methylotrophus,* loses its electron acceptor activity over a period of a few days.[6]

The completely oxidized ETFs from *M. methylotrophus*[1] and bacterium W3A1[6] have absorption spectra very similar to that reported for ETFs from *Paracoccus denitrificans*[11] and those from mammalian sources.[12,13] Both methylotrophic ETFs are rapidly reduced to anionic semiquinone by TMADH in the presence of excess trimethylamine (Fig. 1).[1,6] The reduction reaction shows a high degree of specificity for TMADH. Methanol and methylamine dehydrogenases, the two other major dehydrogenases synthesized by these bacteria, fail to reduce these ETFs. The V_{max} and K_m values determined for this reaction with ETF from bacterium W3A1 as the variable substrate are 2.1 μmol/min/mg and 6.7 μM, respectively.[6]

Continued enzymatic reduction fails to convert the semiquinone to the fully reduced ETF.[1,6] Dithionite titrations and deazaflavin-mediated photoreductions also fail to reduce the methylotrophic ETFs beyond the radical state, which displays a remarkable stability in air.[1]

Complete reduction of *M. methylotrophus* ETF to its two-electron,

[11] M. Husain and D. J. Steenkamp, *J. Bacteriol.* **163,** 709 (1985).
[12] M. Husain and D. J. Steenkamp, *Biochem. J.* **209,** 541 (1983).
[13] R. J. Gorelick, J. P. Mizzer, and C. Thorpe, *Biochemistry* **21,** 6936 (1982).

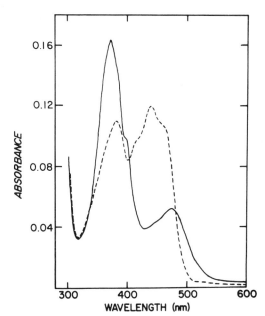

Fig. 1. Reduction of *Methylophilus methylotrophus* electron-transfer flavoprotein by trimethylamine dehydrogenase. Absorption spectra of the oxidized ETF before (---) and after (——) addition of trimethylamine dehydrogenase and trimethylamine are shown.

fully reduced form is, however, achieved during coulometric and potentiometric titrations carried out to determine the oxidation–reduction potentials of the two redox couples.[4] The oxidation–reduction potential for the first electron transfer ($ETF_{ox}/ETF^{\circ-}$) is $+196$ mV and shows no variation with pH. This is the most positive oxidation–reduction potential recorded for an FAD-containing protein. The oxidation–reduction potential for the second redox couple ($ETF^{\circ-}/ETF_{red}H^-$), which involves transfer of both an electron and a proton, is -197 mV. This value is not unusually negative, suggesting that a kinetic rather than a thermodynamic barrier prevents the conversion of $ETF^{\circ-}$ to $ETF_{red}H^-$ during dithionite and photochemical titrations.

These redox data strongly suggest that methylotrophic ETFs cycle between fully oxidized and semiquinone forms during catalysis, and they lend support for the proposed role of ETFs as electron acceptors for TMADH.[4,6] The molar absorptivities for the three oxidation states of *M. methylotrophus* ETF at two selected wavelengths calculated from the redox data are listed in Table II.

Immunological and Enzymatic Cross-Reactivity. Antibodies prepared in rabbits to ETF from bacterium W3A1 cross-react with ETF from *M.*

TABLE II
MOLAR ABSORPTIVITIES FOR THE THREE
OXIDATION STATES OF *Methylophilus*
methylotrophus ELECTRON-TRANSFER
FLAVOPROTEIN

	Molar absorptivity (M^{-1} cm^{-1})		
λ	ETF$_{ox}$	ETF$^{\circ-}$	ETF$_{red}$
438	12,600	4,050	1,360
370	10,240	16,900	4,130

methylotrophus but not with that from pig liver when tested by SDS–PAGE and Western blotting.[1] Further, no immunological reaction is observed between ETFs of *M. methylotrophus* and bacterium W3A1 and antibodies elicited from pig liver ETF.[1] Methylotrophic ETFs are, therefore, antigenically very similar to each other and different from pig liver ETF.

ETFs from *M. methylotrophus* and bacterium W3A1 also cross-react enzymatically with respect to their ability to accept electrons from TMADH.[1] Each of the two methylotrophic ETFs is able to accept electrons from TMADH of both *M. methylotrophus* and bacterium W3A1.

Acknowledgment

This work was supported by Grant HL-16251 from the National Institutes of Health.

[47] Formaldehyde Dehydrogenases from Methylotrophs

By MARGARET M. ATTWOOD

Formaldehyde dehydrogenase catalyzes a key step in the dissimilation of carbon during growth of methylotrophs on reduced C$_1$ compounds and the oxidation of toxic formaldehyde in methylotrophs and nonmethylotrophs. The coenzyme involved in the reaction varies with the organism and the function. The purification of these enzymes has been reported in the literature.[1-3]

[1] J. P. Van Dijken, G. J. Oostra-Demkes, R. Otto, and W. Harder, *Arch. Microbiol.* **111,** 77 (1976).
[2] M. M. Attwood and L. Dijkhuizen, unpublished data (1988).
[3] I. W. Marison and M. M. Attwood, *J. Gen. Microbiol.* **117,** 305 (1980).

Formaldehyde Dehydrogenase (Glutathione) [Formaldehyde:NAD$^+$
Oxidoreductase (Glutathione Formylating), EC 1.2.1.1.]

Formaldehyde + glutathione + NAD$^+$ $\rightleftharpoons$ S-formylglutathione + NADH

The dependence of the glutathione-formylating formaldehyde dehydrogenases on glutathione (GSH) is based on the spontaneous formation of
a hemimercaptal, S-formylglutathione, which is the actual substrate for the
enzyme. These enzymes are characteristically found in yeasts.

Assay Methods

Principle. A spectrophotometric assay based on the reduction of NAD$^+$
in the presence of formaldehyde or the oxidation of NADH in the presence
of S-formylglutathione is used. The change in absorbance at 340 nm is
followed.

Reagents for Assay Method 1

Potassium phosphate buffer, 500 mM, pH 8.0
NAD$^+$, 4 mM
GSH, 30 mM
Formaldehyde, 10 mM, prepared by heating 300 mg of paraformaldehyde in 10 ml of water overnight in a sealed tube at 110°; the
strength ($\sim$ 1 M) should be checked by the procedure of Nash,[4] and
the solution diluted accordingly.

Reagents for Assay Method 2

Potassium buffer, 500 mM, pH 8.0
NADH, 1.4 mM
S-Formylglutathione, 10 mM
Preparation of S-Formylglutathione.[5] S-Formylglutathione is prepared
by mixing 33.2 g of acetic anhydride with 14.5 of formic acid (98 – 100%)
and stirring for 1 hr at 23°. Then 6 g of thioglycolic acid (freshly distilled)
is added and the mixture stirred continuously for a further 1 hr at 23°.
After the addition of 200 ml of water, the mixture is freeze-dried. To the
formyl thioglycolate obtained, 2 g of glutathione dissolved in 20 ml water
is added. The pH of the solution is adjusted to 5.0 with 5 M NaOH with
continuous stirring and cooling. After 1 hr at 23° the mixture is cooled 0°,
acidified with 12 M HCl to pH 3.0, and extracted 3 times with 2 volumes
of ether to remove formyl thioglycolate. The aqueous solution containing
S-formylglutathione is applied to a Sephadex G-10 gel-filtration column

[4] T. Nash, *Biochem. J.* **55**, 416 (1953).
[5] L. Uotila, *Biochemistry* **12**, 3938 (1953).

(70 × 2 cm) to remove impurities. The column is eluted with 10 mM acetic acid. Fractions (4 ml) containing S-formylglutathione are pooled and evaporated to dryness by freeze drying. The preparation of S-formylglutathione (25 mg) contains glutathione.

Detection and Measurement of S-*Formylglutathione.* S-Formylglutathione is detected enzymatically using the purified formaldehyde dehydrogenase and NADH in the assay system described. The same assay is used to determine the concentration of thiol ester in the final preparation. When limiting amounts of S-formylglutathione are added, a linear relationship between the amount of thiol ester added and overall decrease in absorption at 340 nm is seen.

Procedure for Assay Method 1. The assay mixture contains, in a final volume of 1 ml, potassium phosphate buffer, 0.2 ml; NAD, 0.1 ml; reduced glutathione, 0.1 ml; and enzyme. The mixture is equilibrated at 37° and the absorbance at 340 nm followed for several minutes. The reaction is started by the addition of 0.1 ml formaldehyde. The increased rate of absorbance change is measured and this rate minus the formaldehyde-independent rate is taken as the measure of the enzyme activity.

Procedure for Assay Method 2. The assay mixture contains in a final volume of 1 ml, potassium phosphate buffer, 0.2 ml; NADH, 0.1 ml; and enzyme. The mixture is equilibrated for several minutes at 37° and the absorbance at 340 nm measured. The reaction is started by the addition of 0.5 ml S-formylglutathione. The rate of absorbance change is followed, and this rate minus the S-formylglutathione-independent rate is taken as the enzyme activity.

Units. One unit of activity is defined as that amount of enzyme catalyzing the oxidation or reduction of the cofactor per minute at 37°. Specific activity is expressed as units per milligram of protein.

Purification Procedure

Growth of Organism. The strain of *Hansenula polymorpha* de Marais et Maia, CBS 4732, is grown in a mineral salts medium supplemented with vitamins and methanol as the sole source of carbon and energy at 37° in methanol-limited chemostat culture at a dilution rate of 0.1 hr^{-1}. Mineral medium contains, per liter, $(NH)_4SO_4$, 1.5 g; KH_2PO_4, 1 g; $MgSO_4 \cdot 7H_2O$, 0.2 g; and trace elements,[6] 0.2 ml. After heat sterilization and cooling, 5 μg filter-sterilized biotin, 300 μg thiamin-HCl, and 5 g methanol are added. The pH is controlled at 5.0 by automatic adjustment with 1 N NaOH. The actively growing culture is harvested by centrifuga-

---

[6] W. Vishniac and M. Santer, *Bacteriol. Rev.* **21**, 195 (1957).

tion 4500 g for 15 min and washed twice with 0.1 M potassium phosphate buffer, pH 7.5.

Step 1: Preparation of Cell-Free Extract. All procedures are at 0–4°. Cell paste (200 g wet weight) is suspended in 350 ml of the 0.1 M potassium phosphate buffer, pH 7.5. The suspension is disrupted by 4 successive passes through a French press (RMI Sorvall Inc.) at 40,000 psi (2650 atm). The supernatant obtained by centrifugation at 30,000 g for 1 hr is used as the crude cell-free extract.

Step 2: Ammonium Sulfate Fractionation. Solid, finely ground ammonium sulfate is added slowly to the cell-free extract with stirring. During this procedure the pH is adjusted to 7.5 with 10% (v/v) ammonium hydroxide. The protein which precipitates between 40% (1.71 M) and 80% (3.9 M) ammonium sulfate saturation is collected by centrifugation at 18,000 g for 20 min, dissolved in 50 mM potassium phosphate buffer, pH 7.5, containing 1 mM dithiothreitol (DTT), and dialyzed against 100 volumes of the same buffer for 15 hr.

Step 3: DEAE-Cellulose Chromatography. The dialyzed solution is applied to a column (3 × 60 cm) DEAE-cellulose (Whatman DE-52) previously equilibrated with 0.1 M potassium phosphate buffer, pH 7.5, containing 1 mM DTT. Formaldehyde dehydrogenase activity is eluted from the column using a continuous gradient of 0–0.4 M NaCl in 0.1 M potassium phosphate buffer, pH 7.5, containing 1 mM DTT. The active fractions are pooled and concentrated by ultrafiltration (Amicon PM10 ultrafilter; Danvers, MA) and dialyzed against 10 volumes of 50 mM potassium phosphate buffer, pH 7.5, containing 1 mM DTT.

Step 4. Hydroxylapatite Chromatography. The concentrated enzyme from Step 3 is applied to a hydroxylapatite column (1 × 20 cm). Elution is carried out with a linear gradient of potassium phosphate buffer (20–200 mM), pH 7.5, containing 1 mM DTT. The active fractions are pooled and concentrated by ultrafiltration to give a protein concentration of 0.2 mg/ml. The preparation is stored in liquid nitrogen. A typical purification is summarized in Table I.

Properties

Purity. The enzyme preparation is not homogeneous but does not contain methanol oxidase, catalase, formate dehydrogenase, ethanol dehydrogenase, or NADH oxidase.

Substrate Specificity. The enzyme is strictly dependent on glutathione and NAD+ for activity. Glutathione cannot be replaced by cysteine, mercaptoethanol, or thioglycolate. NAD+ cannot be replaced by NADP.+

Stability. The purified enzyme is stable for over 1 year when stored in liquid nitrogen.

TABLE I
PURIFICATION OF FORMALDEHYDE DEHYDROGENASE (GLUTATHIONE) FROM *Hansenula polymorpha*

Fraction	Total volume (ml)	Total protein (mg)	Total activity (units)[a]	Specific activity (units/mg protein)	Yield (%)
Step 1. Cell-free extract	350	12,187	5,728	0.47	100
Step 2. (NH$_4$)$_2$SO$_4$ fraction (40–80% saturation)	50	2,105	6,420	3.05	112
Step 3. Pooled, concentrated fraction after DEAE-cellulose chromatography	360	216	937	4.34	16.4
Step 4. Pooled, concentrated fraction after hydroxylapaitite chromatography	70	136	295	21.22	5.1

[a] Results obtained by Assay Method 1.

pH Optimum. The enzyme has optimum activity at pH 9.0 and 44°.

Michaelis Constants. The apparent K_m values (measured under standard assay conditions for the other reagents) are 1.5×10^{-4}, 1.8×10^{-4}, and 2.1×10^{-4} M for NAD$^+$, GSH, and formaldehyde, respectively. The apparent K_m value for S-formylglutathione (assayed in the reverse direction) is 1 mM. Formylthioglycolate can replace S-formylglutathione in the reverse direction assay.

Formaldehyde Dehydrogenase (Formaldehyde:NAD$^+$ Oxidoreductase, EC 1.2.1.46)

$$\text{Formaldehyde} + \text{NAD}^+ + \text{H}_2\text{O} \rightleftharpoons \text{formate} + \text{NADH}$$

The formaldehyde dehydrogenase catalyzing the above reaction does not depend on glutathione for activity, but some enzymes require glutathione for maximum activity. Such enzymes have been described from methylotrophic bacteria *Methylophilus methylotrophus* (grown on methanol[7]) and *Arthrobacter* P1 (grown on choline, betaine, dimethylglycine, and sarcosine).[8]

Assay Method

Principle. A spectrophotometric assay (as above) follows the reduction of NAD$^+$ at 340 nm.

[7] R. P. Betts, Ph.D thesis, University of Sheffield, England, 1984.
[8] P. R. Levering, L. Tiesma, J. P. Woldendorp, M. Steensma, and L. Dijkhuizen, *Arch. Microbiol.* **146**, 346 (1987).

Reagents

Potassium phosphate buffer, 500 mM, pH 8.0
Magnesium chloride, 30 mM
NAD$^+$, 8 mM
Glutathione, 30 mM
Formaldehyde, 100 mM prepared from paraformaldehyde as described

Procedure. The assay mixture contains, in a final volume of 1 ml, potassium phosphate buffer, 0.1 ml; magnesium chloride, 0.1 ml; NAD$^+$, 0.1 ml; glutathione, if required, 0.1 ml; and enzyme. The mixture is preincubated at 30° and the absorbance at 340 nm measured for several minutes. The reaction is started by the addition of 0.1 ml formaldehyde, and the rate of increase in absorbance is measured. This rate minus the formaldehyde-independent rate is taken as a measure of the enzyme activity.

Purification Procedure

Growth of Organism. Arthrobacter P1 NCIB 11625 was isolated from soil.[9] The organism is grown at 30° in a fermentor as a batch culture in a mineral salts medium containing, per liter, NaH$_2$PO$_4 \cdot$H$_2$O, 0.5 g; K$_2$HPO$_4$, 1.55 g; (NH$_4$)$_2$SO$_4$, 1.0 g; MgSO$_4 \cdot$7H$_2$O, 0.2 g; trace elements,[6] 0.2 ml; and choline, 2 g. The pH of the medium is 7.2. The actively growing culture is harvested by centrifugation at 4500 g for 15 min, washed once, and the cell paste stored at −20°.

Step 1: Preparation of Cell-Free Extract. Cell paste (wet cells) is suspended in 4 volumes of Tris-HCl buffer containing 5 mM KCl. The suspension is disrupted by ultrasonication in an MSE Ultrasonic disintegrator (150 W) using 15 periods of 30-sec sonication interspersed with 30 sec in ice-cold water mixture. The supernatant obtained by centrifugation at 38,000 g for 30 min is the cell-free extract.

Step 2: Ammonium Sulfate Fractionation. All further procedures are carried out at 0–4°. Finely ground solid ammonium sulfate is slowly added to the cell-free extract with stirring. The pH is adjusted throughout to 7.5 with 10% (v/v) ammonium hydroxide. The protein which precipitates between 45% (1.95 M) and 70% (3.3 M) ammonium sulfate saturation is collected by centrifugation at 38,000 g for 15 min. The precipitate is dissolved in 50 mM Tris-HCl containing 5 mM KCl.

Step 3: Ion-Exchange Chromatography. The protein solution from Step 2 is applied to the top of a Sephadex G-25 gel filtration column (Bio-Rad PD10; Richmond, CA); equilibrated and eluted with 20 mM Tris-HCl, pH

[9] P. R. Levering, J. P. Van Dijken, M. Veenhuis, and W. Harder, *Arch. Microbiol.* **129**, 72 (1981).

TABLE II
PURIFICATION OF FORMALDEHYDE DEHYDROGENASE FROM *Arthrobacter* P1

Fraction	Total volume (ml)	Total protein (mg)	Total activity (milli units)	Specific activity (milliunits/) mg protein)	Yield (%)
Step 1. Cell-free extract	2.2	44	873.4	19.85	100
Step 2. (NH$_4$)$_2$SO$_4$ fraction (40–75% saturation)	1.5	4.33	733.5	51.20	84
Step 3. Sephadex G-25, ion-exchange column	4.5	0.38	252	666.7	28.9

7.5, containing 1 mM KCl. The desalted extract is applied to a 25-ml anion-exchange column (Mono Q HR5/5) washed and equilibrated with 20 mM Tris-HCl buffer, pH 7.5, and connected to a Pharmacia fast protein liquid chromatography (FPLC) system. Proteins are eluted at a flow rate of 1.0 ml/min by applying a linear 0–1.0 M KCl gradient in 20 mM Tris-HCl buffer, pH 7.5. Fractions (1.0 ml) are collected, and those containing activity are pooled and stored at −20°. A typical purification is summarized in Table II.

Properties

Purity. Polyacrylamide gel electrophoresis demonstrated that the preparation was not homogeneous.

Substrate Specificity. The enzyme is highly specific for its substrates. Formaldehyde cannot be substituted by a range of aldehydes (1 mM propionaldehyde, n-butyraldehyde, isobutyraldehyde, valeraldehyde, acetaldehyde, or benzaldehyde). NADP$^+$ cannot replace NAD$^+$ in the assay.

Inhibitors. Some metal-chelating agents (2 mM) cause enzyme inhibition: 2,2′-bipyridine (16%), 1,10-phenanthroline (18%), and 8-hydroquinoline (30%). EDTA (40 mM) has no effect on enzyme activity.

Stability. The enzyme is stable to elevated temperatures, showing stability at 45° for 6 min. When stored at −20° activity is maintained for at least 1 month.

pH Optimum. The enzyme has a broad optimum between pH 8.1 and 8.5.

Thiol Reagents. Enzyme activity in crude cell-free extracts is not affected by the addition of GSH. As purification proceeds GSH is required in the assay mixture for maximum activity.

Michaelis Constants. The apparent K_m values are measured in standard assay conditions for the other reagents. The K_m values are 7.4×10^{-5} and 4.5×10^{-4} M for formaldehyde and NAD⁺, respectively.

Molecular Weight and Subunit Structure. The molecular weight of the native enzyme is measured using gel filtration with a Sepharose 16 column (calibrated with marker proteins). The elution position corresponds to a molecular weight of $43,500 \pm 8,000$. Polyacrylamide gel electrophoresis under denaturing conditions using a series of SDS gels ($10-12.5\%$ acrylamide) and marker proteins revealed a subunit molecular weight consistent with 48,000, suggesting that the enzyme is a monomer.

Formaldehyde Dehydrogenase [Formaldehyde:NAD⁺ Oxidoreductase (Novel Cofactor Requirement)]

Another formaldehyde dehydrogenase catalyzes the oxidation of formaldehyde in cells of *Rhodococcus erythropolis* metabolizing compounds which contain methyl groups (methoxybenzoates, methylamine).[11]

Assay Methods

Principle. The spectrophotometric NAD⁺ assay as described above is used.

Reagents

Sodium phosphate buffer, 500 mM, pH 8.0
NAD⁺, 10 mM
Dithiothreitol, 38 mM
Cofactor, prepared by boiling cell-free extract (15 mg protein/ml) for 15 min, clarifying by centrifugation at 10,000 g for 20 min, and applying the supernatant (2.5 ml) to a Sephadex G-25 column (Pharmacia PD10); the column is eluted with water, and fractions eluting between 6 and 9 ml collected
Formaldehyde, 60 mM, prepared from paraformaldehyde as described above

Procedure. The assay mixture contains, in a final volume of 1 ml, sodium phosphate buffer, 0.5 ml; NAD⁺, 0.1 ml; dithiothreitol, 0.1 ml; cofactor, 0.03 ml; and enzyme. The mixture is preincubated at 30° for several minutes and the absorbance at 340 nm measured. The reaction is started by the addition of 0.1 ml formaldehyde, and the increased rate of absorbance change is measured. This rate minus the formaldehyde-independent rate is taken as a measure of enzyme activity.

---

[10] R. J. Mehta, *Antonie van Leeuwenhoek* **39**, 303 (1973).
[11] L. Eggeling and H. Sahm, *Eur. J. Biochem.* **150**, 129 (1985).

Purification Procedure

All procedures are carried out at $0-4°$, and the buffer used throughout is 0.1 M sodium phosphate, pH 7.5, containing 0.01% sodium azide.

Growth of Organism. *Rhodococcus erythropolis* (DSM 1069, Göttingen) is grown at $37°$ in mineral salts medium containing, per liter, KH_2PO_4, 5 g; NH_4Cl, 5 g; $MgSO_4 \cdot 7H_2O$, 0.2 g; yeast extract (Difco, Detroit, MI), 0.5 g; EDTA, sodium salt, 5 mg; $CaCl_2 \cdot 2H_2O$, 10 mg; $FeSO_4 \cdot 7H_2O$, 5 mg; $MnSO_4 \cdot H_2O$, 1 mg; $ZnSO_4 \cdot 7H_2O$, 1 mg; KI, 50 μg; and H_3PO_3, 50 μg. The pH is adjusted to 7.0 with 1 N NaOH. The growth substrate is added at 2 mM. After the consumption of the growth substrate two further additions (each at 2 mM) are added to give a final yield of 12 g wet weight of cells. Actively growing cells are harvested by centrifugation 4500 g for 20 min.

Step 1: Preparation of Cell-Free Extract. A washed cell suspension is frozen in an X-Press (AB-Biox, Naeka, Sweden) to $-18°$ and passed through an orifice of 0.8 mm diameter. After thawing the homogenate is centrifuged at 10,000 g for 30 min and the supernatant used as the cell-free extract.

Step 2: Affinity Chromatography. The cell-free extract (15 ml) is dialyzed against the buffer for 8 hr and then applied to the top of a 5′-AMP-Sepharose column (2.6 × 7 cm). The column is washed with the buffer and the bound material eluted at a flow rate of $5-6$ ml/hr with the buffer containing a linear gradient $0-4$ mM NAD^+ (240 ml). The active fractions are combined and concentrated by ultrafiltration (YM10, Amicon).

Step 3: Sephadex G-200 Gel Filtration. The extract from Step 2 is applied to a Sephadex G-200 column (1.6 × 93 cm) and eluted at a rate of $5-6$ ml/hr with the buffer. The active fractions are collected and concentrated by ultrafiltration. A typical purification scheme is summarized in Table III.

Properties

Purity. The enzyme preparation is homogeneous. A single band is revealed when the sample containing dithiothreitol, to eliminate the oxidation of the enzyme, is subjected to polyacrylamide gel electrophoresis (8–12% polyacrylamide). When the protein is electrofocused using a pH range of $3.5-9.5$ only one band is obtained, corresponding to an isoelectric point of 4.7.

Substrate Specificity. The enzyme shows activity not only with formaldehyde but also with primary alcohols (C_2-C_8). When the activity is measured in the presence of alcohols the cofactor is not required. Methanol is not a substrate for the enzyme but can be used to replace the cofactor. It

TABLE III
PURIFICATION OF FORMALDEHYDE DEHYDROGENASE (NOVEL COFACTOR) FROM *Rhodococcus erythropolis*

Fraction	Total volume (ml)	Total protein (mg)	Total activity (units)	Specific activity (units/mg protein)	Yield (%)
Step 1. Cell-free extract	24	182	15.3	0.09	100
Step 2. Affinity chromatography	0.86	2.4	8.3	3.46	54.1
Step 3. Sephadex G-200 gel filtration and concentration	0.90	1.1	4.6	4.40	30.1

is suggested that a hemiketal of formaldehyde is the true substrate for the enzyme.[11]

pH Optimum. The enzyme has optimum activity at pH 8.0.

Michaelis Constants. The apparent K_m value for formaldehyde is 5×10^{-5} M and for NAD^+ is 1.1×10^{-3} M when enzyme is prepared from cells grown on 3,4-dimethoxybenzoic acid.

Molecular Weight and Subunit Structure. The molecular weight of the native enzyme when measured by gel filtration through a column of Sephadex G-200 corresponds to 120,000–140,000. Subunit molecular weight is determined with polyacrylamide gel electrophoresis in the presence of sodium dodecyl sulfate (SDS). One subunit band was observed with a molecular weight of 44,000–47,000. These values are in accordance with a trimer structure for the native enzyme.

Purified Cofactor. The cofactor is stored either under nitrogen or in the presence of mercaptoethanol or dithiothreitol. Preparative electrofocusing showed that the cofactor is negatively charged at pH values above 6.0. The cofactor behaves as a weak acid with a pK of 6.5.

Formaldehyde Dehydrogenase (Acceptor) [NAD(P)$^+$-Independent Formaldehyde Dehydrogenase (Dye-Linked), EC 1.2.99.3]

Formaldehyde + artificial electron acceptor + $H_2O \rightleftharpoons$ formate + reduced electron acceptor

Formaldehyde dehydrogenase (acceptor) has been observed in non-methylotrophs and in methylotrophs growing on C_1 compounds and multicarbon compounds.[3,10]

Assay Method

Principle. Two assay methods can be used. In Method 1 the activity is measured spectrophotometrically by following the reduction of the artificial electron acceptor added to the reaction mixture. In Method 2 the activity is measured as the increase in oxygen uptake in a Clark-type oxygen electrode.

Reagents for Method 1

Potassium phosphate buffer, 0.8 M, pH 7.0
Phenazine methosulfate (PMS), 20 mM, or phenazine ethosulfate (PES), 20 mM
Dichlorophenolindophenol (DCPIP), 2 mM
Formaldehyde, 0.1 M, prepared from paraformaldehyde as described above
Other dyes can be used, including horse heart cytochrome c (2 mM) and Wurster's blue (2 mM).

Reagents for Method 2

Sodium phosphate buffer, 0.1 M, pH 7.0
PMS or PES, 50 mM
Formaldehyde, 1.0 M, prepared from paraformaldehyde

Procedure for Method 1. The assay mixture contains, in a final volume of 1 ml, sodium phosphate buffer, pH 7.0, 0.1 ml; PMS, 0.1 ml; DCPIP, 0.1 ml; and enzyme. This is preincubated at 30° for several minutes and the absorbance measured at 600 nm. The reaction is started by the addition of 0.1 ml formaldehyde, and the change in absorbance measured. This rate of change in absorbance is a measure of the activity of the dye-linked formaldehyde dehydrogenase. PMS can be replaced by PES (0.1 ml), and PMS and DCPIP can be replaced by horse heart cytochrome c (0.1 ml) or Wurster's blue (0.1 ml). When horse heart cytochrome c or Wurster's blue is used, the absorbance change at 550 or 412 nm, respectively, is measured.

Procedure for Method 2. The assay in the oxygen electrode (Biological Oxygen monitor; Yellow Springs Instruments, Yellow Springs, OH) is carried out in the presence of PMS or PES in the dark. The oxygen electrode is calibrated so that a full-scale deflection of the recorder reflects the difference between the oxygen content of an air-saturated buffer and a solution of zero oxygen content. The oxygen electrode vessel containing, in a final volume of 3.0 ml, 2.4 ml sodium phosphate buffer, pH 7.0, 0.1 ml PMS or PES, and enzyme is preincubated at 30° for at least 3 min. The endogenous oxygen uptake rate is measured, and the reaction is started by the addition of formaldehyde (30 μl). The increase of oxygen uptake is a measure of enzyme activity.

Units. Enzyme activities using Method 1 are expressed as before. The variation in E_{600} of DCPIP with pH must be taken into account.[12] Extinc-

[12] J. McD. Armstrong, *Biochim. Biophys. Acta* **86**, 191 (1964).

tion coefficient values for NADH, cytochrome c, and Wurster's blue are 6.22×10^3, 21.0×10^3, and 8.29×10^3 M^{-1} cm^{-1}. Enzyme activities using Method 2 are expressed as micromoles of oxygen consumed per minute per milligram protein.

Purification Procedure

Growth of Organism. The organism *Hyphomicrobium* X was isolated from natural water.[13] It is grown at 30° in a fermentor as a batch culture in a mineral salts medium containing, per liter, K$_2$HPO$_4$, 1.7 g; NaH$_2$PO$_4 \cdot$2H$_2$O, 1.7 g; (NH$_4$)$_2$SO$_4$, 2.0 g; MgSO$_4 \cdot$7H$_2$O, 0.2 g; and trace elements,[6] 0.2 ml. Filter-sterilized methanol is added (50 mM). The actively growing cells are harvested by centrifugation at 4500 g for 15 min, washed twice with 50 mM NaH$_2$PO$_4$/Na$_2$HPO$_4$ buffer, pH 7.0, containing 4 mM MgCl$_2$, and the cell paste stored at $-20°$.

Step 1: Purification of Cell-Free Extract. Cell paste (1 g) is suspended in 50 mM sodium potassium phosphate buffer, pH 7.5, containing 5 mM MgCl$_2$ (4 ml). The suspension is disrupted by ultrasonication in an MSE ultrasonic disintegrator (150 W) using 10 periods of 30-sec sonication interspersed with 30 sec in ice-cold water mixture. The supernatant obtained by centrifugation at 38,000 g for 30 min is the cell-free extract.

Step 2: Ammonium Sulfate Fractionation. All further procedures are carried out at $0-4°$. Finely ground solid ammonium sulfate is slowly added to the cell-free extract with stirring. The pH is adjusted throughout to 7.0 with 10% (v/v) ammonium hydroxide. The protein which precipitates between 45% (1.95 M) and 85% (4.23 M) ammonium sulfate saturation is collected by centrifugation at 38,000 g for 15 min. The precipitate is dissolved in 50 mM Tris-HCl buffer, pH 7.0.

Step 3: DEAE-Cellulose Chromatography. Any buffer used in the following purification steps contains 10% (v/v) ethanol to stabilize the enzyme. The extract from Step 2 is desalted by applying it to a Sephadex G-25 column (2.5 $\times$ 25 cm) equilibrated with 50 mM Tris-HCl, pH 7.0. The enzyme is eluted in the void volume with the same buffer, and active fractions are collected and applied to the top of a DEAE-cellulose (Whatman DE-52) column (2.5 $\times$ 25 cm) previously equilibrated with the same buffer. The column is washed initially with the buffer, and the protein which remains bound to the column is eluted by the stepwise addition of buffer containing increasing amounts of NaCl. Formaldehyde dehydrogenase activity is eluted with buffer containing 100 mM NaCl. Fractions containing activity are combined and concentrated by ultrafiltration.

Step 4: Hydroxylapitite Chromatography. The concentrated eluate

[13] M. M. Attwood and W. Harder, *Antonie van Leeuwenhoek* **38**, 369 (1972).

from Step 3 is applied to a Sephadex G-25 column equilibrated with 10 mM sodium phosphate buffer, pH 7.0. The column is eluted with the same buffer, and active fractions are combined and applied to an hydroxylapatite (Bio-Rad) column (1.5 × 6 cm) previously equilibrated with the same buffer. The column is eluted in a stepwise manner with increasing concentrations of sodium phosphate buffer, pH 7.0. The enzyme activity is eluted with 100 mM sodium phosphate buffer, pH 7.0. A typical purification is summarized in Table IV.

Properties

Purity. The preparation is not homogeneous, as judged by polyacrylamide gel electrophoresis (7–10.5% acrylamide). Seven protein bands are revealed, and a comparison with gels stained for aldehyde dehydrogenase activity with either formaldehyde or propionaldehyde shows that the major protein band and two minor protein bands have aldehyde dehydrogenase activity. Methanol and formate dehydrogenase activity cannot be detected in the enzyme preparation.

Substrate Specificity. The enzyme shows activity with a wide range of aldehydes (chain length C_1–C_{10}). Other physiological and nonphysiological aldehydes, including the hydrate aldehyde choral, are also substrates for the enzyme.

Electron Acceptors. A range of natural and artificial electron acceptors can replace PMS–DCPIP. These include horse heart cytochrome c, Wurster's blue, DCPIP, PMS–horse heart cytochrome c, and PES–DCPIP in the spectrophotomeric assay system. PES and PMS alone act as primary electron acceptors in the Clark-type oxygen electrode assay.

Stability. In the absence of any stabilizing agent the enzyme loses 33% of its activity after storage at −20° for 10 days. In the presence of 10% (v/v) ethanol the loss of activity after storage at −20° for 3 months is negligible. Stability is not affected by repeated freezing and thawing. The half-life of the enzyme is 168, 89, 8.5, 2.0, and less than 1 min at 30, 45, 55, 60, and 70°, respectively.

pH Optimum. The enzyme has optimum activity at pH 7.2 when cytochrome c or Wurster's blue is used as the primary electron acceptor and at pH 7.6 in the presence of PMS and DCPIP.

Michaelis Constants. The apparent K_m for formaldehyde is 11.2 × 10^{-3} M in presence of PMS–DCPIP and 2.86 × 10^{-3} M when cytochrome c is the primary electron acceptor. The apparent K_m is lowest and relatively constant when the carbon chain length is C_4 to C_7. This K_m value in the presence of PMS–DCPIP is between 7 × 10^{-4} and 3 × 10^{-3} M and in the presence of cytochrome c, 3 × 10^{-6} and 1.5 × $10^{-5}M$.

TABLE IV
PURIFICATION OF DYE LINKED FORMALDEHYDE DEHYDROGENASE FROM *Hyphomicrobium* X

Fraction	Total volume (ml)	Total protein (mg)	Total activity[a] (milliunits)	Specific activity (milliunits/ mg protein)	Yield (%)
Step 1. Cell-free extract	19.5	245.4	3853	15.7	100
Step 2. (NH$_4$)$_2$SO$_4$ fraction (45–85% saturation)	23	146.3	2765	18.9	71.8
Step 3. Sephadex G-25 filtration and DEAE-cellulose chromatography	27.5	50.2	1661	33.1	43.1
Step 4. Sephadex G-25 filtration and hydroxylapatite chromatography	5.0	4.57	800	174.9	20.8

[a] Results obtained by Method 1.

Enzyme Inhibitors. Sodium azide, 1,10-phenanthroline, 2,2′-bipyridyl, thiourea, 8-hydroxyquinoline, 4-chloromercuribenzoate, 5,5′-dithiobis(nitrobenzoic acid), idoacetate, and iodoacetamide (each at 1 mM), preincubated with the purified enzyme for 10 min at 30°, do not inhibit the enzyme activity. EDTA (10 mM) and NaCN (1 mM) do inhibit enzyme activity by 26 and 21.5%, respectively.

Molecular Weight. The location of the purified enzyme in the eluate fractions from a calibrated Sephadex G-150 column indicates a molecular weight of 83,000 ± 2,500. This value is confirmed by the position of the major active aldehyde dehydrogenase bands on polyacrylamide gels after coelectrophoresis with standard marker proteins.

[48] NAD-Linked, Factor-Independent, and Glutathione-Independent Aldehyde Dehydrogenase from *Hyphomicrobium* X

By J. A. DUINE

$$\text{Aldehyde} + \text{NAD}^+ \rightarrow \text{acid} + \text{NADH} + \text{H}^+$$

Aldehyde dehydrogenase oxidizes a wide array of aldehydes and is probably involved in formaldehyde conversion when *Hyphomicrobium* X grows on methanol.[1] Besides NAD$^+$, no other factors are required for

[1] P. A. Poels and J. A. Duine, *Arch. Biochem. Biophys.* **271**, 240 (1989).

activity. Although behaving similar in this respect to formaldehyde dehydrogenase (EC 1.2.1.46) isolated from *Pseudomonas putida* strains, the
enzymes are quite different from each other.

Assay Methods

Principle. The enzyme is assayed by measuring NADH formation at
340 nm. Because the enzyme is oxygen-sensitive, to prevent inactivation
and to restore already existing damage, the assay may be carried out in the
presence of dithiothreitol (DTT) (Method 1). However, as DTT reacts
with the substrate to produce thiohemiketal complexes, in order to perform measurements with reliable substrate concentrations but exclude the
effect of oxygen, an anaerobic assay without DTT can be used (Method 2).

Reagents

Pyrophosphate buffer, 0.1 M tetrasodium pyrophosphate buffer, ad-
justed to pH 9.0 with concentrated HCl
DTT, 0.1 M
NAD⁺, 20 mM
Formaldehyde, 120 mM, prepared by hydrolysis of paraformaldehyde

Procedure for Method 1. To a cuvette (1-cm optical pathway), 600 μl
pyrophosphate buffer (containing 0.17 M KCl), 100 μl DTT, 100 μl
NAD⁺, and 100 μl enzyme solution are added in that order. The mixture is
incubated for 12 min, and the reaction is initiated by addition of 100 μl
formaldehyde solution. The activity is calculated from the rate of absorbance increase at 340 nm (ϵ_{340} for NADH, 6220 M^{-1} cm^{-1}). Protein concentrations are measured with the method of Bradford[2] with bovine serum
albumin as a standard.

Procedure for Method 2. In Method 2, all stock solutions are made
anaerobic by degassing and flushing with nitrogen gas. Manipulations to
fill the cuvette are carried out in an anaerobic cabinet. To restore damage
already present in the enzyme, degassed preparations are dialyzed against
buffers containing 10 mM DTT for 1 hr. Subsequently, dialysis is performed under anaerobiosis against 20 mM potassium phosphate, pH 7.0
(500 volumes, 16 hr). To a cuvette (1-cm optical pathway, stoppered with
Suba seals), 600 μl pyrophosphate buffer, 100 μl NAD⁺, 100 μl enzyme
solution, and 190 μl water are added in that order. The cuvette is transferred to the spectrophotometer compartment, and the reaction is initiated
by injecting 10 μl of an anaerobic formaldehyde solution via a hypodermic

---

[2] M. M. Bradford, *Anal. Biochem.* **72,** 248 (1976).

needle. Measurements and calculations are performed as indicated for Method 1.

Purification Procedure

Growth of Organism. Hyphomicrobium X is grown on a mineral medium supplemented with 0.3% methanol as described.[3] Cell paste is stored frozen.

Step 1: Preparation of Cell-Free Extract. Cell paste (10 g wet weight) is suspended in an equal volume of 0.1 M Tris-HCl buffer, pH 9.0, containing 0.5% (v/v) Triton X-100, and the mixture is passed twice through a French pressure cell at 100 MPa. The suspension (its viscosity is lowered by adding deoxyribonuclease) is centrifuged (48,000 g for 30 min at 4°), giving the cell-free extract. All subsequent manipulations are carried out at 4°.

Step 2: Ammonium Sulfate Precipitation. After adding ammonium sulfate to 45% (w/v) saturation to the cell-free extract from Step 1, the precipitate is collected by centrifugation (48,000 g, 10 min). The precipitate is dissolved and dialyzed against 20 mM potassium phosphate, pH 7.0 (400 volumes, 16 hr).

Step 3: DEAE-Sepharose Chromatography. The solution from Step 2 is centrifuged (48,000 g, 10 min) and applied to a DEAE-Sepharose Fast Flow (Pharmacia, Uppsala, Sweden) column (1 × 10 cm), equilibrated with 20 mM potassium phosphate, pH 7.0. The column is eluted with 0.1 M potassium phosphate, pH 7.0. Active fractions are pooled, dialyzed against 20 mM potassium phosphate, pH 7.0, and concentrated by pressure filtration.

Properties

Purity and Stability. Not unexpectedly, the preparation obtained with Step 3 is not pure. The specific activity of the enzyme (as measured with Method 1) scarcely increases during the purification procedure. This is due in large part to damage by oxygen as DTT treatment of the enzyme preparations and assay with Method 2 lead to a purification factor of 5. However, not all damage can be restored; preparing the cell-free extract under anaerobic conditions gives even higher yields. For these reasons, preparation of homogeneous enzyme should be performed with the exclusion of oxygen.

No other NAD+-linked aldehyde dehydrogenase seems to be present because the properties of the enzyme do not change during purification (in

---

[3] J. Frank and J. A. Duine, this volume [32].

addition, methanol dehydrogenase and dye-linked aldehyde dehydrogenase are absent in the final preparation).

Assay Methods. NAD^+ addition prior to that of aldehyde is crucial for activity in the assays. Restoration and prevention of damage can be achieved by including thiol compounds in Method 1 (DTT and 2-mercaptoethanol are effective whereas glutathione and cysteine are much less effective). These compounds are not required in the anaerobic assay with DTT-treated enzyme. Pyrophosphate buffer gives the highest activities, whereas CHES [2-(N-cyclohexylamino)ethanesulfonic acid], ammonium chloride, and Tris buffers are less active; borate buffer is not active at all.

Activators and Inhibitors. Glutathione is not required for activity. Addition of factor (required for factor-dependent formaldehyde dehydrogenase)[4] does not increase activity. KCl is an activator only in the aerobic assay with DTT (Method 1). Hydrogen peroxide is an inhibitor. However, if NAD^+ is added prior to hydrogen peroxide, significant protection is observed. $HgCl_2$ is also an inhibitor.

Substrate Specificity. A wide range of aldehydes appear to be substrates, but alcohols, even at high concentrations, are not. Propionaldehyde is the best substrate (as demonstrated with Method 2). The aldehydes tested show severe substrate inhibition. $NADP^+$ cannot replace NAD^+. Attempts to demonstrate the reverse reaction have failed so far.

Comparison with Other Enzymes. From its properties, the enzyme can be considered as an NAD^+-linked, factor- and glutathione-independent aldehyde dehydrogenase, just like formaldehyde dehydrogenase (EC 1.2.1.46) from strains of *Pseudomonas putida.* However, there are large differences between these enzymes with respect to substrate specificity, stability, requirement of a thiol compound, the effect of the order of NAD^+ and substrate addition to the assay, and the effect of hydrogen peroxide. On the other hand, bakers' yeast aldehyde dehydrogenase [$NAD(P^+)$] (EC 1.2.1.5)[5] shares several properties with the enzyme (substrate specificity, requirement of a thiol compound, and stimulation by potassium salts in the assay).

---

[4] L. Eggeling and H. Sahm, *Eur. J. Biochem.* **150**, 129 (1985).
[5] S. Black, *Arch. Biochem. Biophys.* **34**, 86 (1951).

[49] Formate Dehydrogenase from *Methylosinus trichosporium* OB3b

By DAVID R. JOLLIE and JOHN D. LIPSCOMB

$$\text{Formic acid} + NAD^+ \rightleftharpoons \text{carbon dioxide} + NADH + H^+$$

Formate dehydrogenase (EC 1.2.1.2) catalyzes the final step in the methane oxidation pathway of methanotrophic bacteria.[1] The NADH produced in this reaction is required by methane monooxygenase which catalyzes the initial step in the pathway, the conversion of methane to methanol.[2,3] Previous attempts to purify formate dehydrogenase from methanotrophs[4] have not been successful owing to the perceived instability of the enzyme. However, the addition of stabilizers to the purification buffers allows the high-yield purification of a high specific activity formate dehydrogenase described here.[5]

Assay Method

Principle. The assay system represents a slight modification of the method presented earlier by Johnson *et al.*[6] It is based on the rate of increase in the optical density at 340 nm owing to the appearance of the product, NADH, using formic acid as the reductant.

Reagents

Buffer, 50 m*M* MOPS,[7] pH 6.5
Formic acid, 0.1 *M* in buffer adjusted to pH 6.5 with KOH
NAD⁺, 50 m*M* in buffer

Procedure. The assay mixture contains, in a final volume of 0.5 ml, 0.44 ml buffer, 25 µl formic acid, 10 µl NAD⁺, and 25 µl of the sample to be analyzed. The reaction rate can be determined from the initial rate of

[1] J. R. Quayle, *Adv. Microb. Physiol.* **7**, 119 (1972).
[2] B. G. Fox, W. A. Froland, D. R. Jollie, and J. D. Lipscomb, this volume [31].
[3] S. J. Pilkington and H. Dalton, this volume [30].
[4] C. Anthony, "The Biochemistry of Methylotrophs," p. 194. Academic Press, London, 1982.
[5] D. R. Jollie and J. D. Lipscomb, unpublished data.
[6] P. A. Johnson, M. C. Jones-Mortimer, and J. R. Quayle, *Biochim. Biophys. Acta* **89**, 351 (1964).
[7] MOPS, 3-(*N*-Morpholino)propanesulfonic acid; PAGE, polyacrylamide gel electrophoresis; DCPIP, 2,6-dichlorophenolindophenol; SDS, sodium dodecyl sulfate; FAD, flavin adenine dinucleotide; FNM, flavin mononucleotide; DTT, dithiothreitol.

increase in the optical density at 340 nm at 23° using an ϵ_{340} value for NADH of 6200 M^{-1} cm^{-1}.

Purification Procedure[5]

The instability of formate dehydrogenase is overcome by the addition of 0.2 mM Fe^{II}(NH$_4$)$_2$(SO$_4$)$_2 \cdot$ 6H$_2$O and 2.0 mM cysteine to all of the purification buffers. All protein concentrations are determined by the method of Bradford,[8] using bovine serum albumin as the standard.

Growth of Organism. The methods for maintenance, growth, and harvesting of *Methylosinus trichosporium* OB3b are described elsewhere in this volume.[2]

Step 1: Preparation of Cell-Free Extract. Cell paste (200 g wet weight) is thawed in 200 ml of 25 mM MOPS buffer, pH 7.0. The cells are broken by sonication (Branson Model 350) in 1-min intervals at an output of 70% in a stainless steel vessel. A saturated CaCl$_2$ slush bath is used to maintain the temperature of the cell suspension at or below 5° at all times. The total time of sonication is 10 min. The broken cells are diluted with an additional 200 ml of 25 mM MOPS, pH 7.0, and centrifuged at 48,000 g for 60 min to pellet the cell wall and other particulate materials.

Step 2: First DEAE Anion-Exchange Column. This and subsequent steps are carried out at 4°. The cell-free extract is diluted with an additional 200 ml of 25 mM MOPS, pH 7.0, and loaded onto a column (4 × 30 cm) of DEAE-Sepharose CL-6B Fast Flow anion-exchange resin (Pharmacia, Piscataway, NJ) equilibrated in 25 mM MOPS, pH 7.0. The loaded column is washed with 600 ml of the equilibration buffer and then developed with a linear gradient of 0 to 0.4 M NaCl over a total volume of 2000 ml at a linear flow rate of 12 cm/hr.

Step 3: Second DEAE Anion-Exchange Column. The fractions that contain at least 20% of the activity of the peak fraction from Step 2 are pooled, the pH is adjusted to 6.5 using a saturated solution of MOPS free acid, and the solution is diluted with 1 volume of cold distilled water. The prepared sample is then loaded onto a column (2.5 × 20 cm) of DEAE-Sepharose CL-6B Fast Flow equilibrated with buffer containing 0.12 M KCl. The loaded column is washed with 150 ml of equilibration buffer, and the protein is eluted with a linear gradient from 0.12 to 0.3 M KCl over a total volume of 800 ml at a linear flow rate of 7 cm/hr.

Step 4: Phenyl-Sepharose Hydrophobic Interaction Column. The fractions that contain at least 20% of the activity of the peak fraction from Step 3 are pooled, and an additional 0.15 M KCl is added. The sample is then

---

[8] M. Bradford, *Anal. Biochem.* **72**, 248 (1976).

loaded onto a column (1 × 18 cm) of Phenyl-Sepharose CL-4B (Pharmacia) equilibrated in buffer containing 0.35 M KCl and 1.0 mM dithiothreitol (DTT). The column is washed with 35 ml of equilibration buffer, and the protein is subsequently eluted with buffer containing 1.0 mM DTT at a linear flow rate of 20 cm/hr.

Step 5: Sepharose CL-6B Gel-Filtration Column. The active fractions from Step 4 are pooled and loaded on to a column (2.5 × 50 cm) of Sepharose CL-6B (Pharmacia) equilibrated in buffer containing 0.1 M KCl and 1.0 mM DTT. The column is developed in equilibration buffer at a linear flow rate of 4 cm/hr. The fractions containing at least 30% of the activity in the peak are pooled and concentrated on a YM30 (Amicon, Danvers, MA) ultrafiltration membrane. The sample is frozen in liquid nitrogen and stored at −80°. A typical purification is summarized in Table I.

Properties

Purity and Stability. Following cell lysis, ferrous ion and a reductant are imperative for the maintenance of active formate dehydrogenase. Without these reagents, the activity decreases noticeably in 1–2 hr at 0° and is unstable even at −80°. In the presence of the stabilizers, however, formate dehydrogenase is stable without detectable loss of activity for 4–5 hr at 0° in dilute solution and longer if it is concentrated to at least 100 units/ml. It is stable indefinitely at −80° when frozen with stabilizers. Repeated freeze–thaw cycles are damaging to the enzyme activity. Denaturing polyacrylamide gel electrophoresis (PAGE) in the presence of sodium dodecyl sulfate (SDS) suggests that the preparation is at least 90% pure. Nondenaturing electrophoresis of the enzyme exhibits just one band that can be observed by either protein or activity-based stains.

Molecular Weight. The molecular weight of formate dehydrogenase determined by gel-permeation chromatography and nondenaturing electrophoresis with different polyacrylamide concentrations is approximately 400,000. Four subunits with molecular weights of 98,000, 56,000, 19,000, and 11,500 are detected by denaturing electrophoresis in the presence of SDS.

Activity. The pH optimum for the enzyme is 6.5, but activity can readily be detected in the pH range of 5 to 9.5. The Michaelis constants at pH 6.5 are 80 and 100 μM for formate and NAD$^+$, respectively. The turnover number is estimated to be 240 sec^{-1}.

Cofactors. Acid precipitation of the protein shows that there is a flavin cofactor present in formate dehydrogenase; however, it appears that it is neither FAD, FMN, nor riboflavin. The iron content of the protein is

TABLE I
PURIFICATION OF FORMATE DEHYDROGENASE FROM *Methylosinus trichosporium* OB3b

Step	Volume (ml)	Activity (units/ml)[a]	Protein (mg)	Specific Activity (units/mg)	Yield (%)
1. Cell-free extract	600	2.11	11,060	0.115	100
2. First DEAE-Sepharose	182	4.26	712	1.09	61.2
3. Second DEAE-Sepharose	109	6.8	92.5	8.06	56.0
4. Phenyl-Sepharose	4.8	122	32.5	18.0	46.3
5. Sepharose CL-6B gel filtration	20.0	22.6	16.0	28.2	35.7

[a] Unit is defined as 1 μmol of NADH produced in 1 min.

20–26 irons per flavin as determined colorimetrically with *o*-bathophenanthroline[9] and inductively coupled plasma emission spectroscopy. No other metals are detected in greater than trace amounts. Optical, electron paramagnetic resonance (EPR), and Mössbauer spectroscopies show that the iron is present primarily as $[Fe-S]_x$ centers. Reductive EPR titrations reveal that there are at least five different types of $[Fe-S]_x$ centers.

Alternative Substrates. Formate and NADH, the natural substrates, are the only biological reductants that have been identified; however, sodium dithionite can be used as a chemical reductant. Many general oxidants have been identified including NAD^+, ferricyanide, dichlorophenolindophenol (DCPIP), and oxygen as well as the endogenous *c*-type cytochrome c_{554}.

Inhibitors. Formate dehydrogenase is inhibited by fundamentally different types of compounds; these include small metal ligands such as azide, cyanide, nitrite, and cyanate as well as irreversible inhibitors that alter or destroy the structure of Fe–S centers, such as mercury, copper, EDTA, and *o*-bathophenanthroline. The role of the individual cofactors in the formate oxidation reaction have not yet been determined, but the inhibition studies suggest that they are involved and necessary for the enzyme to perform the formate oxidation reaction.

Acknowledgments

This work was supported by a grant from the National Institutes of Health (GM 40466).

[9] B. Zak, *Am. J. Clin. Pathol.* **29**, 590 (1958).

[50] Glucose-6-phosphate Dehydrogenase and 6-Phosphogluconate Dehydrogenase from *Methylobacillus flagellatum*

By LUDMILA V. KLETSOVA, MICHAEL Y. KIRIUKHIN, ANDREY Y. CHISTOSERDOV, and YURI D. TSYGANKOV

Glucose-6-phosphate Dehydrogenase

Glucose 6-phosphate + H_2O + $NAD(P)^+$ → 6-phosphogluconate + $NAD(P)H$ + H^+

The enzyme catalyzes the oxidation of glucose 6-phosphate to 6-phosphogluconate, by way of 6-phosphogluconolactone, with the concomitant formation of NADH or NADPH. In obligate methylotrophs using the ketodeoxyphosphogluconate aldolase/transaldolase variant of the ribulose monophosphate cycle, the enzyme participates both in assimilation and cyclic oxidation of formaldehyde. The properties of the enzyme described below have been the subject of a previous report.[1] The purification and properties of similar enzymes from some obligate methylotrophs have been described.[2]

Assay Method

Principle. The enzyme activity is measured continuously by following the increase in NAD(P)H absorbance.

Reagents

Tris-HCl buffer, 50 mM, pH 8.0
Glucose 6-phosphate, 200 mM
NAD or NADP, 16 mM

Procedure. The assay mixture contains, in a final volume of 1 ml, Tris-HCl buffer, 925 µl; NAD or NADP, 25 µl; and cell extract, 25 µl. The mixture is equilibrated at 40° for several minutes, then glucose 6-phosphate (25 µl) is added and the rate of absorbance increase followed at 340 nm.

Units. One unit of glucose-6-phosphate dehydrogenase is defined as that amount of enzyme catalyzing the formation of 1 µmol of NADH or NADPH per minute at 40°. Specific activity is expressed as units per milligram of protein.

[1] M. Y. Kiriuchin, L. V. Kletsova, A. Y. Chistoserdov, and Y. D. Tsygankov, *FEMS Microbiol. Lett.* **52,** 199 (1988).
[2] C. Anthony, "The Biochemistry of Methylotrophs." Academic Press, London, 1982.

Properties

Growth of Organism. The strain *Methylobacillus flagellatum* used originated from Govorukhina *et al.*[3] and the temperature-sensitive mutants
T623 and T903, defective in glucose-6-phosphate dehydrogenase, originated from Kletsova *et al.*[4] The strains are cultivated in a mineral salts
medium containing, per liter, Na_2HPO_4, 6 g; KH_2PO_4, 3 g; NaCl, 0.5 g;
NH_4Cl, 1 g; $MgSO_4$, 120 mg; $CaCl_2$, 11 mg; supplemented with 1% (v/v)
methanol. In the case of mutants, methionine (100 μg/ml) is supplied.
Wild-type *M. flagellatum* is grown at 42°; mutant strains T623 and T903
are grown at 30°.

Preparation of Cell-Free Extracts. Cells are harvested in mid log phase
by centrifugation at 6000 g for 10 min. Cell paste (1 g wet weight) is
suspended in 2 ml of Tris-HCl buffer, pH 8.0, containing 10 mM $MgCl_2$.
The suspension is disrupted by ultrasonication for 2 min at 0–4° in an
MSE ultrasonic disintegrator (150 W) and centrifuged at 30,000 g for 60
minutes, yielding the cell-free extract as the supernatant.

Polyacrylamide Gel Electrophoresis. Vertical electrophoresis in a 7.5%
polyacrylamide gel is run in nondenaturing conditions at 7–9° in slabs
(14 × 14 cm) without a concentrating gel.[5] Cell-free extracts are applied
with an equal volume of a 10 times diluted electrode buffer, pH 8.5, that
contains 10% (v/v) glycerol and 0.01% bromphenol blue. Electrophoresis
(150 W, 14 mA) is run for 9 hr. Finally, the slab is cut into slices that are
incubated in the corresponding reaction mixture for visualizing the enzyme by staining. The reaction mixture for glucose-6-phosphate dehydrogenase contains Tris-HCl, 100 mM, pH 8.0; glucose 6-phosphate, 2 mM;
NAD or NADP, 0.8 mM; phenazine methosulfate. 0.5 mM; and nitrotetrazolium blue, 1 mM.

Two Forms of Glucose-6-Phosphate Dehydrogenase. After staining at
30°, glucose-6-phosphate dehydrogenase is localized in two bands in both
the wild-type strain and the mutants defective in glucose-6-phosphate
dehydrogenase. When the gel is first incubated at 42° for 30 min, both
bands, corresponding to T623 and T903, are absent as a result of inactivation of defective proteins. Thus, the mutation in the *gpd* gene leads to
defects in both forms of glucose-6-phosphate dehydrogenase simulta-
neously. Forms of enzyme differing in mobility in an electric field are
eluted from the corresponding region of the gel with 100 mM Tris-HCl
buffer, pH 8.0, containing 10 mM $MgCl_2$ for 24 hr. Both forms have
specificity to NAD^+ and $NADP^+$ and exhibit the same temperature depen-

[3] N. I. Govorukhina, L. V. Kletsova, Y. D. Tsygankov, Y. A. Trotsenko, and A. I. Netrusov,
Mikrobiologija **56**, 849 (1987).
[4] L. V. Kletsova, E. S. Chibisova, and Y. D. Tsygankov, *Arch. Microbiol.* **149**, 441 (1988).
[5] L. Ornstein and B. J. Davis, *Ann. N.Y. Acad. Sci.* **121**, 321 (1964).

dence of activity either in the wild-type strain or in the mutants T623 and T903. By isoelectrofocusing of the wild-type strain cell-free extract, the glucose-6-phosphate dehydrogenase is localized in two bands, with isoelectric points of 6.5 and 6.3, respectively. Both bands are stained with NAD^+ and $NADP^+$ and correspond to the two forms differing in mobility in the electric field.

Michaelis Constants. The apparent K_m values are determined for both slow and rapid forms of glucose-6-phosphate dehydrogenase: K_m for glucose 6-phosphate, 2.5×10^{-4} and 2.3×10^{-4} M with NAD^+ (0.8 mM) as cofactor; 1.4×10^{-4} and 1.3×10^{-4} M with $NADP^+$ (0.8 mM) as cofactor; K_m for NAD, 2.2×10^{-5} and 3.5×10^{-5} M (at 5 mM glucose 6-phosphate); and for NADP, 2.5×10^{-5} and $4.0 \times 10^{-5}M$. The temperature-sensitive proteins of T623 and T903 mutants have the same affinity for glucose 6-phosphate, NAD^+, and $NADP^+$ as wild-type glucose-6-phosphate dehydrogenase.

Inhibitors. Both forms of glucose-6-phosphate dehydrogenase are effectively inhibited by reduced pyridine nucleotides. NADH (0.2 mM) inhibits 50% of the activity, NADPH (also 2 mM) inhibits 40% of the activity. ATP (10 mM) has an inhibitory effect at high concentrations (up to 45–50% with NAD and 15–20% with NADP).

6-Phosphogluconate Dehydrogenase

6-Phosphogluconate + $NAD(P)^+$ → ribulose 5-phosphate + $NAD(P)H$ + H^+ + CO_2

6-Phosphogluconate dehydrogenase catalyzes the irreversible oxidative decarboxylation of 6-phosphogluconate to ribulose 5-phosphate. In methylotrophs using the ribulose monophosphate cycle, the enzyme is involved in the cyclic oxidation of formaldehyde. The properties of the enzyme described below have been the subject of a previous report.[1] The purification and properties of similar enzymes from some methylotrophs have been described.[2]

Assay Method

Principle. The enzyme activity is measured continuously by following the increase in NAD(P)H absorbance.

Reagents

Tris-HCl buffer, 50 mM, pH 8.0
6-Phosphogluconate, 200 mM
$MgCl_2$, 200 mM
NAD^+ or $NADP^+$, 16 mM
Procedure. The assay mixture contains, in a final volume of 1 ml,

Tris-HCl buffer, 900 μl; MgCl$_2$, 25 μl; NAD$^+$ or NADP$^+$, 25 μl; and cell extract, 25 μl. The mixture is equilibrated at 40° for several minutes; 6-phosphogluconate (25 μl) is then added, and the rate of increase in absorbance is followed at 340 nm.

Units. One unit of 6-phosphogluconate dehydrogenase is defined as the amount of enzyme catalyzing the formation of 1 μmol of NADH or NADPH per minute at 40°. Specific activity is expressed as units per milligram of protein.

Properties

The growth of the organism (wild-type strain) and the preparation of the cell-free extract are the same as for glucose-6-phosphate dehydrogenase.

Isoelectrofocusing. Cell-free extract is focused from pH 4 to 9 in slabs of 5% polyacrylamide gel containing 2% ampholines for 12 hr at 250 W, temperature 7–9°. Aspartate (10 mM) is used as anolyte and 10 mM NaOH as catholyte. After focusing, the slab is incubated in 100 mM Tris-HCl buffer, pH 8.0, containing 10 mM MgCl$_2$ for 15 min at 30°, then cut into slices in order to visualize the enzymes by specific staining. The reaction mixture for visualizing 6-phosphogluconate dehydrogenase contains Tris-HCl buffer, pH 8.0, 100 mM; 6-phosphogluconate, 2 mM; MgCl$_2$, 10 mM; NAD$^+$ or NADP$^+$, 0.8 mM; phenazine methosulfate, 0.5 mM; and nitrotetrazolium blue, 1 mM. After staining the reaction mixture with NAD$^+$, 6-phosphogluconate dehydrogenase is localized in two bands; with NADP$^+$, it is localized in one band.

Isoenzymes of 6-Phosphogluconate Dehydrogenase. In the cell extract there are two isoenzymes, having different isoelectric points and specificity. The enzyme with isoelectric point 6.5 reveals a double specificity (the activity with NAD$^+$ is 10% of that with NADP$^+$). The enzyme with isoelectric point 6.3 is specific for NAD$^+$ only.

Partial Purification and Separation of Isoenzymes. The ammonium sulfate fraction (0–70% saturation) is gel filtered on a column (1.5 × 90 cm) of TSK HW55 (Toyosoda). The activities of 6-phosphogluconate dehydrogenase and glucose-6-phosphate dehydrogenase are found in the same peak. This fraction is applied to a column (1.6 × 20 cm) of DEAE TSK (Toyosoda). Elution of the proteins from an anion exchanger with a linear gradient concentration of NaCl (0–0.5 M) does not improve the separation of the glucose-6-phosphate and 6-phosphogluconate dehydrogenases, but in this case the latter is present in two partially overlapping peaks. The 6-phosphogluconate dehydrogenase with the higher isoelectric point has a lower binding capacity to the anion exchanger; the enzyme with lower isoelectric point is more strongly bound to the anion exchanger.

Stability. The enzyme with an isoelectric point of 6.5 retains virtually all its activity (with NAD$^+$ and NADP$^+$) after storing at 4° for 30 days. The enzyme with an isoelectric point of 6.3 loses about 90% activity after storage at 4° for 3 days.

Inhibitors. NADPH (0.2 mM) inhibits both isoenzymes to a similar extent (65–70%). NADH (0.2 mM) has a strong influence on the enzyme specific to NAD (99%), whereas the enzyme with double specificity is inhibited to a lesser extent (60% with NADP$^+$ and 70% with NAD$^+$ as cofactor). A high ATP concentration (10 mM) is also inhibitory (up to 30% for both isoenzymes).

[51] Glucose-6-phosphate Dehydrogenase and 6-Phosphogluconate Dehydrogenase from *Arthrobacter globiformis*

By A. P. SOKOLOV and Y. A. TROTSENKO

Glucose-6-Phosphate Dehydrogenase

D-Glucose 6-phosphate + NADP$^+$ $\rightleftharpoons$ 6-phospho-D-gluconate + NADPH + H$^+$

Arthrobacter globiformis VKM B-175 is a facultative methylotroph growing on methylated amines as sources of carbon and energy. Enzymes involved in the direct oxidation of formaldehyde to CO_2 (formaldehyde and formate dehydrogenases) have very low activities in this methylotroph.[1] In this bacterium formaldehyde is mainly oxidized via the dissimilatory hexulose phosphate cycle. Methods for the isolation of the first and last oxidative enzymes of this cycle (i.e., glucose-6-phosphate dehydrogenase and 6-phosphogluconate dehydrogenase) reported by Sokolov and Trotsenko[2] are described here.

Assay Method

Principle. Glucose-6-phosphate dehydrogenase is measured spectrophotometrically by following the rate of NADPH formation at 340 nm in the presence of saturating amounts of glucose 6-phosphate and NADP.

Procedure. The assay is carried out at 30° in a recording spectrophotometer. In a total volume of 2 ml the reaction mixture contains 50 mM Tris–acetate buffer, pH 7.5, 10 mM MgSO$_4$, 1 mM NADP$^+$, 2 mM glu-

---

[1] N. V. Loginova and Y. A. Trotsenko, *Microbiology (Engl. Transl.)* **45**, 196 (1976).
[2] A. P. Sokolov and Y. A. Trotsenko, *Biochemistry (Engl. Transl.)* **50**, 1080 (1985).

cose 6-phosphate, and a suitable amount of enzyme to cause an absorbance change of 0.05 to 0.4 A/min. The rate of absorbance change at 340 nm is followed, and activities are calculated by using a ϵ value of 6.22 cm^2/μmol for NADPH at 340 nm.

Units. One unit of glucose-6-phosphate dehydrogenase is defined as the amount of enzyme catalyzing the formation of 1 μmol of NADPH per minute at 30°. Specific activity is expressed as units per milligram of protein.

Growth of Organism

Arthrobacter globiformis VKM B-175 is grown in the following basal medium:[3] 2 g of KH_2PO_4, 2 g of $(NH_4)_2SO_4$, 25 mg of $MgSO_4 \cdot 7H_2O$, and 0.5 g of NaCl in 1 liter of distilled water. The basal medium is supplemented with 3 ml of trace elements solution per 1 liter. Trace elements solution[4] contains 5 g of EDTA, 2.2 g of $ZnSO_4 \cdot 7H_2O$, 0.61 g of $MnSO_4 \cdot 5H_2O$, 0.5 g of $FeSO_4 \cdot 7H_2O$, 0.161 g of $CoCl_2 \cdot 5H_2O$, 0.157 g of $CuSO_4 \cdot 5H_2O$ and 0.151 g of $Na_2MoO_4 \cdot 2H_2O$ per 1 liter of distilled water. The medium is titrated with 10% NaOH to pH 7.2 before sterilization. A sterile solution of trimethylamine is added to the medium after sterilization to a concentration of 0.3% (w/v) as a source of carbon and energy. For large-scale cultivation the organism is grown in a 100-liter fermentor at a stirrer speed of 500 rpm, a temperature of 30°, and an aeration rate of 0.5 liter of air per liter of working volume per minute. The pH is maintained at 7.1 by automatic addition of 10% NaOH. The cells are harvested at the end of the exponential growth phase by a continuous-flow centrifuge and stored at −20° until use.

Synthesis of UTP-Agarose

The periodate method of activating the carrier and ligand is employed in the synthesis of UTP-agarose.[5] One hundred milliliters of 0.2 M $NaIO_4$ is added to 100 ml of settled BioGel A-0.5 M. The mixture is shaken for 2 hr at room temperature, and the gel is washed in a Büchner funnel with 2 liters of water. One hundred milliliters of 0.1 M adipic acid dihydrazide is added to the washed gel, the pH is adjusted to 5.0, and the mixture is shaken for 2 hr at room temperature. The pH of the mixture is adjusted to 9.0 using dry Na_2CO_3, and 27 g of KBH_4 is added to the suspension in small portions at a temperature of 4°. The product is washed with 2 liters

[3] T. Kaneda and J. H. Roxburgh, *Can. J. Microbiol.* **5,** 187 (1959).
[4] J. D. Owens and R. M. Keddy, *J. Appl. Bacteriol.* **32,** 338 (1969).
[5] M. Wilchek and R. Lamed, this series, Vol. 34, p. 475.

of 1 M NaCl, incubated in an equal volume of 1 M NaCl for 15 hr, and again washed with 2 liters of 1 M NaCl. Crystalline $NaIO_4$ is added to an aqueous solution of UTP (10 mM) to a concentration of 9.5 mM, and the mixture is kept in the dark at 0° and pH 7.5 for 1 hr. An equal volume of 0.2 M acetate buffer, pH 5.0, is added to the solution of activated UTP. Fifty milliliters of a solution of activated UTP is mixed with a suspension of 20 ml of settled hydrazide-agarose in 20 ml of 0.1 M acetate buffer, pH 5.0. The mixture is shaken for 4 hr at 4°, then 13.5 g of dry KBH_4 is added in small portions over a period of 12 hr. The UTP-agarose is washed on a Büchner funnel first with 1 M NaCl and then with water.

Purification Procedure

Tris–acetate buffer (15 mM, pH 6.6), containing 5 mM $MgCl_2$ and 10% glycerol (buffer A) is used throughout the purification unless otherwise stated. All operations are carried out at 4°.

Step 1: Preparation of Cell-Free Extract. Cell paste (50 g wet weight) is suspended in 50 ml of 15 mM Tris–acetate buffer, pH 6.6, containing 5 mM $MgCl_2$ and 50% glycerol. The suspension is disrupted by ultrasonication 12 times for 45 sec at 0–4° in an MSE ultrasonic disintegrator (150 W) and centrifuged at 20,000 g for 60 min, yielding the cell-free extract as the supernatant.

Step 2: Streptomycin Sulfate Treatment. To the supernatant solid streptomycin sulfate is added with stirring up to a concentration of 1.5%. After 40 min at 4° the precipitate is collected by centrifugation at 20,000 g for 60 min and discarded.

Step 3: Polyethylene Glycol Fractionation. The supernatant from Step 2 is fractionated with a 40% (w/v) solution of polyethylene glycol (PEG) 6000. The protein precipitating between PEG 6000 concentrations of 7 and 12% is collected by centrifugation at 20,000 g for 60 min and dissolved in buffer A.

Step 4: DEAE-BioGel Chromatography. The protein solution from Step 3 is applied to a column (2.6 × 34 cm) of DEAE-BioGel A previously equilibrated with buffer A. Glucose-6-phosphate dehydrogenase activity is eluted from the column with the equilibration buffer, close to the void volume of the column. Fractions containing dehydrogenase activity are combined.

Step 5: CM-Sepharose Chromatography. The protein solution from Step 4 is applied to a column (2.6 × 40 cm) of CM-Sepharose CL-6B previously equilibrated with buffer A. Glucose-6-phosphate dehydrogenase activity is eluted from the column with the gradient of $(NH_4)_2SO_4$ (0–0.5 M) in the equilibration buffer. The active fractions are combined and dialyzed overnight against buffer A.

Step 6: Polyethyleneimine-Agarose Chromatography. The dialyzed enzyme from Step 5 is further purified by chromatography on polyethyleneimine (PEI)-agarose, prepared by the method of Bystrykh and Trotsenko.[6] The protein solution is applied to a column (1.6 × 10 cm) of PEI-agarose previously equilibrated with buffer A. Glucose-6-phosphate dehydrogenase activity is eluted from the column with a gradient of $(NH_4)_2SO_4$ (0–0.5 M) in the equilibration buffer.

Step 7: UTP-Agarose Chromatography. The active fractions from Step 6 are combined and applied to a column (1.6 × 10 cm) of UTP-agarose previously equilibrated with buffer A. The enzyme is eluted from the column with a gradient of $(NH_4)_2SO_4$ (0–1.0 *M*) in the equilibration buffer. The active fractions are combined, and the glucose-6-phosphate dehydrogenase contained therein is stored at 4°. A typical purification is summarized in Table I.

Properties

Purity. Analysis of the enzyme preparation by polyacrylamide gel electrophoresis in the presence of sodium dodecyl sulfate showed one band on staining with Coomassie Brilliant Blue G-250.

Substrate Specificity. The enzyme is absolutely specific with respect to $NADP^+$. No reaction can be detected in the presence of NAD^+.

Effect of Metal Ions. Glucose-6-phosphate dehydrogenase is activated by bivalent metal ions, Mg^{2+} and Ba^{2+} being most effective, Co^{2+} and Ca^{2+} much less effective. Zn^{2+} at 0.1 m*M* inhibits the enzyme completely.

Stability. The purified enzyme stored at 4° in buffer A remains active for at least 1 month.

pH Optimum. The enzyme has maximum activity at pH 7.5.

Michaelis Constants. The K_m for NADP is 16 µ*M*; for glucose 6-phosphate it is 25 µ*M*.

Molecular Weight and Subunit Structure. The molecular weight of glucose-6-phosphate dehydrogenase in buffer A, as measured by gel permeation through a column of Sepharose CL-6B, corresponds to 142,000. During electrophoresis in polyacrylamide gels in the presence of sodium dodecyl sulfate, a single band is detected at a position corresponding to a molecular weight of approximately 77,000. This suggests that the enzyme is a dimer.

Effectors. Glucose 6-phosphate is inhibited by NADH, NADPH, ATP, 2-oxoglutarate, 3-phosphoglycerate, and fructose 1,6-bisphosphate. Acetyl-CoA has a stimulatory effect. Enzyme activity remains practically un-

---

[6] L. V. Bystrykh and Y. A. Trotsenko, *Biochemistry (Engl. Transl.)* **48,** 1386 (1983).

TABLE I
PURIFICATION OF GLUCOSE-6-PHOSPHATE DEHYDROGENASE FROM *Arthrobacter globiformis*

Fraction	Total protein (mg)	Total activity (units)	Specific activity (units/mg protein)	Yield (%)
Step 1. Cell-free extract	3046	116	0.038	100
Step 2. Streptomycin sulfate treatment	1903	145	0.076	125
Step 3. PEG 6000 fraction (7–12%)	489	77	0.157	67
Step 4. Pooled fractions after DEAE-BioGel	150	72	0.475	62
Step 5. Pooled fractions after CM-Sepharose	87	67	0.77	58
Step 6. Pooled fractions after PEI-agarose	8.5	34	3.9	29
Step 7. Pooled fractions after UTP-agarose	1.1	20	18.5	17

changed in the presence of 1 mM oxaloacetate, citrate, malate, pyruvate, phosphoenolpyruvate, glyoxylate, ADP, cyclic 3′,5′-AMP, and fructose 6-phosphate.

6-Phosphogluconate Dehydrogenase

6-Phospho-D-gluconate + NADP$^+$ ⇌ D-ribulose 5-phosphate + CO$_2$ + NADPH + H$^+$

Assay Method

Principle. The principle is the same as that for glucose-6-phosphate dehydrogenase.

Procedure. The assay procedure is the same as that for glucose-6-phosphate dehydrogenase, except with 2 mM 6-phosphogluconate.

Units. One unit of 6-phosphogluconate dehydrogenase is defined as that amount of enzyme catalyzing the formation of 1 μmol of NADPH per minute at 30°. Specific activity is expressed as units per milligram of protein.

Purification Procedure

Tris-HCl buffer (50 mM, pH 7.5), containing 2.5 mM MgCl$_2$, 0.1 mM EDTA, and 0.1 mM dithioerythritol (buffer B) is used throughout the purification unless otherwise stated. All operations are carried out at 4°.

Step 1: Preparation of Cell-Free Extract. Cell paste (20 g wet weight) is suspended in 20 ml of buffer B. The suspension is disrupted by ultrasonication 6 times for 30 sec at 0–4° in an MSE ultrasonic disintegrator (150 W) and centrifuged at 20,000 g for 40 min, yielding the cell-free extract as supernatant.

Step 2: Streptomycin Sulfate Treatment. To the supernatant 0.1 volume of 10% (w/v) streptomycin sulfate is added with stirring. After 40 min at 4° the precipitate is collected by centrifugation at 20,000 g for 40 min and discarded.

Step 3: Ammoniun Sulfate Fractionation. To the supernatant from Step 2 solid $(NH_4)_2SO_4$ is added with stirring. The protein precipitating between $(NH_4)_2SO_4$ concentrations of 1.6 and 3.2 M is collected by centrifugation at 20,000 g for 40 min and dissolved in buffer B.

Step 4: Sepharose Gel Permeation. The protein solution from Step 3 is applied to a column (1.6 × 80 cm) of Sepharose CL-6B previously equilibrated with buffer B. 6-Phosphogluconate dehydrogenase activity is eluted from the column with the equilibration buffer.

Step 5: Cibacron Blue-Agarose Chromatography. The enzyme from Step 4 is further purified by chromatography on Cibacron Blue-agarose, prepared by the method of Heyns and de Moor[7] using Sepharose CL-6B as the matrix and Cibacron Blue F3-GA as the ligand. The protein solution is applied to a column (1.6 × 15 cm) of Cibacron Blue-agarose previously equilibrated with buffer B. 6-Phosphogluconate dehydrogenase activity is eluted from the column with a gradient of NaCl (0–0.5 M) in the equilibration buffer.

Step 6: Procion Red-Agarose Chromatography. The enzyme from Step 5 is further purified by chromatography on Procion Red-agarose, prepared essentially as a Cibacron Blue-agarose using Procion Red HE-3B as a ligand. The protein solution is applied to a column (1.6 × 10 cm) of Procion Red-agarose previously equilibrated with buffer B. 6-Phosphogluconate dehydrogenase activity is eluted from the column with a gradient of NaCl (0–0.5 M) in the equilibration buffer. Active fractions are combined, diluted with an equal volume of buffer B, and applied to the same column. The enzyme is eluted from the column with buffer B containing 1 M NaCl. Active fractions are combined, dialyzed against buffer B, diluted with an equal volume of glycerol, and stored at −20°. A typical purification is summarized in Table II.

Properties

Purity. Analysis of the enzyme preparation by polyacrylamide gel electrophoresis in the presence of sodium dodecyl sulfate showed one band on staining with Coomassie Brilliant Blue G-250.

[7] W. Heyns and P. de Moor, *Biochim. Biophys. Acta* **358**, 1 (1974).

TABLE II

PURIFICATION OF 6-PHOSPHOGLUCONATE DEHYDROGENASE FROM *Arthrobacter globiformis*

Fraction	Total protein (mg)	Total activity (units)	Specific activity (units/mg protein)	Yield (%)
Step 1. Cell-free extract	629	104	0.17	100
Step 2. Streptomycin sulfate treatment	336	86	0.26	83
Step 3. $(NH_4)_2SO_4$ fraction (1.6–3.2 M)	258	55	0.21	53
Step 4. Pooled fractions after Sepharose CL-6B	108	36	0.34	35
Step 5. Pooled fractions after Cibacron Blue-agarose	2.86	31	11.0	30
Step 6. Pooled, concentrated fractions after Procion Red-agarose	0.65	17	25.7	16

Substrate Specificity. The enzyme is absolutely specific with respect to NADP+. No reaction can be detected in the presence of NAD.+

Effect of Metal Ions. 6-Phosphogluconate dehydrogenase is activated by Mg^{2+}, Mn^{2+}, and Ca^{2+}. Ni^{2+}, Zn^{2+}, Co^{2+}, and Fe^{2+} are inhibitory to the enzyme.

Stability. The purified enzyme stored at 4° in buffer B containing 50% glycerol remains active for at least 3 months.

pH Optimum. The enzyme has maximum activity at pH 7.5.

Michaelis Constants. The K_m for NADP is 35 μM; for 6-phosphogluconate it is 110 μM.

Molecular Weight and Subunit Structure. The molecular weight of 6-phosphogluconate dehydrogenase in buffer B, as measured by gel permeation through a column of Sepharose CL-6B, corresponds to 229,000. During electrophoresis in polyacrylamide gels in the presence of sodium dodecyl sulfate a single band is detected at a position corresponding to a molecular weight of approximately 57,000. This suggests that the enzyme is a tetramer.

Effectors. 6-Phosphogluconate dehydrogenase is inhibited by NADPH, NADH, and, to a lesser extent, by ATP. The enzyme activity remains practically unchanged in the presence of 1 mM oxaloacetate, citrate, malate, 2-oxoglutarate, 3-phosphoglycerate, fructose 1,6-bisphosphate, phosphoenolpyruvate, pyruvate, glyoxylate, ADP, cyclic 3′,5′-AMP, and fructose 6-phosphate.

[52] Glucose-6-phosphate Dehydrogenase from *Pseudomonas* W6

By DIETMAR MIETHE and WOLFGANG BABEL

D-Glucose 6-phosphate + NAD(P)$^+$ ⇌ D-6-phosphogluconate + NAD(P)H + H$^+$

Glucose-6-phosphate dehydrogenase catalyzes the oxidation of glucose 6-phosphate to 6-phosphogluconate in the presence of NAD$^+$ or NADP$^+$. It is involved in the assimilation of formaldehyde via the ketodeoxyphosphogluconate variant of the ribulose monophosphate pathway and in the cyclic oxidation of formaldehyde to CO_2 with ribulose monophosphate-type methylotrophs.[1] The purification and characterization of this enzyme from the obligately methylotrophic *Pseudomonas* W6 have been the subject of a previous report.[2] The purification and properties of glucose-6-phosphate dehydrogenases from *Methylomonas* M15,[3,4] *Pseudomonas* C,[5] *Methylophilus methylotrophus*,[6] and *Methylobacillus flagellatum* KT[7] have been reported.

Assay Method

Principle. Glucose-6-phosphate dehydrogenase is measured spectrophotometrically by following the rate of NADH or NADPH formation at 340 nm in the presence of saturating amounts of glucose 6-phosphate and NAD$^+$ or NADP$^+$.

Reagents

Triethylamine (TEA) buffer, 115 mM, pH 7.6
Glucose 6-phosphate, 27.4 mM
NAD$^+$, 38.5 mM
NADP$^+$, 11.0 mM
MgCl$_2$·6H$_2$O, 55.7 mM
Procedure. The assay mixture contains 1.7 ml TEA buffer, 0.1 ml glucose 6-phosphate, 0.1 ml NAD$^+$ or NADP$^+$, 50 μl MgCl$_2$·6H$_2$O, and 50 μl

[1] C. Anthony, "The Biochemistry of Methylotrophs." Academic Press, London, 1982.
[2] D. Miethe and W. Babel, *Z. Allg. Mikrobiol.* **16**, 289 (1976).
[3] R. A. Steinbach, H. Sahm, and H. Schütte, *Eur. J. Biochem.* **87**, 409 (1978).
[4] R. A. Steinbach, H. Schütte, and H. Sahm, this series, Vol. 89, p. 271.
[5] A. Ben-Bassat, I. Goldberg, and R. I. Mateles, *J. Gen. Microbiol.* **116**, 213 (1980).
[6] A. J. Beardsmore, P. N. G. Aperghis, and J. R. Quayle, *J. Gen. Microbiol.* **128**, 1423 (1982).
[7] M. Y. Kiriuchin, L. V. Kletsova, A. Y. Chistoserdov, and Y. D. Tsygankov, *FEMS Microbiol. Lett.* **52**, 199 (1988).

enzyme solution in a total volume of 2.0 ml. Prior to assay, the reaction mixture is brought to 30°. NADH or NADPH formation is followed at this temperature in a recording spectrophotometer. The reaction is initiated by the addition of glucose 6-phosphate. The rate of absorbance change is followed at 340 nm for 10 min.

Units. One unit of glucose-6-phosphate dehydrogenase activity is defined as that amount of enzyme causing the reduction of 1 μmol of NAD$^+$ or NADP$^+$ per minute. Specific activity is expressed as units per milligram of protein.

Growth of Bacteria

Pseudomonas W6 is grown aerobically at 30° in a mineral medium containing, per liter of doubly distilled water, NH$_4$Cl, 1.0 g; KH$_2$PO$_4$, 0.3 g; MgSO$_4$·7H$_2$O, 0.2 g; CuSO$_4$·5H$_2$O, 24 mg; ZnCl$_2$, 12 mg; CoSO$_4$·7H$_2$O, 3 mg; MnSO$_4$·4H$_2$O, 3 mg; H$_3$PO$_4$, 34 mg; FeCl$_3$·6H$_2$O, 2 mg; and 10 ml of methanol. The cells are inoculated into 100 ml of the medium in 500-ml shake flasks and harvested at the end of logarithmic growth phase (after around 16 hr) by centrifugation at 5000 g and 0–4°. The biomass is washed twice with ice-cold water and stored at this temperature until use.

Purification Procedure

All operations are carried out in the cold room at around 4°.

Step 1: Preparation of Crude Extract. The cell paste is suspended in 50 mM TEA buffer, pH 7.6, at 100 mg dry weight per milliliter. The cells are disrupted by two passages through an Aminco French pressure cell press at 96 MPa. The resulting homogenate is centrifuged at 23,000 g for 20 min at 0° to remove unbroken cells and cell debris. The glucose-6-phosphate dehydrogenase activity is present in the supernatant solution.

Step 2: Protamine Sulfate Treatment. Under mechanical stirring, a 1% solution of protamine sulfate is added dropwise to the crude extract. After an additional 20 min, the precipitate is removed by centrifugation.

Step 3: Ammonium Sulfate Fractionation. To the supernatant from Step 2, 43 g of solid ammonium sulfate per 100 ml (65% saturation) is added. The suspension is centrifuged at 20,000 g after stirring for 30 min at 0°. Another 7 g of ammonium sulfate (75% saturation) is added, and the centrifugation is repeated after stirring for 30 min. The collected precipitate is dissolved in 50 mM TEA buffer, pH 7.6 and dialyzed for 18 hr at 0–4° against a 200-fold volume of the buffer.

Step 4: Chromatography on DEAE-Cellulose. The dialyzed enzyme solution is applied to a DEAE-cellulose column (2.5 × 20 cm) equilibrated with 50 mM TEA buffer, pH 7.6. The column is washed with 500 ml of the same buffer and then with 500 ml of the buffer containing 0.1 M of KCl. Enzyme elution is performed with a linear gradient from 0.1 to 1.0 M KCl in a total volume of 2 liters of TEA buffer. The fractions containing high specific activity are concentrated by ammonium sulfate treatment as described above. The precipitate is dissolved in 50 mM TEA buffer, pH 7.6, and dialyzed for 12 hr at 0–4° against a 200-fold volume of 50 mM TEA buffer. To the dialyzed extract solid ammonium sulfate is added to a final concentration of 1 M. The dialyzate is stored at −20°. A typical purification is summarized in Table I.

Properties

Purity. The described procedure results in an about 80-fold enrichment with a yield of around 10%. The enzyme preparation is not homogeneous. Polyacrylamide gel electrophoresis of the enzyme solution revealed several bands of protein. However, the enzyme preparation is free of activities that might interfere with the assay of glucose-6-phosphate dehydrogenase.

Kinetic Properties. Glucose-6-phosphate dehydrogenase from *Pseudomonas* W6 catalyzes the oxidation of glucose 6-phosphate with NAD$^+$ as well as NADP$^+$ as the electron acceptor. The affinity for NADP$^+$ is approximately 13 times higher than that for NAD$^+$, but the catalytic activity is higher at greater NAD$^+$ concentrations that at saturating NADP$^+$ concentrations. The ratio of maximum reaction rates $V_{max,NAD}/V_{max,NADP}$ is 1.6. Since the ratio of the specific activities of the NAD$^+$ and NADP$^+$-linked reaction does not change in the course of purification, only one enzyme protein appears to be responsible for the catalysis. The K_m values in the presence of 1 mM MgCl$_2$ are 0.24 mM for NAD$^+$ and 0.018 mM for NADP,$^+$ and $s_{0.5}$ values with respect to glucose 6-phosphate are 0.18 and 0.13 mM, respectively.

pH Optimum. The enzyme is active in the pH range between 7 and 9 and exhibits maximum activity at around pH 8.8.

Stability. The purified enzyme shows no loss of activity over a period of 12 months when stored as a solution of 1 M ammonium sulfate at −20°.

Inhibitors. Glucose-6-phosphate dehydrogenase is a key enzyme in the assimilation of glucose. It controls the flux of hexoses to pentoses or to trioses. Therefore, it is plausible that the activity is regulated by various metabolites. In *Pseudomonas* W6 this enzyme is involved in the assimilation and dissimilation of methanol, but it does not immediately control the distribution of the carbon flux because glucose 6-phosphate is not a branch point. Therefore, its regulatory properties are of particular interest, and the

TABLE I

PURIFICATION OF GLUCOSE-6-PHOSPHATE DEHYDROGENASE FROM *Pseudomonas W6*

Fraction	Volume (ml)	Total protein (mg)	Total activity (units)		Specific activity[a]		Yield (%)		Purification (-fold)	
			NAD+	NADP+	NAD+	NADP+	NAD+	NADP+	NAD+	NADP+
Step 1. Crude extract	125	3812	1900	1410	0.5	0.37	100	100	1	1
Step 2. Protamine sulfate	198	2772	1745	1110	0.63	0.4	92	79	1.3	1.1
Step 3. Ammonium sulfate	13.5	97.2	565	438	5.8	4.5	29.8	31	11.6	12.1
Step 4. DEAE-cellulose	12	4.6	182	132	40.1	28.9	10.4	10.7	80	78

[a] Specific activity values are units per milligram protein.

effect of a number of metabolites has been studied. With respect to ATP and NADH inhibition, the *Pseudomonas* W6 glucose-6-phosphate dehydrogenase resembles those of *Hydrogenomonas* H16,[8] *Pseudomonas fluorescens*,[9] and *Rhodopseudomonas sphaeroides*.[10] NADH competitively inhibits only the NAD⁺-linked reaction; the NADP⁺-linked oxidation is not influenced by NADH. NADPH does not inhibit the catalysis at all. ATP (1 mM) causes a 37% inhibition of the NAD⁺-linked glucose-6-phosphate dehydrogenase and a 13% inhibition of the NADP⁺-linked enzyme. Intermediates of carbon metabolism at a concentration of 2 mM hardly decrease the enzyme activity; only acetyl-coenzyme A at a concentration of 1 mM inhibits the activity (by 50%).

[8] F. Blackkolb and H. G. Schlegel, *Arch. Microbiol.* **63**, 177 (1968).
[9] J. Schindler and H. G. Schlegel, *Arch. Microbiol.* **66**, 69 (1969).
[10] E. Ohmann, R. Borriss, and K. R. Rindt, *Z. Allg. Mikrobiol.* **10**, 37 (1970).

[53] Citrate Synthases from Methylotrophs

By GABRIELE MÜLLER-KRAFT and WOLFGANG BABEL

Oxaloacetate + acetyl-CoA $\longrightarrow$ citrate + CoASH

Citrate synthase (EC 4.1.3.7) is the key regulatory enzyme of the amphibolic tricarboxylic acid cycle and is inhibited by energy metabolites.[1-3] In methylotrophic nutrition the energy-supplying function of the tricarboxylic acid cycle is dispensable.

Assay Method

Principle. Citrate synthase activity is measured according to the principle described by Srere *et al.*[4] The CoASH formed reacts with DTNB (5,5′-dithiobis-2-nitrobenzoic acid) giving thionitrobenzoate which adsorbs at 412 nm. Because the citrate synthase is inhibited by DTNB and the oxaloacetate can prevent this inhibition, the reaction is started by the addition of enzyme. Deacylase activity is estimated by omission of oxaloacetate.

Reagents

Tris-HCl buffer, pH 7.6, 0.2 M
DTNB, 18 mM

[1] P. D. J. Weitzman and D. Jones, *Nature (London)* **219**, 270 (1968).
[2] V. R. Flechtner and R. S. Hanson, *Biochim. Biophys. Acta* **222**, 253 (1970).
[3] P. D. J. Weitzman, *Soc. Appl. Bacteriol. Symp. Ser.* **8**, 107 (1980).
[4] P. A. Srere, H. Brazil, and L. Gonen, *Acta Chem. Scand.* (Suppl. 1) **17**, 129 (1963).

Acetyl-CoA, 10 mM

Oxaloacetate, 7.57 mM

Procedure. Activity is determined at 30° in a final volume of 2 ml containing 150 mM Tris-HCl buffer, 0.16 mM DTNB (0.2 ml stock solution), 1 mM acetyl-CoA (20 μl stock solution), 0.38 mM oxaloacetate (0.1 ml stock solution), 0.1 ml of appropriately diluted enzyme solution, and water to 2 ml.

Growth of Bacteria

The organisms *Pseudomonas* MA (ATCC 23819), *Pseudomonas* MS (ATCC 25262), *Methylobacterium* AM1 (ATCC 1418), *Pseudomonas* M27 (NCIB 986), *Methylocystis curvatus* (IBT B385), and *Pseudomonas methylica* and *Arthrobacter globiformis* obtained from Pushchino (USSR) are grown at pH 6.9, (*Acetobacter methanolicus* MB58 at pH 4.5) at 30° in a mineral medium containing, per liter; NH$_4$Cl, 0.76 g; KH$_2$PO$_4$, 68 mg; K$_2$HPO$_4$, 87 mg; CaCl$_2$, 5.5 mg; MgSO$_4 \cdot$7H$_2$O, 15 mg; ZnSO$_4 \cdot$7H$_2$O, 0.44 mg; MnSO$_4 \cdot$4H$_2$O, 0.81 mg; CuSO$_4 \cdot$5H$_2$O, 0.78 mg; Na$_2$MoO$_4 \cdot$2H$_2$O, 0.25 mg; FeSO$_4 \cdot$7H$_2$O, 5 mg.

Cultures are grown in 500-ml Erlenmeyer flasks containing 100 ml of medium. For the cultivation of *A. globiformis* and *Microcyclus aquaticus,* 0.1% (w/v) yeast extract is added to the medium for *Pseudomonas* MA, *Pseudomonas* MS, and *Ac. methanolicus,* 0.02% (w/v) pantothenic acid is added. The following sources of carbon and energy are used: 0.5% methanol (v/v) for *Acetobacter methanolicus, Methylobacterium* AM1, *Pseudomonas* M27, *Ps. methylica,* and *Mc. aquaticus;* 0.5% (v/v) methylamine for *Pseudomonas* MA and *Pseudomonas* MS; 0.5% (v/v) trimethylamine for *A. globiformis;* and methane for *Mtc. curvatus* in special flasks with a methane–air mixture of 1 : 1.5 at 37°.

Purification Procedure

Preparation of Crude Extract. The cells are harvested near the end of the logarithmic phase of growth by centrifugation at 5000 g for 15 min at 0–4° and then washed twice in 50 mM phosphate buffer, pH 7.5. Then cells are disrupted by two passages through an Aminco French pressure cell at 96 MPa. Unbroken cells and cell debris are removed by centrifugation at 23,000 g at 0–4°.

Affinity Chromatography on Sephadex G-200-Linked Cibacron Blue F3G-A. The binding of the dye to the carrier is accomplished by stirring the

TABLE I
BEHAVIOR OF CITRATE SYNTHASE AFTER AFFINITY CHROMATOGRAPHY[a]

| | Nutrition | | | |
| | Chemoorganoheterotrophic | | Methylotrophic | |
Strain	A units (%)	B units (%)	A units (%)	B units (%)
Pseudomonas MS	80	4	40	35
Pseudomonas MA	95	1	33	56
Pseudomonas M 27	75	—	74	—
Methylobacterium AM1	98	—	77	—
Ps. methylica	85	—	80	—
Mtc. curvatus			92	—
A. globiformis	82	—	55	—
Ac. methanolicus	2	75	65	—
Mc. aquaticus	62	—	80	—

[a] Dash indicates no activity. A, Fraction not bound to the carrier; B, fraction eluated with 0.2 M KCl.

components in a water bath at 60° according to Böhme *et al.*[5] Cibacron blue F3G-A (0.4 g) dissolved in 12 ml double-distilled water is added to 2 g Sephadex G-200 in 70 ml double-distilled water. After 30 min of stirring, NaCl (9 g) is added, and stirring is continued for an additional 60 min. After increasing the temperature to 80°, adding 0.8 g Na_2CO_3, stirring 120 min, and decreasing the temperature of the carrier to 20° it is washed with double-distilled water until the eluate is colorless.

Purification of Crude Extract. Dialyzed crude extract is applied to a 0.8×15 cm column equilibrated with 50 mM phosphate buffer, pH 7.5. Nonbound protein is eluted with the same buffer at a flow rate of 1 ml/hr. The elution is completed by washing with 0.2 mol KCl/liter working buffer.

Properties

Homogeneity. Only in a few cases is citrate synthase activity bound to the carrier. As seen in Table I, the citrate synthase of *Pseudomonas* M27, *Methylobacterium* AM1, *Ps. methylica, Mtc. curvatus, A. globiformis,* and *Mc. aquaticus* is not bound. This is also true for a part of the citrate

[5] H. J. Böhme, G. Kopperschläger, J. Schulz, and E. Hofmann, *J. Chromatogr.* **69**, 209 (1972).

TABLE II
INFLUENCE OF EFFECTORS (2 mM) ON THE ACTIVITY OF CITRATE SYNTHASE AFTER AFFINITY CHROMATOGRAPHY[a]

Strain	Fraction	Methylotrophic nutrition							Chemoorganoheterotrophic nutrition						
		ATP	AMP	NADH	NAD⁺	NADPH	NADP⁺	α-KGl	ATP	AMP	NADH	NAD⁺	NADPH	NADP⁺	α-KGl
Pseudomonas MS	A	84	100	65	100	100	100	n.e.	90	100	76	100	110	83	100
	B	71	n.e.	83	n.e.	n.e.	n.e.	n.e.	100	100	90	95	113	84	97
Pseudomonas MA	A	76	n.e.	45	n.e.	n.e.	n.e.	n.e.	84	85	65	90	110	98	n.e.
	B	86	n.e.	91	n.e.	n.e.	n.e.	n.e.	96	n.e.	88	n.e.	n.e.	n.e.	n.e.
Pseudomonas M 27	A	90	90	53	n.e.	107	93	93	91	89	54	91	104	87	93
Methylobacterium AM 1	A	81	92	54	100	100	100	97	86	80	65	n.e.	100	92	91
Ps. methylica	A	91	91	31	97	111	103	103	89	87	38	94	112	80	99
Mtc. curvatus	A	81	n.e.	n.e.	n.e.	n.e.	n.e.	n.e.	n.e.	n.e.	n.e.	n.e.	n.e.	n.e.	n.e.
A. globiformis	A	160	180	160	182	80	150	118	131	140	120	180	95	141	128
Ac. methanolicus	A	73	94	100	n.e.	75	n.e.	n.e.	79	100	100	87	74	84	100
	B	54	91	100	110	85	92	116	45	n.e.	100	100	n.e.	n.e.	n.e.
Mc. aquaticus	A	91	91	79	91	106	85	96	98	110	88	98	110	98	98

[a] n.e., not estimated. Data are given as the percentage of citrate synthase activity in the absence of effector.

synthase of *Pseudomonas* MA, *Pseudomonas* MS, and *Ac. methanolicus* MB58. But for these bacteria additional citrate synthase activity is eluated with 0.2 *M* KCl. The portion of the citrate synthase bound to the carrier depends on cultivation. In the "ICl⁺ serine pathway bacteria" *Pseudomonas* MS and *Pseudomonas* MA the nonbound part of citrate synthase (A) is more abundant after chemoorganoheterotrophic nutrition (80–95%) than after methylotrophic (30–40%), and the bound part (B) after chemoorganoheterotrophic nutrition is smaller (5%) when compared with methylotrophic nutrition (35–56%). The citrate synthase of *Ac. methanolicus* behaves in a reverse manner. After affinity chromatography the citrate synthase fractions are free of deacylase and malate dehydrogenase which allows for the investigation of the influence of NADH on citrate synthase activity.

Effectors. The influence of effectors on the activity of citrate synthase after affinity chromatography is estimated in concentrated fractions; the citrate synthase eluted with KCl-containing buffer is first dialyzed against 50 m*M* phosphate buffer, pH 7.5.

The activity of citrate synthase in the presence of different effectors at a concentration of 2 m*M* is summarized in Table II. Excluding *A. globiformis*, AMP, NAD⁺, NADP⁺, and α-ketoglutarate show no influence on citrate synthase activity. The citrate synthase of fraction A from bacteria with the ICl⁺ serine pathway and the citrate synthase of all other facultatively methylotrophic microorganisms with the exception of *A. globiformis* and *Ac. methanolicus* is inhibited by NADH. The bacteria with the ICl⁺ serine pathway possess an additional citrate synthase which is not inhibited by NADH or by other effectors. The methylotrophic acetic acid bacterium *Ac. methanolicus* also possesses a citrate synthase which is not inhibited by effectors (fraction A). Remarkably, the citrate synthase fraction B of *Ac. methanolicus* is inhibited by ATP. The citrate synthase of *A. globiformis* is activated by the effectors tested, namely, ATP, AMP, NAD⁺, NADP⁺, and α-ketoglutarate.

[54] Dichloromethane Dehalogenase from *Hyphomicrobium* DM2

By THOMAS LEISINGER and DORIS KOHLER-STAUB

Dichloromethane + $H_2O \longrightarrow$ formaldehyde + 2 HCl

The first step in the utilization of dichloromethane as a carbon and energy source by methylotrophic bacteria is a hydrolytic dehalogenation yielding formaldehyde and inorganic chloride.[1] This reaction is catalyzed by dichloromethane dehalogenase, a strongly inducible enzyme, and is dependent on reduced glutathione (GSH) as a cofactor. By analogy with the GSH-dependent metabolism of dichloromethane in rat liver cytosol,[2] the formation of formaldehyde is thought to proceed through the following intermediates:

$$CH_2Cl_2 \xrightarrow[\text{HCl}]{\text{GSH}} [GS-CH_2Cl] \xrightarrow[\text{HCl}]{H_2O} GS-CH_2OH \rightleftharpoons CH_2O + GSH$$

Dichloromethane dehalogenase has been purified from *Hyphomicrobium* strain DM2[3] as well as from four other facultative methylotrophic bacteria.[4,5] In dichloromethane-grown cells of these organisms the enzyme represents 8 to 20% of the total soluble protein. The nucleotide sequence of the dichloromethane dehalogenase structural gene from *Methylobacterium* strain DM4 has been established, and analysis indicated that the dichloromethane dehalogenase structural gene is a member of the glutathione *S*-transferase gene family.[6]

Assay Method

Principle. The assay of dichloromethane dehalogenase is based on either the discontinuous or the continuous measurement of formaldehyde production from dichloromethane. The discontinuous colorimetric mea-

[1] G. Stucki, R. Gälli, H.-R. Ebersold, and T. Leisinger, *Arch. Microbiol.* **130**, 366 (1981).
[2] A. E. Ahmed and M. W. Anders, *Biochem. Pharmacol.* **27**, 2021 (1978).
[3] D. Kohler-Staub and T. Leisinger, *J. Bacteriol.* **162**, 676 (1985).
[4] D. Kohler-Staub, S. Hartmans, R. Gälli, F. Suter, and T. Leisinger, *J. Gen. Microbiol.* **132**, 2843 (1986).
[5] R. Scholtz, L. P. Wackett, C. Egli, A. M. Cook, and T. Leisinger, *J. Bacteriol.* **170**, 5698 (1988).
[6] S. La Roche and T. Leisinger, *J. Bacteriol.* **172**, 164 (1990).

surement of formaldehyde formation is a convenient routine assay for measuring enzyme activity in crude extracts and column eluates. With a detection limit for formaldehyde of 50 nM it is relatively insensitive. Increased sensitivity is provided by an enzymatic assay involving the continuous measurement of the rate of formaldehyde formation. Commercially available NAD$^+$-dependent formaldehyde dehydrogenase is used as a coupling enzyme, and a detection limit of 1 nM formaldehyde is attained. The coupled assay is suitable for determining the kinetic properties of dichloromethane dehalogenase.

Reagents for Colorimetric Assay

200 mM Tris sulfate, pH 8.2
100 mM reduced glutathione
100 mM dichloromethane

Procedure for Colorimetric Assay. The reaction mixture contains 1 ml Tris buffer, 0.1 ml reduced glutathione, and up to 0.2 units of dichloromethane dehalogenase. The volume is made to 1.9 ml, and the reaction is started with 0.1 ml dichloromethane. Wheaton tubes of 4.0 ml total volume closed with gas-tight mininert stoppers (Precision Sampling Corp., Baton Rouge, LA) to prevent volatilization of dichloromethane are used for incubating the mixture at 30°. Timed samples of 0.1 ml are removed with a gas-tight syringe over a period up to 30 min. The samples are mixed with 4.9 ml distilled water to which 0.5 ml iodine solution (40 mM I$_2$ in acetone) is immediately added. The concentration of formaldehyde in the diluted samples is determined colorimetrically by the method of Nash[7] as modified by Sawicki and Sawicki.[8]

Reagents for Coupled Enzymatic Assay

100 mM potassium phosphate, pH 8.2
100 mM reduced glutathione
10 mM NAD$^+$
Formaldehyde dehydrogenase (EC 1.2.1.1; from *Pseudomonas putida;* Sigma Chemical Co., St. Louis, MO)
50 mM dichloromethane

Procedure for Coupled Enzymatic Assay. The assay mixture in a cuvette (10-mm light path) contains 250 μl potassium phosphate buffer, 25 μl reduced glutathione, 25 μl NAD$^+$, 45 units formaldehyde dehydrogenase, and extract containing approximately 0.01 units dichloromethane dehalo-

---

[7] T. Nash, *Biochem. J.* **55**, 416 (1953).
[8] E. Sawicki and C. R. Sawicki, "Aldehydes: Photometric Analysis," Vol. 5, p. 153. Academic Press, New York, 1978.

genase. Water is added to bring the volume to 0.45 ml, and the mixture is preincubated for several minutes at 30°. Fifty microliters dichloromethane is added, the cuvette is tightly closed with a Teflon stopper, and the rate of absorbance increase is followed at 339 nm. Blank reactions with assay mixtures lacking either dichloromethane dehalogenase or formaldehyde dehydrogenase, or both, are not significant.

Units. One unit of dichloromethane dehalogenase is the quantity of enzyme catalyzing the formation of 1 μmol of formaldehyde per minute at 30°. Specific activity is expressed as units per milligram of protein.

Growth of Organism

Hyphomicrobium strain DM2 (ATCC 43129) is grown at 30° in continuous culture under conditions of carbon limitation. A 2-liter fermentor with a working volume of 1.7 liters is run at a dilution rate of 0.03 hr^{-1}, and the pH is maintained at 7.0 by a pH controller using 3.0 M NaOH as the neutralizing agent. Air is supplied at a rate of 0.1 liter/min. The minimal medium contains, per liter, KH_2PO_4, 1.36 g; $(NH_4)_2SO_4$, 1.0 g; and $MgSO_4 \cdot 7H_2O$, 1.0 g. After sterilization the medium is supplemented with 1 ml of a trace element solution,[1] and dichloromethane is added to give a final concentration of 120 mM in the storage vessel. It should be noted that dichloromethane concentrations above 20 mM are growth inhibitory for dichloromethane-utilizing bacteria. The effluent from the chemostat over a 24-hr period is collected in an ice-cooled flask. Cells are harvested by centrifugation, washed once with 50 mM Tris hydrochloride (pH 7.5), and stored at −20°. The cell yield is about 5 g wet weight per liter.

An alternative to continuous culture for obtaining large quantities of cells is provided by batch cultures of the organism on a rotary shaker at 120 rpm. Five-liter Erlenmeyer flasks tightly closed with rubber stoppers and containing 1 liter of medium are used. A total of 24 mmol dichloromethane per flask is added sequentially in three portions of 8 mmol, each after the pH has been adjusted with 1 M NaOH to 7.15. The minimal medium used for batch cultures is more strongly buffered and contains, per liter, KH_2PO_4, 2.0 g; $Na_2HPO_4 \cdot 2H_2O$, 4.0 g; $(NH_4)_2SO_4$, 1.0 g; $MgSO_4 \cdot 7H_2O$, 1.0 g; plus 1 ml of trace element solution.

Purification of Enzyme

Buffers

Buffer A: 50 mM Tris hydrochloride, pH 7.5, 5 mM dithiothreitol, 0.1 mM ethylenediaminetetraacetic acid (EDTA), 25% (v/v) glycerol

Buffer B: 25 mM Tris hydrochloride, pH 8.2, 2.5 mM dithiothreitol, 0.1 mM EDTA, 25% (v/v) glycerol

Buffer C: 1 M potassium phosphate, pH 7.5, 2.5 mM dithiothreitol, 0.1 mM EDTA

Buffer D: 100 mM potassium phosphate, pH 7.5, 2.5 mM dithiothreitol, 0.1 mM EDTA

Buffer E: 50 mM potassium phosphate, pH 7.5, 2.5 mM dithiothreitol, 0.1 mM EDTA, 25% (v/v) glycerol

Preparation of Cell-Free Extract. All procedures are performed at 0–5°. Frozen wet cells are thawed in 200 ml buffer A and disrupted by passing 40-ml portions of the suspension 3 times through a French pressure cell at 35 MPa. The suspension of broken cells is centrifuged at 30,000 *g* for 30 min, yielding 200 ml cell-free extract as supernatant.

Protamine Sulfate Treatment. Forty-four milliliters of a 2% (w/v) protamine sulfate solution is added slowly to the cell-free extract. After stirring for 30 min the precipitate is removed by centrifugation at 30,000 *g* for 30 min.

DEAE-Cellulose Chromatography. The supernatant is diluted with 1 volume of a 25% (v/v) glycerol solution containing 0.1 mM EDTA, and the pH is adjusted to 8.2 by the addition of KOH. The protein solution is applied to a column of DEAE-cellulose (Whatman DE-52), 5 × 60 cm, equilibrated with 1 volume of buffer B. After washing with 1 volume of buffer B, dichloromethane dehalogenase is eluted with a linear gradient (4 liters) of 0 to 100 mM potassium phosphate in buffer B at a flow rate of 250 ml/hr. Active fractions are pooled. To remove glycerol, which interferes with the next purification step, the protein solution (560 ml) is diluted with 1 volume of 2 M potassium phosphate, pH 7.5, containing 2.5 mM dithiothreitol and 0.1 mM EDTA. The volume of the solution is reduced one-fifth in a hollow-fiber concentrator (Amicon Corp., Lexington, MA), and the preparation is diluted 5-fold in buffer C and concentrated again.

N-Pentyl-Sepharose Chromatography. N-Pentyl-Sepharose is prepared[9] by linking 1-aminopentane covalently to cyanogen bromide-activated Sepharose 4B (Pharmacia Fine Chemicals, Uppsala, Sweden). Enzyme solution from the previous step (265 ml) is applied to a column of N-pentyl-Sepharose, 1.6 × 40 cm, equilibrated with buffer C. After washing the column with 700 ml buffer C, dichloromethane dehalogenase is eluted at a flow rate of 80 ml/hr with buffer D. Fractions containing enzyme activity are pooled and concentrated by ultrafiltration with an Amicon PM10 membrane to 60 ml. The concentrate is desalted by passage through columns of Sephadex G-25M, equilibrated with buffer E.

[9] S. Shaltiel, this series, Vol. 34, p. 126.

Hydroxylapatite Chromatography. The desalted sample is added to a hydroxylapatite column, 1.6×60 cm, equilibrated with buffer E. The column is washed with 250 ml of buffer E, and the enzyme is eluted at 40 ml/hr by a linear gradient (1.0 liter) from 50 to 300 mM potassium phosphate in buffer E. Fractions containing enzyme activity are pooled, concentrated by ultrafiltration to 170 ml, and stored in portions at $-20°$. Some preparations of purified dichloromethane dehalogenase contain minor contaminants, probably degradation products of the enzyme. These are removed by precipitation in $(NH_4)_2SO_4$ of 55% saturation. A typical purification is summarized in Table I.

Comments on Purification Procedure

The purification procedure described is suited for the preparation of a large quantity of more than 98% pure dichloromethane dehalogenase from *Hyphomicrobium* DM2. The same procedure, without column chromatography on N-pentyl-Sepharose, can be used for the preparation of approximately 96% pure dichloromethane dehalogenases from *Hyphomicrobium* DM2, *Hyphomicrobium* GJ21, *Pseudomonas* DM1, and *Methylobacterium* DM4.[4] Dichloromethane dehalogenase from the unidentified methylotrophic bacterium strain DM11 differs in its chromatographic properties from the dichloromethane dehalogenases of the other methylotrophs. On a small scale it is conveniently purified to electrophoretic homogeneity in a one-step fast-protein liquid chromatography procedure.[5] Cell-free extract (2.3 ml, 20 mg protein/ml) prepared in 50 mM Tris sulfate, pH 7.0, containing 25% (v/v) glycerol (buffer F) is loaded on a Mono Q HR16/10 column (Pharmacia). The column is eluted at 6 ml/min with buffer F for 20 min and then with a linear gradient of 0 to 0.4 M KCl in buffer F for 67 min. The chromatography is performed at $25°$.

TABLE I
PURIFICATION OF DICHLOROMETHANE DEHALOGENASE FROM *Hyphomicrobium* DM2

Step	Total volume (ml)	Total protein (mg)	Total activity (units)	Specific activity (units/mg protein)	Yield (%)
Cell-free extract	220	5770	1194.0	0.207	100
Protamine sulfate-treated extract	275	4140	1278.0	0.309	112
DEAE-cellulose pool	560	1540	1101.6	0.715	96
N-Pentyl-Sepharose pool	183	970	819.6	0.845	72
Hydroxylapatite eluate	170	690	718.2	1.041	63

TABLE II
APPARENT K_m VALUES FOR DIHALOMETHANES
AND GLUTATHIONE

Substrate or cofactor	Apparent K_m (μM)	V_{max} (units/mg protein)
CH_2Cl_2	30	1.014
CH_2BrCl	15	0.918
CH_2Br_2	13	0.948
CH_2I_2	3	0.246
Glutathione	320	—

Properties

Substrates. Dichloromethane dehalogenase is highly specific for dihalomethanes. The apparent K_m values for the four dihalomethanes tested are given in Table II. 1,2-Dichloroethane is dehalogenated at a 1,000 times lower rate than dichloromethane; 1,1-dichloroethane, 1,1,1-trichloroethane, 1,1,2-trichloroethane, tetrachloroethane, 1,1-dichloroethene, trichloroethene, tetrachloroethene, 1-chloropropane, and chlorobenzene are not substrates for the enzyme. Sulfhydryl compounds such as 2-mercaptoethanol, dithiothreitol, or cysteine cannot substitute for reduced glutathione as a cofactor.

Inhibitors. The enzyme is strongly inhibited by biterminally monochlorinated alkanes and moderately by monoterminally monochlorinated alkanes. The extent of inhibition decreases as the chain length of the inhibiting chloroalkane increases. It appears that the inhibition is of the competitive type. The K_i values for 1,2-dichloroethane and 1-chloropropane are 3 and 56 μM, respectively.

pH Optimum. The enzyme has optimum activity at pH 8.5.

Stability. The pure enzyme, stored at a concentration of 4 mg/ml in 100 mM potassium phosphate, pH 7.5, 2.5 mM dithiothreitol, 0.1 mM EDTA, 25% (v/v) glycerol, is stable for 6 months at $-20°$. Fifty percent of the activity is lost within 6 days at room temperature.

Structure. The molecular weight of dichloromethane dehalogenase under native conditions, as measured by gel filtration on Sephadex G-200, corresponds to $195,000 \pm 10,000$. Sodium dodecyl sulfate (SDS)–polyacrylamide gel electrophoresis[10] of pure dichloromethane dehalogen-

---

[10] U. Laemmli, *Nature (London)* **227**, 680 (1970).

ase gives a single band corresponding to a molecular weight of 33,000 ± 1,000. This suggests that the enzyme is a hexamer composed of identical subunits. Subunit cross-linking with dimethyl suberimidate[11] and subsequent SDS–polyacrylamide gel electrophoresis confirms the hexameric quaternary structure.

[11] G. E. Davies and G. R. Stark, *Proc. Natl. Acad. Sci. U.S.A.* **66**, 651 (1970).

[55] Assay of Assimilatory Enzymes in Crude Extracts of Serine Pathway Methylotrophs

By Patricia M. Goodwin

It is often necessary to assay key assimilatory enzymes when characterizing C_1 negative mutants of serine pathway methylotrophs. Assays for serine hydroxymethyltransferase and hydroxypyruvate reductase are given elsewhere in this volume.[1] In this chapter methods for the detection of serine–glyoxylate aminotransferase, glycerate kinase, acetyl-CoA-independent phosphoenolpyruvate carboxylase, and malyl-CoA lyase in crude extracts are described.

Preparation of Crude Extracts of Bacteria

The crude extract is preferably prepared from bacteria grown to midexponential phase on a reduced C_1 compound. Cells are harvested, washed in cold 20 mM phosphate buffer, pH 7.0, and resuspended in the same buffer to approximately 0.5 g wet weight/ml. After cell breakage (sonication for 15 times, 30 sec each, at an amplitude of 12 μm, with 30-sec cooling periods, is suitable for the pink-pigmented facultative methylotrophs) the suspension is centrifuged at 90,000 g for 1 hr to remove NADH oxidase activity. In several facultative methylotrophs the serine pathway enzymes are inducible, and some are repressed by multicarbon compounds such as succinate.[2] Therefore, if mutants which are unable to utilize any C_1 compound as the sole carbon source are under investigation, they are grown on a permissive carbon source, and methanol (or another suitable C_1 substrate) is included in the growth medium. The bacteria are harvested aseptically and incubated overnight in medium containing methanol to ensure maximum induction of the enzymes under investigation.[3]

[1] M. E. Lidstrom, this volume [56]; C. Krema and M. E. Lidstrom, this volume [57].
[2] T. McNerney and M. L. O'Connor, *Appl. Environ. Microbiol.* **40**, 370 (1980).
[3] S. Stone and P. M. Goodwin, *J. Gen. Microbiol.* **135**, 227 (1989).

Assay Methods

Serine–Glyoxylate Aminotransferase (EC 2.6.1.45)

$$\text{Serine} + \text{glyoxylate} \rightleftharpoons \text{hydroxypyruvate} + \text{glycine} \qquad (1)$$

Principle. Serine-glyoxylate aminotransferase is measured using a coupled assay which depends on the presence of excess hydroxypyruvate reductase in the extract.[4] The glyoxylate-dependent rate of formation of hydroxypyruvate from serine is determined by estimating the rate of oxidation of NADH resulting from the subsequent reduction of hydroxypyruvate to glycerate. In wild-type bacteria the activity of hydroxypyruvate reductase is usually considerably greater than that of serine–glyoxylate aminotransferase, but if a mutant deficient in the reductase is under investigation exogenous enzyme must be added to the crude extract. A suitable source is a crude extract of *Paracoccus denitrificans* grown on succinate, which contains an NADPH-linked hydroxypyruvate reductase.[4] In this case NADPH is substituted for NADH in the assay.

Reagents

Potassium phosphate buffer, pH 7.1, 500 mM
Pyridoxal phosphate, 1 mM
Sodium glyoxylate, 50 mM
L-Serine, 50 mM
NADH, 1.5 mM

Procedure. The assay mixture contains, in a final volume of 1 ml, phosphate buffer, 0.1 ml; pyridoxal phosphate, 10 μl; NADH, 0.1 ml; sodium glyoxylate, 0.1 ml; and varying volumes of extract and water. The endogenous rate of NADH oxidation at 30° is measured at 340 nm against a blank containing all the reagents except NADH. The reaction is then started by the addition of 0.1 ml of L-serine. The rate of NADH oxidation is measured and corrected for the endogenous rate.

Units. One unit of activity is defined as that amount of enzyme which catalyzes the oxidation of 1 μmol of NADH in 1 min at 30°. Specific activity is expressed as units per milligram protein.

Glycerate Kinase (EC 2.7.1.31)

$$\text{Glycerate} + \text{ATP} \rightarrow \text{2-phosphoglycerate} + \text{ADP} \qquad (2)$$

Principle. Glycerate kinase activity is measured by estimating the rate of ADP formation from glycerate and ATP using a coupled assay. Excess

[4] M. A. Blackmore and J. R. Quayle, *Biochem. J.* **118**, 53 (1970).

pyruvate kinase, lactate dehydrogenase, and phosphoenolpyruvate are added, and the rate of ADP formation is estimated by measuring the rate of NADH oxidation, using the method of Heptinstall and Quayle[5] as modified by Harder et al.[6]

Reagents

Tris-HCl, pH 7.5, 500 mM
EDTA, pH 7.3, 100 mM
NADH, 1 mM
Tetrasodium ATP, 50 mM
MgCl$_2$, 1 M
Sodium glycerate, approximately 10 mM
Phosphoenolpyruvate, 250 mM
Pyruvate kinase, 7 units
Lactate dehydrogenase, 10 units (a suitable combined preparation of pyruvate kinase and lactate dehydrogenase is available commercially)

Sodium glycerate is made from calcium glycerate as follows: Dowex 50W (H$^+$ form) is added to a solution of 100 mM calcium DL-glycerate in water. The slurry is stirred and the pH monitored until it has reached about 2.3 and there is no further change. The solution of sodium glycerate is then decanted off and neutralized with NaOH. It can be stored at $-20°$ for at least 1 month and diluted as required.

Procedure. The assay mixture contains, in a final volume of 1 ml, Tris-HCl buffer, 0.1 ml; EDTA, 10 μl; NADH, 0.1 ml; ATP, 10 μl; phosphoenolpyruvate, 10 μl; 7 units pyruvate kinase; 10 units lactate dehydrogenase; MgCl$_2$, 10 μl; and varying volumes of extract and water. The endogenous rate of NADH oxidation at 30° is measured at 340 nm against a blank containing all the assay components except NADH. The reaction is then started by the addition of 0.1 ml of sodium glycerate.

Units. One unit is defined as that amount of enzyme which catalyzes the oxidation of 1 μmol of NADH in 1 min at 30°. Specific activity is expressed as units per milligram protein.

Acetyl-CoA-Independent Phosphoenolpyruvate Carboxylase (EC 4.1.1.31)

$$\text{Phosphoenolpyruvate} + CO_2 \rightarrow \text{oxaloacetate} \qquad (3)$$

Principle. The rate of formation of oxaloacetic acid is measured using a

[5] J. Heptinstall and J. R. Quayle, *Biochem. J.* **117**, 563 (1970).
[6] W. Harder, M. M. Attwood, and J. R. Quayle, *J. Gen. Microbiol.* **78**, 155 (1973).

coupled assay; excess malate dehydrogenase is added, and the rate of oxidation of NADH is estimated.[7]

Reagents

Tris-HCl buffer, pH 8.5, 400 mM
Sodium bicarbonate, 200 mM
NADH, 2 mM
MgCl$_2$, 20 mM
Phosphoenolpyruvate, 20 mM
Malate dehydrogenase from bovine heart suspension (if this is stored in ammonium sulfate it must be dialyzed before use as NH$_4^+$ inhibits the reaction; it may not be necessary to add exogenous enzyme to crude extracts as these often have excess malate dehydrogenase activity)

Procedure. The assay mixture contains, in a final volume of 1 ml, Tris-HCl buffer, 0.1 ml; sodium bicarbonate, 0.1 ml; NADH, 0.1 ml; MgCl$_2$, 0.1 ml; 100 units malate dehydrogenase; and varying volumes of extract and water. The rate of endogenous oxidation of NADH at 30° is measured at 340 nm against a blank containing all the assay components except NADH, then the reaction is started by the addition of 50 μl phosphoenol pyruvate.

Units. One unit is defined as that amount of enzyme which catalyzes the oxidation of 1 μmol of NADH in 1 min at 30°. Specific activity is expressed as units per milligram protein.

Malyl-CoA Lyase

Malyl-CoA lyase activity[8] cannot easily be measured since the substrate, malyl-CoA, is not commerically available. However, the presence of this enzyme in crude extracts of the pink-pigmented facultative methylotrophs can be ascertained by measuring the apparent malate synthase activity [Eq. (6)],[3] which is due to the concerted action of malyl-CoA lyase [Eq. (4)] and malyl-CoA hydrolase [Eq. (5)].[9]

$$\text{Acetyl-CoA} + \text{glyoxylate} \rightleftharpoons \text{malyl-CoA} \qquad (4)$$

$$\text{Malyl-CoA} + \text{H}_2\text{O} \rightarrow \text{malate} + \text{CoA} \qquad (5)$$

$$\text{Acteyl-CoA} + \text{glyoxylate} \rightarrow \text{malate} + \text{CoA} \qquad (6)$$

Principle. The apparent malate synthase activity is measured by deter-

[7] P. J. Large, D. Peel, and J. R. Quayle, *Biochem. J.* **85**, 243 (1962).
[8] See A. J. Hacking and J. R. Quayle, this volume [58].
[9] R. B. Cox and J. R. Quayle, *J. Gen. Microbiol.* **95**, 121 (1976).

mining the glyoxylate-dependent rate of disappearance of acetyl-CoA, using a modification of the method of Dixon and Kornberg.[10]

Reagents

Tris-HCl, pH 8.0, 200 mM
MgCl$_2$, 100 mM
Acetyl-CoA, 2 mM
Sodium glyoxylate, 20 mM

Procedure. The assay mixture contains, in a final volume of 0.9 ml, Tris-HCl, 0.4 ml; MgCl$_2$, 30 μl; acetyl-CoA, 20 μl; and varying volumes of extract and water. The endogenous rate of disappearance of acetyl-CoA is measured at 232 nm at 30°, against a blank containing all the assay components except acetyl-CoA. Then 20 μl sodium glyoxylate is added to start the reaction. The rate of cleavage of acetyl-CoA is linear for only about 2 min after the addition of glyoxylate.

Units. One unit is defined as the activity (due to the concerted activity of the two enzymes) which catalyzes the cleavage of 1 μmol acetyl-CoA in 1 min. A ΔE_{232} value of 0.1 min^{-1} is equivalent to 9.35 milliunits.

[10] G. H. Dixon and H. L. Kornberg, this series, Vol. 5, p. 633.

[56] Serine Hydroxymethyltransferases from *Methylobacterium organophilum* XX

By Mary E. Lidstrom

Glycine + N^5,N^{10}-methylene tetrahydrofolate $\rightleftharpoons$ serine + tetrahydrofolate

Serine hydroxymethyltransferase (serine transhydroxymethylase, EC 2.1.2.1) catalyzes the interconversion of serine and glycine using the C$_1$ carrier tetrahydrofolate. In most organisms, the physiological role of this enzyme is to synthesize glycine from serine and generate methylene tetrahydrofolate for the C$_1$ pool.[1] However, in methylotrophic bacteria containing the serine cycle for formaldehyde assimilation, during growth on C$_1$ compounds the enzyme functions to generate serine from glycine and methylene tetrahydrofolate, and it is the enzyme that incorporates C$_1$ units derived from the growth substrate into the serine cycle.[2] *Methylobacterium* strains contain two different serine hydroxymethyltransferases, one induced during growth on C$_1$ compounds, involved in the serine cycle, and

[1] S. H. Mudd and G. J. Cantoni, *in* "Comprehensive Biochemistry" (M. Forlein and E. H. Stolz, eds.), p. 1. Elsevier, Amsterdam, 1964.
[2] C. Anthony, "The Biochemistry of Methylotrophs." Academic Press, London, 1982.

the other present during growth on multicarbon compounds, apparently serving the more common heterotrophic function.[3,4] The C_1-specific enzyme is strongly activated by glyoxylate, whereas the heterotrophic enzyme is largely unaffected. This difference can be used to distinguish between the two activities in both crude cell extracts and purified preparations.[3] Both enzymes have been purified from *Methylobacterium organophilum* XX,[3] and the methanol-inducible enzyme has been purified from *Hyphomicrobium methylovorum*.[5] This chapter describes the purification and characteristics of the two *M. organophilum* XX enzymes.

Assay Methods

Principle. Serine hydroxymethyltransferase can be assayed in the direction of glycine synthesis with a continuous or a discontinuous assay. The continuous assay involves coupling the methylene tetrahydrofolate produced to NADP+ reduction using commercially available methylenetetrahydrofolate dehydrogenase.[6] The reduction of NADP+ is monitored spectrophotometrically. The discontinuous assay involves incubating enzyme with specifically radiolabeled serine and then trapping and counting the radiolabeled C_1 unit derived from the serine. This is accomplished via the nonenzymatic exchange reaction between formaldehyde and methylene tetrahydrofolate, which occurs rapidly under the conditions used. Nonlabeled formaldehyde is used to chase the labeled C_1 unit out of the methylene tetrahydrofolate, and the formaldehyde is trapped using dimedon (5,5-dimethyl-1,3-cyclohexadione). The C_1–dimedon complex can then be extracted into toluene and counted in a scintillation counter.

Although the continuous assay is more convenient, it cannot be used to distinguish between the two enzymes because methylenetetrahydrofolate dehydrogenase is inhibited by glyoxylate. The discontinuous assay must be used in this case, and for all studies involving effectors. Another disadvantage of the continuous assay is that it is carried out at suboptimal pH, owing to the lower pH optimum for the coupling enzyme, methylenetetrahydrofolate dehydrogenase. Therefore, it is not useful for kinetic measurements. A more convenient variant of the discontinuous assay has been described recently which involves binding the radiolabeled methylene tetrahydrofolate to DEAE-cellulose paper.[7] Serine hydroxymethyltransferase can be measured in the direction of serine synthesis by measuring glycine-

[3] M. L. O'Connor and R. S. Hanson, *J. Bacteriol.* **124**, 985 (1975).
[4] T. McNerney and M. L. O'Connor, *Appl. Environ. Microbiol.* **40**, 370 (1980).
[5] S. S. Miyazaki, S. Toki, Y. Izumi, and H. Yamada, *Eur. J. Biochem.* **148**, 786 (1986).
[6] J. Heptinstall and J. R. Quayle, *Biochem. J.* **117**, 563 (1970).
[7] A. M. Geller and M. Y. Kotb, *Anal. Biochem.* **180**, 120 (1989).

dependent formaldehyde disappearance,[5] an assay that also relies on the nonenzymatic exchange reaction between methylene tetrahydrofolate and formaldehyde. The first two assays noted above are described here.

Procedure for Continuous Assay

Reagents

0.5 M KPO$_4$ buffer, pH 7.5
0.1 M NADP$^+$
2 mM pyridoxal phosphate
5 mM tetrahydrofolate in 0.1 M 2-mercaptoethanol, stored anaerobically under N$_2$ or Ar
5 units methylenetetrahydrofolate dehydrogenase (may not be necessary in crude cell extracts)
1.0 M L-serine

Assay. One-tenth milliliter of buffer, 10 μl NADP$^+$, 10 μl pyridoxal phosphate, 50 μl tetrahydrofolate, the methylenetetrahydrofolate dehydrogenase, and the enzyme sample are added to a cuvette containing water to make a total volume of 0.98 ml. The reaction is started with 20 μl L-serine, and the production of NADPH is followed at 340 nm. Assays are routinely run at 30°.

Procedure for Discontinuous Assay

Reagents

0.1 M bicine buffer, pH 8.5
2 mM pyridoxal phosphate
0.1 M 2-mercaptoethanol
80 mM tetrahydrofolate in 0.1 M 2-mercaptoethanol, stored anaerobically under N$_2$
DL-[3-^{14}C]Serine (Amersham, 55 mCi/mmol), made to 10 mM with DL-serine and diluted to 10^5 dpm/μmol (5×10^4 dpm/50 μl)
1.0 M sodium acetate buffer, pH 4.5
0.1 M formaldehyde (made from paraformaldehyde by heating)[8]
0.4 M dimedon (5,5-dimethyl-1,3-cyclohexadione) in 50% (v/v) ethanol
Toluene

Assay. To a test tube is added 0.25 ml bicine buffer, 50 μl pyridoxal phosphate, 50 μl 2-mercaptoethanol, 10 μl tetrahydrofolate, and 25–50 μl enzyme sample. The test tube is preincubated at 30° for 5 min, then 25 μl

---

[8] M. M. Attwood, this volume [47].

[^{14}C]serine is added to start the reaction. The tube is incubated at 30° for 15 min, then the reaction is stopped with 0.5 ml sodium acetate buffer. Two-tenths milliliter formaldehyde and 0.3 ml dimedon are added immediately, and the tubes are mixed. The tubes are then heated for 5 min in a boiling water bath and cooled for 5 min on ice. Five milliliters toluene is added, the tubes are stoppered and vortexed at maximum speed for 30 sec, and the tubes are then centrifuged at 5000 g for 2 min. Three milliliters of the upper phase is removed and counted in a water-absorbing cocktail, such as Aquasol-II. Controls are carried out with no enzyme and with mixtures to which the sodium acetate buffer is added before the labeled serine, and the average background readings are subtracted from the total radioactivity. The assay is linear with time for 20–25 min.

The decays per minute obtained are converted to micromoles formaldehyde, correcting for the fraction of the toluene sampled, the efficiency of counting, and the efficiency of extraction of the formaldehyde into the dimedon. The latter is approximately 80% using this procedure, but it can be tested directly using [^{14}C]formaldehyde. The specific activity of the labeled serine must be corrected for the presence of the D-serine, which is assumed to account for 50% of the total.

Definition of Units. One unit of serine hydroxymethyltransferase activity is defined as the amount of enzyme required to generate 1 μmol of product per minute. Specific activity is defined as units per milligram protein.

Growth of *Methylobacterium organophilum* XX

Methylobacterium organophilum XX (ATCC 27886) is grown at 30° in a mineral salts medium on either methanol or succinate, as described for *Methylobacterium extorquens* AM1 in another chapter in this volume.[9] Cells are harvested, washed once with 50 mM KPO$_4$ buffer, pH 7.3, and frozen at −20°.

Purification of Glyoxylate-Activated (C$_1$-Specific) Serine Hydroxymethyltransferase

All purification procedures are performed at 4°. All buffers contain 50 nM pyridoxal phosphate, 5 μM dithiothreitol, and 1 mM EDTA. All buffers also contain 30% (v/v) glycerol unless otherwise noted. For these purifications, the discontinuous assay is used so that the correct isoenzyme may be distinguished.

[9] C. Krema and M. E. Lidstrom, this volume [57].

Step 1: Preparation of Cell-Free Extracts. Frozen methanol-grown cells (10–20 g wet weight) are thawed in 20–40 ml of 50 mM KPO$_4$ buffer, pH 7.3, containing the additions noted above (buffer A) and passed through a French pressure cell 3 times at 110 MPa. The extract is centrifuged at 20,000 g for 20 min, and the supernatant is used for further purification. Magnesium chloride (1 mM), ribonuclease (50 μg/ml), and deoxyribonuclease (50 μg/ml) are added, and the preparation is dialyzed overnight at 4° against 3 changes of buffer A plus 1 mM MgCl$_2$.

Step 2: DEAE-Cellulose Column Chromatography 1. A DEAE-cellulose (Whatman DE-52) column (2 × 10 cm) is prepared and equilibrated with buffer A. The enzyme preparation from Step 1 is layered onto the top and eluted with buffer A at a flow rate of 0.5 ml/min. The enzyme elutes with the buffer.

Step 3: DEAE-Cellulose Column Chromatography 2. The fractions from Step 2 containing the highest specific activity are pooled, diluted 5-fold with distilled water, and layered onto a second DEAE column (2 × 10 cm) equilibrated with 10 mM KPO$_4$ buffer, pH 7.3, containing the same additions as buffer A (buffer B). The column is washed with 50 ml buffer B, and the enzyme is eluted using a linear potassium acetate gradient (0 to 0.25 M in buffer B). The tubes containing the highest specific activity are pooled and concentrated using an Amicon Diaflo ultrafiltration unit (Danvers, MA) equipped with an XM50 membrane.

Step 4: Preparative Electrophoresis. The concentrated enzyme preparation from Step 3 is layered onto a Sephadex G-25 column which has been equilibrated with 60 mM Tris-HCl buffer (pH 8.0; no glycerol; buffer C) and which has been prepared in a stoppered Büchler Fractophorator column. The top of the column is sealed with a 1-cm layer of 0.5% (w/v) agar, and the column is inserted into the Fractophorator apparatus. Electrophoresis is performed at 10 mA with buffer C. After 20 to 34 hr, the enzyme band is discernible near the center of the column. The current is then turned off, the agar seal is broken, and the band is eluted with buffer C. The fractions containing the highest specific activity are pooled and stored at −20°.

Purification of Glyoxylate-Insensitive (Heterotrophic) Serine Hydroxymethyltransferase

Step 1: Preparation of Cell-Free Extracts. Cell free extracts are prepared as described for the glyoxylate-activated enzyme, except that succinate-grown cells are used. All buffers include the additions noted previously.

Step 2: DEAE-Cellulose Column Chromatography 1. The dialyzed extract is loaded onto a DEAE-cellulose column prepared as described for the glyoxylate-activated enzyme. About 10% of the total activity elutes with the buffer, and this activity is activated by glyoxylate. The remainder of the activity is then batch eluted with 0.2 M potassium acetate in buffer A. This activity is glyoxylate-insensitive. The fractions of the glyoxylate-insensitive enzyme containing the highest activity are pooled and dialyzed against 3 changes of 10 mM Tris–citrate buffer (pH 7.8; buffer D).

Step 3: DEAE-Cellulose Column Chromatography 2. The dialyzed preparation from Step 2 is loaded onto a second DEAE-cellulose column (2 × 10 cm) which has been equilibrated with buffer D. The column is washed with 50 ml buffer D, then 50 ml buffer D containing 0.1 M potassium acetate. The enzyme is eluted with buffer D containing 0.2 M potassium acetate. Fractions containing the highest specific activity are pooled and concentrated as noted above.

Step 4: Glycerol Density Gradient. A linear 10 to 30% (v/v) glycerol density gradient is prepared in polyallomer tubes and precooled to 4°. Samples (0.5 ml) of the concentrated preparation from Step 3 are layered onto each tube and the tubes are centrifuged at 38,000 rpm for 24 hr at 4° in a type SW41 rotor in a Beckman Model L-2 centrifuge. The tubes are punctured with an 18-gauge needle and 12-drop fractions are collected. The fractions containing the highest specific activity are pooled and stored at −20°.

The purification of both enzymes is summarized in Table I.[3]

Properties

Both serine hydroxymethyltransferase enzymes are present in cells grown on methanol and on succinate, but the major isoenzyme in each case represents about 90% of the total activity.[3] These are separated on the first DEAE-cellulose column. A summary of the properties of each enzyme is presented in Table II.

Purity and Stability. Both enzymes are pure as judged by denaturing and nondenaturing polyacrylamide gel electrophoresis, staining with Coomassie blue. A minor band is observed on native gels in the case of the glyoxylate-insensitive enzyme, but it shows activity when gels are sliced and assayed. The glyoxylate-activated enzyme is stable at −20°, showing no decrease in activity after 2 months, and also shows no decrease in activity after 24 hr at 4°. The glyoxylate-insensitive enzyme is less stable, however, and shows a 50% reduction in activity after storage for 2 months at −20° and an 80% loss of activity after 24 hr at 4°.

Kinetic Properties. Both enzymes show similar apparent K_m and V_{max} values, as determined with the pooled fractions from the second DEAE-cellulose column in both cases. These are 1.25 mM and 0.1 μmol/min/mg protein for the glyoxylate-activated enzyme and 1.0 mM and 85 nmol/

TABLE I

PURIFICATION OF TWO SERINE HYDROXYMETHYLTRANSFERASE ENZYMES FROM
*Methylobacterium organophilum*XX

Step	Total protein (mg)	Total units	Specific activity	Yield (%)	Purification -fold
Methanol-grown cells					
1. Cell-free extract	366	15.8	0.043	100	—
2. DEAE 1	222	9.3	0.045	59	1.04
3. DEAE 2	69	3.7	0.053	23	1.23
4. Electrophoresis	2	1.2	0.6	7.6	11.2
Succinate-grown cells					
1. Crude extract	430	9.7	0.023	100	—
2. DEAE 1	66	4.2	0.043	44	1.9
3. DEAE 2	32	2.2	0.069	23	3.1
4. Glycerol gradient	2	1.0	0.52	9.4	23.0

min/mg protein for the glyoxylate-insensitive enzyme.

pH Optimum. The pH optimum for the glyoxylate-activated enzyme is 8.7–8.8; for the glyoxylate-insensitive enzyme the pH optimum is slightly lower, 8.5–8.6.

Cofactor Requirements. Both enzymes require pyridoxal phosphate and tetrahydrofolate as cofactors. Maximum activity is obtained for both enzymes at 20 μM pyridoxal phosphate and 200 μM tetrahydrofolate.[3]

Molecular Weight. The molecular weight of the glyoxylate-stimulated

TABLE II

PROPERTIES OF SERINE HYDROXYMETHYLTRANSFERASES FROM *Methylobacterium
organophilum* XX

Property	Enzyme	
	Glyoxylate-activated (C_1-specific)	Glyoxylate-insensitive (heterotrophic)
K_m (mM)	1.25	1.00
V_{max} (μmol/min/mg protein)	0.10	0.085
pH optimum	8.7–8.8	8.5–8.6
Stimulation by cations (1 mM)	Ca^{2+}, K^+, Na^+	Ca^{2+}, K^+, Na^+, Mg^{2+}, Mn^{2+}, Zn^{2+}
Effect of glyoxylate (5 mM) (% of control)	450	90
Molecular weight		
Total	200,000	100,000
Subunit	50,000	100,000

enzyme is 200,000, as determined by polyacrylamide gel electrophoresis at varying acrylamide concentrations, and by sucrose density centrifugation. One band of 50,000 is found on gels containing sodium dodecyl sulfate (SDS), suggesting that the enzyme is a tetramer. The molecular weight of the glyoxylate-insensitive enzyme on both SDS-containing and native polyacrylamide gels is 100,000, suggesting that it is a monomeric enzyme.

Cation Stimulation. Both enzymes are stimulated by 1 mM Ca^{2+}, K$^+$, and Na$^+$, but only the glyoxylate-insensitive enzyme is stimulated by 1 mM Mg^{2+}, Mn^{2+}, and Zn^{2+}.[3]

Stimulation by Small Molecules. A variety of small molecules have been tested for their effects on activity of both enzymes, and in the case of AMP, ATP, methionine, S-adenosylmethionine, phosphoenolpyruvate, guanine, and thymine, little effect is observed. However, glyoxylate stimulates the glyoxylate-activated enzyme 4.5-fold, whereas it has little effect on the glyoxylate-stimulated enzyme. Glycine inhibits both enzymes about 50%. All molecules were tested at a concentration of 5 mM, except guanine (0.25 mM) and thymine (1 mM).

Serine Hydroxymethyltransferases in Other Serine Cycle Methylotrophs

Although both isoenzymes have been purified from only one serine cycle methylotroph, it is possible to assess the presence of the glyoxylate-activated enzyme in cell-free extracts. In this case, the activation is usually 1- to 2-fold.[4] In *Methylobacterium* 3A2 and *M. extorquens* AM1, a glyoxylate-activated activity is present in methanol-grown cells but not in cells grown on multicarbon compounds,[4] suggesting that isoenzymes of serine hydroxymethyltransferase are probably widespread in these bacteria. However, in *Hyphomicrobium* X and *H. methylovorum,* bacteria that grow only on C$_1$ and C$_2$ compounds, current evidence suggests that only the glyoxylate-activated enzyme is present. The enzyme that has been purified from methanol-grown cells of *H. methylovorum* is a dimer of 50,000-MW subunits, and antisera prepared against that enzyme does not cross-react with cell-free extracts of *M. organophilum* XX,[10] suggesting that these enzymes are different.

[10] S. S. Miyazaki, S. Toki, Y. Izumi, and H. Yamada, *Arch. Microbiol.* **147,** 328 (1987).

[57] Hydroxypyruvate Reductase from *Methylobacterium extorquens* AM1

By CINDER KREMA and MARY E. LIDSTROM

$$
\begin{array}{ccc}
\text{COOH} & & \text{COOH} \\
| & & | \\
\text{C}{=}\text{O} \quad + \text{NADH} + \text{H}^+ \longrightarrow \text{H}{-}\text{C}{-}\text{OH} + \text{NAD}^+ \\
| & & | \\
\text{H}{-}\text{C}{-}\text{OH} & & \text{H}{-}\text{C}{-}\text{OH} \\
| & & | \\
\text{H} & & \text{H}
\end{array}
$$

Hydroxypyruvate reductase (EC 1.1.1.81) catalyzes the reduction of hydroxypyruvate to D-glycerate. The enzyme has been isolated from *Pseudomonas acidovorans*,[1] in which it exists as isoenzymes of 75,000 and 85,000, respectively.[1,2] The 85,000 isoenzyme is induced by growth on glyoxylate and exists as a homodimer of two identical subunits.[1] In *Methylobacterium extorquens* AM1 (*Pseudomonas* AM1), hydroxypyruvate reductase is necessary for growth on one-carbon compounds and is a key enzyme in the serine cycle for assimilation of formaldehyde.[3]

Assay Method

Principle. Hydroxypyruvate reductase is assayed spectrophotometrically by measuring the hydroxypyruvate-dependent change in optical density at 340 nm arising from the oxidation of NADH to NAD$^+$.

Reagents

Sodium phosphate buffer, 0.5 M (pH 7.5)
Ammonium sulfate, 38 mM
NADH, 4 mM
Lithium hydroxypyruvate, 2.0 mM

Procedure. One-tenth milliliter each of the first three reagents are added in the above order to 0.5 ml distilled water. Enzyme plus water to make 0.1 ml is then added. Readings are taken to obtain background rates, then 0.1 ml hydroxypyruvate is added to initiate the reaction. The background rate is subtracted from the total to obtain the substrate-dependent rate. The assay is routinely carried out at room temperature, and ammonium sulfate stimulates the activity.

[1] L. D. Kohn and J. M. Utting, this series, Vol. 89, p. 341.
[2] L. D. Kohn and W. B. Jakoby, this series, Vol. 9, p. 229.
[3] C. Anthony, "The Biochemistry of Methylotrophs." Academic Press, London, 1982.

Definition of Unit. One unit of enzyme activity is defined as the amount of enzyme required to oxidize 1 μmol NADH to NAD$^+$ per minute.

Production of Hydroxypyruvate Reductase

Culture Growth. The enzyme is purified from *Methylobacterium extorquens* AM1[4] (NCIMB 9133) grown on a minimal salts medium supplemented with 0.5% (v/v) methanol or 50 m*M* disodium succinate (Sigma Chemical Co., St. Louis, MO). The minimal salts medium is prepared as follows. Solution I contains 170 g K_2HPO_4 and 170 g $NaH_2PO_4 \cdot 2H_2O$ in 1 liter, with distilled water. Solution II is 200 g $(NH_4)_2SO_4$ and 20 g $MgSO_4 \cdot 7H_2O$ in 1 liter, with distilled water. Ten milliliters of Solution I and 10 ml of Solution II are added to 979 ml distilled water. To this is added 1 ml of Vishniac's trace elements consisting of the following, per 500 ml: EDTA, 5 g; $ZnSO_4 \cdot 7H_2O$, 2.2 g; $CaCl_2 \cdot 2H_2O$, 0.733 g; $MgCl_2 \cdot 4H_2O$, 0.506 g; $FeSO_4 \cdot 7H_2O$, 0.499 g; $(NH_4)Mo_7O_{24} \cdot 4H_2O$, 0.11 g; $CuSO_4 \cdot 5H_2O$, 0.157 g; and $CoCl_2 \cdot 6H_2O$, 0.161 g. Methanol is filter-sterilized using a solvent-resistant membrane filter and added to cooled autoclaved medium at a final concentration of 0.5% (v/v).

A fresh culture grown on a minimal salts medium plate supplemented with either 0.5% methanol or 50 m*M* disodium succinate is used to inoculate 40 ml of the same medium. After 3 days of growth at 30° with shaking (270 rpm), this culture is used to inoculate 500 ml of the same medium in a 2-liter flask. After 2 days of growth, the contents of the flask are transferred to a 10-liter fermentor containing the same medium and vigorously aerated with compressed air and constant stirring. This culture is grown at 30° to late log or stationary phase. The cells are harvested by centrifugation at 10,000 *g* for 10 min at 4°. The pellets are washed once in the minimal salts medium without substrate, centrifuged, and resuspended in a minimal volume of the medium. The enzyme remains active for at least 4 months when the cells are stored at −20°.

Purification Procedure

In methanol-grown cells, hydroxypyruvate reductase can be separated into two active fractions by DEAE-cellulose chromatography, which are referred to as the LS (low-salt) and HS (high-salt) activities. Succinate-grown cells contain only the HS fraction.

Step 1: Preparation of Crude Extract. Frozen cells (usually 80–100 g wet weight) are thawed at room temperature, and a few grains of DNase (Sigma) are added during thawing. The cells are lysed by passing the cell

---

[4] J. Heptinstall and J. R. Quayle, *Biochem. J.* **117**, 563 (1970).

solution through a French pressure cell precooled at 4° (SLM Aminco), 3 times under 20,000 psi. The extracts are then centrifuged at 13,000 g for 10 min. The pellets are discarded, and the supernatant is centrifuged at 150,000 g for 1 hr. The resulting supernatants are collected and stored on ice. All procedures are carried out at 4° unless otherwise noted.

Step 2: Ammonium Sulfate Precipitation. To the combined supernatant from Step 1, ammonium sulfate [$(NH_4)_2SO_4$] is gradually added to 25% of saturation and dissolved with constant stirring for at least 1 hr. The solution is then centrifuged for 15 min at 10,000 g. The pellet is discarded, and the supernatant is brought up to 50% of saturation with $(NH_4)_2SO_4$, dissolved, and centrifuged as above. After the pellet is discarded, the concentration of $(NH_4)_2SO_4$ in the supernatant is raised to 75% of saturation, and the solution is stirred and centrifuged as noted above. The pellet is resuspended in a minimal volume of buffer A [50 mM Tris-chloride (pH 8.0), 25% glycerol (v/v), 2 mM dithiothreitol (DTT)], and this solution is dialyzed against three 2-liter changes of the same buffer.

Step 3: DEAE-Cellulose Chromatography. The dialyzed solution is loaded onto a DEAE-cellulose (Sigma) column (2.5 × 44 cm) that has been equilibrated with buffer A, and the column is washed with 500 ml of this buffer. In preparations from methanol-grown cells, the major hydroxypyruvate reductase activity peak (LS activity) is eluted during the first wash with buffer A. The remaining activity (HS) is eluted with buffer B which consists of buffer A with 200 mM potassium chloride (KCl). In preparations from succinate-grown cells, the column is washed first with buffer A, and the activity is eluted with buffer B. This preparation has not been further purified, but it appears identical to the HS activity found in the methanol-grown cells. Fractions containing more than 10% of the total activity for each peak are pooled, each pool is concentrated separately in a stirred cell (Amicon Corp., series 200) containing a YM10 filter, and the fraction eluted with buffer B is dialyzed against buffer A.

Step 4: DEAE-Toyopearl Chromatography. The LS fraction is applied to a DEAE-Toyopearl (Supelco, Inc., Bellefonte, PA) column (1.5 × 26 cm), equilibrated with buffer A. The column is washed with 100 ml buffer A, and the activity is eluted with a 100 ml wash of buffer C (buffer A with 75 mM KCl). The column is then washed with 2 bed volumes of buffer A, and the HS fraction is loaded onto the column and washed with 200 ml of buffer A. A 200 ml wash of buffer D (buffer A with 100 mM KCl) elutes the HS fraction. Both of the eluted portions are then separately concentrated in microconcentrators with 10,000-MW cutoff membranes (Centricon 10, Amicon, Danvers, MA), to a volume of 5 ml each and dialyzed against buffer A.

Step 5: Isoelectric Focusing. A 10 × 20 cm flat-bed isoelectric focusing unit (LKB 2117-003 Multiphor II system, Pharmacia LKB, Piscataway,

NJ) is set at a constant temperature of 4° and loaded with a 4% dextran gel (Ultrodex 165, Pharmacia LKB) including 2% carrier ampholites (LKB Ampholine), pH 3.5–10, and diluted with buffer A. The LS and HS activity fractions are run on separate but identical gels. The concentrated enzyme solution eluted from the DEAE-Toyopearl column is loaded into the gel, and the unit is run at an initial voltage of 500 V for 14–18 hr. The gel is sectioned with a LKB-Pharmacia Preparative Kit after electrophoresis is complete, and each section is tested for pH in a 10 mM KCl solution and for activity in buffer A. The isoelectric point of hydroxypyruvate reductase is approximately pH 5.0 for both the LS and HS activity fractions. Each sample is then microconcentrated (YM10 membrane filters) and dialyzed against buffer A.

Step 6: Gel-Filtration Chromatography. A 1.5 × 58 cm Sephadex G-150 column (Sigma) is equilibrated and washed with buffer A. The LS fraction from isoelectric focusing is loaded onto the column and eluted with the same buffer at a flow rate of about 0.25 ml/min. Fractions showing activity are collected and concentrated in the stirred cell with a YM10 membrane. A 1.5 × 28 cm Sephadex G-200 (Sigma) column is prepared in the same manner as the Sephadex G-150, and the HS activity sample is eluted from it and concentrated under the same conditions as for the LS activity sample.

Step 7: Polyacrylamide Gel Electrophoresis. A zymogram is obtained by polyacrylamide slab gel electrophoresis as follows. A 10% nondenaturing polyacrylamide gel (16 × 13.5 × 0.4 cm) is used to separate the proteins following the buffer system of Laemmli[5] with the exceptions that sodium dodecyl sulfate (SDS) is omitted, 2-mercaptoethanol is omitted from the sample buffer, and 25% glycerol (v/v) and 2 mM DTT are added to the gel buffer. The gel is run at 10–20 mA at 4° with bromphenol blue as the front marker. The concentrated samples taken from each molecular weight column are run separately. After electrophoresis is complete, the slab gel is removed from the glass plates, rinsed with distilled water at room temperature, and incubated with the following staining mixtures for detection of active enzyme. Solution A consists of nitro blue tetrazolium (Sigma), 30 mg, and 50 mM sodium phosphate buffer, pH 7.5, in a final volume of 100 ml. Solution B contains 10 mM hydroxypyruvate, lithium salt (Sigma), 2 mg phenazine methosulfate (Sigma), 50 mg NADH (Sigma), 406 mg MgCl$_2$·6H$_2$O (J. T. Baker Chemical Co., Phillipsburg, NJ), and 50 mM sodium phosphate buffer, pH 7.5, to a final volume of 100 ml. The gel is covered with foil and soaked in solution A at room

[5] U. K. Laemmli, *Nature (London)* **227**, 680 (1970).
[6] L. D. Kohn and W. B. Jakoby, *J. Biol. Chem.* **243**, 2494 (1968).

temperature for 40 min. This solution is removed and solution B added. The gel is removed after the background becomes dark blue. The lighter, active region is excised from the gel, and the protein is removed from the gel strip by an electroelution system (Elutrap, Schleicher & Schuell, Keene, NH) with a constant power of 100 V for 14 hr at room temperature. The eluate is removed from the apparatus and dialyzed against buffer A. Both fractions show only one major polypeptide band on Coomassie blue-stained polyacrylamide gels, although a few minor bands are present on silver-stained gels.

Properties

The properties of the LS and HS activities are summarized in Table I.

Molecular Weight and Subunit Structure. The molecular weight of the hydroxypyruvate reductase as determined by both gel filtration chromatography and polyacrylamide gel electrophoresis is approximately 50,000 ± 5,000 for both the LS and HS enzymes. A similar size was obtained with polyacrylamide gel electrophoresis in the presence of SDS, indicating that both enzymes are most likely composed of a single subunit.

Specificity. Studies of substrate specificity on the active fractions derived from DEAE-cellulose chromatography have shown that both fractions exhibit strong specificity for hydroxypyruvate. Neither enzyme showed detectable activity with glyoxylate, glycolate, or oxaloacetate. The LS fraction did utilize dihydroxyfumarate (DHF) at a final concentration of 30 mM, although this activity was 3.75-fold less than that for hydroxypyruvate. When hydroxypyruvate and DHF were used together as substrates, the activity increased 7-fold in the LS enzyme fraction only. This activation may be due to the inhibitory effect of DHF on other glycerate-producing enzymes such as glyoxylate reductase.[6] Alternatively, DHF may have a direct stimulatory effect on the enzyme itself. Glycolate (Sigma), at a concentration of 2 mM, appeared to have a stimulatory effect on the activity of the LS enzyme for hydroxypyruvate, increasing it 2-fold. This effect was greatly diminished after the enzyme underwent freezing and thawing. Glyoxylate (Sigma) at the same concentration also seemed to have a small but similar effect on this enzyme.

Kinetics. Complete purification of the active enzyme was not possible due to precipitation after extraction of the enzyme from a native polyacrylamide gel. Therefore, apparent K_m values were calculated using the separated DEAE-cellulose fractions. The apparent K_m for hydroxypyruvate was 4.0×10^{-2} M for the LS enzyme fraction, and 1.0×10^{-2} M for both the HS enzyme fractions. In both the LS and HS enzymes, the activity toward hydroxypyruvate was enhanced in the presence of ammonium ions

TABLE I

Properties of Two Active Hydroxypyruvate Reductase
Fractions Separated by DEAE-Cellulose
Chromatography

Property	Fraction	
	LS (low-salt)[a]	HS (High-salt)[b]
Size (daltons)[c]	50,000 ± 5,000	50,000 ± 5,000
Number of subunits	1	1
Substrates		
Hydroxypyruvate	+++	+++
Dihydroxyfumarate	+	−
Oxaloacetate	−	−
Glyoxylate	−	−
Glycolate	−	−
Apparent K_m (mM)	40	10

[a] Found only in methanol-grown cells.

[b] Found in both methanol and succinate-grown cells; properties are identical for the HS fraction from cells grown on either substrate.

[c] As judged by zymograms and Sephadex column chromatography.

(NH_4^+). The stimulation by ammonium ions was saturated at $1.6 \times 10^{-2}\ M$ for the LS enzyme and $1.2 \times 10^{-3}\ M$ for both the HS enzyme fractions.

The data presented here suggest that *M. extorquens* AM1 contains two forms of hydroxypyruvate reductase, one of which (eluting at low salt from a DEAE-cellulose column) is induced during growth on methanol. However, as the native molecular sizes of the two enzymes are similar, it is not yet clear whether these two activities represent different enzymes or whether the LS enzyme may be a modified form of the HS enzyme. Further studies will be necessary to determine which of these possibilities is correct.

[58] Malyl-CoA Lyase from *Methylobacterium extorquens* AM1

By A. J. HACKING and J. R. QUAYLE

Malyl-CoA $\rightleftharpoons$ acetyl-CoA + glyoxylate

Malyl-CoA lyase has been found in a number of bacteria able to utilize C_1 compounds as the sole source of carbon and energy.[1] The specific activity is severalfold higher in *Methylobacterium extorquens* AM1 (previously known as *Pseudomonas* AM1) when grown on C_1 compounds than on C_2, C_3, or C_4 compounds. A mutant of *Methylobacterium extorquens* AM1 which lacks this enzyme is unable to grow on C_1 compounds unless supplemented with substrate quantities of glyoxylate or a precursor such as glycolate. Revertants of this mutant which have regained the ability to grow on C_1 compounds have also regained this activity.[2] The enzyme is routinely assayed in the direction of cleavage to acetyl-CoA and glyoxylate with added phenylhydrazine.

Assay Method

Reagents. (S)-$(\beta$-hydroxysuccinyl)-N-octanoylcysteamine[(S)-$(\beta$-hydroxy-3-carboxypropionyl)-N-octanoylcysteamine] is prepared from S-malate, octanoic acid, and cysteamine using the methods described by Eggerer and Grünewälder[3] and is used to make $(2S)$-4-malyl-CoA by ester interchange with CoA. (S)-$(\beta$-Hydroxysuccinyl)-N-octanoylcysteamine (16 mg; 50 μmol) and CoA (20 mg; 26 μmol) are dissolved in 4 ml of 0.2 M $KHCO_3$, 1 ml of diethyl ether (peroxide-free) is added, and the mixture is shaken at room temperature for 1 hr.

The mixture is then extracted with ether in a continuous extraction apparatus for 1 hr with that part of the extraction apparatus containing the aqueous phase kept at $0°$ in an ice bath. The aqueous phase is then withdrawn and acidified to pH 4 with 5 M HCl at $0°$. The acidified mixture is again continuously extracted with ether for 3 hr, the aqueous layer being maintained at $0°$. At the end of this time, the aqueous phase is withdrawn and the dissolved ether removed from it with a stream of N_2. The resulting solution of $(2S)$-4-malyl-CoA (~ 3.5 mM) can be stored

---

[1] A. R. Salem, A. J. Hacking, and J. R. Quayle, *Biochem. J.* **136**, 89 (1973).

[2] A. R. Salem, A. J. Hacking, and J. R. Quayle, *J. Gen. Microbiol.* **81**, 525 (1974).

[3] H. Eggerer and C. Grünewälder, *Justus Liebigs Ann. Chem.* **677**, 200 (1964).

METHODS IN ENZYMOLOGY, VOL. 188

either at 0° or at −15°, and is stable under these conditions for several weeks. (2R)-4-Malyl-CoA is similarly prepared from (R)-(β-hydroxysuc-cinyl)-N-octanoylcysteamine. S-Citryl-CoA is prepared from S-citryl-N-octanoylcysteamine by the method of Eggerer.[4]

Acetyl-CoA and succinyl-CoA are prepared by reaction of their respective acid anhydrides with CoA, and DL-3-hydroxybutyryl-CoA is prepared by reaction of the mixed anhydride of ethyl hydrogen carbonate and DL-3-hydroxybutyric acid with CoA as described by Stadtman.[5] Oxalyl-CoA is prepared by ester interchange between thiocresyloxalic acid and CoA as described by Quayle.[6,7] Propionyl-CoA, n-butyryl-CoA, and 3-hydroxymethylglutaryl-CoA can be purchased from P-L Biochemicals Inc. (Milwaukee, WI). Succinyl-CoA, 3-hydroxybutyryl-CoA, 4-malyl-CoA, and oxalyl-CoA are assayed chemically from the decrease in absorbance at 232 nm (succinyl-CoA and 3-hydroxybutyryl-CoA), 234 nm (4-malyl-CoA), or 270 nm (oxalyl-CoA), associated with their hydrolysis in 0.1 M NaOH at 30° for 10 min. Citryl-CoA is incubated in 0.1 M NaOH at 30° until no further decrease in E_{235} can be detected (~ 8 hr). The differences in molar extinction coefficients of the acyl-CoA derivatives and their products are taken to be 4.5×10^3 liter mol^{-1} cm^{-1} for succinyl-CoA, 3.7×10^3 liter mol^{-1} cm^{-1} for 3-hydroxybutyryl-CoA, 4.8×10^3 liter mol^{-1} cm^{-1} for 4-malyl-CoA, 2.3×10^3 liter mol^{-1} cm^{-1} for oxalyl-CoA, and 4.7×10^3 liter mol^{-1} cm^{-1} for citryl-CoA.

Acetyl-CoA is assayed enzymatically by arsenolysis in the presence of phosphate acetyltransferase (acetyl-CoA orthophosphate acetyltransferase, EC 2.3.1.8) and arsenate[5] or by conversion to citrate and CoA in the presence of oxaloacetate and citrate synthase [citrate oxaloacetate-lyase (CoA-acetylating), EC 4.1.3.7] followed by assay of CoA with 5,5′-dithiobis(2-nitrobenzoate).[8]

Calcium phosphate gel is prepared by the method described by Keilin and Hartree.[9] DEAE-cellulose (preswollen DE-52) is obtained from Whatman Biochemical Ltd., Maidstone, Kent, UK. Purified enzymes, nucleotides, and CoA are obtained from Boehringer Corp., Lewes, East Sussex, BN7 1LG, UK.

Procedure. The reaction mixture for the routine assay of 4-malyl-CoA lyase contains the following, in a final volume of 1 ml: potassium phosphate buffer, pH 7.4 (100 μmol), MgCl₂ (10 μmol), phenylhydrazine-HCl (5 μmol), and enzyme extract (2–100 μg protein). The reaction is started

[4] H. Eggerer, *Justus Liebigs Ann. Chem.* **666,** 192 (1963).
[5] E. R. Stadtman, this series, Vol. 3, p. 931.
[6] J. R. Quayle, *Biochim. Biophys. Acta* **57,** 398 (1962).
[7] J. R. Quayle, *Biochem. J.* **87,** 368 (1963).
[8] P. K. Tubbs and P. B. Garland, this series, Vol. 13, p. 535.
[9] D. Keilin and E. I. Hartree, *Proc. R. Soc. London, Ser. B* **124,** 397 (1938).

by the addition of 0.2 μmol of 4-malyl-CoA, and the E_{324} is recorded at 30°
against a reference mixture lacking 4-malyl-CoA. For measurement of the
pH optimum of the enzyme, the glyoxylate produced is determined discontinuously.

The enzyme can also be assayed discontinuously in the direction of
4-malyl-CoA synthesis by measuring the disappearance of acetyl-CoA or
glyoxylate. Reaction mixtures contain, in a 2.0 ml final volume, potassium
phosphate buffer, pH 7.4 (200 μmol), MgCl$_2$ (20 μmol), acetyl-CoA (1
μmol) and potassium glyoxylate (1 μmol). The reaction is started by the
addition of enzyme extract (1 – 5 μg protein). At convenient time intervals,
samples (250 μl) are withdrawn and 50 μl of 3.0 M HClO$_4$ are added to
each. After incubation for 10 min at room temperature, the samples are
adjusted to neutral pH by the addition of 2.5 M K$_2$CO$_3$, and the precipitate of KClO$_4$ is removed by centrifugation. A sample (200 μl) of the
supernatant is then added to a cuvette containing (in 800 μl) potassium
phosphate buffer, pH 7.4 (100 μmol), and phenylhydrazine (5 μmol). The
E_{324} is measured against a similar reaction mixture to which 200 μl of
water has been added in place of the sample. When necessary, the 4-malyl-
CoA concentration can also be measured at this stage by adding purified
malyl-CoA lyase and MgCl$_2$ (10 μmol) and recording the subsequent
increase in E_{324}. Alternatively, the malyl-CoA concentration may be measured in the supernatant after precipitation of KClO$_4$ by hydrolysis with
citrate synthase.[10] Reaction mixtures contain, in 1 ml, Tris-HCl buffer,
pH 8.0 (100 μmol), 5,5'-dithiobis(2-nitrobenzoate) (0.1 μmol), and sample
(200 μl). The increase in E_{412} after the addition of citrate synthase (100 μg)
is measured. Acetyl-CoA in the same sample is then determined by measuring the subsequent increase in E_{412} on the addition of oxaloacetate (0.1
μmol).

Source of Enzyme

Methylobacterium extorquens AM1 (NCIB 9133) can be obtained from
the National Collection of Industrial and Marine Bacteria, Torry Research
Station, P.O. Box 31, Aberdeen, AB9 8DG, Scotland. The organism is
maintained in slope culture on an inorganic salts medium[11] containing
methylamine hydrochloride (50 mM) and agar (1.5%, w/v). It is subcultured monthly and stored at 4° after incubation for 3 – 4 days at 30°.
Liquid cultures are prepared as described by Heptinstall and Quayle[12] with
methanol (0.5%, v/v) as the carbon source.

[10] H. Eggerer, U. Remberger, and C. Grünewälder, *Biochem. Z.* **339**, 436 (1964).
[11] G. C. N. Jayasuria, *J. Gen. Microbiol.* **12**, 419 (1955).
[12] J. Heptinstall and J. R. Quayle, *Biochem. J.* **117**, 563 (1970).

Purification Procedure

Step 1: Preparation of Cell-Free Extract. Methanol-grown bacteria (60 g wet weight) are suspended in 300 ml (final volume) of 0.1 M potassium phosphate buffer, pH 7.0, and are circulated twice through a Rapidis F400 continuous sonicator (Ultrasonics Ltd.) at a flow rate of 8 ml/min with power and tuning both at setting 7. The suspension is cooled on ice and circulated at 0–10°. The resulting extract is centrifuged at 30,000 g for 40 min at 4° and the pellet discarded.

Step 2: Protamine Sulfate Treatment. To the supernatant is added protamine sulfate (15 mg/ml in 0.1 M potassium phosphate buffer, pH 7.0) dropwise and with continuous stirring to a final concentration of 1 mg/10 mg of protein. The solution is equilibrated for 30 min at room temperature, and the resultant precipitate is removed by centrifugation (30,000 g for 20 min) and discarded.

Step 3: Ammonium Sulfate Precipitation. To the supernatant is added, with stirring, $(NH_4)_2SO_4$ to 45% saturation at 4°. The pH of the solution is maintained at 7.0 by addition of 2 M NH_3. After equilibration at 4° for 30 min, the precipitate is removed by centrifugation and discarded. To the supernatant is added $(NH_4)_2SO_4$ to 55% saturation. After equilibration as above, the precipitate is collected by centrifugation and the supernatant discarded. The precipitate is dissolved in 50 ml of 0.1 M potassium phosphate buffer, pH 7.0, and dialyzed overnight against 2 liters of 10 mM potassium phosphate buffer containing 10 mM $MgCl_2$, pH 7.0.

Step 4: Adsorption on Calcium Phosphate Gel. To the dialyzed $(NH_4)_2SO_4$ fraction is added calcium phosphate gel in a ratio 1:1 (w/w) to protein. The suspension is equilibrated at 4° with stirring for 30 min. The gel is removed by low-speed centrifugation (2000 g for 10 min) and discarded. A second addition of gel is made in a 5:1 (w/w) ratio to protein, and after equilibration and centrifugation the supernatant is discarded. The gel is resuspended in 40-ml batches of 0.1 M potassium phosphate buffer, pH 7.0, and, after equilibration and centrifugation, the supernatants are assayed for malyl-CoA lyase activity. Normally, two extractions are sufficient to elute 80–90% of the enzyme activity from the gel.

Step 5: DEAE-Cellulose Chromatography. The combined eluates from the calcium phosphate gel are diluted with water until they are 50 mM with respect to phosphate buffer (pH 7.0), and they are then applied to a DEAE-cellulose column (12 × 6.0 cm) which has been equilibrated with this buffer. The column is washed with 50 ml of 50 mM potassium phosphate buffer, then eluted by a linear gradient of 50–150 mM potassium phosphate buffer, pH 7.0 (600 ml total). Fractions (7 ml) are collected, and malyl-CoA lyase activity is eluted at a concentration of approximately 100 mM phosphate in fractions 32–51. The purification procedure is summarized in Table I.

TABLE I
PURIFICATION OF MALYL-CoA LYASE FROM *Methylobacterium extorquens* AM1[a]

Step	Enzyme activity (units/ml)	Volume (ml)	Total activity (units)	Recovery (%)	Protein (mg/ml)	Specific activity (units/mg)	Purification
Cell-free extract	15.8	255	4029	100	11.5	1.37	1
Protamine sulfate precipitation	13.6	280	3819	94.8	8.6	1.58	1.15
45–55% saturated (NH$_4$)$_2$SO$_4$ fraction (after dialysis)	37.9	60	2276	56.5	7.3	5.2	3.8
Calcium phosphate gel	13.9	80	1112	27.6	0.95	14.6	10.7
DEAE-cellulose chromatography	2.8	140	392	9.7	0.105	26.6	19.4
	(5.34)	—	—	—	(0.19)	(28.1)	(20.5)

[a] Experimental details are given in the text. A unit of enzyme activity is defined as that quantity which will catalyze the cleavage of 1 μmol of substrate per minute at 30°. Values in parentheses refer to the peak fraction from DEAE-cellulose.

Concentration of Protein Samples. Dilute fractions of the DEAE-cellulose column effluent containing malyl-CoA lyase activity are concentrated by filtration under pressure [N$_2$ gas at 10^5 Pa (1 kg/cm^2) through PM30 membranes (Diaflo ultrafiltration membranes; Amicon Corp., Lexington, MA)]. Where protein solutions are required at concentrations of approximately 5 mg/ml for sedimentation velocity analysis, protein is applied to a small (0.6 × 2.5 cm) column of DEAE-cellulose equilibrated with 50 mM potassium phosphate buffer, pH 7.0, and eluted with 150 mM buffer. The recovery of enzyme activity is 90–95%.

Protein Determinations. Protein is measured by the Folin–Ciocalteau method as described by Lowry et al.[13] with bovine serum albumin as standard.

This procedure gives a preparation which is greater than 95% pure malyl-CoA lyase as determined by gel electrophoresis and sedimentation velocity experiments.

Properties

Molecular Weight. The molecular weight of malyl-CoA lyase is estimated to be 190,000 ± 4,000 using the meniscus depletion method of Yphantis[14] and assuming a partial specific volume of 0.73 for the protein.

[13] O. H. Lowry, N. J. Rosebrough, A. L. Farr, and R. J. Randall, *J. Biol. Chem.* **193**, 265 (1951).
[14] D. A. Yphantis, *Biochemistry* **3**, 297 (1964).

Stability. Purified enzyme preparations retain over 95% of their initial activity during storage periods of up to several months at $-15°$. The enzyme is also stable when stored at $0-4°$ for $2-3$ weeks, but activity is rapidly and irreversibly lost at temperatures above $55°$.

Specificity. Purified malyl-CoA lyase preparations possess no detectable malate synthase or citrate synthase activities when assayed by the standard procedures for these enzymes.

Substitution of $(2S)$-4-malyl-CoA by an equimolar amount of the optical isomer $(2R)$-4-malyl-CoA in the cleavage assay system may give a slight initial rise in E_{324}, but this can be shown to be due to contamination by the $(2S)$ isomer. The purity of preparations may be tested by hydrolysis of the thioester bond under mild conditions (0.1 M NaOH at $25°$) followed by determination of $(2S)$-malate in the presence of NAD, acetyl-CoA, malate dehydrogenase, and citrate synthase.[15] With preparations of the $(2R)$ isomer that are pure by this method, no lyase activity can be measured, and any activity in impure preparations is proportional to the amount of $(2S)$ isomer detected. The enzyme has no activity toward DL-3-hydroxybutyryl-CoA or 3-hydroxy-3-methylglutaryl-CoA, and citryl-CoA does not function as a substrate either in cleavage to oxaloacetate and acetyl-CoA or in hydrolysis of the thioester bond.

If the enzyme is assayed in the direction of synthesis, glyoxylate cannot be substituted by glycolate, glycoaldehyde, or pyruvate with acetyl-CoA as cosubstrate. With glyoxylate, however, there is a slow reaction if acetyl-CoA is replaced by propionyl-CoA, but not n-butyryl-CoA, succinyl-CoA, or oxalyl-CoA.

Metal Ion Requirement. Malyl-CoA lyase has an absolute requirement for a bivalent metal ion. The relative effectiveness of various ions is shown in Table II.

Inhibition. $(2R)$-4-Malyl-CoA causes inhibition of $(2S)$-4-malyl-CoA cleavage, such that at equimolar concentrations (0.2 mM) 28% inhibition is observed, and this is increased to 45% by doubling the concentration of the $(2R)$ form to 0.4 mM. Activity is also affected by the buffer system used; at pH 7, 0.1 M Tris buffers cause $45-50\%$ inhibition relative to equimolar phosphate at the same pH, but activities are 10% higher in 0.1 M N-glycylglycine than in 0.1 M phosphate. The enzyme is not inhibited by the thiol group binding reagents N-ethylmaleimide, iodoacetate, or p-chloromercuribenzoate at 1 mM concentrations.

pH Optimum. Malyl-CoA lyase has a broad optimum range between pH 7.2 and 8.2 with maximum activity being observed at pH 7.8.

---

[15] J. W. Cornforth, J. W. Redmond, H. Eggerer, W. Buckel, and Ch. Gutschow, *Eur. J. Biochem.* **14,** 1 (1970).

TABLE II
METAL ION REQUIREMENT OF MALYL-CoA
LYASE[a]

Metal ion	Rate (μmol/min/ml enzyme)	Rate relative to Mg^{2+} (%)
Mg^{2+}	12.9	—
Co^{2+}	12.9	100
Mn^{2+}	11.4	88
Zn^{2+}	5.3	41
Ni^{2+}	1.2	9
Ca^{2+}	0	0

[a] Reaction mixtures contain, in a final volume of 1.0 ml; N-glycylglycine buffer, pH 7.5 (100 μmol), 4-malyl-CoA (0.2 μmol), phenylhydrazine-HCl (5 μmol), and EDTA (0.2 μmol). The reaction is initiated by the addition of metal ion (10 μmol), and the E_{324} is recorded at 30° with a reference mixture lacking metal ion. The enzyme preparation contains 0.5 mg protein/ml.

TABLE III
MICHAELIS CONSTANTS OF MALYL-CoA LYASE[a]

Substrate	$K_m(M)$
(2S)-4-Malyl-CoA (10 mM Mg^{2+})	6.6×10^{-5}
Acetyl-CoA (10 mM glyoxylate; 10 mM Mg^{2+})	1.5×10^{-5}
Glyoxylate (0.4 mM acetyl-CoA; 10 mM Mg^{2+})	1.7×10^{-3}
Mg^{2+} (0.2 mM 4-malyl-CoA)	1.2×10^{-3}

[a] Concentrations of additional reactants are shown in parentheses. The Michaelis constant for Mg^{2+} is determined in the direction of 4-malyl-CoA cleavage. All the reactions are carried out in 0.1 M potassium phosphate buffer (pH 7.4) at 30°

Equilibrium Constant. The equilibrium constant

$$K = \frac{[\text{acetyl-CoA}][\text{glyoxylate}]}{[\text{4-malyl-CoA}]}$$

has a value of $4.7 \times 10^{-4}\ M$ at pH 7.4 and 30°.
Michaelis Constants. The Michaelis constants determined from double-reciprocal plots of initial velocity versus substrate concentration are given in Table III.

[59] Synthesis of L-4-Malyl Coenzyme A

By PEGGY J. ARPS

Introduction

The serine pathway for the bacterial assimilation of formaldehyde was proposed in the early 1960s by Quayle and co-workers when they found that *Pseudomonas* AM1 (now called *Methylobacterium extorquens* AM1) grown on radioactive methanol produced malate and amino acids (mainly serine) as the first labeled products.[1] The pathway has three phases. In the first, formaldehyde and CO_2 are used to produce acetyl-CoA and glyoxylate through a series of reactions. Malyl-CoA is the final intermediate in this series, and it is the substrate that is cleaved by the enzyme malyl-CoA lyase to yield acetyl-CoA and glyoxylate:

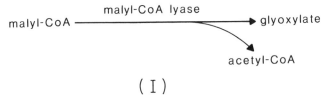

In the second phase various reactions proceed to convert acetyl-CoA to glycine. Finally, glycine and a second molecule of formaldehyde are converted to cell material by way of 3-phosphoglycerate. This is the only route for acetyl-CoA production in bacteria such as obligate methylotrophs and restricted facultative methylotrophs (e.g., *Hyphomicrobium*) that lack pyruvate dehydrogenase.[2]

Malyl-CoA lyase (EC 4.1.3.24) is a key cleavage enzyme of the serine

---

[1] P. J. Large, D. Peel, and J. R. Quayle, *Biochem. J.* **81**, 470 (1961).
[2] C. Anthony, "The Biochemistry of Methylotrophs," p. 98. Academic Press, New York, 1982.

pathway since it is responsible for cleaving L-4-malyl-CoA [also called (S)-3-hydroxysuccinyl-CoA], a C_4 compound, into two C_2 units. The enzyme is inducible by growth on C_1 compounds, and the reaction is readily reversible. The lyase is specific for L-4-malyl-CoA; the alternative optical isomer D-4-malyl CoA is not cleaved but is able to inhibit the reaction.[3-5] For more details concerning malyl-CoA lyase, see the chapter in this volume by Hacking and Quayle.[6]

L-4-Malyl-CoA, the substrate for the above reaction, is not available commercially, but it can be chemically synthesized following a procedure that was first described by Eggerer and Grünewälder[7] and later modified in the laboratory of Quayle.[3] The following protocol is an adaptation of the latter method. The synthesis of L-4-malyl-CoA is carried out in five steps, as shown in Scheme 1. L-Malic acid is first reacted with chloral hydrate to produce the L-4-malylchloralide (1), in which the 1-carboxylate and 2-hydroxyl functional groups are masked. The L-4-malic acid chloralide is converted to the L-4-malic acid chloralide chloride (2) using thionylchloride, then treated with N-octanoylcysteamine (RSH) to give the thioester, L-4-malic acid choralidyl-N-octanoylcysteamine (3). Deprotection with hydrochloric acid produces L-4-malyl-N-octanoylcysteamine (4). In order to prove that the deprotection step does not racemize the L-malyl moiety, Eggerer and Grünewälder[7] reduced the L-4-malyl-N-octanoylcysteamine and showed that the optically active 2,4-dihydroxybutanoic acid was produced. In the final step, the N-acylcysteamine thioester undergoes thioester interchange with coenzyme A to produce the coenzyme A ester, L-4-malyl-CoA (5), and N-octanoylcysteamine. The L-4-malyl-CoA product is used directly to test for the presence of malyl-CoA lyase activity in bacterial cell extracts.[6]

L-Malic Acid Chloralide (1)

A 13 g (0.10 mol) quantity of L-malic acid (Sigma Chemical Co., St. Louis, MO) is added to 16 g (0.12 mol) of chloral hydrate[8] (Sigma) in a 1-liter double-neck, ice-cooled flask fitted with a $CaCl_2$ drying tube and magnetic stirring bar. A 25 ml volume of concentrated H_2SO_4 (0°) is

[3] A. R. Salem, A. J. Hacking, and J. R. Quayle, *Biochem. J.* **136,** 89 (1973).
[4] L. B. Hersh, *J. Biol. Chem.* **248,** 7295 (1973).
[5] A. J. Hacking and J. R. Quayle, *Biochem. J.* **139,** 399 (1974).
[6] A. J. Hacking and J. R. Quayle, this volume [58].
[7] H. Eggerer and C. Grünewälder, *Justus Liebigs Ann. Chem.* **677,** 200 (1964).
[8] In the United States, chloral hydrate is classified by the Drug Enforcement Administration (DEA) as a Class IV controlled substance, requiring the laboratory/institution to have a DEA registration number before ordering.

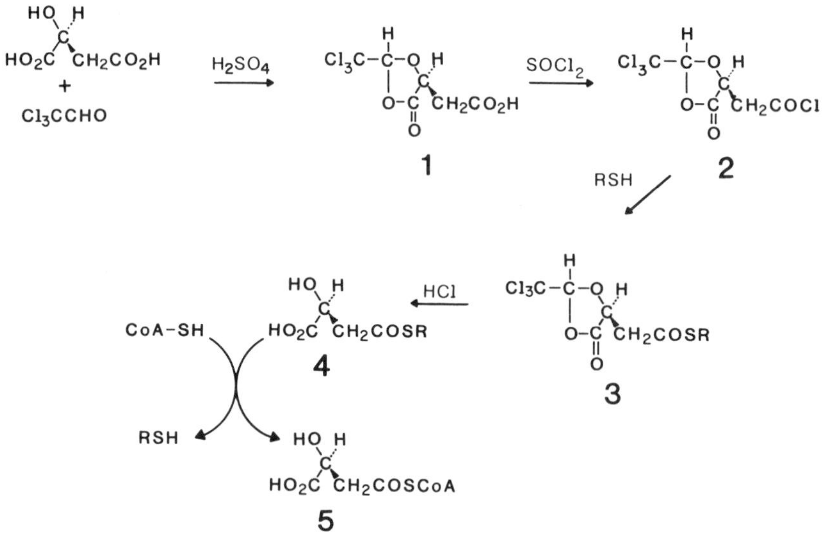

SCHEME 1. Synthesis of L-4-malyl-CoA. L-malic acid, $HO_2CCH_2CH(OH)CO_2H$; chloral hydrate, Cl_3CCHO; **1**, L-4-malylchloralide; **2**, L-4-malic acid chloralide chloride; **3**, L-4-malic acid chloralidyl-N-octanoylcysteamine; **4**, L-4-malyl-N-octanoylcysteamine; **5**, L-4-malyl-CoA; R, $CH_2CH_2NHCO(CH_2)_6CH_3$; CoASH, coenzyme A.

cautiously added, and the mixture is stirred for 2 hr at 0° then allowed to stand overnight at room temperature. The resulting solid is mixed with 200 g of ice, transferred to a separatory funnel, and extracted with four 50-ml volumes of ethyl acetate. The organic phase is washed once with water and dried by adding Na_2SO_4 (30 g) to the solution and allowing it to sit for several hours. The Na_2SO_4 is removed by filtering, washed with ethyl acetate, and added to the organic phase. Ethyl acetate is removed under vacuum using a rotary evaporator. The malic acid chloralide is recrystallized by dissolving the crystals in hot toluene or ethyl acetate (10 ml) and allowing the solution to cool at room temperature for a few hours. Excess solvent is removed using a Büchner funnel, and the crystals are transferred from the filter paper to a watch glass to air dry. Crystals of L-malic acid chloralide have a melting point of 140–141°. The yield is 21.2 g (82%).

L-Malic Acid Chloralide Chloride (2)

L-Malic acid chloralide crystals (13.2 g) are mixed with 21.5 ml of fresh thionyl chloride (Aldrich Chemical Co., Milwaukee, WI) and refluxed (with exclusion of moisture), for a period of 50 hr. (Although it has not been tested in our laboratory, addition of dimethylformamide may accelerate this reaction substantially.)[9] The thionyl chloride is then distilled off in a fume hood using a small distillation apparatus to remove the majority of thionyl chloride. The last residues of thionyl chloride are removed by vacuum distillation. Petroleum ether is used to recrystallize the L-malic acid chloralide chloride; a small amount is added to the flask, which is then warmed to dissolve the crystals, and the solution is transferred to a clean flask and allowed to cool for several hours. The crystals (needles) are filtered on a Büchner funnel and dried under vacuum in a desiccator containing $CaCl_2$ (to exclude moisture). Crystals with a melting point of 70–72° are obtained. Note: Waste thionyl chloride may be disposed of by adding it to a beaker of ice and water in the fume hood and allowing it to decompose over several hours.

N-Octanoylcysteamine

The compound N-octanoylcysteamine is required in the next stage of the synthesis. Its preparation is based on the thesis work of Eggerer[10] and is outlined in detail elsewhere in this series.[11] In summary, octanoyl chloride (Aldrich) is reacted with cysteamine (Aldrich) to produce the disulfide N,N'-dioctanoylcystamine, which is then reduced to N-octanoylcysteamine by sodium amalgam.

L-Malylchloralidyl-N-octanoylcysteamine (3)

N-Octanoylcysteamine (1.02 g, 5 mmol) and L-malic acid chloralide chloride (1.48 g, 5.25 mmol) are added to a 250-ml round-bottomed flask and dissolved in 5.0 ml of fresh tetrahydrofuran (peroxide free). To this solution is added 1.0 ml of dry pyridine (12.5 mmol), and the solution is tightly stoppered and allowed to sit overnight at room temperature. A reddish brown crystalline precipitate forms. This material is collected by suction filtration, dissolved in 40 ml of ethyl acetate, then washed in a separatory funnel with two 50-ml volumes of 1 N HCl, two 50-ml volumes of 1 N $KHCO_3$, and finally two 50-ml volumes of water. The aqueous washes are then sequentially back-extracted with one 50-ml volume of ethyl acetate; this series is repeated using another 50 ml of ethyl acetate. The combined ethyl acetate fractions are dried with 5 g of anhydrous

[9] L. F. Feiser and M. Feiser, "Reagents for Organic Synthesis," Vol. 1, p. 286. Wiley, New York, 1967.

[10] H. Eggerer, Dissertation, University of Munich, 1957.

[11] F. Lynen, this series, Vol. 5, p. 443.

Na_2SO_4 for 1 hr, filtered to remove the Na_2SO_4, and a second 5 g is added to the solution, which is left at room temperature overnight. The Na_2SO_4 is again filtered off and the solution evaporated to an oily liquid with a rotary evaporator. The residue is crystallized by adding toluene–petroleum ether (40%/60%), producing crystals with a melting point of 47–50°. The yield is 2.03 g (89%).

L-Malyl-N-octanoylcysteamine (4)

L-4-Malic acid chloralidyl-N-octanoylcysteamine (989 mg) is added to a 250-ml conical flask and dissolved in 20 ml of dimethylformamide. Concentrated HCl (10 ml) is added to the flask and the solution kept at 50°. Additional 5-ml volumes of HCl are added after 30 and 90 min, and the solution is incubated at 50° for another 5.5 hr. The mixture is poured into a beaker containing about 50 g of ice, and solid NH_4Cl is added with stirring until the solution is saturated. The mixture is extracted with three 15-ml volumes of ethyl acetate, and the ethyl acetate layers are pooled and washed with two 45-ml volumes of water. The second water wash may require the addition of a small amount of NaCl to assist separation and prevent emulsions from forming. The organic phase is extracted by adding 3-ml volumes to a 250-ml separatory funnel containing 50 ml of saturated $KHCO_3$ solution, with careful mixing after each addition. The extract ($KHCO_3$ layer) is acidified (0°) with 30–35 ml of 5 N HCl to pH 3–4. The acidified $KHCO_3$ solution is then extracted with one 50-ml, then two 25-ml volumes of ethyl acetate. The ethyl acetate layers are pooled and washed with two 50-ml volumes of water and the solution dried over anhydrous Na_2SO_4 overnight at room temperature. The Na_2SO_4 is filtered off and the ethyl acetate removed by rotary evaporation under vacuum. The product is recrystallized from ethyl acetate–petroleum ether (40%/60%) to give needles with a melting point of 114–115°. The yield is 452 mg (71%).

L-4-Malyl-CoA (5)

L-4-Malyl-CoA is prepared from L-malyl-N-octanoylcysteamine by a modification[4] of the original ester interchange method of Eggerer and Grünewälder.[7] In a stoppered 16 × 20 mm test tube, 16 mg (50 μmol) of the L-4-malyl cysteamine derivative (4) is dissolved in 4.0 ml of 0.2 M $KHCO_3$ (pH 8.1). Coenzyme A (20 mg, 26 μmol) is added, followed by 1 ml of fresh diethyl ether (peroxide free). The mixture is shaken at room temperature for 1 hr. The pH is 7.5 at this point. The mixture is then extracted with ether (peroxide free) in a continuous-extraction apparatus

(Cat. 6840, Ace Glass, Inc., Vineland, NJ) for 1 hr, keeping the part of the extraction apparatus containing the aqueous phase at 0° in an ice bath. The aqueous phase is then withdrawn, acidified to pH 4.0 at 0° with 5 M HCl (~0.1 ml), and extracted with diethyl ether for 3 hr to remove unreacted L-malyl-N-octanoylcysteamine. The dissolved ether is removed from the aqueous phase by blowing a stream of nitrogen or hydrogen through the solution. The resulting solution of L-4-malyl-CoA (~5.0 mM) is stored either at 0 or at −15° and is stable under these conditions for several weeks.

4-Malyl-CoA is assayed chemically by measuring the decrease in absorbance at 234 nm associated with alkaline hydrolysis of the thioester.[3] Enzymatic hydrolysis by citrate synthase, which is specific for the L isomer, may also be utilized. Measuring thioester absorption at 234 nm or determining the liberated coenzyme A results in the quantitative determination of L-4-malyl-CoA.[12] The assay of malyl-CoA lyase activity using L-4-malyl-CoA as the substrate is described elsewhere in this volume.[6]

Acknowledgments

The author is grateful to Dr. John Keller for helpful discussions and comments, and to Dr. Ken Kokjer for technical assistance.

[12] H. Eggerer, U. Remberger, and C. Grünewälder, *Biochem. Z.* **339**, 436 (1964).

---

[60] 3-Hexulose-6-phosphate Synthase from Thermotolerant Methylotroph *Bacillus* C 1

By N. Arfman, L. Bystrykh, N. I. Govorukhina, and L. Dijkhuizen

D-Ribulose 5-phosphate + formaldehyde ⇌ D-*arabino*-3-hexulose 6-phosphate

Thermotolerant methylotrophic *Bacillus* strains, like various other methylotrophic bacteria, employ the ribulose monophosphate cycle of formaldehyde fixation during growth on methanol.[1,2] The first step, condensation of formaldehyde and D-ribulose 5-phosphate to give D-*arabino*-3-hexulose 6-phosphate, is catalyzed by 3-hexulose-6-phosphate synthase. Enzymes with a similar function but different properties have been purified from *Methylomonas capsulatus,*[3] *Methylomonas* M15,[4] *Methylo-*

[1] L. Dijkhuizen, N. Arfman, M. M. Attwood, A. G. Brooke, W. Harder, and E. M. Watling, *FEMS Microbiol. Lett.* **52**, 209 (1988).

[2] N. Arfman, E. M. Watling, W. Clement, R. J. van Oosterwijk, G. E. de Vries, W. Harder, M. M. Attwood, and L. Dijkhuizen, *Arch. Microbiol.* **152**, 280 (1989).

[3] J. R. Quayle, this series, Vol. 90, p. 314.

[4] H. Sahm, H. Schütte, and M.-R. Kula, this series, Vol. 90, p. 319.

monas aminofaciens 77a,[5] *Mycobacterium gastri* MB19,[6] and *Methylophilus methylotrophus.*[7]

Assay Methods

Principle. Two assay methods can be used, namely, the discontinuous colorimetric measurement of the rate of ribulose 5-phosphate-dependent removal of formaldehyde (Method 1) and the continuous measurement of the rate of ribulose 5-phosphate- and formaldehyde-dependent formation of hexulose 6-phosphate (Method 2). Method 2 is a modified version of the assay described by van Dijken *et al.*[8] and depends on hexulose-6-phosphate isomerase as a coupling enzyme. The direct assay (Method 1) is less convenient but is more suitable for studies of pH and temperature effects, metal ion requirement, and substrate specificity.

Instead of using commercial ribulose 5-phosphate, which is known to partially inhibit the enzyme,[9] this compound is generated *in situ* from ribose 5-phosphate by ribose-5-phosphate isomerase.

Procedure for Assay Method 1. The assay is performed as described by Quayle.[3] Formaldehyde is measured by the procedure of Nash.[10]

Reagents for Assay Method 2

Potassium phosphate buffer, pH 7.0, 100 mM
Magnesium chloride, 250 mM
NADP+, 8 mM
Formaldehyde, 100 mM (prepared from paraformaldehyde[3])
D-Ribose 5-phosphate (sodium salt), 100 mM
Glucose-6-phosphate dehydrogenase (grade I, from yeast; Boehringer Mannheim)
Glucose-6-phosphate isomerase (phosphoglucoisomerase) (from yeast; Boehringer Mannheim)
Ribose-5-phosphate isomerase (phosphoriboisomerase) (type III, from yeast; Sigma Chemical Co., St. Louis, MO)
Hexulose-phosphate isomerase (phosphohexuloisomerase) (partially purified from *Bacillus* C1, see below)
Procedure for Assay Method 2. The assay mixture in a quartz cuvette

---

[5] N. Kato, H. Ohashi, Y. Tani, and K. Ogata, *Biochim. Biophys. Acta* **523**, 236 (1978).
[6] N. Kato, this volume [61].
[7] A. J. Beardsmore, Ph.D. thesis, University of Sheffield, 1983.
[8] J. P. van Dijken, W. Harder, A. J. Beardsmore, and J. R. Quayle, *FEMS Microbiol. Lett.* **4**, 97 (1978).
[9] T. Ferenci, T. Strøm, and J. R. Quayle, *Biochem. J.* **144**, 477 (1974).
[10] T. Nash, *Biochem. J.* **55**, 416 (1953).

(10-mm light path) contains the following, in a final volume of 1 ml: potassium phosphate buffer, 0.5 ml; magnesium chloride, 20 μl; NADP$^+$, 50 μl; ribose 5-phosphate, 50 μl; glucose-6-phosphate dehydrogenase, 1 unit; phosphoglucoisomerase, 1 unit; phosphoriboisomerase, 1.75 units; phosphohexuloisomerase, 1.3 units; hexulose-6-phosphate synthase sample, to give a change in absorbance of 0.1–0.2 per min; and water to adjust the volume to 0.95 ml. The mixture is preincubated at 50° for temperature equilibration and to ensure formation of an equilibrium mixture of ribose 5-phosphate and ribulose 5-phosphate. After a time interval of at least 5 min, the reaction is started by adding 50 μl of formaldehyde. The resulting rate of absorbance increase at 340 nm is taken to be due to hexulose-6-phosphate synthase activity. When necessary, formaldehyde-independent activity is corrected for by running parallel assays containing water instead of formaldehyde.

Assay Method for Phosphohexuloisomerase. Phosphohexuloisomerase can be assayed using a slightly modified version of Method 2. The reaction mixture contains 0.6 units of purified hexulose-6-phosphate synthase instead of 1.3 units of phosphohexuloisomerase. Formaldehyde is included during the preincubation period, and the reaction is started with the phosphohexuloisomerase sample.

Units of Enzyme Activity. One unit of hexulose-6-phosphate synthase activity is defined as the amount of enzyme catalyzing the formation of 1 μmol of hexulose 6-phosphate per minute at 50° under the reaction conditions described. Specific activity is expressed as units per milligram of protein. Protein is determined by the Bradford method,[11] with bovine serum albumin as a standard. One unit of phosphohexuloisomerase activity is defined as the amount of enzyme catalyzing the conversion of 1 μmol of hexulose 6-phosphate per minute at 50°.

Purification Procedure

Growth of Organism. Growth of *Bacillus* C1 is described elsewhere in this volume.[12] Cells are harvested by centrifugation at 3800 g for 10 min at 4° and washed twice with 50 mM potassium phosphate buffer, pH 7.5, containing 5 mM MgSO$_4$. The resulting pellet (3 g wet weight) is resuspended in 20 ml of the same buffer and stored at −20° until used for purification. Extracts and samples obtained in various purification steps (see below) are stored at −80° in the presence of 5 mM dithiothreitol

[11] M. M. Bradford, *Anal. Biochem.* **72,** 248 (1976).
[12] N. Arfman and L. Dijkhuizen, this volume [35].

(DTT). Hexulose-6-phosphate synthase activity is stabilized by adding ribose 5-phosphate (5 mM) and phosphoriboisomerase (1.75 units/ml).

Step 1: Preparation of Cell-Free Extract. Cells are disrupted in the presence of 5 mM DTT by passage through a French pressure cell operating at 1.4×10^5 kN/m^2. Unbroken cells and debris are removed by centrifugation at 25,000 g for 20 min at 4°. The supernatant thus obtained contains approximately 10 mg protein/ml. The viscosity of the extract is lowered by adding protamine sulfate (10% solution) to a final concentration of 1 mg/10 mg of protein. The resulting precipitate (nucleic acids) is removed by centrifugation at 25,000 g for 20 min at 4°. The supernatant is desalted in buffer A (20 mM Tris-HCl, pH 7.5, 5 mM MgSO$_4$, 5 mM 2-mercaptoethanol) by Sephadex G-25 gel filtration (PD10 column, Bio-Rad, Richmond, CA). Further purification steps are performed at room temperature with a Pharmacia fast protein liquid chromatography (FPLC) system.

Step 2: Anion-Exchange Chromatography. The desalted cell-free extract (320 mg of protein) is applied to a Fast Flow Q-Sepharose (Pharmacia) column (volume 25 ml, washed and equilibrated with buffer A) and eluted by applying a linear $0-1.0$ M potassium chloride gradient in 120 ml of buffer A, at a flow rate of 2.0 ml/min. Fractions of 4 ml are collected and analyzed for enzyme activity. Hexulose-6-phosphate synthase activity (assayed by Method 1) and phosphohexuloisomerase activity elute from the column at KCl concentrations of 150 and 250 mM, respectively. Synthase and isomerase peak fractions are pooled separately and prepared for hydrophobic interaction chromatography by adding ammonium sulfate to each pool to a final concentration of 1.7 M.

Step 3: Hydrophobic Interaction Chromatography. The hexulose-6-phosphate synthase sample (Q-Sepharose pool, 13.6 mg of protein) is applied to a FPLC Phenyl-Superose HR5/5 column (volume 1 ml; Pharmacia), equilibrated with 50 mM Tris-HCl, pH 7.0, 5 mM MgSO$_4$, 5 mM 2-mercaptoethanol containing 1.7 M (NH$_4$)$_2$SO$_4$ (buffer B). Bound protein is eluted by applying a linear $1.7-0$ M (NH$_4$)$_2$SO$_4$ gradient in 20 ml of buffer B, at a flow rate of 0.4 ml/min. Fractions of 0.5 ml are collected and analyzed for enzyme activity. Hexulose-6-phosphate synthase activity elutes from the column at (NH$_4$)$_2$SO$_4$ concentrations of around 680 mM. The peak fractions are combined and used for further characterization of enzyme properties. A typical purification is summarized in Table I.

Phosphohexuloisomerase is also further purified by hydrophobic interaction chromatography. Samples (16–17 mg of protein) from the isomerase-containing Q-Sepharose pool are applied onto a Phenyl-Superose HR5/5 column, which is run as described above. Phosphohexuloisomerase activity elutes from the column at (NH$_4$)$_2$SO$_4$ concentrations of around

TABLE I

PURIFICATION OF HEXULOSE-6-PHOSPHATE SYNTHASE FROM *Bacillus* C1

Fraction	Total volume (ml)	Total protein (mg)	Total activity[a] (units)	Specific activity (units/mg)	Purification (-fold)	Yield (%)
Step 1. Desalted extract	32	320	1133	3.5	1.0	100
Step 2. Q-Sepharose	7.5	13.6	804 (339)[b]	59.1 (24.9)	16.9 (7.1)	71 (30)
Step 3. Phenyl-Superose	2.6	6.11	703	115	32.9	62
Step 4. Superose 12	2.5	0.84	276[c]	64	18	24[c]

[a] Results obtained by Assay Method 2.

[b] Data in parentheses reflect the enzyme activity of the Q-Sepharose pool when stored in absence of phosphoriboisomerase and ribose 5-phosphate.

[c] The gel-filtration run is performed with a 500-μl sample from the Phenyl-Superose pool. To estimate the yield, the total activity has been corrected for the total Phenyl-Superose pool (factor 2.6/0.5).

1 M. The peak fractions from two separate runs, containing neglectable amounts of hexulose-6-phosphate synthase activity, are pooled and used for assaying hexulose-6-phosphate synthase activity in samples obtained in various purification steps. The partial purification of phosphohexuloisomerase is summarized in Table II.

Step 4: Gel-Filtration Chromatography. To check the homogeneity of the hexulose-6-phosphate synthase preparation and to determine the native molecular weight of the enzyme, a 500-μl sample (1.2 mg of protein) of the Phenyl-Superose pool is applied to a Pharmacia FPLC Superose 12 HR 10/30 gel-filtration column equilibrated with 100 mM Tris-HCl, pH 7.5, 5 mM MgSO$_4$, 5 mM 2-mercaptoethanol. Proteins are eluted with the same buffer, at a flow rate of 0.5 ml/min. Fractions of 0.5 ml are collected and analyzed for enzyme activity.

Properties

Purity. The homogeneity of the hexulose-6-phosphate synthase preparation obtained by the procedure described above has been checked by the following criteria: (1) gel-filtration chromatography showed only a single symmetric peak, and (2) analysis of approximately 5 μg of enzyme preparation by sodium dodecyl sulfate–polyacrylamide gel electrophoresis (SDS–PAGE) with silver staining gave only a single band, both in case of the Phenyl-Superose pool and the Superose 12 pool.

TABLE II

PURIFICATION OF PHOSPHOHEXULOISOMERASE FROM *Bacillus* C1

Fraction	Total volume (ml)	Total protein (mg)	Total activity (units)	Specific activity (units/mg)	Purification (-fold)	Yield (%)
Step 1. Desalted extract	32	320	1235	3.9	1.0	100
Step 2. Q-Sepharose	15	34.2	908	26.6	6.8	74
Step 3. Phenyl-Superose	2.2	1.18	566	480	123	46

Molecular Weights. The molecular weight of native hexulose-6-phosphate synthase as determined by gel filtration chromatography, using Bio-Rad gel-filtration standards (1,350–67,000 range) as a reference, corresponds to 32,000. Under denaturing conditions (SDS–PAGE) a single 27,000 protein species is observed, which indicates that the enzyme is monomeric.

Substrate Specificity. Formaldehyde fixation (determined with Method 1) is observed with ribulose 5-phosphate, generated from ribose 5-phosphate by phosphoriboisomerase (100%), commercial ribulose 5-phosphate (83%), and fructose 6-phosphate (37%). No activity is detected with the following compounds: ribose 5-phosphate, dihydroxyacetone phosphate, xylulose 5-phosphate, and sedoheptulose 7-phosphate.

Effect of Metal Ions. Hexulose-6-phosphate synthase activity (determined with Method 1) of desalted enzyme preparations is strongly stimulated by a variety of bivalent metal ions. The following relative activities are found (at 1 mM metal ion concentrations): Co^{2+}, 100%; Mn^{2+}, 100%; Mg^{2+}, 98%; Ni^{2+}, 97%; Zn^{2+}, 39%; Cu^{2+}, 33%; Ca^{2+}, 28%; no metals, 25%. No further decrease is observed when 1 mM EDTA is added to the control assay (without metal ions). Either hexulose-6-phosphate synthase has a residual (metal ion-independent) activity or it strongly binds metal ions.

Stability. The purified enzyme, in the presence of 5 mM MgSO$_4$, 5 mM ribose 5-phosphate, and 1.75 units/ml phosphoriboisomerase, is stable for at least 5 months at −80°. At 55°, the enzyme retains more than 50% of its activity ($t_{0.5}$) after 1 hr. At 65°, the $t_{0.5}$ is 30 min.

Michaelis Constants. The apparent K_m value for formaldehyde (at 5 mM ribulose 5-phosphate concentration) is 14.7×10^{-5} M, for commercial ribulose 5-phosphate (at 5 mM formaldehyde concentration) it is 70×10^{-5} M, and for ribulose 5-phosphate generated from ribose 5-phosphate (5 mM formaldehyde) it is 45×10^{-5} M (assuming a 100% conversion, the true value will be even lower). Clearly, the inhibitory component present in commercial ribulose 5-phosphate not only lowers the reaction rate but also affects the affinity for the substrate. The apparent K_m value for Mg^{2+} (after correcting for Mg^{2+}-independent activity) is 13×10^{-5} M.

pH Optimum. The enzyme has optimal activity (determined with Method 1) at pH 7.0.

Immunology. Polyclonal antibodies directed against purified hexulose-6-phosphate synthase from *Methylophilus methylotrophus,* which show cross-reactivity toward hexulose-6-phosphate synthases from various methylotrophic organisms, do not recognize the synthase described here.[13]

[13] J. M. Burton and C. W. Jones, unpublished.

[61] 3-Hexulose-6-phosphate Synthase from *Mycobacterium gastri* MB19

By NOBUO KATO

D-Ribulose 5-phosphate + formaldehyde $\rightleftharpoons$ D-*arabino*-3-hexulose 6-phosphate

3-Hexulose-6-phosphate synthase catalyzes the aldol condensation of formaldehyde with D-ribulose 5-phosphate and participates in the ribulose monophosphate pathway for formaldehyde fixation in some methylotrophic bacteria. This enzyme has so far been purified from a methane utilizer (*Methylomonas capsulatus*[1]), obligate methanol utilizers (*Methylomonas* M15,[2] *Methylomonas aminofaciens,*[3] and *Methylophilus methylotrophus*[4]), and facultative methanol utilizers (*Pseudomonas oleovolans*[5] and *Mycobacterium gastri*[6]). The enzyme from *M. gastri* also catalyzes the condensation of glycolaldehyde with D-ribulose 5-phosphate to yield a 4-heptulose 7-phosphate.

Assay Method

Principle. The enzyme is assayed by measuring the rate of the D-ribulose 5-phosphate-dependent disappearance of formaldehyde. D-Ribulose 5-phosphate is derived from D-ribose 5-phosphate by ribose-5-phosphate isomerase (phosphoriboisomerase) in the reaction mixture.

[1] J. R. Quayle, this series, Vol. 90, p. 314.
[2] H. Sahm, H. Schütte, and M.-R. Kula, this series, Vol. 90, p. 319.
[3] N. Kato, H. Ohashi, Y. Tani, and K. Ogata, *Biochim. Biophys. Acta* **523**, 236 (1978).
[4] A. J. Beardsmore, P. N. G. Aperghis, and J. R. Quayle, *J. Gen. Microbiol.* **128**, 1423 (1982).
[5] A. P. Sokolov and Y. A. Trotsenko, *Biochemistry (Engl. Transl.)* **43**, 782 (1978).
[6] N. Kato, N. Miyamoto, M. Shimao, and C. Sakazawa, *Agric. Biol. Chem.* **52**, 2659 (1988).

Reagents

Potassium phosphate buffer, 500 mM, pH 7.5

Magnesium chloride, 50 mM

Formaldehyde, 20 mM, prepared by heating 0.5 g of paraformaldehyde in 5 ml of water at 100° in a sealed tube for 15 hr; the
formaldehyde solution is standardized with glutathione-independent formaldehyde dehydrogenase[7] (Grade II, from *Pseudomonas
putida;* Nacalai Tesque, Inc., Kyoto), and the solution is diluted
accordingly

D-Ribose 5-phosphate, sodium salt, 50 mM

Ribose-5-phosphate isomerase (phosphoriboisomerase) (type I, Sigma
Chemical Co., St. Louis, MO), 10 units/ml in distilled water.

Procedure. The assay mixture consists of 50 μl of potassium phosphate
buffer, 50 μl of magnesium chloride, 50 μl of D-ribose 5-phosphate, 50 μl
of phosphoriboisomerase, and a limited amount of the enzyme (0.05 to 0.2
units), in a final volume of 0.45 ml. After a 10-min preincubation at 30°,
the reaction is started by the addition of 50 μl of formaldehyde and allowed
to proceed for 5 min at 30°. The reaction is stopped by adding 0.1 ml of
0.5 N HCl, and the precipitate (if one forms) is removed by centrifugation
at 16,000 g for 10 min. To 0.1 ml of the clear supernatant, 1.9 ml of water
is added (20-fold dilution). The formaldehyde remaining is determined
according to Nash.[8] A blank test, namely, a parallel assay without D-ribose
5-phosphate, is needed when a crude enzyme preparation is used.

Definition of Unit and Specific Activity. One unit of the enzyme activity
is defined as the amount of enzyme catalyzing the removal of 1 μmol
formaldehyde per minute at 30°. Specific activity is defined as the number
of units per milligram of protein.

Growth of the Organism

Mycobacterium gastri MB19 is cultivated at 30° for 3 days in a 2-liter
shake flask containing 1 liter of medium with the following composition
(per liter): 10 ml of methanol, 2 g of NaNO$_3$, 2 g of (NH$_4$)$_2$SO$_4$, 2 g of
K$_2$HPO$_4$, 1 g of KH$_2$PO$_4$, 0.2 g of MgSO$_4 \cdot$7H$_2$O, 1 g of yeast extract, a
vitamin mixture,[9] and trace metal salts,[9] pH 7.0. Methanol is added aseptically to the medium just before inoculation. The cells are harvested by
centrifugation at 6700 g for 20 min and then washed twice with 10 mM
potassium phosphate buffer (pH 7.0). The cell paste is stored at −20° until use.

---

[7] M. Ando, T. Yoshimoto, S. Ogushi, K. Rikitake, S. Shibata, and D. Tsuru, *J. Biochem.* **85,**
1165 (1979).

[8] T. Nash, *Biochem. J.* **55,** 416 (1953).

Purification Procedure

All procedures are performed at $0-5°$. To the buffer solution used are added 1 mM MgCl$_2$, 1 mM dithiothreitol, and 0.5 mM phenylmethylsulfonyl fluoride.

Step 1: Preparation of Cell-Free Extract. Frozen cell paste (100 g wet weight) is suspended in 1000 ml of 10 mM potassium phosphate buffer. The cell suspension is disrupted in 200-ml portions, separately, by ultrasonication for 30 min with a ultrasonic oscillator (19 kHz), followed by centrifugation at 14,000 g for 20 min. The resultant supernatant is used as the cell-free extract.

Step 2: Phenyl-Sepharose CL-4B Chromatography. To the cell-free extract is added solid NaCl to a concentration of 3 M under gentle stirring. The protein solution is put on a phenyl-Sepharose column (3 × 57 cm) previously equilibrated with 10 mM potassium phosphate buffer (pH 7.0) containing 3 M NaCl. After washing the column with 1.2 liters of the equilibration buffer, elution is carried out with a gradient of decreasing NaCl concentration and increasing ethylene glycol concentration (the final concentrations being 0 and 50%, respectively; total volume, 2 liters) at a flow rate of 15 cm/hr. The active fractions are collected, concentrated to about 100 ml by ultrafiltration, using an Amicon YM10 membrane (Danvers, MA), and then dialyzed against 10 mM Tris-HCl buffer (pH 8.2).

Step 3. DEAE-Sephacel Chromatography. The dialyzed enzyme solution is applied to a DEAE-Sephacel column (1.6 × 50 cm) previously equilibrated with 10 mM Tris-HCl buffer (pH 8.2), then the column is washed with 500 ml of the buffer. The enzyme is eluted with a linear gradient between 300 ml of 10 mM and 300 ml of 100 mM Tris-HCl buffer (pH 8.2). The flow rate is 10 cm/hr. The active fractions are collected, concentrated as described above, and dialyzed against 10 mM potassium phosphate buffer (pH 7.0).

Step 4: Second Phenyl-Sepharose Chromatography. To the dialyzed solution is added solid NaCl to a concentration of 3 M, followed by application to a Phenyl-Sepharose column (2.6 × 30 cm) equilibrated with 10 mM potassium phosphate buffer (pH 7.0) containing 3 M NaCl. The enzyme is eluted with a concentration gradient of NaCl (3 to 0 M) and ethylene glycol (0 to 50%) in 10 mM potassium phosphate buffer, pH 7.0 (total volume, 800 ml), at a flow rate of 10 cm/hr. The active fractions are concentrated and then dialyzed against 50 mM potassium phosphate buffer (pH 7.0).

The concentrated enzyme solution ($\sim$2 mg/ml) can be stored at $0°$ for

[9] N. Kato, Y. Yamagami, Y. Kitayama, M. Shimao, and C. Sakazawa, *J. Biotechnol.* **1**, 295 (1984).

TABLE I

PURIFICATION OF 3-HEXULOSE-6-PHOSPHATE SYNTHASE FROM *Mycobacterium gastri* MB19

Procedure	Total protein (mg)	Total activity (units)	Specific activity (units/mg protein)	Purification (-fold)	Yield (%)
Cell-free extract	3850	12,320	3.2	1	110
First Phenyl-Sepharose	730	10,290	14.1	4.4	84
DEAE-Sephacel	103	3510	34.1	11	28
Second Phenyl-Sepharose	52	3856	74.2	23	31

at least 1 month without loss of activity, or it can be kept in 20% glycerol at −70° for a longer period. A typical purification is summarized in Table I.

Properties

Purity. The purified enzyme preparation gives a single band on polyacrylamide (7%) gel electrophoresis at pH 8.3 and on gels containing sodium dodecyl sulfate (SDS).

Molecular Weight and Subunit Structure. The relative molecular weight of the 3-hexulose-6-phosphate synthase is determined to be 43,000 by high-performance liquid chromatography on a column of TSK gel G-3000SW (Tosoh, Tokyo). The SDS gel profile of the enzyme comprises a single band, and a molecular weight of 24,000 is estimated for the subunit. Only methionine is found as the N-terminal residue. The native enzyme is a molecule composed of two, possibly identical, subunits.

Substrate Specificity and Kinetic Constants. The enzyme is highly specific for D-ribulose 5-phosphate as an acceptor of aldehydes. The enzyme shows no detectable formaldehyde fixation activity with the following compounds: D-ribulose 1,5-bisphosphate, D-ribose 5-phosphate, D-erythrose 4-phosphate, D-fructose 6-phosphate, D-fructose 1,6-bisphosphate, D-glucose 6-phosphate, D-glucose 1-phosphate, and D-xylulose 5-phosphate. The enzyme utilizes glycolaldehyde and methylglyoxal in the presence of D-ribulose 5-phosphate as the acceptor, as well as formaldehyde. The activities toward glycolaldehyde and methylglyoxal are assayed by measuring the rate of the D-ribulose 5-phosphate-dependent disappearance of the aldehydes by the MBTH (3-methyl-2-benzothiazolinone hydrazone) method.[10] The product of aldol condensation between glycolaldehyde and D-ribulose 5-phosphate is isolated from the reaction mixture by several chromatographic steps and identified as a 4-heptulose 7-phosphate by [1]H

[10] E. Sawicki, T. R. Hauser, T. W. Stanley, and E. Elbert, *Anal. Chem.* **33**, 93 (1961).

NMR, ^{13}C NMR, and GC–MS analyses.[6] The apparent K_m values for aldehydes (determined at 5 mM D-ribulose 5-phosphate) are as follows: formaldehyde, 1.4 mM; glycolaldehyde, 4.3 mM; methylglyoxal, 5.7 mM. The V_{max} values for glycolaldehyde and methylglyoxal are 239 and 344%, respectively, of that for formaldehyde.

Effect of pH. The enzyme exhibits optimum activity at pH 7.5 to 8.0 (potassium phosphate buffer) and is stable in the pH range of 6.5 to 7.5 at 30° for 2 hr.

Effects of Metal Ions. The enzyme shows an essential requirement for Mg^{2+} or Mn^{2+} (1 mM each). These cations are also essential for the enzyme's stability. The enzyme is completely inhibited by the presence of EDTA, tiron, and *o*-phenanthroline (2 mM each) and is inactivated irreversibly on dialysis against buffer without Mg^{2+}.

[62] 3-Hexulose-6-phosphate Synthase from *Acetobacter methanolicus* MB58

By ROLAND H. MÜLLER and WOLFGANG BABEL

D-Ribulose 5-phosphate + formaldehyde ⇌ D-*arabino*-3-ketohexulose 6-phosphate

3-Hexulose-6-phosphate synthase catalyzes the condensation of ribulose 5-phosphate with formaldehyde to give D-*arabino*-3-ketohexulose 6-phosphate in a key step of the pathway for assimilation of reduced C_1 compounds. The purification and characterization of hexulose-phosphate synthase from *Acetobacter methanolicus* MB58 were the subject of a previous report.[1] Purification procedures of related enzymes from other sources have been described.[2–4]

Assay Methods

Principle. Determination of activity is based on the ribulose 5-phosphate-dependent decrease of formaldehyde. Ribulose 5-phosphate is generated from ribose 5-phosphate by the action of ribose-5-phosphate isomerase (phosphoriboisomerase) because of the inhibitory properties of commercial ketopentose phosphate.[5]

[1] R. H. Müller and W. Babel, *Wiss. Z. Karl-Marx-Univ. Leipzig, Math.-Naturwiss. Reihe* 27, 67 (1978).
[2] N. Kato, H. Ohashi, Y. Tani, and K. Ogata, *Biochim. Biophys. Acta* 523, 236 (1978).
[3] J. R. Quayle, this series, Vol. 90, p. 314.
[4] H. Sahm, H. Schütte, and M.-R. Kula, this series, Vol. 90, p. 319.
[5] T. Ferenci, T. Strøm, and J. R. Quayle, *Biochem. J.* 144, 477 (1974).

Reagents

Sodium potassium phosphate buffer, 100 m*M*, pH 7.2

Magnesium chloride, 60 m*M*

Formaldehyde, 50 m*M* (prepared by heating 500 mg of paraformaldehyde in 20 ml water in a sealed tube for 24 hr at 100°; the
formaldehyde concentration is checked by the method of Nash[6] and
the solution diluted accordingly)

Ribose 5-phosphate, sodium salt (Boehringer, Mannheim, FRG),
50 m*M*

Ribose-5-phosphate isomerase (phosphoriboisomerase) (type II,
Sigma Chemical Co., St. Louis, MO)

Procedure. The assay mixture contains the following, in a final volume
of 1.5 ml: sodium potassium phosphate buffer, 1.2 ml; magnesium chloride, 0.1 ml; ribose 5-phosphate, 0.1 ml; phosphoriboisomerase, 0.4 units;
and up to 0.06 units of hexulose-phosphate synthase. This mixture is
preincubated for 30 min at 30° to generate ribulose 5-phosphate in equilibrium with ribose 5-phosphate. Then the reaction is started by adding
0.1 ml of formaldehyde. Its consumption at 30° is followed by taking
aliquots of 50 μl from the assay mixture and adding them to 0.2 ml of 5%
(w/v) trichloroacetic acid solution. The remaining formaldehyde is determined according to the method of Nash.[6] It should be mentioned that
starting the reaction with hexulose-phosphate synthase does not significantly alter the properties.

Units. One unit of hexulose-phosphate synthase is defined as that
amount of enzyme catalyzing the consumption of 1 μmol of formaldehyde
per minute at 30°. Specific activity is expressed as units per milligram of
protein.

Purification Procedure

Growth of Organism. Acetobacter methanolicus MB58 is cultivated in a
methanol-limited chemostat at a dilution rate of 0.15 hr^{-1}, a pH of 4.0, and
a temperature of 30° on a mineral salts medium containing the following
per liter: NH_4Cl, 3.4 g; KH_2PO_4, 0.31 g; $CaCl_2 \cdot 2H_2O$, 20 mg;
$MgSO_4 \cdot 7H_2O$, 15 mg; $FeSO_4 \cdot 7H_2O$, 5 mg; $ZnSO_4 \cdot 7H_2O$, 0.44
mg; $MnSO_4 \cdot 4H_2O$, 0.82 mg; $CuSO_4 \cdot 5H_2O$, 0.78 mg; $Na_2MoO_4 \cdot 2H_2O$,
0.25 mg; and pantothenic acid, 0.5 mg. The effluent medium is collected
in an ice bath. Cells are harvested by centrifugation for 20 min at 5000 *g* at
4° and washed 2 times with 10 m*M* sodium potassium phosphate buffer
containing 5 m*M* magnesium chloride. The pellet is stored at 0–4°, if

[6] T. Nash, *Biochem. J.* **55**, 416 (1953).

consecutive steps are done immediately, or frozen at $-20°$ for later use. If not otherwise stated, all consecutive steps are performed at $0-4°$.

Step 1: Preparation of Cell-Free Extract. Cell paste is resuspended in precooled 10 mM sodium potassium phosphate buffer, pH 7.2, containing 5 mM magnesium chloride at a cell concentration of 24 mg dry weight/ml. This suspension is treated by ultrasonification at 150 W for 5 min at $0-4°$. Cell-free supernatant is obtained after centrifugation for 20 min at 12,000 g. The supernatant is used for further treatment.

Step 2: Ammonium Sulfate Fractionation. To the supernatant from Step 1, solid ammonium sulfate is added slowly with stirring. Stirring is continued for 1 hr after the last addition. The protein precipitating between 1.65 and 2.85 M ammonium sulfate is collected by centrifugation at 12,000 g for 30 min and dissolved in a small amount of 10 mM sodium potassium phosphate buffer, pH 7.2, containing 5 mM MgCl$_2$. This solution is then dialyzed against 2 liters of the same buffer for 20 hr.

Step 3: DEAE-Cellulose Chromatography. The dialyzed protein solution of Step 2 is applied to a column (10 × 2.5 cm) of DEAE-cellulose (DEAE-SS, Serva, Heidelberg, FRG), previously equilibrated with 10 mM sodium potassium phosphate buffer, pH 7.2, containing 5 mM magnesium chloride. Protein is eluted with a flow rate of 50 ml/hr by a discontinuous sodium chloride gradient increased in 50 mM concentration steps after the respective protein peak measured at 254 nm has appeared. Hexulose-phosphate synthase activity is eluted at 0.1 M sodium chloride. Peak fractions with the highest specific activity are pooled and concentrated by ultrafiltration through a Diaflo M10 membrane filter (Amicon, Danvers, MA).

Step 4: Gel Filtration. The concentrated protein solution from Step 3 is applied to a column (2.5 × 100 cm) of Sephadex G-100 superfine (Pharmacia, Uppsala, Sweden), previously equilibrated with 10 mM sodium potassium phosphate buffer, pH 7.2, containing 5 mM magnesium chloride. Protein is eluted at a rate of 10 ml/hr, and fractions of 6.2 ml are collected. Hexulose-phosphate synthase activity appears after about 340 ml. Seven fractions with the highest specific activity are pooled and concentrated by ultrafiltration through a Diaflo M10 membrane filter. The concentrated protein solution is stored either at $-20°$ or saturated with 3.2 M ammonium sulfate at $0-4°$. A typical purification procedure is summarized in Table I.

Properties

Purity. Analysis of 20 μg protein by polyacrylamide gel electrophoresis (7.5% gel according to gel system 1 in Maurer,[7] separation with 4 mA per

---

[7] H. R. Maurer, "Theorie und Praxis der Diskontinuierlichen Polyacrylamidgel-Elektrophorese." de Gruyter, Berlin, 1968.

TABLE I
PURIFICATION OF 3-HEXULOSE-PHOSPHATE SYNTHASE FROM *Acetobacter methanolicus* MB58

Fraction	Total volume (ml)	Total protein (mg)	Total activity (units)	Specific activity (units/mg protein)	Yield (%)
Step 1. Cell-free extract	224	2550	3650	1.4	1
Step 2. (NH$_4$)$_2$ SO$_4$ fraction (1.65–2.85 M)	42	1380	4450	3.2	2.3
Step 3. Pooled, concentrated fractions after DEAE-cellulose chromatography	7	61	1880	31.0	22
Step 4. Pooled, concentrated fraction after gel filtration	3.25	12.7	810	64.0	46

gel) and of 4 μg protein in the presence of sodium dodecyl sulfate[8] (separation with 8 mA per gel) shows only one protein band on staining with Coomassie blue. Enzymatic analysis, however, reveals traces of phosphoriboisomerase and hexulose-phosphate isomerase.

Molecular Weight and Subunit Structure. The molecular weight of hexulose-phosphate synthase from *Acetobacter methanolicus* MB58 was found to be 80,000 ± 2,000 as estimated by gel permeation with cytochrome c as the reference on Sephadex G-150 superfine according to the method of Jarowek.[9] Sodium dodecyl sulfate–polyacrylamide gel electrophoresis gave a molecular weight for the subunits of 20,400. This suggests a tetrameric structure for the hexulose-phosphate synthase from *A. methanolicus.*

Effect of Metal Ions. Bivalent cations are essential for hexulose-phosphate synthase activity. Most effective is Mg^{2+} and by about 20% less is Mn^{2+}. Co^{2+}, Fe^{2+}, and Ca^{2+} in 1 mM concentration show less than one-half the activity obtained with Mg^{2+}. Cu^{2+} is inhibitory, and the effect is partly abolished by Mg^{2+}.

Stability. The purified enzyme is stable for at least 1 year at $-20°$ or at $0-4°$ in the presence of 3.2 M ammonium sulfate. At 40°, in the presence of 5 mM MgCl$_2$, no decrease in activity is observed within 6 hr. At 50 and 60° the enzyme is stable for 2.5 and 0.5 hr, respectively; one-half the initial activities are found after 6 and 1.5 hr, respectively.

pH Optimum. The enzyme has optimum activity in the range between pH 7 and 8.

Temperature Optimum. The optimum reaction temperature is 60°.

[8] K. Weber, J. R. Pringle, and M. Osborn, this series, Vol. 26, p. 3.
[9] D. Jarowek, *Chromatographia* 3, 414 (1970).

From an Arrhenius plot, activation energies of 47.0 and 37.8 kJ/mol are found with a break at 37°.

Kinetic Properties. The progress curve of hexulose-phosphate synthase activity is biphasic. An initial phase of high activity (burst) is transformed to a lower, again linear, phase. Accumulation of a certain amount of the reaction product D-*arabino*-3-hexulose 6-phosphate seems to be responsible.[10] The kinetics of hexulose-phosphate synthase from *Acetobacter methanolicus* MB58 show a complex property in that intermediary plateau regions are evident, indicative of the presence of multiple enzyme forms.[11] This was found also with the enzyme from *Pseudomonas* W6 in permeabilized cells,[12] but indications of the existence of this property were found also with purified enzymes from other sources.[13]

[10] R. H. Müller and W. Babel, *Acta Biol. Med. Ger.* **40**, 123 (1981).
[11] R. H. Müller and W. Babel, *Acta Biol. Med. Ger.* **40**, 137 (1981).
[12] W. Babel and R. H. Müller, *Z. Allg. Mikrobiol.* **17**, 175 (1977).
[13] R. H. Müller and W. Babel, *Z. Allg. Mikrobiol.* **20**, 325 (1980).

[63] Transaldolase Isoenzymes from *Arthrobacter* P1

By P. R. LEVERING and L. DIJKHUIZEN

Fructose 6-phosphate + erythrose 4-phosphate → sedoheptulose 7-phosphate + glyceraldehyde 3-phosphate

The enzyme transaldolase (EC 2.2.1.2) has multiple functions in methylotrophic bacteria. In *Arthrobacter* P1[1] transaldolase is a key enzyme in the ribulose monophosphate (RuMP) cycle during growth on methylated amines.[2] Moreover, as in other prokaryotic and eukaryotic organisms, the enzyme functions in the nonoxidative pentose phosphate pathway to furnish precursors for the biosynthesis of aromatic amino acids and nucleic acids.[3] Characterization of transaldolase mutants of *Arthrobacter* P1 revealed that the enzyme also plays an indispensable role in the catabolism of compounds such as gluconate and xylose. The latter mutants (i.e., strain Art 98) were still able to metabolize methylamine, displaying high activity levels of transaldolase, which led to the identification of transaldolase isoenzymes in *Arthrobacter* P1.[4] Kinetic studies on the purified isoenzymes

[1] P. R. Levering, J. P. van Dijken, M. Veenhuis, and W. Harder, *Arch. Microbiol.* **129**, 72 (1981).
[2] P. R. Levering, L. Dijkhuizen, and W. Harder, *FEMS Microbiol. Lett.* **14**, 257 (1982).
[3] O. Tsolas and B. L. Horecker, *in* "The Enzymes" (P. D. Boyer, ed.), 3rd Ed., Vol. 7, p. 259. Academic Press, New York, 1972.
[4] P. R. Levering and L. Dijkhuizen, *Arch. Microbiol.* **144**, 116 (1986).

showed that both are functionally similar but that their synthesis is regulated differently. One of the enzymes is synthesized constitutively, whereas the other is induced only during growth on C_1 compounds. The procedures followed for the separation and purification of both isoenzymes from *Arthrobacter* P1 wild type and mutant strain Art 98 are described below.

Transaldolase (isoenzymes) has been purified from various yeast sources (brewers' yeast,[5] bakers' yeast,[6] and *Candida utilis*[7,8]) and from mammalian tissue.[9] To our knowledge the characterization of the transaldolase isoenzymes from *Arthrobacter* P1[4] represents the first detailed report on the enzyme of prokaryotic origin.

Assay Method

Principle. Glyceraldehyde 3-phosphate produced in the reaction with fructose 6-phosphate and erythrose 4-phosphate as substrates is converted to dihydroxyacetone phosphate by triose-phosphate isomerase. The formation of this compound is measured with NADH and α-glycerol-phosphate dehydrogenase. The reaction is followed spectrophotometrically by measuring the change in absorbance at 340 nm. The observed rate of NADH oxidation is, under appropriate conditions, proportional to the amount of transaldolase added.

Reagents

TEA–EDTA buffer (triethanolamine–EDTA buffer), pH 8.0: a solution containing 200 mM triethanolamine and 40 mM EDTA
NADH, 3 mM
D-Fructose 6-phosphate, disodium salt, 100 mM
D-Erythrose 4-phosphate, sodium salt (grade II), 50 mM
α-Glycerol-phosphate dehydrogenase (from rabbit muscle), 35 units/ml
Triose-phosphate isomerase (from rabbit muscle), 200 units/ml

All biochemicals and enzymes were purchased from Boehringer (Mannheim, FRG), with the exception of erythrose 4-phosphate (Sigma Chemical Co., St. Louis, MO).

Procedure. The assay mixture in a quartz cuvette (10-mm light path) contains TEA–EDTA buffer, 0.5 ml; NADH, 50 μl; fructose 6-phosphate,

[5] B. L. Horecker and P. Z. Smyrniotis, *J. Biol. Chem.* **212**, 811 (1955).
[6] R. Venkataraman and E. Racker, *J. Biol. Chem.* **236**, 1876 (1961).
[7] O. Tsolas and B. L. Horecker, *Arch. Biochem. Biophys.* **136**, 287 (1970).
[8] O. Tsolas and L. Joris, this series, Vol. 42, p. 290.
[9] E. Kuhn and K. Brand, *Biochemistry* **11**, 1767 (1972).

50 μl; α-glycerol-phosphate dehydrogenase, 50 μl; triose phosphate isomerase, 50 μl; transaldolase sample (0.01–0.1 ml), to give a change in absorbance of 0.1–0.2 per min; and water to adjust the volume to 0.99 ml. This mixture is incubated at 30° for several minutes to record any endogenous NADH oxidation. The transaldolase reaction is then started by the addition of 10 μl of erythrose 4-phosphate. The resulting rate of absorbance decrease at 340 nm, minus the erythrose 4-phosphate-independent rate, is taken as transaldolase activity.

Unit of Enzyme Activity. One unit is defined as the amount of enzyme catalyzing the formation of 1 μmol of glyceraldehyde 3-phosphate per minute at 30°. Specific activity is expressed as units per milligram of protein. Protein is determined by the Lowry method,[10] with bovine serum albumin as a standard.

Purification Procedure

Growth of Organism. *Arthrobacter* P1 (NCIB 11625)[1] and the gluconate-negative mutant strain Art 98[4] are grown at 30° in 3-liter conical flasks containing 1 liter of a mineral salts medium of the following composition (per liter): $NaH_2PO_4 \cdot H_2O$, 0.5 g; K_2HPO_4, 1.55 g; $(NH_4)_2SO_4$, 1.0 g; $MgSO_4 \cdot 7H_2O$, 0.2 g; and trace element solution,[11] 0.2 ml. Heat-sterilized solutions of potassium gluconate or methylammonium chloride are added at final concentrations of 20 and 50 mM, respectively. Incubation is on a rotary shaker at 250 rpm. Cultures are harvested in the late-exponential growth phase by centrifugation at 6000 g for 10 min at 4° and washed once with 50 mM potassium phosphate buffer, pH 7.0. The resulting pellet is resuspended in the same buffer, to a concentration of 50–75 mg dry weight/ml. These suspensions are stored at −20° until required for purification.

Step 1: Preparation of Cell-Free Extracts. The concentrated cell suspensions are sonicated in aliquots of 3 ml at 20 kHz for 15 times, 15 sec each, at 0–4° in an MSE 100 W ultrasonic disintegrator (MSE Ltd., Crawley, Sussex, UK), with 45-sec intermittent cooling periods. Unbroken cells and debris are removed by centrifugation at 40,000 g for 10 min at 4°, and the supernatant is again centrifuged at 40,000 g for 30 min at 4°. The supernatants thus obtained contain approximately 25 mg protein/ml. The further purification steps are performed at room temperature with a high-performance liquid chromatography (HPLC) system (Waters Associates, Milford,

[10] O. H. Lowry, N. H. Rosebrough, A. L. Farr, and R. J. Randall, *J. Biol. Chem.* **193**, 265 (1951).
[11] W. Vishniac and M. Santer, *Bacteriol. Rev.* **21**, 195 (1957).

MA), consisting of a Model 6000A solvent delivery system equipped with a U6K injector and a Model 441 absorbance detector.

Step 2: DEAE-Polyol Chromatography. Protein samples (0.25 – 1.0 ml) are injected in a DEAE-Si 100 polyol column (9.5 × 500 mm; Serva, Heidelberg, FRG), equilibrated with 25 mM potassium phosphate, pH 7.0 (buffer A), and eluted with a nonlinear gradient of 0 – 100% solvent B (25 mM potassium phosphate plus 1 M sodium chloride, pH 7.0) in buffer A at a flow rate of 3.0 ml/min. The stepwise gradient (0 – 35% B in 3 min, 35 – 60% B in 15 min, and 60 – 100% in 1 min) is controlled by an Acorn Atom microcomputer (Acorn Computer Ltd., Cambridge, UK), interfaced with a three-way solenoid valve (type LFYA, Lee Company, Westbrook, CT) as described previously.[12] The profile of the gradient is checked by measuring the conductivity of eluate samples. Fractions of 1.5 ml are collected and analyzed for enzyme activity. The transaldolase isoenzymes typically elute from the column at NaCl concentrations between 400 and 525 mM (Fig. 1). Depending on the source of the cell-free extract employed, three different separation profiles are recognized:[4]

a. Extracts from methylamine-grown cells of wild-type *Arthrobacter* P1 contain both the C_1-inducible and the constitutive transaldolase activity. Application of a protein sample of 6.3 mg (0.25 ml) to the column results in separation of transaldolase in two distinct activity peaks (at elution volumes around 58.5 and 67.5 ml, respectively; Fig. 1A). Although the isoenzymes can be separated from each other at this step, preference is given to the purification methods outlined below (Steps 2b and 2c) which exclude any cross-contamination.

b. Extracts from gluconate-grown cells of wild-type *Arthrobacter* P1 contain only the constitutive transaldolase activity. A protein sample of 25 mg (1.0 ml) is run and transaldolase peak fractions are pooled (total 6 ml; eluate between 64.5 and 70.5 ml; Fig. 1B). Several runs are performed and the pooled peak fractions combined.

c. Extracts from methylamine-grown cells of mutant strain Art 98 contain only the C_1-inducible transaldolase activity. A protein sample of 25 mg (1.0 ml) is run and transaldolase peak fractions are pooled (total 6 ml; eluate between 57 and 63 ml; Fig. 1C). Several runs are performed and the pooled peak fractions combined.

Step 3: Protein Concentration. Combined pooled peak fractions from the DEAE column (Step 2b or 2c) are concentrated (to 1/25 or 1/30 the volume) with a Minicon B-15 ultrafiltration system (Amicon, Danvers, MA).

Step 4: Gel-Permeation Chromatography. Samples (0.1 ml) of the con-

[12] R. van der Zee and G. W. Welling, *J. Chromotogr.* **292**, 412 (1984).

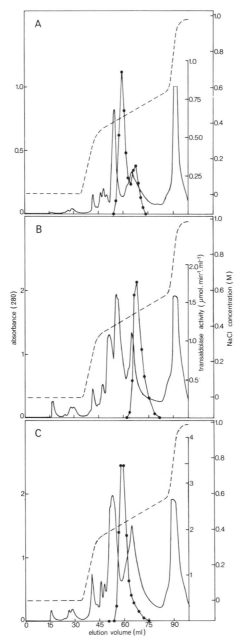

Fig. 1. DEAE-polyol chromatography elution profiles of cell-free extracts of (A) methyla-mine- or (B) gluconate-grown cells of wild-type *Arthrobacter* P1 and (C) methylamine-grown cells of mutant strain Art 98. The transaldolase specific activities in these cell-free extracts were (A) 1.4, (B) 0.64, (C) 1.0 units/mg protein. _____, absorbance at 280 nm; – – –, sodium chloride gradient; ●, transaldolase. Reproduced (with permission) from P. R. Levering and L. Dijkhuizen, *Arch. Microbiol.* **144,** 116 (1986).

centrated enzyme preparations from Step 3, containing 0.5–1.0 mg protein, are applied to a Protein Pak 300-SW gel permeation column (7.5 × 300 mm; Waters Associates) and eluted with 200 mM potassium phosphate buffer, pH 7.0, at a flow rate of 0.5 ml/min. Fractions (0.25 ml) are collected and analyzed for transaldolase activity.

a. For material obtained in Steps 2b and 3, activities of the constitutive transaldolase are recovered in the 8.25–10.25 ml elution volume range. Again, several runs are performed, and the fractions containing the highest transaldolase activities are pooled (1 ml per run), combined, and stored at −20°.

b. For material obtained in Steps 2c and 3, activities of the C_1-inducible transaldolase are recovered in the 9.25–11.25 ml elution volume range. The peak fractions of several runs (1 ml per run) are pooled, combined, and stored at −20°.

The final preparations obtained (Steps 1–4) contain 0.09 and 0.1 mg protein/ml for the constitutive and C_1-inducible isoenzyme, respectively. The overall yield (total units) for both enzymes is approximately 20%.

Properties

Purity. The specific activities of the partially purified constitutive and C_1-inducible transaldolases are 16.0 and 27.5 units/mg protein, corresponding to overall purifications of 25- and 27.5-fold, respectively. Activities of the other enzymes involved in the nonoxidative branch of the pentose phosphate pathway are also detected in these preparations (expressed as percent activity of the constitutive and C_1-inducible transaldolase, respectively): transketolase (0.7, 0.1); ribulose-5-phosphate 3-epimerase (2.0, 3.7), and ribose-5-phosphate isomerase (4.7, 2.2). These activities are low, however, compared to the levels originally present in cell-free extracts of *Arthrobacter* P1.[2,4]

pH Optima. Both transaldolase isoenzymes display maximum activities at pH 8.0, falling off very little in value between pH 7.8 and 8.2. This is similar to transaldolase purified from yeast.[3,6]

Michaelis Constants. Transaldolase is inhibited reversibly by the high concentrations of orthophosphate present in the samples, as has been reported for the yeast enzyme.[3] Orthophosphate is precipitated by adding sufficient amounts of calcium acetate, and the precipitate is removed by centrifugation. Calcium acetate itself does not affect transaldolase activities. The supernatants obtained are used for the determination of apparent K_m values. The constitutive and C_1-inducible transaldolase isoenzymes display similar apparent K_m values for fructose 6-phosphate (2.1 and

2.3 mM, respectively) and erythrose 4-phosphate (70 and 100 μM, respectively).

Molecular Weights. The molecular weights of the purified isoenzymes are estimated with the HPLC gel-permeation technique described above.[13] The molecular weights of the isoenzymes differ little, but significantly, from each other, namely, 47,500 ± 2,000 for the C_1-inducible transaldolase and 52,000 ± 2,000 for the constitutive enzyme. When using the same method a molecular weight of about 64,000 is found for bakers' yeast transaldolase (Sigma) which is close to published values.[14] Sodium dodecyl sulfate (SDS)–polyacrylamide gelelectrophoresis[15,16] [with the calibration proteins Combithek (104558, Boehringer) as references] yields one major band in the 50,000 region with both isoenzymes, indicating that they are monomeric.

Stability. Both partially purified transaldolase isoenzymes are stable at −20° for at least 2 months. A remarkable difference is observed after incubation at elevated temperatures. The C_1-inducible enzyme loses 90% of its activity within 2 min at 50° (totally inactivated after 10 min), whereas the constitutive transaldolase remains fully active under these conditions. Similar observations are made when using crude cell-free extracts.

[13] Samples of 50 μl (~5 μg protein) are run together with the following marker proteins: chrymotrypsinogen A, molecular weight 25,000; albumin, 45,000; and albumin, 68,000 (5 μg of each; Boehringer, Mannheim).
[14] R. Venkataraman and E. Racker, *J. Biol. Chem.* **236,** 1883 (1961).
[15] U. K. Laemmli and K. Favre, *J. Mol. Biol.* **80,** 575 (1973).
[16] W. Wray, T. Boulikas, V. P. Wray, and R. Hancock, *Anal. Biochem.* **118,** 197 (1981).

[64] Cytochemical Staining Methods for Localization of Key Enzymes of Methanol Metabolism in *Hansenula polymorpha*

By Marten Veenhuis and Ida J. van der Klei

During growth of cells of the yeast *Hansenula polymorpha* on methanol, peroxisomes fulfill an indispensable role because these organelles harbor the key enzymes involved in the metabolism of this substrate. These enzymes include (1) alcohol oxidase, which catalyzes the oxidation of methanol into formaldehyde and hydrogen peroxide, (2) catalase, which decomposes the hydrogen peroxide produced into water and dioxygen, and (3) dihydroxyacetone synthase (formaldehyde transketolase), which is involved in the assimilation of formaldehyde.[1-5] The localization of these enzymes has been established both biochemically by cell fractionation studies and *in situ* by cytochemical and immunocytochemical methods.

General Outline of Cytochemical Staining Procedures

Methods for detection of the subcellular sites of enzyme activities are generally based on trapping of product(s) of enzyme reaction(s); enzyme proteins, but not activities, can be localized immunocytochemically. The usefulness of modern immunocytochemistry has been considerably enhanced by the recent development of the protein A/gold method[6] and the possibility of embedding biological material in resins at temperatures below zero.[7] Low-temperature embedding methods, together with improved fixation techniques, have significantly reduced the negative effects of conventional preservation methods on protein antigenicity and thus improved the possibility of detecting antigens at the subcellular level.

Catalase activity is localized using 3,3'-diaminobenzidine (DAB) as an artificial electron acceptor in the peroxidative reaction of catalase;[8] the localization of alcohol oxidase activity is based on trapping the H_2O_2 produced in the oxidase reaction, either directly in a reaction with $CeCl_3$ or indirectly during incubations with DAB and the oxidase substrate for the endogenous production of H_2O_2.[9] Immunocytochemical and DAB-based methods can be applied to intact cells; when $CeCl_3$ is used in cytochemical experiments, protoplasts have to be prepared to facilitate the penetration of Ce^{3+} ions into the cells.

[1] A. C. Douma, M. Veenhuis, W. de Koning, M. E. Evers, and W. Harder, *Arch. Microbiol.* **143**, 237 (1985).

[2] M. Veenhuis and W. Harder, *Yeast* **5**, S517 (1989).

[3] W. Harder and M. Veenhuis, *in* "The Yeasts" (A. H. Rose and H. J. Harrison, eds.), Vol. 3, p. 289. Academic Press, London, New York, 1989.

[4] M. Veenhuis and W. Harder, *in* "The Yeasts" (A. H. Rose and H. J. Harrison, eds.), Vol. 4, in press. Academic Press, London, New York, 1990.

[5] M. Veenhuis and W. Harder, *in* "Peroxisomes in Biology and Medicine" (H. D. Fahimi and H. Sies, eds.), p. 436. Springer-Verlag, Berlin, 1987.

[6] M. Bendayan, *in* "Electron Microscopy 1982; Proceedings of the 10th International Congress on EM," Vol 3, p. 427. Deutsche Gesellschaft fur EM, 1982.

[7] E. Carlemalm, R. M. Garavito, and W. Villiger, *J. Microsc. (Oxford)* **126**(2), 123 (1982).

[8] J. P. van Dijken, M. Veenhuis, C. A. Vermeulen, and W. Harder, *Arch. Microbiol.* **105**, 261 (1975).

[9] M. Veenhuis, J. P. van Dijken, and W. Harder, *Arch. Microbiol.* **111**, 123 (1976).

Growth of Organism

Hansenula polymorpha CBS 4732 can be grown in either batch or continuous cultures as described.[10-11]

Protoplast Formation

Reagents

Buffer A: 50 mM potassium phosphate, pH 7.0
Buffer B: 50 mM potassium phosphate, pH 7.0, supplemented with 3 M sorbitol
2-Mercaptoethanol
Zymolyase 20T

Procedure. Intact cells are harvested by centrifugation and washed once in buffer A, followed by preincubation in the same buffer supplemented with 0.2 M 2-mercaptoethanol at 37° for 10 min. The cells are again washed in phosphate buffer with 3 M sorbitol as an osmotic stabilizer (buffer B) to remove excess 2-mercaptoethanol, then resuspended (60 mg cells wet weight/ml) in buffer B containing 1 mg/ml Zymolyase and incubated for 1–3 hr at 37°. Digestion of the cell wall is followed by light microscopy. Protoplasts are harvested by centrifugation (10 min at 5000 g) and washed once by carefully resuspending them in buffer B.

Cytochemical Staining Procedure for Localization of Catalase Activity

Principle. Catalase activity is localized by the DAB method, which is based on the peroxidative reaction of catalase:

$$H_2O_2 + DAB \rightarrow DAB_{ox} + H_2O$$

Upon incubation the accumulating oxidized DAB forms polymers which are subsequently transformed into an electron-dense product by treatment (postfixation) of the incubated samples with either osmium tetroxide (OsO_4) or potassium permanganate ($KMnO_4$).

Reagents

Prefixative: either 3 or 6% glutaraldehyde in 0.1 M sodium cacodylate buffer, pH 7.2

[10] M. Veenhuis, J. P. van Dijken, S. A. F. Pilon, and W. Harder, *Arch. Microbiol.* **117**, 153 (1979).
[11] I. J. van der Klei, L. V. Bystrykh, and W. Harder, this volume [65].

Incubation mixture: 0.1 M Tris-HCl, pH 8.5, containing 2 mg/ml
DAB and 0.05% H_2O_2

Osmium tetroxide postfixative: 0.5% OsO_4 plus 2.5% $K_2Cr_2O_7$ in
0.1 M sodium cacodylate buffer, pH 7.2

Potassium permanganate postfixative: 1.5% $KMnO_4$ in distilled water

Procedure. Intact cells are fixed in 3% glutaraldehyde and protoplasts in
6% glutaraldehyde in the cacodylate prefixative buffer for 60 min at 0°.
After fixation the cells are washed for 30 min in 0.1 M Tris-HCl buffer, pH
8.5, and subsequently transferred into 5 ml of the incubation mixture and
incubated at 37° in the dark. Incubation times are 30–90 min for protoplasts and 3–4 hr for intact cells. The incubation medium should be
refreshed every 60 min. After incubation the cells are washed 30 min in
distilled water (intact cells) or 0.1 M sodium cacodylate buffer, pH 7.2
(protoplasts), and postfixed in either 1.5% $KMnO_4$ for 20 min at room
temperature (intact cells) or in the mixed $OsO_4/K_2Cr_2O_7$ solution for 90
min at 0° (protoplasts). After postfixation the cells are washed with distilled water to remove excess fixative, dehydrated, and embedded in epoxy
resin (e.g., Epon 812) by established procedures.[1,9]

Controls. Controls are performed in the absence of substrate (H_2O_2) or
in the presence of 50 mM 3-amino-1,2,4-triazole (as an inhibitor of catalase activity) or 6 mM KCN (as an inhibitor of peroxidase activities).[8,9]

Cytochemical Staining Procedure for Localization of Alcohol Oxidase Activity

Indirect Method

Principle. The localization of alcohol oxidase activity is based on the
peroxidative reaction of catalase with DAB as described above but differs
from this method in that the H_2O_2 required for this reaction is not added
exogenously but generated inside the peroxisome by the oxidase reaction:

$$O_2$$
$$\text{Alcohol oxidase: } CH_3OH \xrightarrow{\quad} CH_2O + H_2O_2$$

$$\text{Catalase: } H_2O_2 + DAB \longrightarrow DAB_{ox} + H_2O$$

Reagents

Prefixative: 3% glutaraldehyde in 0.1 M sodium cacodylate buffer, pH
7.2

Incubation mixture: 0.1 M Tris-HCl, pH 8.0, supplemented with 2
mg/ml DAB and 50 mM methanol

Postfixative: 1.5% $KMnO_4$ in distilled water

Procedure. Intact cells are washed once with distilled water, fixed in 3% glutaraldehyde for 30 min at 0°, washed 30 min in 0.1 M Tris-HCl, pH 8.0, to remove excess fixative, and subsequently transferred into 5 ml of the incubation mixture. Incubation is performed at 37° in the dark under continuous aeration for 30–120 min. After incubation the cells are washed for 30 min in distilled water. Postfixation in $KMnO_4$ and processing for electron microscopy are as described above. Controls are performed in the absence of substrate (methanol) or under anaerobic conditions.

Comments on DAB-Based Methods. Reaction products accumulate into peroxisomes and mitochondria (Figs. 1 and 2). Mitochondrial staining is due to cytochrome-c peroxidase and can be prevented by KCN.[8,12] In experiments performed on cells from a methanol-limited chemostat grown at low dilution rates ($D < 0.07$ hr^{-1}), not all the peroxisomes are completely stained (Fig. 2). It has been shown that this absence of reaction products is not due to the absence of catalase activity at these sites.[9] From these and other observations it is concluded that the staining patterns obtained reflect the actual sites of alcohol oxidase activity;[2,3] therefore, this method enables discrimination between active and inactive alcohol oxidase protein inside one organelle.

Direct Method

Principle. Alcohol oxidase activity can be localized inside the cell by a direct method based on trapping of H_2O_2 produced during the oxidase reaction by cerous ions to form an insoluble, electron-dense reaction product, cerium perhydroxide. This method was initially used for the detection of extracellularly produced H_2O_2[13] and has been subsequently adapted for the localization of intracellular sites of H_2O_2-producing enzymes such as alcohol oxidase.[9]

$$CH_3OH \xrightarrow{\;O_2\;} CH_2O + H_2O_2$$

$$H_2O_2 + Ce^{3+} \longrightarrow Ce(OH)_2OOH$$

The method is predominantly applied to protoplasts to overcome the very slow penetration rates (compared to the other reagents used) of $CeCl_3$ in intact cells, but it has the advantage over DAB-based methods because its usage is not dependent on a simultaneous catalase activity.

---

[12] F. Roels, *J. Histochem. Cytochem.* **22**(6), 442 (1973).

[13] R. T. Briggs, B. D. Draht, M. L. Karnovsky, and M. J. Karnovsky, *J. Cell Biol.* **67**, 566 (1975).

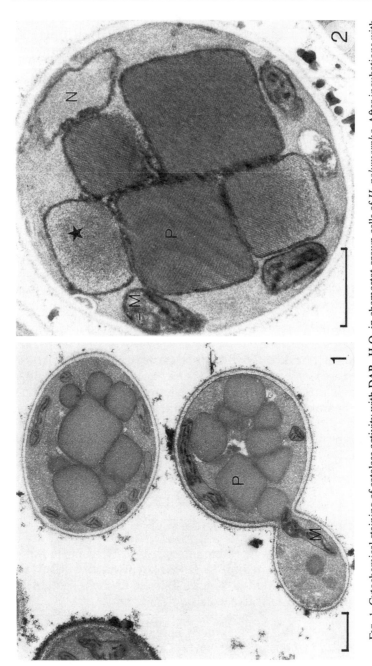

Fig. 1. Cytochemical staining of catalase activity with DAB–H_2O_2 in chemostat-grown cells of *H. polymorpha*. After incubations with DAB–H_2O_2, peroxisomes (P) are homogeneously stained. M, Mitochondrion. Bar, 0.5 μm.

Fig. 2. Cytochemical staining of alcohol oxidase activity by the indirect method with DAB–methanol in chemostat-grown cells of *H. polymorpha*. Some of the peroxisomes (P) are partly unstained (★) after DAB–methanol incubations. N, Nucleus. Bar, 0.5 μm.

Reagents

Prefixative: 6% glutaraldehyde in 0.1 M sodium cacodylate buffer, pH 7.2

Preincubation mixture: 0.1 M Tris–maleate, pH 7.5, containing 5 mM CeCl$_3$ and 50 mM 3,3′-amino-1,2,4-triazole

Incubation mixture: preincubation mixture supplemented with 50 mM methanol

Postfixative: 0.5% OsO$_4$ plus 2.5% K$_2$Cr$_2$0$_7$ in 0.1 M sodium cacodylate buffer, pH 7.2; 0.1 M sodium cacodylate buffer, pH 6.0

Procedure. Protoplasts are prefixed in 6% glutaraldehyde for 30 min at 0°, subsequently washed 30 min in 0.1 M Tris–maleate, pH 7.5, followed by preincubation in the same buffer supplemented with 50 mM aminotriazole (to inhibit catalase activity) and 5 mM CeCl$_3$ for 30–60 min at room temperature. After preincubation the cells are transferred into 10 ml of complete incubation mixture, supplemented with the substrate (methanol), and incubated for 15–90 min at 37° under continuous aeration. After incubation the cells are washed in 5 ml of 0.1 M sodium cacodylate buffer, pH 6.0, to remove eventually the precipitated cerium hydroxide [Ce(OH)$_3$].[9] Postfixation in the mixed OsO$_4$–K$_2$Cr$_2$O$_7$ solution and processing for electron microscopy are as described above.

Comments on Cerium Method. Cerous ions easily precipitate as the carbonate salt [Ce(CO$_3$)$_3$], especially at alkaline pH values. This necessitates that all cerium-containing solutions be freshly prepared prior to use in order to obtain clear and stable incubation media.

Preincubations of the yeast protoplasts with CeCl$_3$ in the absence of substrate are essential for optimal results. Compared to the methanol substrate, Ce^{3+} ions diffuse into the cells relatively slowly; if preincubation is omitted this may result in premature generation of H$_2$O$_2$, which will diffuse from the site where it is formed (the peroxisomal matrix) to be subsequently trapped in the cytosol, thus leading to artificial staining patterns.

The final reaction product, cerium perhydroxide, diffuses within a cellular compartment but does not traverse intact membranes. Aerobic incubations of cells with CeCl$_3$ and methanol result in an overall, homogeneous staining of the peroxisomal matrix (Fig. 3). The method therefore does not allow discrimination between active/inactive alcohol oxidase protein inside one organelle as does the DAB-based method.

Immunocytochemical Detection of Alcohol Oxidase Protein

Principle. Immunocytochemical detection differs from the cytochemical metal salt method in that it (1) localizes alcohol oxidase protein instead

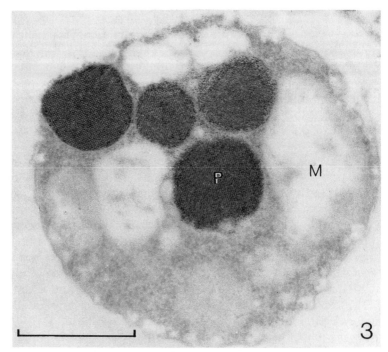

FIG. 3. Cytochemical demonstration of alcohol oxidase activity by the direct method with CeCl$_3$ and methanol (protoplasts prepared from cells from a batch culture grown on 0.5% methanol). M, Mitochondrion. Bar, 0.5 μm.

of activity and (2) is not applied *in situ* but on ultrathin sections prepared from embedded cells.

Reagents

Fixative: 3% glutaraldehyde in 0.1 M cacodylate buffer, pH 7.2
Ethanol solutions of 50, 70, 96, and 100%
Lowicryl K$_4$M resin
Monospecific antibodies against alcohol oxidase
Protein A/gold (commercially available)

Procedure. Intact cells are washed once in distilled water, subsequently fixed in 3% glutaraldehyde for 60 min at 0°, washed in distilled water to remove excess fixative, and pelleted. Dehydration is started at room temperature by incubation of the intact pellet in 50% ethanol, followed by 70% ethanol (30 min each). Subsequently, the pellet is fragmentated (blocks should not exceed 1 mm^3 in size), transferred to 96% ethanol, and gradually adapted to the cold ($-35°$). Table I shows the scheme for further dehydration and impregnation of the samples in Lowicryl K$_4$M. Polymeri-

TABLE I
GENERAL SCHEME FOR DEHYDRATION AND IMPREGNATION

Time	Temperature (°C)	Solution
Approximately		
60 min	$+20 \rightarrow -35$	96% ethanol
30 min	-35	100% ethanol
60 min	-35	100% ethanol
12 hr	-35	Ethanol : Lowicryl K_4M (1 : 1)
12 hr	-35	Ethanol : Lowicryl K_4M (1 : 3)
24 hr	-35	Pure Lowicryl K_4M

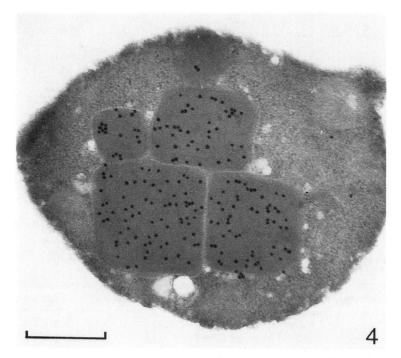

FIG. 4. Immunocytochemical demonstration of alcohol oxidase protein in peroxisomes of a methanol-limited cell of *H. polymorpha* (Lowicryl K_4M; anti-alcohol oxidase–protein A/gold). Bar, 0.5 μm.

zation is performed with UV light[14] at $-35°$. Ultrathin sectioning and immunolabeling of alcohol oxidase protein are performed by established methods.[1,6,15] A typical result of an immunocytochemical experiment for the detection of alcohol oxidase protein, performed on Lowicryl-embedded cells of methanol-grown *H. polymorpha*, is shown in Fig. 4.

Comments. Immunocytochemical methods have been successfully used for the localization of various enzymes in cells of *H. polymorpha*, including integral membrane proteins.[1,16] In addition to low-temperature embedded cells, ultrathin cryosections of frozen cells have also been shown to be very useful for immunocytochemical purposes.[15]

[14] J. Zagers, K. A. Sjollema, and M. Veenhuis, *Lab. Pract.* **35**, 114 (1986).
[15] J. W. Slot and H. J. Geuze, *in* "Immunolabeling in Electron Microscopy" (J. M. Polak and J. M. Varndell, eds.), p. 129. Elsevier, Amsterdam, 1984.
[16] A. C. Douma, M. Veenhuis, H. R. Waterham, and W. Harder, *Yeast* **6**, 45 (1990).

[65] Alcohol Oxidase from *Hansenula polymorpha* CBS 4732

By Ida J. van der Klei, Leonid V. Bystrykh, and Wim Harder

$$CH_3OH + O_2 \rightarrow CH_2O + H_2O_2$$

Alcohol oxidase (EC 1.1.3.13) is a key enzyme in methanol metabolism in yeasts. It is a flavoprotein containing FAD as a prosthetic group and catalyzes the oxidation of methanol, using dioxygen as the electron acceptor, to yield formaldehyde and hydrogen peroxide. Alcohol oxidase has been encountered in the methylotrophic yeasts *Hansenula polymorpha*, *Pichia pastoris*, *Candida boidinii*, and *Torulopsis candida*.[1-9] It is also present in some fungi[10] but not in prokaryotes or higher eukaryotes.

[1] R. H. Patel, C. T. Hou, A. I. Laskin, and P. Derelanko, *Arch. Biochem. Biophys.* **210**, 481 (1981).
[2] T. R. Hopkins and F. Muller, *in* "Proceedings of the International Symposium on Growth of Microorganisms on C₁ Compounds" (H. W. van Verseveld and J. A. Duine, eds.), p. 150. Martinus Nijhof, Dordrecht, 1986.
[3] R. Couderc and J. Baratti, *Agric. Biol. Chem.* **44**, 2279 (1980).
[4] T. R. Hopkins, U.S. Patent 4,619,898 (1986).
[5] N. Kato, Y. Omori, Y. Tani, and K. Ogata, *Eur. J. Biochem.* **36**, 250 (1976).
[6] J. P. van Dijken, R. Otto, and W. Harder, *Arch. Microbiol.* **111**, 137 (1976).
[7] H. Sahm and F. Wagner, *Eur. J. Biochem.* **36**, 250 (1973).
[8] H. Yamada, K.-G. Sgin, N. Kato, S. Shimizu, and Y. Tani, *Agric. Biol. Chem.* **43**, 877 (1979).
[9] Y. Tani, T. Miya, H. Nishikawa, and K. Ogata, *Agric. Biol. Chem.* **36**, 68 (1972).
[10] S. Bringer, B. Sprey, and H. Sahm, *Eur. J. Biochem.* **101**, 563 (1979).

The inactive monomeric alcohol oxidase protein is synthesized in the yeast cytosol and posttranslationally translocated into special cell organelles, called peroxisomes, where the active octameric FAD-containing enzyme is assembled.[11] Alcohol oxidase activity *in vivo* is thus confined to the peroxisomal matrix[12] where it may be present in a crystalline form.[13] Alcohol oxidase synthesis is regulated at the transcriptional level by repression/derepression. Under methanol-limiting growth conditions, synthesis of alcohol oxidase is greatly enhanced, and the protein may constitute up to 35% of the soluble protein of the cells.[3,6,14]

Assay Methods

Principle. Alcohol oxidase activity can be measured using either discontinuous[9] or continuous procedures. The latter assays are more convenient and precise and are therefore described below. These involve either determination of the rate of oxygen consumption using an oxygen electrode[6,15] or measurement of the rate of hydrogen peroxide production using a coupling reaction with peroxidase and an indicator dye.[15] One unit of activity is defined as the amount of enzyme that catalyzes the oxidation of 1 μmol of methanol per minute.

Reagents for Method 1

Potassium phosphate buffer, 50 mM, pH 7.0
Methanol, 10 M (40%, v/v, in water)

Reagents for Method 2

Potassium phosphate buffer, 50 mM, pH 7.0, containing 6 mM 2,2'-azinobis(3-ethylbenzthiazoline 6-sulfonate) (ABTS)
Methanol 10 M (40%, v/v, in water)
Peroxidase from horseradish (Boehringer, Mannheim), 1000 U/ml, dissolved in 50 mM potassium phosphate buffer, pH 7.0

Procedure for Method 1. The assay mixture (5 ml final volume) consists of the potassium phosphate buffer and 5–50 μl sample containing alcohol oxidase. The reaction is monitored for several minutes at 37° until a constant rate of oxygen consumption is established. Then 20 μl of the

[11] J. M. Goodman, C. W. Scott, P. W. Donahue, and J. P. Atheron, *J. Biol. Chem.* **259,** 8485 (1984).

[12] M. Veenhuis, J. P. van Dijken, and W. Harder, *Arch. Microbiol.* **111,** 123 (1976).

[13] M. Veenhuis, W. Harder, J. P. van Dijken, and F. Mayer, *Mol. Cell. Biol.* **1,** 949 (1981).

[14] R. Roggenkamp, Z. Janowicz, Z. Stanikowski, and C. P. Hollenberg, *Mol. Gen. Genet.* **194,** 489 (1984).

[15] H. Sahm, *Arch. Microbiol.* **105,** 179 (1975).

methanol solution is added, and the reaction is measured until the next constant rate is observed. The difference between the rates of oxygen consumption in the presence and absence of substrate is used to calculate alcohol oxidase activity.

General Comments on Method 1. It is useful to include catalase (1–2 U/ml) in the assay mixture if catalase-free preparations of alcohol oxidase are used for the analysis. This prevents possible inhibition of the enzyme by hydrogen peroxide upon prolonged incubation. In the presence of catalase, alcohol oxidase activity is calculated by multiplying the observed rate by a factor of 2 because of the decomposition of H_2O_2 into dioxygen and water.

Procedure for Method 2. The assay mixture consists of 10 μl peroxidase, 5–50 μl sample containing alcohol oxidase, and potassium phosphate buffer containing ABTS in a final volume of 0.99 ml. The absorbance at 420 nm is monitored in a spectrophotometer equipped with a thermostatted cuvette holder at 37° until a constant rate is observed. Then 10 μl of the methanol solution is added, and the reaction is followed until the next constant rate is obtained. The difference between the rates in the presence and absence of substrate is used to calculate the activity of alcohol oxidase. The molar extinction coefficient for ABTS is 43,200 M^{-1} cm^{-1}. Note that the stoichiometry of ABTS peroxidation is 2 mol of ABTS oxidized per mole of H_2O_2.[16]

General Comments on Method 2. *o*-Dianisidine can be used instead of ABTS[9,16,17] in which case the reaction reaches a constant rate sooner. However, the oxidized product easily forms dimers and precipitates, which is a disadvantage, particularly during prolonged incubations. Also, *o*-dianisidine is a carcinogenic compound.

Purification Procedure

Alcohol oxidase has been purified and characterized from various strains of *Candida*,[5,7,9,17] *Hansenula*,[5,6] and *Pichia*.[1-4] Two different methods have been described for enzyme purification. The first has only been successfully applied for the oxidase from *P. pastoris*[4] and is based on direct crystallization of the enzyme from the crude extract at a slightly acidic pH (close to the isoelectric point). The second procedure is apparently more universal and has been commonly used (with slight modifications) for several yeast strains. It generally consists of three consecutive

[16] G. Michal, H. Mollering, and J. Siedel, *in* "Methods of Enzymatic Analysis" (J. Bergmeyer and M. Grassl, eds.), 3rd Ed., Vol. 1, p. 197. Verlag Chemie, Weinheim, 1987.
[17] T. Fujii and K. Tonomura, *Agric. Biol. Chem.* **39,** 2325 (1975).

steps, namely, ammonium sulfate precipitation, ion-exchange chromatography, and gel filtration. An example of this procedure for the enzyme from *H. polymorpha* is given below (Table I).

Growth of Organism. For optimal results *Hansenula polymorpha* CBS 4732 is grown in a methanol-limited chemostat culture in a mineral medium which contains the following (per liter): $(NH_4)_2SO_4$, 2.5 g; KH_2PO_4, 1.0 g; $MgSO_4 \cdot 7H_2O$, 0.2 g; yeast extract, 0.5 g; 0.2 ml of a trace element solution;[18] thiamin, 0.5 mg; biotin, 50 μg; and methanol 1% (v/v) as a carbon and energy source.

The alcohol oxidase content of the cells increases with decreasing dilution rate.[6] In view of the fact that the efficiency of cell breakage decreases considerably at low dilution rates, the optimal dilution rate for enzyme purification is approximately $0.06-0.1$ hr^{-1}. Glucose–methanol or glucose–formaldehyde mixtures can also be used for carbon-limited cultivation of cells.[19] Cells are grown at 37° at an oxygen concentration corresponding to 50% of air saturation, and the pH of the culture is kept at 5.0 by automatic titration with 1 N NaOH. The cells are harvested by centrifugation at 5000 g for 10 min, washed twice with 50 mM potassium phosphate buffer, pH 7.5, and the cell paste used immediately or stored at $-20°$.

Step 1: Preparation of Cell-Free Extract. For the preparation of a crude extract approximately 1 g cell paste (wet weight) is suspended in 5 ml of 50 mM potassium phosphate buffer, pH 7.5, and the cells are disintegrated by passing the suspension five times through a French pressure cell at 1.4×10^5 kN/m^2 or by grinding with glass beads using a Bead Beater, 5 times for 1 min with intervals of 1 min at 0°. The resulting extract is centrifuged at 30,000 g for 30 min at 4° to remove unbroken cells and debris.

Step 2: Ammonium Sulfate Fractionation. The crude extract is fractionated by ammonium sulfate precipitation at 0°. After addition of solid ammonium sulfate to 45% saturation, the solution is kept on ice for 30 min, then centrifuged for 20 min at 10,000 g at 4°. To the supernatant a further portion of ammonium sulfate is added, and the fraction precipitating between 45 and 55% saturation, which contains most of the alcohol oxidase activity, is collected. This precipitate is resuspended in a minimal volume of the phosphate buffer and desalted by gel filtration on a Sephadex G-25 column (PD10, Pharmacia, Uppsala, Sweden) or by dialysis (overnight, 4°) against the phosphate buffer. Further purification is per-

[18] W. Vishniak and M. Santer, *Bacteriol. Rev.* **21,** 195 (1957).
[19] M. L. F. Giuseppin, H. M. L. van Eijk, and B. S. M. Bes, *Biotechnol. Bioeng.* **32,** 577 (1988).

TABLE I
PURIFICATION OF ALCOHOL OXIDASE FROM *Hansenula polymorpha*

Step	Total protein (mg)	Total units (U)	Specific activity (U/mg)	Purification (-fold)
1. Crude extract	51.4	265.6	5.2	1
2. Ammonium sulfate fraction, 45–55%	14.9	169.9	11.4	2.2
3. Mono Q chromatography	5.4	105.9	19.5	3.8
4. Gel filtration on Superose 6	3.5	81.6	23.2	4.5

formed using a fast protein liquid chromatography system (FPLC, Pharmacia).

Step 3: Mono Q Chromatography. The protein fraction obtained from Step 2 is applied to a Mono Q ion-exchange column (HR 5/5, Pharmacia) which has been equilibrated with 20 mM Tris-HCl buffer, pH 7.5. Alcohol oxidase is eluted using a linear gradient of 0–0.5 M ammonium sulfate in the same buffer (flow rate 1 ml/min, 25 min, room temperature). The fractions containing most of the alcohol oxidase activity are pooled and concentrated by ammonium sulfate precipitation (60% saturation) at 0°. The pellet is resuspended in a minimal volume of the same phosphate buffer (about 0.5 ml) and used for gel filtration.

Step 4: Gel Filtration on Superose 6. The fraction obtained in Step 3 is loaded on a Superose 6 column (HR 10/30, Pharmacia) equilibrated with 50 mM potassium phosphate buffer, pH 7.5, and eluted using a flow rate of 0.4 ml/min at room temperature. Alcohol oxidase activity is present in a symmetric peak after 12–13 ml of elution. The fractions containing the enzyme are pooled and stored at 4° as a sterile solution which is obtained by filtration through a 0.25-μm membrane.

Properties

Purity, Molecular Weight, and Structure. The enzyme preparation is homogeneous as judged by sodium dodecyl sulfate (SDS)–polyacrylamide gel electrophoresis followed by Coomassie blue or silver staining. The mass calculated from the amino acid sequence derived from the nucleotide sequence of the structural gene is 74 kDa.[20] According to gel filtration, the molecular mass of the native enzyme is approximately 600 kDa, which

---

[20] A. M. Ledeboer, L. Edens, J. Maat, G. Visser, J. W. Bos, C. T. Verrips, Z. Janowicz, M. Eckart, and C. P. Hollenberg, *Nucleic Acids Res.* **13**, 3063 (1985).

corresponds to an octameric structure for the enzyme. The isoelectric point (pI) is 6.2.[3]

Alcohol oxidase contains noncovalently bound flavin adenine dinucleotide (FAD) as a prosthetic group. It possesses one FAD-binding site per subunit. FAD may be present in a modified form; this modified FAD differs from the normal or classic form by having a different absorbance ratio at 280 and 450 nm and shorter retention time during HPLC analysis.[21,22] The ratio of classic and modified FAD in alcohol oxidase depends on the growth conditions of the cells. In alcohol oxidase purified from methanol-limited chemostat-grown *H. polymorpha,* the amount of modified FAD may be higher than 90% of the total. The commercially available preparation from *Pichia pastoris* (Sigma, St. Louis, MO) also contains modified FAD at the same level. Alcohol oxidase purified from cells grown in batch culture on methanol contains mainly classic FAD.[22] Except for FAD no other coenzymes or metal cations were found in the active enzyme.[5]

Stability. Alcohol oxidase is apparently stable for at least 6 months in sterile buffer solutions at 4°. In activity assays it is stable for at least 1 hr at 45° and at room temperature, at least 24 hr. The SH-stabilizing agent 2-mercaptoethanol has no significant stabilizing effect. Addition of this compound leads to its oxidation because it is a substrate analog. The enzyme from *P. pastoris* (Sigma) is available in sucrose–buffer solution and can be frozen and thawed several times. However, it is easily inactivated by freezing and thawing in buffer solution in the absence of stabilizers. Alcohol oxidase can also be stored as a precipitate or crystalline suspension in 60–80% ammonium sulfate in potassium phosphate buffer, pH 7.5. However the enzyme gradually loses activity (about 20% per month).

Substrate Specificity. Short-chain primary alcohols with up to four carbons and their halogenated derivatives are oxidized by alcohol oxidase. The specific activity and affinity decrease with increasing carbon chain length.[5] The most effective substrates are methanol and ethanol. Depending on the enzyme preparation, the K_m for methanol varies from 0.2 to 2.0 mM. The K_m for oxygen is 0.4 mM.[6] The enzyme can also oxidize formaldehyde (K_m 2.6 mM) but no other aldehydes. Probably this anomalous specificity can be explained by the formation of methylene glycol from formaldehyde in aqueous solutions. In some enzyme preparations an activation effect has been observed during incubation of the enzyme at 30° at pH values between 7.5 and 8.5.[22] The observed kinetics is Michaelian; the

[21] B. Sherry and R. H. Abeles, *Biochemistry* **24,** 2594 (1985).
[22] L. V. Bystrykh, V. P. Romanov, J. Steczko, and Y. A. Trotsenko, *Biotechnol. Appl. Biochem.* **11,** 184 (1989).

reaction involves a Ping-Pong mechanism and operates via a two-electron transfer from substrate to oxygen.[23]

Catalysis involves changes in the redox state of the FAD, where dioxygen acts as an electron acceptor and the substrate as a reducing agent. The occurrence of a semiquinone form of FAD has been suggested for the enzyme from *H. polymorpha* DL1, *P. pastoris*,[21,24] and *Candida boidinii*.[24,25] However, there is no evidence for its significance as an intermediate in the catalytic process.

Inhibitors and Inactivators. The reaction product H_2O_2, at a concentration of 5–10 mM, inactivates alcohol oxidase.[2,5] Tightly bound formaldehyde has been found in purified preparations of the enzyme,[2,26] and this may inhibit the enzyme activity.[23]

Alcohol oxidase is irreversibly inactivated by the acetylenic alcohols propargyl alcohol and 1,4-butynediol.[27] This inactivation is caused by covalent modification of amino acid residues in the active center of the enzyme by the oxidation products of these alcohols, propynal and 4-hydroxy-2-butynal.

The enzyme catalyzes a suicide reaction with cyclopropanol and cyclopropanone. In the latter case covalent modification of FAD by this substrate has been suggested.[28] The oxidase also catalyzes oxidation of allyl alcohol yielding acrolein. *In vivo* this reaction leads to cell intoxication. Alcohol oxidase is inactivated by SH-inactivating agents (*p*-chloromercuribenzoate).

Alcohol oxidase is partially inactivated by incubation at 30° for 1–2 hr with 6 M urea or 3.5 M KBr and is completely inactivated in a mixture of these agents. The mechanism of enzyme inactivation by urea is complex and includes both dissociation of the octamer into monomers and dissociation of FAD from the protein. KBr effects mainly removal of FAD from the holoenzyme; KCN exhibits the same effect.[29]

Sodium azide affects enzyme activity (at 1–10 mM concentrations) and also turns the color of the enzyme solution from yellow to red.[2,10,24] The effectiveness of the azide binding appears to vary with the source of enzyme. However, in most of the cases the azide can be removed by dialysis for 20–30 hr.

[23] J. Geissler, S. Gishla, and P. M. H. Kroneck, *Eur. J. Biochem.* **160**, 93 (1986).
[24] T. Mincey, G. Tayrien, A. S. Mildvan, and R. H. Abeles, *Biochemistry* **77**, 7099 (1980).
[25] J. Geissler and P. Hemmerich, *FEBS Lett.* **126**, 152 (1981).
[26] Y. Sakai and Y. Tani, *Agric. Biol. Chem.* **52**, 227 (1988).
[27] C. S. Nichols and Th. H. Cromartie, *Biochem. Biophys. Res. Commun.* **97**, 216 (1980).
[28] Th. H. Cromartie, *Biochem. Biophys. Res. Commun.* **105**, 785 (1982).
[29] I. J. van der Klei, M. Veenhuis, K. Nicolay, and W. Harder, *Arch. Miocrobiol.* **151**, 26 (1989).

pH Optimum. The enzyme is characterized by a broad pH optimum both for activity and stability (5.5–8.5). *In vivo* it functions at pH values around 6.0 owing to acidic nature of the peroxisomal matrix.[30]

[30] K. Nicolay, M. Veenhuis, A. C. Douma, and W. Harder, *Arch. Microbiol.* **143**, 37 (1987).

[66] Amine Oxidases from Methylotrophic Yeasts

By Peter J. Large and Geoffrey W. Haywood

$$RCH_2NH_3^+ + O_2 + H_2O \rightarrow RCHO + NH_4^+ + H_2O_2$$

Amine oxidases catalyze the liberation of ammonia from primary amines and play an essential role in growth of aerobic yeasts on amines as the sole nitrogen source. The purifications described below have been the subject of earlier reports.[1,2] Most yeasts are capable of producing two different amine oxidases, similar in subunit molecular weight, but differing in substrate specificity and heat stability.[1] Whether one of the two enzymes or both is expressed depends on the amine used as nitrogen source, and on the yeast species. The methylotrophic yeasts *Candida boidinii* and *Pichia pastoris* are convenient sources of amine oxidases, but because during growth of the yeasts on methanol very high levels of methanol (alcohol) oxidase (EC 1.1.3.13) are produced,[3] which may interfere with both assay and purification, it is more convenient to use cells grown on glucose with the appropriate amine as the sole nitrogen source. The purification of three amine oxidases is described here, the methylamine and benzylamine oxidases from *Candida boidinii*[1] and the rather different benzylamine/putrescine oxidase of *Pichia pastoris*.[2] In addition, *Candida boidinii,* when grown on spermidine or putrescine as nitrogen source, elaborates a polyamine oxidase that cleaves acetylspermidine derivatives at the secondary amino group.[4,5] For the purification of this enzyme, see Ref. 4.

Assay Method

Principle. The hydrogen peroxide generated in the above equation is coupled to the oxidation of 2,2′-azinobis(3-ethylbenzthiazoline 6-sulfonate) (ABTS) with horseradish peroxidase:

[1] G. W. Haywood and P. J. Large, *Biochem. J.* **199**, 187 (1981).
[2] J. Green, G. W. Haywood, and P. J. Large, *Biochem. J.* **211**, 481 (1983).
[3] H. Sahm, *Adv. Biochem. Eng.* **6**, 77 (1977).
[4] G. W. Haywood and P. J. Large, *J. Gen. Microbiol.* **130**, 1123 (1984).
[5] G. W. Haywood and P. J. Large, *Eur. J. Biochem.* **148**, 277 (1985).

$$H_2O_2 + 2\ ABTS \rightarrow 2\ H_2O + 2\ ABTS_{ox}$$

to form a highly colored radical cation with absorption maxima at 405, 420, 623, and 660 nm.[6] Any of these wavelengths can be used to measure the product, but 405 nm is the most satisfactory.

Reagents

Potassium phosphate buffer, 0.1 M, pH 7.0

Methylammonium chloride, 0.1 M; putrescine dihydrochloride, 0.1 M; benzylamine hydrochloride, 10 mM

Disodium ABTS, 50 mM, prepared from the commercial diammonium salt by passage down a column of Zeo-Karb 225 (Na$^+$ form) (Permutit, London) and standardized by using a millimolar absorbance coefficient of 43 liter mmol^{-1} cm^{-1} at 343 nm (for less critical work in which the possibility of product inhibition is not important, the diammonium salt may be used)

Horseradish peroxidase (type I from Sigma, St. Louis, MO), 10 mg/ml distilled water (800 units/ml)

Procedure. The assay mixture contains, in a final volume of 1 ml in polystyrene cuvettes (light path, 10 mm), 0.3 ml potassium phosphate buffer, 17 μl disodium ABTS, 3 μl horseradish peroxidase, enzyme, and water. The cuvettes are incubated for 2–3 min at 25° and the absorbance at 405 nm followed. The reaction is then initiated by addition of the amine hydrochloride substrate (30 μl of methylamine, 10 μl of putrescine, or 15 μl of benzylamine), and the rate of increase in absorbance at 405 nm is measured against a blank containing all the assay components except the amine hydrochloride.

Most yeast extracts contain high levels of catalase, but catalase does not interfere with this assay until a concentration of more than 3×10^4 units of catalase per unit of amine oxidase is reached.

The assay is generally applicable in crude extracts without any problems, unless the extracts contain thiol compounds. Such extracts may exhibit a lag of up to 30 min before a significant rate of ABTS oxidation can be measured. The lag can be abolished by precipitation of the extract with an 80% saturation of $(NH_4)_2SO_4$ at 0° followed by dissolving the precipitate in the original volume of buffer and a short (2-hr) dialysis against the same buffer.

Units. One unit of amine oxidase is defined as that amount of enzyme catalyzing the formation of 1 μmol of ABTS radical cation per minute at 25°, using a millimolar absorbance coefficient of 18.4 liter mmol^{-1} cm^{-1}.[6]

[6] W. Werner, H.-G. Rey, and H. Wielinger, *Fresenius' Z. Anal. Chem.* **252,** 224 (1970).

It follows from the equations above that the oxidation of 1 mol of substrate is accompanied by the formation of 2 mol of ABTS radical cation. Specific activity is expressed as units per milligram protein (measured by the method of Bradford[7]).

Purification Procedures

Growth of Organisms. The yeast strains used are *Candida boidinii* CBS 5777 (ATCC 56897) and *Pichia pastoris* CBS 704 (ATCC 28485). These are grown at 30° in shake flasks or a fermentor in the following medium, containing (per liter) 10 g D-glucose, 0.2 g $MgSO_4 \cdot 7H_2O$, 3 g KH_2PO_4, a nitrogen source (see below), 20 μg (+)-biotin, 4 mg D-pantothenate (calcium salt), 0.4 mg thiamin hydrochloride, 0.2 mg 4-aminobenzoic acid, 2 μg folic acid, 0.4 mg nicotinic acid, 10 mg *myo*-inositol, 0.4 mg pyridoxine hydrochloride, 0.2 mg riboflavin, 10 mg disodium EDTA, 4.4 mg $ZnSO_4 \cdot 7H_2O$, 1.47 mg $CaCl_2 \cdot 2H_2O$, 0.22 mg $MnCl_2 \cdot 4H_2O$, 0.3 mg $CuSO_4 \cdot 5H_2O$, and 0.3 mg $CoCl_2 \cdot 6H_2O$. (Bacto-yeast carbon base from Difco, Detroit, MI, is an alternative but more expensive medium.) The pH of the medium is adjusted to 6.0 with NaOH. The glucose, vitamins, and nitrogen source are sterilized by filtration through a 0.22-μm pore size membrane filter; the other components may be autoclaved. The filtered components are added aseptically after the rest of the medium is cool. The nitrogen source concentrations (per liter) are as follows: 2 g methylammonium chloride or 0.73 g *n*-butylamine (adjusted to pH 6.0 with concentrated HCl). Cells are harvested at an absorbance (663 nm) of not more than 2.5, centrifuged at 5000 *g* and 4° for 20 min, and washed with 50 mM potassium phosphate buffer, pH 7.0, before storage at $-15°$.

Methylamine oxidase is best prepared from cells grown on methylammonium chloride as the nitrogen source. Although both methylamine and benzylamine oxidases can be obtained from cells grown on *n*-butylamine, yields of the former enzyme are lower.[1,8] In the procedures described below, to obtain homogeneous material at the end of the purification, it is necessary when combining fractions from column chromatographic procedures to include only those fractions with more than 50% of the maximum activity.

Methylamine Oxidase from Candida boidinii

Step 1: Preparation of Cell-Free Extract. Cell material (about 20 g wet weight of cells grown on methylammonium chloride) is suspended in 2

[7] M. M. Bradford, *Anal. Biochem.* **72,** 248 (1976).
[8] J. Green, G. W. Haywood, and P. J. Large, *J. Gen. Microbiol.* **128,** 991 (1982).

volumes of 50 mM potassium phosphate, pH 7.0, and passed once through a French pressure cell at 8.6 MPa. The material is then centrifuged for 15 min at 30,000 g at 4°. The pellet obtained is resuspended in 1 volume of the same buffer and passed a second time through the pressure cell. After centrifuging for 15 min at 30,000 g, the supernatant is combined with the previous one.

Step 2: Ammonium Sulfate Fractionation. The combined supernatants are adjusted to 35% saturation by addition of solid $(NH_4)_2SO_4$, and the precipitated protein separated by centrifugation is discarded. The supernatant is then taken to 70% saturation and centrifuged at 50,000 g for 20 min. The precipitate is collected.

Step 3: Sepharose Gel-Permeation Chromatography. The precipitate from Step 2 is dissolved in not more than 7.5 ml of 10 mM potassium phosphate buffer pH 7.0 and applied to a column (2.5 cm $\times$ 1 m) of Sepharose 4B or 6B equilibrated with the same buffer. The column is eluted at 15 ml/hr overnight and 4-ml fractions collected.

Step 4: 5-Aminopentyl-Agarose Chromatography. The combined active fractions from Step 3 are applied to a column (2.5 $\times$ 10 cm) of 5-aminopentyl-agarose (Sigma, A5018) and washed with 10 mM potassium phosphate, pH 7.0. A linear gradient (10–250 mM potassium phosphate, total volume 300 ml) is then applied. The activity elutes at about 80 mM phosphate. Fractions are combined and concentrated by using an Amicon (Danvers, MA) concentration cell with a Diaflo PM30 membrane to about one-tenth of the original volume.

Step 5: Hydroxyapatite Chromatography. The concentrated material from Step 4 is diluted to a final concentration of 20 mM phosphate with distilled water and applied to a column (1 $\times$ 12 cm) of hydroxyapatite (Bio-Rad, Richmond, CA; BioGel HTP) equilibrated in 20 mM potassium phosphate, pH 7.0. After washing with the same buffer, the enzyme is eluted with a linear gradient of 20–250 mM potassium phosphate, pH 7.0 (total volume 150 ml). The activity elutes at about 63 mM phosphate. Fractions are combined and concentrated as in Step 4.

Step 6: Ion-Exchange Chromatography. Pooled concentrated fractions from Step 5 are diluted with distilled water to a concentration of 20 mM phosphate and applied to a column (2 $\times$ 25 cm) of DEAE-Sepharose CL-6B and washed with 20 mM potassium phosphate, pH 7.0. The enzyme is then eluted with a linear gradient of 0–300 mM phosphate, pH 7.0 (total volume 300 ml). Fractions are combined and concentrated as in Step 4.

Step 7. To obtain homogeneous material, it is sometimes necessary to dilute the material from Step 6 to 10 mM in phosphate and repeat Step 4.

A typical purification of methylamine oxidase is summarized in Table I.

TABLE I

PURIFICATION OF METHYLAMINE OXIDASE FROM *Candida boidinii*

Fraction	Total volume (ml)	Total protein (mg)	Total activity (units)	Specific activity (units/mg protein)	Yield (%)
Step 1. Crude extract	91	1074	203	0.188	100
Step 2. 35–70% (NH$_4$)$_2$ SO$_4$ precipitate	34	741	167	0.225	83
Step 3. Sepharose 4B chromatography	155	270	157	0.58	78
Step 4. 5-Aminopentyl-agarose chromatography	84	99	130	1.31	64
Step 5. Hydroxyapatite chromatography	14.6	44	94	2.15	47
Step 6. DEAE-Sepharose CL-6B chromatography	8.56	15	59	3.95	29
Step 7. 5-Aminopentyl-agarose chromatography (homogeneous fractions)	11.6	3.8	13	3.39	6

Benzylamine Oxidase from Candida boidinii

Cells grown on *n*-butylamine are used as the source of the enzyme. Steps 1–3 are identical with those above except that the ammonium sulfate fraction precipitating between 45 and 85% saturation at 0° is used. Benzylamine, acetylputrescine, or isobutylamine (Table IV) may be used as the assay substrate.

Step 4: Chromatography on DEAE-Sepharose. The combined active fractions from Step 3 are applied to a column (2 × 25 cm) of DEAE-Sepharose CL-6B and washed with 20 mM potassium phosphate buffer, pH 7.0. The enzyme is eluted with a linear gradient of 20–300 mM phosphate, pH 7.0 (total volume 300 ml). This step separates methylamine oxidase (which elutes between 55 and 85 mM phosphate) from benzylamine oxidase (which elutes at 105 mM phosphate).[2] Fractions are combined and concentrated.

Step 5: Chromatography on DEAE-Cellulose. The pooled concentrate from Step 4 is diluted with distilled water to 20 mM phosphate and applied to a column of DEAE-cellulose (Whatman DE-11) previously equilibrated in 20 mM phosphate, pH 7.0. After washing with the same buffer, the column is eluted with a linear gradient of 20–250 mM phosphate (total volume 300 ml). The peak of activity elutes at about 70 mM phosphate. Fractions are combined and concentrated.

TABLE II

PURIFICATION OF BENZYLAMINE OXIDASE FROM *Candida boidinii*

Fraction	Total volume (ml)	Total protein (mg)	Total activity (units)	Specific activity (units/mg protein)	Yield (%)
Step 1. Crude extract	67	992	28	0.029	100
Step 2. 45–85% (NH$_4$)$_2$ SO$_4$ precipitate	15	390	16	0.041	57
Step 3. Sepharose 6B chromatography	83	101	16	0.157	57
Step 4. DEAE-Sepharose CL-6B chromatography	28	18	15	0.810	54
Step 5. DEAE-cellulose chromatography	7	4.6	7	1.52	25
Step 6. Hydroxyapatite chromatography	5	1.1	5	4.55	18

Step 6: Hydroxyapatite Chromatography. Hydroxyapatite chromatography is performed essentially as for Step 5 of methylamine oxidase above, except that the column is eluted with a gradient of 20–300 m*M* phosphate, pH 7.0 (in a total volume of 200 ml). The peak of activity emerges at 130 m*M* phosphate. Fractions are combined and concentrated.

A typical purification of benzylamine oxidase is summarized in Table II.

Benzylamine/Putrescine Oxidase from Pichia pastoris

Pichia pastoris grown on *n*-butylamine is used as the source of the enzyme. Steps 1–3 are identical with those above except that the active fraction from the (NH$_4$)$_2$SO$_4$ treatment is that precipitated between 30 and 80% saturation at 0°. Benzylamine or putrescine may be used as the assay substrate (Table IV).

Step 4: Chromatography on DEAE-Sepharose. The combined active and concentrated fractions from Step 3 are brought to 250 m*M* in phosphate by addition of 1 *M* potassium phosphate buffer, pH 7.0, and then applied to a column (2 × 40 cm) of DEAE-Sepharose CL-6B. The column is washed with 250 m*M* potassium phosphate buffer, pH 7.0, and then eluted with 1 *M* potassium phosphate buffer, pH 7.0. The peak active fractions are combined, and the phosphate concentration is reduced to about 50 m*M* by addition of distilled water to the concentration cell. The solution is concentrated to about 2–3 ml before the protein concentration is determined.

A typical purification of benzylamine/putrescine oxidase is summarized in Table III.

TABLE III
PURIFICATION OF BENZYLAMINE/PUTRESCINE OXIDASE FROM *Pichia pastoris*

Fraction	Total volume (ml)	Total protein (mg)	Total activity (units)[a]	Specific activity (units/mg protein)	Yield (%)
Step 1. Crude extract	47	322	105	0.33	100
Step 2. 30–80% $(NH_4)_2 SO_4$ precipitate	9.5	238	86	0.36	82
Step 3. Sepharose 6B chromatography	36	21	55	2.62	53
Step 4. DEAE-Sepharose CL-6B chromatography	33	10	48	4.80	46

[a] Putrescine as substrate.

Properties

Purity. If side fractions of low activity are rigorously discarded, all three enzymes described above can be obtained homogeneous as judged by dissociating or nondissociating polyacrylamide gel electrophoresis.

Substrate Specificity. There are significant differences in substrate specificity between the three enzymes. Table IV gives apparent K_m values for the most important substrates in which differences are observed. Methylamine oxidase is really a short-chain primary amine oxidase, whereas benzylamine oxidase is an aromatic and long-chain aliphatic primary amine oxidase. A compound like β-phenylethylamine with characteristics of both aliphatic and aromatic compounds is a good substrate for both enzymes. The enzyme from *Pichia pastoris* is really an α,ω-diamine and lysyl-peptide oxidase. Recent studies by Tur and Lerch[9] have shown that the enzyme has a high affinity for peptides containing lysine and also shows activity with elastin, collagen, and histones. In this respect it resembles mammalian lysyl oxidase.[10]

Stability. There is a significant difference in stability between the methylamine oxidase group of enzymes and the benzylamine/putrescine oxidase group.[1,2] In general, at temperatures in excess of 45°, the methylamine oxidases are much less stable.[1,2] The enzymes store well in buffer for up to 1 year at −15°.

Michaelis Constants and Enzyme Mechanism. Kinetic studies with the *Candida boidinii* methylamine oxidase suggested a sequential mechanism

[9] S. S. Tur and K. Lerch, *FEBS Lett.* **238,** 74 (1988).
[10] H. M. Kagan and K. A. Sullivan, this series, Vol. 82, p. 637.

TABLE IV
Substrate Specificity of Three Amine Oxidases

Substrate	Apparent K_m values (mM)		
	Candida boidinii methylamine oxidase	Candida boidinii benzylamine oxidase	Pichia pastoris benzylamine/ putrescine oxidase
Methylamine	0.23	1.83	0.27
Ethylamine	0.77	16.74	0.71
Propylamine	1.54	2.04	0.21
Butylamine	0.24	0.18	0.083
Hexylamine	2.50	0.044	0.038
Decylamine	0.20	0.008	0.078
Isobutylamine	Inactive	1.13	0.093
1,2-Diaminoethane	Inactive	Inactive	a
1,3-Diaminopropane	Inactive	Inactive	a
1,4-Diaminobutane (putrescine)	Inactive	Inactive	5×10^{-4}
1,5-Diaminopentane (cadaverine)	Inactive	1.35	8×10^{-4}
1,6-Diaminohexane	Inactive	0.50	a
Acetylputrescine	Inactive	0.33	0.011
Ornithine	Inactive	Inactive	0.80^b
Lysine	Inactive	Inactive	0.063
Spermine	Inactive	Inactive	a
Spermidine	Inactive	Inactive	a
Ethanolamine	Inactive	130.00	a
Δ^1-Pyrroline	1.43	0.45	0.16
Benzylamine	Inactive	0.003	0.035
β-Phenylethylamine	0.11	0.044	0.023
Histamine	Inactive	a	a

[a] Active but K_m value not determined.
[b] Data of S. S. Tur and K. Lerch, FEBS Lett. 238, 74 (1988).

for the enzyme, with absolute K_m values of 0.20 mM for methylamine and 0.01 mM for oxygen, with noncompetitive product inhibition by NH_4^+ ions (K_i[slope] 7.6 mM), uncompetitive inhibition by formaldehyde (K_i[intercept] 31 mM), and competitive inhibition by hydrogen peroxide (K_i[slope] 22.5 mM).[1] Detailed kinetic studies have not been performed with the benzylamine oxidases owing to technical difficulties.[1,2]

Inhibitors. The enzymes are all thought to contain copper and pyrroloquinoline quinone (PQQ),[1,2] although this has been demonstrated experimentally only in the case of the *Pichia pastoris* benzylamine/putrescine oxidase.[9] Accordingly, they are extremely sensitive to carbonyl reagents such as semicarbazide, izoniazid, hydroxylamine, and potassium cyanide. Cuprizone, mersalyl, and aminoacetonitrile are also potent inhibitors.[1,2]

Molecular Weight and Subunit Structure. The two enzymes from *Candida boidinii* each contain only one type of subunit, M_r in both cases about 80,000. The methylamine oxidase gives rise quite readily to dimers of M_r 150,000 and on gel permeation behaves as if it is a tetramer with M_r 286,000. The benzylamine oxidase behaves as if it is usually a dimer of M_r 136,000, which under certain conditions aggregates to a tetramer of M_r 286,000. The *Pichia pastoris* enzyme has a higher subunit molecular weight (116,000) and behaves consistently as a dimer of M_r 256,000.

pH Optimum. The pH optimum tends to vary slightly with the amine substrate but is in the region of 7.0 for all three enzymes.

Applications. The enzymes have potential applications for the removal of amines from food materials[11] and for the estimation of amines. In particular we have found the *Pichia pastoris* enzyme very useful in the estimation of *n*-butylamine, spermine, spermidine, and putrescine.[2,4]

[11] P. J. Large and C. W. Bamforth, "Methylotrophy and Biotechnology," p. 168. Longman Scientific and Technical, Harlow, 1988.

[67] Dihydroxyacetone Synthase from *Candida boidinii* KD1

By Leonid V. Bystrykh, Wim de Koning, and Wim Harder

$$CH_2O + \text{xylulose 5-phosphate} \rightarrow \text{dihydroxyacetone} + \text{glyceraldehyde 3-phosphate}$$

Dihydroxyacetone synthase catalyzes the first step in the assimilation of formaldehyde via the xylulose monophosphate (dihydroxyacetone) cycle during growth of yeasts on methanol. The enzyme differs from transketolase with respect to both substrate specificity and its subcellular localization in peroxisomes.

Assay Methods

Principle. Several methods are available for the assay of dihydroxyacetone synthase; these are summarized in Table I. The methods are based on either determining the rate of product formation from formaldehyde and xylulose 5-phosphate or measuring the xylulose 5-phosphate-dependent rate of formaldehyde consumption. In crude extracts a number of interfering reactions may occur. In assays using coupling enzymes that enable the determination of reaction products by way of monitoring concentration

TABLE I

METHODS FOR DIHYDROXYACETONE SYNTHASE ASSAY WITH FORMALDEHYDE AND XYLULOSE
5-PHOSPHATE AS SUBSTRATES [a]

Compounds detected	Mode of assay	Assay system	Interfering reaction(s)	Ref.
Formaldehyde	Discontinuous	Chemical	AO	d, e
DHA and GAP (Method 1)	Discontinuous	Chemical	TK	f
DHA and GAP (Method 2)	Continuous	GK/TPI/ GPDH; ATP, NADH	TK[b], FR[c]	g
DHA (Method 3)	Continuous	GDH: NADH	FR[c]	This publication
GAP (Method 4)	Continuous	GaPDH; NAD+	TK[b]	This publication h, i

[a] DHA, Dihydroxyacetone; AO, alcohol oxidase; GAP, glyceraldehyde 3-phosphate; TK, transketo-lase; TPI, triose-phosphate isomerase; GK, glycerol kinase; GPDH, glycerol-3-phosphate dehydro-genase; GDH, glucose dehydrogenase; GaPDH, glyceraldehyde-3-phosphate dehydrogenase; FR, formaldehyde reductase.

[b] Detected as endogenous activity.

[c] Eliminated by using a two cuvette assay.

[d] N. Kato, T. Nishizawa, C. Sakazawa, Y. Tani, and H. Yamada, *Agric. Biol. Chem.* **43**, 2013 (1979).

[e] T. Nash, *Biochem, J.* **55**, 416 (1953).

[f] L. V. Bystrykh, A. P. Sokolov, and Y. A. Trotsenko, *Dokl. Akad. Nauk SSSR* **258**, 499 (1981).

[g] M. L. O'Connor and J. R. Quayle, *J. Gen. Microbiol.* **120**, 219 (1980).

[h] G. A. Kochetov, L. I. Nikitushkina, and N. N. Chernov, *Biochem. Biophys. Res. Commun.* **40**, 873 (1970).

[i] G. A. Kochetov, this series, Vol. 90, p. 209.

changes of NADH, the consumption of NADH by formaldehyde reductase is usually important. Another source of interference is the combined action of ribulose-phosphate 3-epimerase, ribose-5-phosphate isomerase (usually present at very high levels in crude extracts), and transketolase. The net result of these reactions is the formation of sedoheptulose 7-phosphate and glyceraldehyde 3-phosphate. Transketolase activity can be measured in an independent assay.[1] Detection of low activities of dihydroxyacetone syn-thase is complicated by the fact that transketolase is also able to catalyze the reaction between formaldehyde and xylulose 5-phosphate, although only at a low rate and with low affinity for formaldehyde.[2,3] Conversely, dihydroxyacetone synthase also can catalyze the various transketolase reac-tions (see below).

[1] T. Ozawa, S. Saitou, and I. Tomita, *Chem. Pharm. Bull.* **20**, 2723 (1972).

[2] M. J. Waites and J. R. Quayle, *J. Gen. Microbiol.* **124**, 309 (1981).

[3] W. de Koning, K. Bonting, W. Harder, and L. Dijkhuizen, in preparation.

One procedure (Table I), is discontinuous and measures the rate of xylulose 5-phosphate-dependent consumption of formaldehyde by a colorimetric method.[4] It is not described in any detail here; the main disadvantages are its apparent low sensitivity and the discontinuous mode of assay. However, it can be used to assay various formaldehyde-converting enzymes.

Method 1 involves the discontinuous measurement of the rate of formaldehyde- and xylulose 5-phosphate-dependent formation of both dihydroxyacetone and glyceraldehyde 3-phosphate, using 2,4-dinitrophenyl hydrazine.

Method 2 is based on the continuous measurement of formaldehyde- and xylulose 5-phosphate-dependent formation of dihydroxyacetone and glyceraldehyde 3-phosphate.[5] Both compounds are detected via the NADH-dependent reduction of dihydroxyacetone phosphate by purified glycerol-3-phosphate dehydrogenase. Dihydroxyacetone and glyceraldehyde 3-phosphate are converted to dihydroxyacetone phosphate by way of purified glycerol kinase and purified triose-phosphate isomerase, respectively. Although the sensitivity is enhanced by the detection of both dihydroxyacetone and glyceraldehyde 3-phosphate, the assay can be simplified by omitting glycerol kinase and ATP or triose-phosphate isomerase.

Method 3 involves the use of purified glycerol dehydrogenase to detect the rate of formaldehyde- and xylulose 5-phosphate-dependent formation of dihydroxyacetone.

Method 4 is based on the continuous measurement of the rate of formaldehyde- and xylulose 5-phosphate-dependent formation of glyceraldehyde 3-phosphate using purified glyceraldehyde-3-phosphate dehydrogenase as has been described for transketolase. This method, however, provides lower activities than Methods 1–3.

One unit of activity is defined as the amount of enzyme catalyzing the formation of 1 μmol of dihydroxyacetone or glyceraldehyde 3-phosphate per minute.

General Reagents

Potassium phosphate buffer, 50 mM, pH 6.8 (except for Method 4)
Magnesium chloride, 500 mM
Thiamin pyrophosphate, 50 mM
Formaldehyde 250 mM, prepared by heating 75 mg of paraformaldehyde in 10 ml of water overnight in a sealed tube at 100° (for good reproducibility the concentration of the solution should be checked

[4] T. Nash, *Biochem. J.* **55**, 416 (1953).
[5] M. J. Waites and J. R. Quayle, *J. Gen. Microbiol.* **129**, 935 (1983).

by the procedure of Nash[4]); the solution can be used for approximately 2 weeks when stored at 4°

Xylulose 5-phosphate, 250 mM; *Note:* because of the hygroscopic nature of xylulose 5-phosphate the entire contents of the vial (25 or 100 mg) should be dissolved in distilled water and stored at −20°

Reagents for Method 1

2,4-Dinitrophenylhydrazine 0.1% (w/v) in 2 N HCl
Mixture of 20% (w/v) KOH in water and ethanol (4:7, v/v)

Reagents for Method 2

NADH, 15 mM
ATP, 200 mM, pH adjusted to 6.8 with 5 N KOH
Glycerol kinase from *Candida mycoderma* (Boehringer, Mannheim, FRG), 200 U/ml
Glycerol-3-phosphate dehydrogenase from rabbit muscle (Boehringer), 80 U/ml
Triose-phosphate isomerase from rabbit muscle (Boehringer), 1000 U/ml

Reagents for Method 3

NADH, 15 mM
Glycerol dehydrogenase from *Enterobacter aerogenes* (Boehringer), 200 U/ml
EDTA, 200 mM (for discontinuous assay only)

Reagents for Method 4

1,4-Piperazinediethanesulfonic acid (PIPES)–Tris buffer, 50 mM, pH 6.8
NAD$^+$, 40 mM
Sodium arsenate, 5 mM
Glyceraldehyde-3-phosphate dehydrogenase from rabbit muscle (Boehringer), 500 U/ml

In case crude extracts are used for the assay of dihydroxyacetone synthase, low molecular weight compounds should be removed by a minimum of 2 hr of dialysis against potassium phosphate buffer (see Purification Procedure) or by passage through a Sephadex G-25 column equilibrated with the same buffer.

Procedure for Method 1. The assay mixture contains, in a final volume of 2 ml, potassium phosphate buffer, 1.85 ml; magnesium chloride, 20 μl; thiamin pyrophosphate, 10 μl; formaldehyde, 10 μl; xylulose 5-phosphate, 4 μl; and 100 μl of the protein solution containing 0.1–1 mU of dihydroxyacetone synthase activity. The reaction is started by the addition of enzyme; an assay mixture with acid-inactivated protein is used as a reference. The mixture is incubated at 30° for 3–10 min. During the incubation 0.2-ml samples are removed from the reaction mixture at 1- to 2-min intervals and added to 0.5 ml of the 2,4-dinitrophenylhydrazine solution in Pyrex tubes. The tubes are stoppered and heated for 30 min in a boiling water bath. Then the tubes are rapidly cooled to room temperature, and 10 ml of the alkaline ethanol–water mixture is added. The optical density at 560 and 435 nm is measured; enzyme activity is derived from the difference ($A_{560} - A_{435}$) between the two readings. For calibration, a solution of dihydroxyacetone of known concentration is used.

Comments on Method 1. Because both dihydroxyacetone and glyceraldehyde 3-phosphate react with the hydrazine and the specific extinction of the hydrazones at 560 and 435 nm is the same, calculation of the activity is based on the apparent formation of two molecules of dihydroxyacetone. The spectrum of the reaction mixture containing inactivated protein should show two apparently equal maxima in the visible region. The maximum at 435 nm corresponds to the hydrazone resulting from the condensation of formaldehyde and 2,4-dinitrophenylhydrazine, the maximum at 560 nm to ozazones formed in the reaction with ketoses (xylulose 5-phosphate, dihydroxyacetone, glyceraldehyde 3-phosphate, etc.). As formaldehyde and xylulose 5-phosphate are consumed in an equimolar ratio, the difference between the absorbance at 560 and 435 nm corresponds to formation of dihydroxyacetone and glyceraldehyde 3-phosphate. This method is convenient when large numbers of samples are to be assayed (e.g., during the purification of dihydroxyacetone synthase). The advantages of the method are that it is sensitive and less expensive than those discussed below. It is not suitable, however, for the determination of kinetic parameters, and substrate concentrations are below saturation of the enzyme.

Procedure for Method 2. For crude extracts the assay is performed in a double-beam spectrophotometer. Both reference and measuring cuvettes contain, in a final volume of 1 ml, potassium phosphate buffer, 0.91 ml; magnesium chloride, 10 μl; thiamin pyrophosphate, 10 μl; NADH, 10 μl; glycerol kinase, 10 μl; glycerol-3-phosphate dehydrogenase, 10 μl; and triose-phosphate isomerase, 10 μl. Xylulose 5-phosphate (10 μl) is added to one cuvette and water (10 μl) to the other. To achieve an increase in the absorbance reading, the xylulose 5-phosphate-containing cuvette is placed in the reference cuvette holder. After equilibration at the desired temperature (30°), 10 μl of the protein solution containing 0.2–0.8 mU dihydroxyacetone synthase activity is added to both cuvettes. The reaction is followed at 340 nm for 3 to 4 min to determine the xylulose 5-phosphate-dependent NADH oxidation. If this activity is high, the protein concentration should be reduced. The reaction is started by the addition of 10 µl formaldehyde to both cuvettes.

Comments on Method 2. By subtracting the xylulose 5-phosphate-dependent NADH oxidation, interfering reactions are eliminated. However, care should be taken when the sample contains large amounts of formaldehyde reductase: this may cause a fast exhaustion of NADH in both cuvettes and severely limits the accuracy of the assay. Also, a relatively small difference in its activity in the two cuvettes, owing, for instance, to inaccuracy in the pipetting of the protein solution, can result in a significant change in the apparent dihydroxyacetone synthase activity found. For the calculation of activity it should be realized that 2 molecules of NADH (ϵ_{NADH} 6.22 mM^{-1} cm^{-1}) are oxidized per molecule of formaldehyde consumed. When (partly) purified enzyme is assayed the use of two cuvettes is not essential.

Procedure for Method 3. The assay mixture contains, in a final volume of 1 ml, potassium phosphate buffer, 0.94 ml; magnesium chloride, 10 µl; thiamin pyrophosphate, 10 µl; formaldehyde, 10 µl; xylulose 5-phosphate, 10 µl; glycerol dehydrogenase, 10 µl; and 10 µl of the protein solution containing 0.4–2 mU dihydroxyacetone synthase activity. After temperature equilibration (30°), the reaction is started by the addition of formaldehyde and the change in absorbance at 340 nm recorded. When crude extracts are used, a two-cuvette assay as used in Method 2 eliminates interfering reactions.

Comments on Method 3. Low activities of dihydroxyacetone synthase in crude extracts are best measured using a discontinuous modification of this assay. The same reaction mixture can be used, except that NADH and glycerol dehydrogenase are omitted and replaced by water. After equilibration to the desired temperature, the reaction is initiated by addition of the protein solution or formaldehyde. Samples of 200 µl are removed from the reaction mixture at $t = 0$ and at regular intervals and pipetted into a precooled tube containing 50 µl EDTA. The samples are analyzed for dihydroxyacetone in a 1-ml assay mixture containing 780 µl potassium phosphate buffer, 10 µl NADH, and 200 µl sample. The reaction is started by the addition of 10 µl glycerol dehydrogenase. In case the sample contains interfering enzyme activities, protein can be removed by ultrafiltration (Amicon micropartition system) or by stopping the reaction with 1 M perchloric acid instead of EDTA. After the protein is removed by centrifugation, samples are neutralized with 10 M KOH at 0° to precipitate the perchloric acid and, after centrifugation, analyzed as described above.

Procedure for Method 4. The assay mixture contains, in a final volume of 1 ml, PIPES–Tris buffer, 0.92 ml; magnesium chloride, 10 µl; thiamin pyrophosphate, 10 µl; sodium arsenate, 10 µl; xylulose 5-phosphate, 10 µl;

glyceraldehyde-3-phosphate dehydrogenase, 10 μl; NAD$^+$, 10 μl; and 10 μl of the protein solution containing 0.2–2 mU dihydroxyacetone synthase activity. After temperature equilibration (30°), the reaction is started by the addition of 10 μl formaldehyde and the change in absorbance at 340 nm recorded. Any transketolase activity present may be corrected by subtracting the activity observed prior to the addition of formaldehyde.

Comments on Method 4. If Method 4 is used for the enzyme assay, the samples should not contain orthophosphate.

Purification Procedure

Growth of Organism. Candida boidinii KD1 may be obtained from the collection of methylotrophic microorganisms at The Institute of Biochemistry and Physiology of Microorganisms, USSR Academy of Science, Pushchino, USSR. *Hansenula polymorpha* DL1 is available elsewhere.[6]

Yeast cells are grown in chemostat culture in a fermentor at 29° (*Candida* and *Pichia* strains) or 37° (*Hansenula* strains) at a dissolved oxygen tension of 30–50% of air saturation, pH 5.5, and dilution rate 0.1 hr^{-1} using the following mineral medium[7] (g/liter): NH$_4$Cl, 7.63; KH$_2$PO$_4$, 2.81; MgSO$_4 \cdot 7H_2O$, 0.59; microelements (mg/liter) as follows: H$_3$BO$_3$, 4.0; CuSO$_4 \cdot 5H_2O$, 4.0; KI, 0.6; FeCl$_3 \cdot 6H_2O$, 37.5; MnSO$_4 \cdot H_2O$, 17.0; ZnSO$_4 \cdot 7H_2O$, 22.0; Na$_2$MoO$_4 \cdot 2H_2O$, 2.6; CaCl$_2 \cdot 2H_2O$, 55.0; CoCl$_2 \cdot 6H_2O$, 2.8; EDTA, 450.0; and vitamins (mg/liter): thiamin 5.0; biotin, 0.05. The cells are harvested from steady-state cultures by centrifugation at 5000 g for 10 min, washed 1 or 2 times with 50 mM PIPES–Tris (or potassium phosphate) buffer, pH 6.8, containing 0.5 mM EDTA, 5 mM MgCl$_2$, and 0.5 mM dithiothreitol, and used either immediately for enzyme purification or stored at $-20°$ for up to 1 month. All steps for the enzyme purification are carried out at 4° (Table II).

Step 1: Preparation of Cell-Free Extract. Frozen cells (50 g wet weight) are suspended in 100 ml of 50 mM PIPES–Tris buffer, pH 6.8, containing 0.5 mM EDTA, 5 mM MgCl$_2$, and 0.5 mM dithiothreitol (buffer A) with 1 ml of dodecanethiol (Fluka). Cells are disintegrated using a French press at 1.5×10^5 kN/m^2 and centrifuged at 10,000 g for 30 min, yielding the cell-free extract as supernatant.

Note: Dodecanethiol is a water-immiscible liquid which floats on the surface of the buffer. Its stabilizing effect is probably due to a high reactivity toward SH-inactivating impurities possibly present in the buffer as well as maintenance of the SH groups of the enzyme and dithiothreitol in the

[6] D. W. Levine and C. L. Cooney, *Appl. Microbiol.* **26**, 982 (1973).
[7] Th. Egli and A. Fiechter, *J. Gen. Microbiol.* **123**, 365 (1981).

TABLE II
PURIFICATION OF DIHYDROXYACETONE SYNTHASE FROM *Candida boidinii* KD1

Fraction	Total volume (ml)	Total protein (mg)	Total activity (units)	Specific activity (units/mg)	Yield (%)
Step 1. Cell-free extract	100	216	32.4	0.15	100
Step 2. Phenyl-Sepharose chromatography	24	21.8	24.0	1.1	74
Step 3. Procion Red-Sepharose chromatography	12	1.6	6.4	4.1	20

reduced state. To avoid problems during chromatography the buffer level above the gel should be high enough to prevent penetration of dodecanethiol.

Step 2: Phenyl-Sepharose CL-6B Chromatography. The cell-free extract is fractionated using conventional ammonium sulfate precipitation, and the 30–60% precipitate is collected by centrifugation. This is then resuspended in a minimal volume of buffer A and applied to a column (1.6 × 20 cm) of Phenyl-Sepharose CL-6B (Pharmacia, Uppsala, Sweden) equilibrated with buffer A containing 30% ammonium sulfate. On the surface of the buffer solution in the column, 2 ml of dodecanethiol is applied. Fractions containing dihydroxyacetone synthase activity are eluted with a descending convex semilogarithmic gradient of ammonium sulfate, 30–0%, 40 ml/hr for 4 hr. Usually, fractions containing dihydroxyacetone synthase activity are eluted at 0% ammonium sulfate.

Note: The shape of the gradient is important. If the residual concentration of salts in the dihydroxyacetone synthase-containing fractions is too high to use the following step, the curvature of the gradient can be increased.

Step 3: Procion Red-Sepharose Chromatography. Procion Red-Sepharose is prepared as follows.[8] One hundred fifty milliliters of Sepharose CL-6B (Pharmacia) is washed with distilled water on a glass funnel and resuspended in 50 ml of distilled water. Then a solution of 2.5 g Na_2CO_3 and 1 g Procion Red HE-3B (ICI, England) in 50 ml of distilled water is added. The suspension is carefully mixed and incubated for 40 hr at 45°. The gel is washed on a glass funnel with 10 volumes of distilled water, then with 1 *M* KCl, and finally with water. The freshly prepared gel is placed in

[8] W. Heyns and P. de Moor, *Biochim. Biophys. Acta* **358**, 1 (1974).

a column of proper size, equilibrated with 50 mM potassium phosphate buffer, and 5 ml of a 20 mg/ml bovine serum albumin solution is applied to the column and eluted with 1 M KCl. The gel is stored at 4° in a 20% buffered solution of ethanol.

For chromatography, the dihydroxyacetone synthase fraction from Step 2 is applied to a column (1.6 × 20 cm) filled with the Procion Red-Sepharose prepared as above and equilibrated with buffer A. Dodecanethiol is also added on top of the buffer in the column as described in Step 2. Fractions containing dihydroxyacetone synthase activity are eluted with a linear gradient of ammonium sulfate, 0–10%, flow rate 40 ml/hr for 4 hr. Active fractions are pooled and concentrated to 3–5 ml final volume by dialysis against buffer A containing 20% polyethylene glycol 6000. On the top of the dialysis tube 200 μl of dodecanethiol is placed. The enzyme preparation is kept in a stoppered tube under 1 ml of dodecanethiol at 4°. Under these conditions the enzyme remains active for at least 2–3 weeks. The enzyme from *H. polymorpha* DL1 as well as from *C. boidinii* KD1 is unstable in the absence of dodecanethiol and loses activity in a few hours. According to previously published procedures,[9,10] hydroxylapatite chromatography can also be used as an additional step for the purification.

Properties

Purity, Molecular Weight, and Structure. Following the method described above, dihydroxyacetone synthase is obtained as a single protein band by polyacrylamide gel electrophoresis in either the presence or absence of sodium dodecyl sulfate (SDS) followed by Coomassie blue staining. Sepharose CL-6B gel filtration and nondenaturating gradient (5–20%) polyacrylamide gel electrophoresis revealed a native molecular mass of 145 kDa for *C. boidinii*. The molecular mass of the subunit, as determined by SDS–polyacrylamide gel electrophoresis, is 76 kDA,[9] making a dimer the most likely native configuration. Similar results were found with the enzyme from *H. polymorpha* DL1. The molecular mass of the enzyme purified from other *Candida* strains is, however, different.[5,10] The enzyme from *C. boidinii* KD1 has an isoelectric point of 7.1, and it contains tightly bound thiamin pyrophosphate as a cofactor but no metal cations.

Stability. The most effective stabilizer is dodecanethiol. EDTA and Mg^{2+} also stabilize the enzyme but to a lesser extent. A homogeneous preparation of dihydroxyacetone synthase is partially inactivated in 60%

[9] L. V. Bystrykh, A. P. Sokolov, and Y. A. Trotsenko, *FEBS Lett.* **132**, 324 (1981).
[10] N. Kato, C. Higuchi, C. Sakazawa, T. Nishizawa, Y. Tani, and H. Yamada, *Biochim. Biophys. Acta* **715**, 134 (1982).

TABLE III
SUBSTRATE SPECIFICITY OF DIHYDROXYACETONE
SYNTHASE FROM *Candida boidinii* KD1

Substrate	V_{max}	K_m	V_{max}/K_m
Donors (1 mM formalde-hyde)			
Hydroxypyruvate	2.5	0.25	10
Xylulose 5-phosphate	5.2	0.35	15
Fructose 6-phosphate	1.6	0.9	1.8
Sedoheptulose 7-phos-phate	0.9	1.2	0.8
Acceptors (2.5 mM xylulose 5-phosphate)			
Formaldehyde	5.7	0.5	12
Glycolaldehyde	8.3	0.4	20
D-Glyceraldehyde	5.2	0.65	8
L-Glyceraldehyde	0.6	0.85	0.6
Ribose 5-phosphate	2.9	0.75	4

ammonium sulfate solution; however, it is stable for several weeks when the crude extract is fractionated as described in Step 2.

Substrate Specificity. The enzyme operates by the way of a Ping-Pong mechanism and transfers the glycolaldehyde moiety from the substrate–donor (phosphorylated keto sugars or hydroxypyruvate) to the substrate–acceptor (phosphorylated aldo sugars or formaldehyde). The most effective substrate–donor is xylulose 5-phosphate (Table III). The enzyme activity and affinity decrease proportionally to the carbon chain length of the substrate donor. The highest affinity and activity among substrate donors are displayed by glycolaldehyde, followed by formaldehyde and D-glyceraldehyde. Similar data have been reported for dihydroxyacetone synthases from other *Candida* strains.[5,10]

Inhibitors. Dihydroxyacetone synthase is irreversibly inactivated by 1 mM SH-inactivating reagents (*p*-chloromercuribenzoate, *N*-ethylmalei-mide). Excess formaldehyde also causes an inhibitory effect, with an apparent K_i of 8.2 mM. Reduced glutathione (GSH) inhibits the enzyme formally by a competitive mechanism with respect to formaldehyde, but not with any other substrate acceptor. Actually, the mechanism of enzyme inhibition by GSH is based on chemical trapping of the formaldehyde by the formation of the adduct hydroxymethylglutathione. The equilibrium constant for adduct formation, 1.5 mM,[11] agrees with this inhibitory effect.

[11] L. Uotila and M. Koivusalo, *J. Biol. Chem.* **249,** 7653 (1974).

Effect of Metal Cations. Mg^{2+} stabilizes the enzyme and is also necessary in the enzyme assay. Other bivalent cations do not influence dihydroxyacetone synthase activity (Ca^{2+}) or inactivate the enzyme (Mn^{2+}, Zn^{2+}, and Co^{2+}) at 0.1 mM concentrations.

pH Optimum. The pH optimum of both activity and stability is apparently narrow, namely, pH 6.8–7.1.

[68] Triokinase from *Candida boidinii* KD1

By Leonid V. Bystrykh, Wim de Koning, and Wim Harder

Dihydroxyacetone + MgATP → dihydroxyacetone phosphate + MgADP

Triokinase catalyzes the second step in the assimilation of formaldehyde via the xylulose monophosphate (dihydroxyacetone) cycle during growth of yeasts on methanol. It is one of the key enzymes in the assimilatory pathway.[1-3] According to mutant studies the enzyme is not required for the assimilation of glycerol by methylotrophic yeast.[2,4] Since this triokinase is specific for the phosphorylation of dihydroxyacetone, the designation "dihydroxyacetone kinase" is also used in the literature. No other triokinases are found in methylotrophic yeast.[5,6]

Assay Methods

Principle. Two assay methods may be used. Both methods involve the determination of the rate of dihydroxyacetone phosphate production from dihydroxyacetone and MgATP. Method 1 involves the continuous measurement of dihydroxyacetone phosphate formation using glycerol-3-phosphate dehydrogenase as a coupling enzyme.[3] Method 2 is discontinuous and is based on dihydroxyacetone phosphate cleavage in sodium hydroxide

[1] Y. A. Trotsenko, L. V. Bystrykh, and V. M. Ubiyvovk, *in* "Proceedings of the 4th International Symposium on Microbiol Growth on C$_1$ Compounds" R. L. Crawford and R. S. Hanson, eds.), p. 118. American Society for Microbiology, Washington, D.C., 1984.

[2] W. Harder, Y. A. Trotsenko, L. V. Bystrykh, and Th. Egli, *in* Proceedings of the 5th International Symposium on Microbial Growth on C$_1$ Compounds" (H. W. van Verseveld and J. A. Duine, eds.), p. 139. Martinus Nijhof, Dordrecht, 1987.

[3] J. P. van Dijken, W. Harder, A. J. Beardsmore, and J. R. Quayle, *FEMS Microbiol. Lett.* **4**, 97 (1978).

[4] W. de Koning, W. Harder, and L. Dijkhuizen, *Arch. Microbiol.* **148**, 314 (1987).

[5] W. de Koning, M. A. G. Gleeson, W. Harder, and L. Dijkhuizen, *Arch. Microbiol.* **147**, 375 (1987).

[6] L. V. Bystrykh, L. R. Aminova, and Y. A. Trotsenko, *FEMS Lett.* **51**, 89 (1988).

solution, yielding methylglyoxal and inorganic phosphate.[7,8] The phosphate is then determined by way of a malachite green/molybdate reagent.[9]

Reagents for Method 1

Tris-HCl buffer, 50 mM, pH 8.0
Magnesium chloride, 500 mM
ATP, 500 mM; the pH of this solution is adjusted to pH 6.0–8.0 by addition of solid tris(hydroxymethyl)aminomethane or NaHCO$_3$
Dihydroxyacetone, 10 mM
NADH, 15 mM
Glycerol-3-phosphate dehydrogenase from rabbit muscle (Boehringer, Mannheim), 200 U/ml, desalted to remove ammonium sulfate and suspended in 0.1 M potassium phosphate buffer, pH 7.0

Reagents for Method 2

Tris-HCl buffer, 50 mM, pH 8.0
Magnesium chloride, 500 mM
ATP, 500 mM (for preparation, see above)
Dihydroxyacetone, 10 mM
Malachite green/molybdate reagent, prepared by mixing 1 volume of 3% (NH$_4$)$_6$MoO$_{24}$·4H$_2$O in a 2% aqueous solution of Triton X-100 and 1 volume of 0.06% malachite green in 6 N HCl
1 N NaOH

In case crude extracts are used for the assay of triokinase, low molecular weight compounds should be removed by a minimum of 2 hr of dialysis against the Tris-HCl buffer (see Purification Procedure) or by passage through a Sephadex G-25 column equilibrated with the same buffer.

Procedure for Method 1. The assay mixture contains the following, in a final volume of 1 ml: Tris-HCl buffer, 0.94 ml; magnesium chloride, 10 μl; dihydroxyacetone, 10 μl; glycerol-3-phosphate dehydrogenase, 10 μl; and NADH, 10 μl. The mixture is preincubated at 30° in glass cuvettes for temperature equilibration, then 10 μl of the triokinase-containing solution (0.1–1 mU) is added. Changes in the NADH concentration are monitored at 340 nm using a recording spectrophotometer equipped with a thermostatted cell holder at 30° to establish a control rate. Then 10 μl of the ATP solution is added and the measurement continued. The enzyme activity is calculated from the difference between the rates of NADH oxidation in the

[7] L. V. Bystrykh and Y. A. Trotsenko, *Biokhimia* **48,** 1611 (1983).
[8] R. Iyengar and I. Rose, *Biochemistry* **20,** 1223 (1981).
[9] Y. Tashima and N. Yoshimura, *J. Biochem.* **78,** 1161 (1975).

presence and absence of ATP and expressed as units (micromoles per minute per milligram protein).

General Comments on Method 1. Glycerol dehydrogenase is usually present in crude extracts of methylotrophic yeasts. This enzyme catalyzes the NADH-dependent reduction of dihydroxyacetone, and the activity can be as high as several units per milligram of protein. The interference by this enzyme can be reduced by decreasing the dihydroxyacetone concentration in assay mixture, because glycerol dehydrogenase possesses a much lower apparent affinity for dihydroxyacetone than does the triokinase (about $\frac{1}{1000}$). Another possibility is to use the alternative, Method 2. MgATP complex is the true substrate for triokinase; excess of either component of the complex inhibits the enzyme. To avoid this inhibitory effect, the ATP/Mg ratio should be carefully adjusted. This was demonstrated for the kinase from *Candida boidinii* KD1[10] and *C. methylica*;[11] however, it was not found for the enzyme from *Hansenula polymorpha* CBS 4732.[12]

Procedure for Method 2. The assay mixture contains the following, in a final volume of 2 ml: Tris-HCl buffer, 1.9 ml; magnesium chloride, 20 μl; ATP, 20 μl; and dihydroxyacetone, 20 μl. The same sample without dihydroxyacetone is used as a reference to correct for any liberation of orthophosphate from ATP by unspecific phosphatases which may be present in crude extracts. Reaction mixtures are preincubated for 3–15 min at 30°, the reaction is started by adding enzyme solution, and 0.2-ml samples are withdrawn from each mixture at 1 to 2-min intervals, mixed with 0.2 ml of 1 N NaOH, and incubated at 37° for 10 min. The alkaline treatment leads to hydrolysis of the dihydroxyacetone phosphate and release of inorganic orthophosphate. After incubation, 1.6 ml of distilled water and 1 ml of the malachite green/molybdate reagent are added sequentially; the mixtures are incubated at room temperature for 15 min and the optical density read at 614 nm.

General Comments on Method 2. The final mixture should be transparent. A green sediment in the mixture indicates an excess of orthophosphate. In this case it is necessary to decrease the triokinase activity in the assay mixture or the time of incubation. A white sediment in the final mixture indicates an excess of protein. The method works well within a concentration range of 0.5–13 nM inorganic phosphate in the assay mixture.

[10] L. V. Bystrykh, *Biokhimia* **50**, 1012 (1985).
[11] K. H. Hoffman and W. Babel, *Z. Allg. Mikrobiol.* **21**, 219 (1982).
[12] N. Kato, H. Yoshikawa, K. Tanaka, M. Shimao, and C. Sakazawa, *Arch. Microbiol.* **150**, 155 (1988).

Purification Procedure

Growth of Organism. Candida boidinii KD1 may be obtained from the collection of methylotrophic microorganisms at the Institute of Biochemistry and Physiology of Microorganisms, USSR Academy of Science, Puschino, USSR. *Hansenula polymorpha* DL1 is available elsewhere.[13] For growth of the organism in batch culture, a mineral medium containing 0.5–1% (v/v) of methanol can be used.[14] It contains the following (g/liter): NH_4Cl, 4.0; KH_2PO_4, 1.0; K_2HPO_4, 1.0; $MgSO_4 \cdot 7H_2O$ 0.5; thiamin, 1.0 mg/liter; and biotin, 0.01 mg/liter. However, a better yield of cells and of enzyme is obtained by chemostat cultivation in a fermentor using the following mineral salts solution[15] (g/liter): NH_4Cl, 7.63; KH_2PO_4, 2.81; $MgSO_4 \cdot 7H_2O$, 0.59; microelements (mg/liter): H_3bO_3, 4.0; $CuSO_4 \cdot 5H_2O$, 4.0; KI, 0.6; $FeCl_3 \cdot 6H_2O$, 37.5; $MnSO_4 \cdot H_2O$, 17.0; $ZnSO_4 \cdot 7H_2O$, 22.0; $Na_2MoO_4 \cdot 2H_2O$, 2.6; $CaCl_2 \cdot 2H_2O$, 55.0; $CoCl_2 \cdot 6H_2O$, 2.8; EDTA, 450.0; and vitamins (mg/liter): thiamin, 5.0; biotin, 0.05.

Cells are grown in chemostat culture at 29° (*Candida* and *Pichia* strains) or 37° (*Hansenula* strains) at a dissolved oxygen concentration of 30–50% of air saturation, pH 5.5, and a dilution rate 0.1–0.15 hr^{-1}. The level of triokinase is proportional to the growth rate of the culture. Cells are harvested by centrifugation at 5000 g for 10 min, washed 1 or 2 times with distilled water or 50 mM Tris-HCl buffer, pH 8.0, containing 0.5 mM EDTA, and used either immediately or stored at −20° for up to 2 months.

All steps for enzyme purification are carried out at 4° (see Table I).

Step 1: Preparation of Cell-Free Extract. Frozen cells (500 g wet weight) are suspended in 1000 ml of 50 mM Tris-HCl buffer, pH 8.0, containing 0.5 mM EDTA, 5 mM $MgCl_2$, and 15% (v/v) glycerol (buffer A), disintegrated using a French press (1.5×10^5 kN/m^2), and centrifuged at 10,000 g for 30 min, yielding the cell-free extract as the supernatant.

Step 2: Fractionation by Polyethylene Glycol. A 40% (w/v) solution of polyethylene glycol 6000 (BDH, Poole, England) in buffer A is used. The solution is added to the cell-free extract to a concentration of 15% (v/v) with stirring, and the resulting mixture is centrifuged at 25,000 g for 40 min. To the supernatant a further portion of the polyethylene glycol solution is added to a final concentration of 20%. The mixture is centrifuged as above, and the sediment is carefully resuspended in a minimal volume of buffer A to obtain a transparent solution.

Step 3: Chromatography on Polyethyleneimine-BioGel. Polyethyleneimine is immobilized on BioGel A 0.5m (Bio-Rad, Richmond, CA). This

[13] D. W. Levine and C. L. Cooney, *Appl. Microbiol.* **26**, 982 (1973).
[14] K. Ogata, H. Nishikawa, and M. Ohsugi, *Agric. Biol. Chem.* **33**, 1519 (1969).
[15] Th. Egli and A. Fiechter, *J. Gen. Microbiol.* **123**, 365 (1981).

TABLE I
PURIFICATION OF TRIOKINASE FROM *Candida boidinii* KD1

Fraction	Total volume (ml)	Total protein (mg)	Total activity (units)	Specific activity (units/mg)	Yield (%)
Step 1. Cell-free extract	950	9880	2865	0.29	100
Step 2. Polyethylene glycol fraction 15–20%	31	524	1231	2.35	43
Step 3. Polyethyleneimine-BioGel chromatography	60	38	798	21.00	28
Step 4. DEAE-Sepharose CL-6B chromatography	6.8	27	684	25.32	24
Step 5. Gel filtration on Sepharose CL-6B	40	24	488	20.33	17

gel provides essentially better purification of the triokinase than commercially available DEAE gels. A procedure for its synthesis is as follows.[11] Fifty grams of wet BioGel A 0.5m is placed in a funnel equipped with a glass filter and washed with 1 volume of cold distilled water followed by cold solutions of 10, 20, and 40% acetone in water. Finally, the gel is washed with pure acetone (250 ml), suspended in 100 ml of acetone, transferred to a flask with a stopper, and cooled again to 4°. Then 2 g of cyanuric chloride (Aldrich, Milwaukee, WI) is added to the gel suspension and carefully mixed. (*Note:* cyanuric chloride is very poisonous.) After a 5-min incubation, 1 ml of pure freshly distilled triethanolamine is added while the mixture is stirred vigorously. After the flask is stoppered, the mixture is incubated for 16 hr at 4° with gentle shaking. Then the gel is washed on a filter funnel with 300 ml of acetone, suspended in 200 ml of acetone, and transferred to a flask fitted with a stopper. Ten grams of a 50% polyethyleneimine solution (Serva, Heidelberg, Germany) is diluted in 200 ml of water, and the pH is adjusted to 9.0. This solution is mixed with the gel suspension under vigorous shaking and incubated at room temperature for 4 hr. The gel is washed on the funnel with 5 volumes of distilled water, then suspended in 200 ml 1 M ethanolamine-HCl, pH 8.5, and incubated for 4 hr at room temperature. The gel is washed with water and stored, after adding NaN_3 to a final concentration of 0.1 mM, at 4°. For enzyme purification, the gel is poured into a column (1.6 × 10 cm) and equilibrated with buffer A using a flow rate of 25 ml/hr.

The protein solution from Step 2 is applied to the column containing polyethyleneimine-BioGel at a rate of 25 ml/hr. Triokinase activity is eluted using a linear gradient of 0.1–0.5 M NaCl for 8 hr. Fractions containing activity are collected, diluted one-third with buffer A, and used for the following step.

Step 4: DEAE-Sepharose Chromatography. The fraction obtained from Step 3 is applied to a column (1.6 × 20 cm) containing DEAE-Sepharose (Pharmacia) equilibrated with buffer A. Dihydroxyacetone kinase is eluted from the column using a linear gradient of 0–0.25 M NaCl for 2 hr. Fractions containing activity are pooled.

Step 5: Gel Filtration on Sepharose CL-6B. Gel filtration is used only if the preparation obtained after the preceding steps still contains protein impurities. The fractions obtained in Step 4 are applied to a column (1.6 × 100 cm) containing Sepharose CL-6B (Pharmacia) equilibrated with buffer A and eluted using the same buffer with flow rate of 60 ml/hr. The active fractions are pooled and dialyzed for 10–16 hr at 4° against a solution of 40% glycerol in buffer A.

The enzyme is stored at 20° in 40% glycerol solution and remains active for at least 1 month.

Properties

Purity, Molecular Weight, and Structure. The enzyme preparation is homogeneous as judged by polyacrylamide gel electrophoresis either in the presence or absence of sodium dodecyl sulfate (SDS), followed by Coomassie blue staining. The above procedure can also be used for the triokinase from *Hansenula polymorpha* DL1. The enzyme from *Candida boidinii* KD1 is a dimer with a native molecular mass of 139 kDa. The molecular mass of the subunits as determined by SDS gel electrophoresis is 71 kDa.[7] The isoelectric point (pI) is 4.6. The enzyme does not contain tightly bound cofactors or metal cations.

Stability. The most effective stabilizer is glycerol. EDTA and MgCl$_2$ also stabilize the enzyme, albeit to a lesser extent. The kinase is completely inactivated during ammonium sulfate precipitation. However, the enzyme can be stored as a suspension in a buffered solution of polyethylene glycol 6000 (see Step 2). In the absence of stabilizing reagents, the enzyme loses activity within 24 hr.

Substrate Specificity. ATP and dihydroxyacetone are the most effective substrates for the kinase from *C. boidinii* KD1. The activity with other nucleotide triphosphates (ITP, CTP, GTP, UTP) as well as with DL-glyceraldehyde instead of dihydroxyacetone is about 25% of that observed with dihydroxyacetone and ATP. A similar specificity was found for the enzyme from *C. methylica*.[11] However, the triokinase from *H. polymorpha* CBS 4732 is active with ATP only.[12] The enzyme from *C. boidinii* KD1 shows a slight cooperative response toward dihydroxyacetone, with a Hill coefficient of $n_H = 1.56$, $s_{0.5} = 0.006$ mM. For MgATP the K_m is 0.35 mM. A plot of initial velocity *(v)* versus Mg^{2+} is sigmoidal, whereas a plot of *v* versus ATP shows an intermediate plateau. Both components of the MgATP complex inhibit the kinase at excess concentrations. The reaction catalyzed by triokinase is practically irreversible.

Effect of Metal Ions. Triokinase requires Mg^{2+} cations for activity, and no other cations can replace Mg^{2+} for the enzyme from *Candida* strains. Bivalent cations such as Co^{2+}, Zn^{2+}, Be^{2+}, Mn^{2+}, Ca^{2+}, and Cu^{2+} inhibit enzyme activity by 60 to 100%. The enzyme from *H. polymorpha* CBS4732 is also active when the cations of Mn^{2+}, Co^{2+}, or Ca^{2+} are added.[12]

pH Optimum. The pH optimum for activity is 7.8–8.2 and for stability, pH 6.7–8.2.

[69] Glycerone Kinase from *Candida methylica*

By KLAUS H. HOFMANN and WOLFGANG BABEL

Dihydroxyacetone + ATP $\rightleftharpoons$ dihydroxyacetone phosphate + ADP

Glycerone kinase (dihydroxyacetone kinase, EC 2.7.1.29) catalyzes a key step in the assimilation of carbon during growth of yeasts on methanol.[1,2] This enzyme is also involved in the assimilation of glycerol by microorganisms.[3,4] The partial purification described below has been the subject of a previous report.[5] The dihydroxyacetone kinases from *Candida boidinii* KD1[6] and *Hansenula polymorpha* CBS 4732[7] have been purified to homogeneity.

Assay Method

Principle. Enzyme activity is determined in a coupled reaction with glycerol phosphate dehydrogenase and NADH by following the change of optical density at 340 nm at 30°.[8]

[1] J. P. van Dijken, W. Harder, A. J. Beardsmore, and J. R. Quayle, *FEMS Microbiol. Lett.* **4**, 97 (1978).
[2] W. Babel and N. Loffhagen, *Z. Allg. Mikrobiol.* **19**, 299 (1979).
[3] E. C. C. Lin, *Annu. Rev. Microbiol.* **30**, 525 (1976).
[4] W. Babel and K. H. Hofmann, *Arch. Microbiol.* **132**, 179 (1982).
[5] K. H. Hofmann and W. Babel, *Z. Allg. Mikrobiol.* **21**, 219 (1981).
[6] L. V. Bystrykh and Y. A. Trotsenko, *Biokhimiya* **48**, 1611 (1983).
[7] N. Kato, H. Yoshikawa, K. Tanaka, M. Shimao, and C. Sakazawa, *Arch. Microbiol.* **150**, 155 (1988).
[8] F. Heinz and W. Lamprecht, *Hoppe-Seyler's Z. Physiol. Chem.* **324**, 88 (1961).

Reagents

Imidazole-HCl buffer, 200 mM, pH 7.5
Magnesium sulfate, 50 mM
ATP, 50 mM
Dihydroxyacetone, 5 mM
NADH, 3.4 mM
α-Glycerol-phosphate dehydrogenase, 70 units

Procedure. The assay mixture contains, in a final volume of 2.0 ml, imidazole-HCl buffer, 1.0 ml; magnesium sulfate, 0.20 ml; ATP, 0.20 ml; NADH, 0.10 ml; dihydroxyacetone, 0.20 ml; glycerol-phosphate dehydrogenase, 50 μl; and enzyme solution. NADH oxidation independent of dihydroxyacetone phosphate formation is corrected for, when necessary, by running parallel assays lacking ATP.

Units. One unit of dihydroxyacetone kinase is defined as that amount of enzyme catalyzing the formation of 1 μmol of dihydroxyacetone per minute at 30°. Specific activity is expressed as units per milligram of protein.

Growth of Organism

Candida methylica is grown in 500-ml shake flasks at 30°. A mineral salt medium (Reader, 1927) with 0.1% of yeast extract and 1% (v/v) of methanol is used. The initial pH value is adjusted to 4.5 by adding NaOH. Cells from 2-day cultures are harvested by centrifugation at 6,000 g.

Purification Procedure

Step 1: Preparation of Cell-Free Extract. For the preparation of cell-free extracts the pellet is washed twice with 20 mM Tris-HCl buffer (pH 7.5) and resuspended in the same buffer. The cell suspension (about 20 mg wet weight/ml) is passed 3 times through a French pressure cell (9.9 kN/m^2) followed by centrifugation of the homogenate at 20,000 g for 20 min, yielding the cell extract as supernatant. The protein concentration is determined by the method of protein–dye binding,[9] with human serum albumin used as a standard.

Step 2: Streptomycin Sulfate Treatment. Solid streptomycin sulfate is added to the cell-free extract to a concentration of 2%. The precipitate is removed by centrifugation (20,000 g, 30 min) and discarded.

Step 3: Ammonium Sulfate Precipitation. In the following step the enzyme is precipitated by solid ammonium sulfate between 2.0 and 3.0 M.

[9] M. M. Bradford, *Anal. Biochem.* **72,** 248 (1976).

The precipitate is collected by centrifugation and dissolved in a small volume of 20 m*M* Tris-HCl buffer (pH 7.5) containing 20% glycerol (standard buffer).

Step 4: Cibacron Blue F3G-A Sephadex G-200 Chromatography. The protein solution from Step 3 is placed on a column of Cibacron blue F3G-A Sephadex G-200[10] (2.5 × 10 cm) equilibrated with standard buffer. Since the enzyme is not bound by the dye it could be eluted with standard buffer. The active fractions are combined and concentrated by ultrafiltration (Diaflo PM30, Amicon, Danvers, MA).

Step 5: DEAE-Cellulose Chromatography. The concentrated solution from Step 4 is loaded onto a DEAE-cellulose column (2.5 × 20 cm) equilibrated with standard buffer. The column is washed with 0.2 *M* KCl in standard buffer until the eluate is free of protein. Afterward the enzyme is eluted with a linear KCl gradient (0.2 to 0.4 *M*). The active fractions are pooled and concentrated (see Step 4). Glycerol is added to the concentrated enzyme to maintain activity. A typical purification is summarized in Table I.

Properties

Purity. The described procedure results in about a 100-fold enrichment with a yield of around 52%. The enzyme preparation is not homogeneous. Polyacrylamide gel electrophoresis of the purified enzyme revealed several bands of protein. But the enzyme preparation is free of activities disturbing the assay of dihydroxyacetone kinase.

Substrate Specificity and Michaelis Constants. Substrates tested are listed in Table II. The purified enzyme phosphorylates dihydroxyacetone 4 times faster than DL-glyceraldehyde. The reaction rate of dihydroxyacetone kinase shows normal Michaelis–Menten kinetics[11] depending on the concentration of dihydroxyacetone and DL-glyceraldehyde. The apparent Michaelis constants for the two substrates obtained by Lineweaver–Burk plots are 0.011 and 0.024 m*M*. Other C_3 compounds including glycerol are not phosphorylated by the dihydroxyacetone kinase of *Candida methylica.* ITP and UTP are used as phosphate donors with reaction rates of 11.2 and 3.1%, respectively, in relation to ATP, whereas the reaction rates with CTP or GTP are much lower than 1% (Table II).

Effect of Metal Ions. The reaction of dihydroxyacetone kinase from *Candida methylica* depends on the presence of divalent cations in the

[10] H.-J. Böhme, G. Kopperschläger, J. Schulz, and E. Hofmann, *J. Chromatogr.* **69,** 209 (1972).

[11] K. H. Hofmann and W. Babel, *Z. Allg. Mikrobiol.* **24,** 713 (1984).

TABLE I
PURIFICATION OF DIHYDROXYACETONE KINASE FROM *Candida methylica*

Fraction	Protein (mg)	Units (μmol/min)	Specific activity (units/mg protein)	Yield (%)
Step 1. Cell-free extract	2003	737.8	0.369	100
Step 2. Streptomycin sulfate (2% supernatant)	1542	737.6	0.478	100
Step 3. Ammonium sulfate (2.0 to 3.0 M precipitate)	694	544.0	0.784	73.8
Step 4. Cibacron blue F3G-A Sephadex G 200	202	430.2	2.134	58.3
Step 5. DEAE-cellulose eluate (0.2 to 0.4 M KCl)	10.2	382.5	37.646	51.8

TABLE II
SUBSTRATE SPECIFICITY OF DIHYDROXYACETONE KINASE OF
Candida methylica

Substrate	Relative activity (%)	Apparent Michaelis constants (mM)
C$_3$ compounds (0.5 mM of each)		
Dihydroxyacetone	100	0.011
DL-Glyceraldehyde	25.2	0.024
DL-Glycerate	0	
Glycerol	0	
Pyruvate	0	
Hydroxypyruvate	0	
Trinucleotides (5 mM of each)		
ATP	100	
ITP	11.2	
UTP	3.1	
CTP or GTP	1.0	
Divalent cations (5 mM of each)		
MgCl$_2$ or MgSO$_4$	100	
CoSO$_4$	57.3	
CaCl$_2$	30.3	
MnCl$_2$	0	
—	0	

assay. The highest activity was found with Mg^{2+} (Table II). If magnesium sulfate is replaced by cobalt(II) or calcium ions, the reaction rates are only 57.3 and 30.3%, respectively, in relation to the assay with magnesium ions. Manganese chloride in the assay led to a complete loss of activity.

Stability. The purified enzyme, in the presence of 60% glycerol, is stable for at least 4 weeks at $-20°$.

pH Optimum. The enzyme has optimum activity at pH 7.5.

Inhibitors[12]. Among metabolites of the methanol oxidation sequence, glycolysis, and the tricarboxylic acid cycle as well as intermediates of the cyclic oxidation no inhibitors were found, but the enzyme is influenced by the energy charge. In the presence of 5 mM of ADP the enzyme is inhibited by 60%.

[12] K. H. Hofmann and W. Babel, *Z. Allg. Mikrobiol.* **20,** 389 (1980).

[70] Formaldehyde Dehydrogenase from Methylotrophic Yeasts

By NOBUO KATO

Formaldehyde + glutathione + NAD⁺ ⇌ S-formylglutathione + NADH + H⁺

The true substrate of NAD-linked and glutathione-dependent formaldehyde dehydrogenase (glutathione) (EC 1.2.1.1) is a hemimercaptal, S-hydroxymethylglutathione, spontaneously formed from formaldehyde and glutathione, the reaction product being S-formylglutathione. This enzyme catalyzes a key step of the methanol catabolism in yeasts and has so far been purified from several methanol-grown yeasts: *Candida boidinii* (*Kloeckera* sp.) 2201,[1] *Candida boidinii* ATCC 32195,[2] *Hansenula polymorpha*,[3] *Pichia* NRRL-Y-11328,[4] and *Pichia pastoris* IFP 206.[5]

Assay Method

Principle. Formaldehyde dehydrogenase is measured spectrophotometrically by following the rate of NADH formation at 340 nm.

[1] N. Kato, T. Tamaoki, Y. Tani, and K. Ogata, *Agric. Biol. Chem.* **36,** 2411 (1972).
[2] H. Schutte, J. Flossdorf, H. Sahm, and M. R. Kula, *Eur. J. Biochem.* **62,** 151 (1976).
[3] J. P. van Dijken, G. J. Oostra-Demkes, R. Otto, and W. Harder, *Arch. Microbiol.* **111,** 77 (1976).
[4] R. Patel, C. T. Hou, and P. Derelanko, *Arch. Biochem. Biophys.* **221,** 135 (1983).
[5] J. J. Allais, A. Louktibi, and J. Baratti, *Agric. Biol. Chem.* **47,** 1509 (1983).

Reagents

Potassium phosphate buffer, 100 mM, pH 7.5

NAD$^+$, 60 mM, in distilled water

Glutathione (reduced), 120 mM, in distilled water

Formaldehyde, 60 mM, prepared by heating 0.5 g of paraformaldehyde in 5 ml of water at 100° in a sealed tube for 15 hr; the
formaldehyde solution is standardized with glutathione-independent formaldehyde dehydrogenase[6] (Grade II, from *Pseudomonas
putida;* Nacalai Tesque, Inc., Kyoto), and the solution is diluted
accordingly

Procedure. The enzyme activity is measured by reading the increase in
absorbance at 340 nm with a recording spectrophotometer thermostatically maintained at 30°. The complete assay mixture consists of 1.0 ml of
potassium phosphate buffer, 0.2 ml of NAD$^+$, 0.1 ml of glutathione,
0.1 ml of formaldehyde, and a limited amount of the enzyme, in a final
volume of 3.0 ml. The reaction is initiated by the addition of enzyme. A
blank test, that is, a parallel assay, is performed with a reaction mixture
with formaldehyde omitted.

Definition of Unit and Specific Activity. One unit of enzyme activity is
defined as the amount of enzyme catalyzing the formation of 1 μmol of
NADH per minute at 30°. Specific activity is defined as the number of
units per milligram of protein.

Growth of Organism

Candida boidinii 2201 is cultivated at 30° for 3 days in a 2-liter shake
flask containing 500 ml of growth medium with the following composition
(per liter): 10 ml methanol, 5 g (NH$_4$)$_2$SO$_4$, 2 g Na$_2$HPO$_4$, 4 g KH$_2$PO$_4$,
0.6 g MgSO$_4$·7H$_2$O, 0.2 g NaCl, 0.2 g CaCl$_2$·2H$_2$O, and 0.2 g yeast extract, pH 6.0. Methanol is added aseptically to the medium just before
inoculation. The cells are harvested by centrifugation at 6700 g for 15 min
and then washed twice with 10 mM potassium phosphate buffer (pH 7.0).
The cell paste is stored at $-20°$ until use.

Purification Procedure

All procedures are performed at 0 to 5°. The buffer solution comprises
1 mM dithiothreitol, 1 mM EDTA, and 0.5 mM phenylmethylsulfonyl
fluoride.

Step 1: Preparation of a Cell-Free Extract. Frozen cell paste of *C.*

[6] M. Ando, T. Yoshimoto, S. Ogushi, K. Rikitake, S. Shibata, and D. Tsuru, *J. Biochem.* **85,**
1165 (1979).

boidinii 2201 (180 g wet weight) is suspended in 180 ml of 10 m*M* potassium phosphate buffer (pH 7.5). Twenty-milliliter portions of the suspension are disrupted, separately, with a Braun MSK Cell Homogenizer with 20 g of glass beads for 4 min at 0°, followed by centrifugation at 14,000 g for 20 min, the resultant supernatant being used as the cell-free extract, which is dialyzed against 10 m*M* potassium phosphate buffer (pH 7.5).

Step 2: First DEAE-Toyopearl 650 Chromatography. The dialyzed solution is applied to a DEAE-Toyopearl 650 column (2.6 × 25 cm) previously equilibrated with 10 m*M* potassium phosphate buffer (pH 7.5). The column is washed with 500 ml of the buffer, and the enzyme is eluted with a linear gradient between 250 ml of 10 m*M* and 250 ml of 200 m*M* potassium phosphate buffer (pH 7.5) at a flow rate of 50 cm/hr. The active fractions are concentrated and dialyzed against 10 m*M* potassium phosphate buffer (pH 7.0).

Step 3: Second DEAE-Toyopearl 650 Chromatography. The dialyzed solution is applied to a DEAE-Toyopearl 650 column (1.6 × 50 cm) previously equilibrated with 10 m*M* potassium phosphate buffer (pH 7.0), then the column is washed with 100 ml of the buffer. The enzyme is eluted with a linear gradient between 250 ml of 10 m*M* and 250 ml of 150 m*M* phosphate buffer (pH 7.0) at a flow rate of 20 cm/hr. The peak fractions are pooled, concentrated to about 10 ml by ultrafiltration, using an Amicon YM10 membrane (Danvers, MA), and dialyzed against 10 m*M* potassium phosphate buffer (pH 7.5).

Step 4: Hydroxylapatite Chromatography. The dialyzed solution is placed on a hydroxylapatite column (2.6 × 20 cm) previously equilibrated with 50 m*M* potassium phosphate buffer (pH 7.5). Successive elution is carried out with 200-ml volumes of 10, 50, and 100 m*M* potassium phosphate buffer (pH 7.5). The active peak, which appears in the eluate with the 100 m*M* buffer, is concentrated to a volume of 5 ml.

Step 5: TSK Gel Toyopearl HW55 Gel Filtration. The concentrated enzyme solution is subjected to gel filtration on a TSK gel Toyopearl HW55 column (2.6 × 80 cm). Elution is carried out with 50 m*M* potassium phosphate buffer (pH 7.0) at a flow rate of 3 cm/hr. The active peak is concentrated to 5 ml. The concentrated enzyme solution is added to an equal volume of cold glycerol and stored at −20°. Under these conditions, the enzyme can be stored for at least 6 months without loss of activity.

A typical purification is summarized in Table I.

Properties

Purity. The purified enzyme preparation gives a single band on acrylamide (7%) gel electrophoresis at pH 8.3 in the presence of sodium dodecyl sulfate (SDS).

TABLE I
PURIFICATION OF FORMALDEHYDE DEHYDROGENASE FROM *Candida boidinii* 2201

Procedure	Total protein (mg)	Total activity (units)	Specific activity (units/mg protein)	Purification (-fold)	Yield (%)
Cell-free extract	3170	1010	0.32	1	100
First DEAE-Toyopearl	91.5	933	10.2	32	92
Second DEAE-Toyopearl	24.2	568	23.5	73	56
Hydroxylapatite	14.5	458	31.6	99	45
TSK gel Toyopearl HW55	10.8	414	38.3	120	41

Molecular Weight and Subunit Structure. The relative molecular weight of formaldehyde dehydrogenase from *C. boidinii* 2201 is determined to be 82,000 by high-performance liquid chromatography on a column of TSK gel G-3000SW (Tosoh, Tokyo). The molecular weights of the subunits are determined to be 43,000 by SDS gel electrophoresis. The enzyme is assumed to be a dimer.

Substrate Specificity. The enzyme catalyzes the oxidation of formaldehyde and methylglyoxal. The relative activity toward methylglyoxal is 89% of that toward formaldehyde. The enzyme does not use other aliphatic or aromatic aldehydes and is inactive toward $NADP^+$. Glutathione cannot be replaced by other thiol compounds such as cysteine, 2-mercaptoethanol, dithiothreitol, or thioglycolate. The K_m values are 0.29 mM for formaldehyde, 2.8 mM for methylglyoxal, and 25 μM for NAD^+.

Reaction Product and Reverse Reaction. The oxidation product of formaldehyde in the reaction system is *S*-formylglutathione, the formation of which is followed as the increase in absorbance at 240 mM owing to formation of a thiol ester. When purified *S*-formylglutathione hydrolase (EC 3.1.2.12) is added to the reaction mixture, formate formation is detected.[7] In the reverse reaction, the K_m values for *S*-formylglutathione and NADH are 0.12 and 0.025 mM, respectively.[8]

Effect of pH. The enzyme exhibits optimum activity at pH 8.0 in potassium phosphate buffer (Tris-HCl buffer is not suitable for the reaction) and is stable in the pH range of 6.0 to 10 at 20° for 1 hr.

Inhibition. The enzyme is completely inhibited by 1 mM *p*-chloromercuribenzoate and several cations (1 mM each), Cd^{2+}, Cu^{2+}, Hg^{2+}, and Ag^+. EDTA (1 mM) has no effect on the activity.

[7] N. Kato, C. Sakazawa, T. Nishizawa, Y. Tani, and H. Yamada, *Biochim. Biophys. Acta* **611,** 323 (1980).
[8] N. Kato, H. Sahm, and F. Wagner, *Biochim. Biophys. Acta* **566,** 12 (1979).

Kinetics. The results of kinetic studies on formaldehyde dehydrogenase from *C. boidinii* ATCC 32195 suggest an ordered Bi-Bi mechanism, with NAD+ as the first substrate and NADH as the last product.[8] In the forward reaction, NADH is a competitive inhibitor (product inhibition) with respect to NAD+ (K_i 1.1 mM), and *S*-formylglutathione is a noncompetitive inhibitor with respect to NAD+ and formaldehyde (actually, *S*-hydroxymethylglutathione) (K_i 20–80 μM). Nucleoside phosphates are competitive inhibitors with respect to NAD+. The K_i values at pH 6.0 are 0.80 mM for ATP, 0.15 mM for ADP, and 6.0 mM for AMP.

Others. The general properties of formaldehyde dehydrogenase from methylotrophic yeasts[2-5] are very similar to those of that from *C. boidinii* 2201.

[71] Formate Dehydrogenase from Methylotrophic Yeasts

By NOBUO KATO

$$\text{Formate} + \text{NAD}^+ \rightleftharpoons \text{CO}_2 + \text{NADH} + \text{H}^+$$

In methylotrophic yeasts, NAD-linked formate dehydrogenase (EC 1.2.1.2) is the last enzyme in methanol catabolism. This enzyme has so far been purified from several methanol-grown yeasts; *Candida boidinii* (*Kloeckera* sp.) 2201,[1] *Candida boidinii* ATCC 32195,[2] *Hansenula polymorpha* CBS 4732,[3] *Pichia pastoris* NRRL-Y-7556[4] and IFP 206,[5] *Candida methylica*,[6] and *Candida methanolica* ATCC 26175.[7]

Assay Method

Principle. Formate dehydrogenase is measured spectrophotometrically by following the rate of NADH formation at 340 nm.

[1] N. Kato, M. Kano, Y. Tani, and K. Ogata, *Agric. Biol. Chem.* **38**, 111 (1974).
[2] H. Schutte, J. Flossdorf, H. Sahm, and M. R. Kula, *Eur. J. Biochem.* **62**, 151 (1976).
[3] J. P. van Dijken, G. J. Oostra-Demkes, R. Otto, and W. Harder, *Arch. Microbiol.* **111**, 77 (1976).
[4] C. T. Hou, R. N. Patel, A. I. Laskin, and N. Barnabe, *Arch. Biochem. Biophys.* **216**, 296 (1982).
[5] J. J. Allais, A. Louktibi, and J. Baratti, *Agric. Biol. Chem.* **47**, 2547 (1983).
[6] T. V. Avilova, O. A. Egorova, L. S. Ioanesyan, and A. M. Egorov, *Eur. J. Biochem.* **152**, 657 (1985).
[7] Y. Izumi, H. Kanzaki, S. Morita, and H. Yamada, *FEMS Microbiol. Lett.* **48**, 139 (1987).

Reagents

Potassium phosphate buffer, 100 mM, pH 7.5
NAD$^+$, 60 mM, in distilled water
Sodium formate, 500 mM, in distilled water

Procedure. The enzyme activity is measured as the increase in absorbance at 340 nm with a recording spectrophotometer thermostatically maintained at 30°. The complete assay mixture consists of 1.0 ml of potassium phosphate buffer, 0.1 ml of NAD$^+$, 1.0 ml of sodium formate, and a limited amount of enzyme, in a final volume of 3.0 ml. The reaction is initiated by the addition of enzyme. A blank test, that is, a parallel assay, is performed with a reaction mixture with sodium formate omitted.

Definition of Unit and Specific Activity. One unit of the enzyme activity is defined as the amount of enzyme catalyzing the formation of 1 μmol NADH per minute at 30°. Specific activity is defined as the number of units per milligram of protein.

Growth of the Organism

Candida boidinii 2201 is grown on methanol as the sole carbon and energy source under the same conditions as given in the preceding chapter.[8]

Purification Procedure

All procedures are performed at 0 to 5°. The buffer solution comprises 1 mM dithiothreitol and 0.5 mM phenylmethylsulfonyl fluoride.

Step 1: Preparation of a Cell-Free Extract. Frozen cell paste of *C. boidinii* 2201 (340 g wet weight) is suspended in 3 liters of 10 mM potassium phosphate buffer (pH 7.5). Three hundred-milliliter portions of the cell suspension are disrupted, separately, by ultrasonication for 2 hr with an ultrasonic oscillator (19 kHz), followed by centrifugation at 14,000 g for 20 min, the resultant supernatant being used as the cell-free extract.

Step 2: Heat Treatment. The cell-free extract, dispensed in 20-ml portions into test tubes (18 × 180 mm), is heated for 10 min in a water bath at 55°, cooled in an ice bath, and then centrifuged to remove denatured protein. The clear supernatant is dialyzed against 10 mM potassium phosphate buffer (pH 7.5).

Step 3: DEAE-Sephacel Chromatography. The dialyzed solution is applied to a DEAE-Sephacel column (2.6 × 50 cm) previously equilibrated with 10 mM potassium phosphate buffer (pH 7.5). The column is washed

---

[8] N. Kato, this volume [70].

with 750 ml of the equilibration buffer, and the enzyme is eluted with 50 mM potassium phosphate buffer (pH 7.5) at a flow rate of 10 cm/hr. The peak fractions are pooled, concentrated to about 50 ml by ultrafiltration, using an Amicon (Danvers, MA) YM10 membrane, and then dialyzed against 50 mM potassium phosphate buffer (pH 7.5).

Step 4: Hydroxylapatite Chromatography. The dialyzed solution is placed on a hydroxylapatite column (2.6 × 20 cm) previously equilibrated with 50 mM potassium phosphate buffer (pH 7.5), then the column is washed with 200 ml of the buffer. The enzyme is eluted with a linear gradient between 250 ml of 10 mM and 250 ml of 200 mM potassium phosphate buffer (pH 7.5) at a flow rate of 0.8 cm/hr. The active fractions are concentrated and dialyzed against 10 mM potassium phosphate buffer (pH 7.5). The concentrated enzyme solution (~5 mg/ml) is added to an equal volume of cold glycerol and then stored at −20°. Under these conditions, the enzyme can be stored for at least 1 year without loss of activity.

A typical purification is summarized in Table I.

Comments. Yeast cells are disrupted more effectively with a Dyno-Mill Disintegrator in the case of larger scale purification. The enzyme yield is greatly improved by the use of a buffer containing dithiothreitol and phenylmethylsulfonyl fluoride, compared with the originally described method.[1]

Properties

Purity. The purified enzyme preparation gives a single band on acrylamide (7%) gel electrophoresis at pH 8.3 in the presence of sodium dodecyl sulfate (SDS).

Molecular Weight and Subunit Structure. The relative molecular weight of formate dehydrogenase from *C. boidinii* 2201 is determined to be 76,000 by high-performance liquid chromatography on a column of TSK gel G-3000SW (Tosoh, Tokyo). The molecular weights of the subunits are determined to be 37,000 by SDS gel electrophoresis. The enzyme is assumed to be a dimer.

Substrate Specificity. The enzyme shows strict substrate specificity for formate and NAD$^+$. The K_m values for formate and NAD$^+$ are 0.10 and 22 mM, respectively. No *S*-formylglutathione hydrolase activity is detected in the purified enzyme preparation.

Effects of pH and Temperature. The enzyme exhibits optimum activity at pH 7.5 to 8.0 (potassium phosphate buffer) and is stable in the pH range of 7.0 to 8.0 at 30° for 1 hr. The maximum activity is observed at 50°, most of the activity remains after 10 min at 50°, and 60% of the activity is lost at 60° for 10 min.

TABLE I

PURIFICATION OF FORMATE DEHYDROGENASE FROM *Candida boidinii* 2201

Procedure	Total protein (mg)	Total activity (units)	Specific activity (units/mg protein)	Purification (-fold)	Yield (%)
Cell-free extract	1280	212	0.166	1	100
Heat treatment	876	242	0.276	1.7	114
DEAE-Sephacel	45.4	143	3.15	19	67.4
Hydroxylapatite	26.1	90.0	3.45	21	42.5

Inhibition. The enzyme is completely inhibited by KCN (1 mM), *p*-chloromercuribenzoate (1 mM), and sodium azide (5 μM), and by several cations (1 mM each), Cu^{2+} (80% inhibition), Hg^{2+} (100%), and Pb^{2+} (45%). EDTA, *o*-phenanthroline, and α,α'-dipyridyl at a concentration of 1 mM have no effect on the activity.

Kinetics. The kinetic data are consistent with an ordered Bi-Bi mechanism for a reaction in which NAD^+ is bound first to the enzyme and NADH released last.[9] NADH is a competitive inhibitor with respect to NAD^+ (K_i 0.30 mM). Nucleoside phosphates inhibit the enzyme at acidic pH. The K_i values are 70 μM for ATP, 0.15 mM for ADP, and 0.4 mM for AMP.

Others. The specific activities of the purified enzymes from other methylotrophic yeasts vary, as follows: the enzyme from *C. boidinii* ATCC 32195,[2] 2.4; *H. polymorpha* CBS 4732,[3] 2.83; *P. pastoris* NRRL-Y-7556,[4] 8.2; *P. pastoris* IFP 206,[5] 2.80; *C. methylica,*[6] 16.0 (at 37°); and *C. methanolica* ATCC 26175,[7] 7.52 units/mg, under the respective assay conditions. The general properties of formate dehydrogenase from methylotrophic yeasts are similar to those of that from *C. boidinii* 2201. However, there are some differences between the enzymes from *C. boidinii* 2201 and *P. pastoris* NRRL-Y-7556 with regard to molecular weight (that of the latter is 94,000) and optimum temperature (25°).

[9] N. Kato, H. Sahm, and F. Wagner, *Biochim. Biophys. Acta* **566,** 12 (1979).

[72] Catalase from *Candida boidinii* 2201

By MITSUYOSHI UEDA, SABIHA MOZAFFAR, and ATSUO TANAKA

$$2 H_2O_2 \rightarrow O_2 + 2 H_2O$$
$$(\text{Donor} + H_2O_2 \rightarrow \text{oxidized donor} + 2 H_2O)$$

Catalase or H_2O_2:H_2O_2 oxidoreductase (EC 1.11.1.6) catalyzes the decomposition of hydrogen peroxide. The enzyme is a typical marker enzyme of peroxisomes, subcellular organelles which proliferate, for example, in the liver of hypolipidemic drug-treated rats, germinating seeds, and *n*-alkane- or methanol-assimilating yeasts. The enzyme from *n*-alkane-grown *Candida tropicalis* has already been purified,[1] although the enzyme was partially purified from methanol-grown cells of *Candida* N-16[2] and *Hansenula polymorpha*.[3] The enzyme of *Candida boidinii* (*Kloeckera* sp.) 2201 was the first to be purified and crystallized.[4] Catalase exhibits a peroxidatic activity as well as a catalatic activity, via coupling with alcohol oxidase, so that these enzymes can catalyze the formation of 2 mol of formaldehyde from 2 mol of methanol with the consumption of 1 mol of molecular oxygen.[5] Furthermore, this enzyme forms a crystalloid structure with alcohol oxidase in peroxisomes.[6-8]

Assay Methods

Principle. The catalatic activity of catalase is most conveniently measured by following the decomposition of hydrogen peroxide spectrophotometrically at 240 nm.[1] The peroxidatic activity of catalase is assayed by the method of Srivastava and Ansari[9] with slight modifications[1] or by the method of van Dijken *et al.*[3]

[1] T. Yamada, A. Tanaka, and S. Fukui, *Eur. J. Biochem.* **125**, 517 (1982).
[2] T. Fujii and K. Tonomura, *Agric. Biol. Chem.* **39**, 1891 (1975).
[3] J. P. van Dijken, R. Otto, and W. Harder, *Arch. Microbiol.* **106**, 221 (1975).
[4] S. Mozaffar, M. Ueda, K. Kitatsuji, S. Shimizu, M. Osumi, and A. Tanaka, *Eur. J. Biochem.* **155**, 527 (1986).
[5] A. Tanaka, S. Yasuhara, M. Osumi, and S. Fukui, *Eur. J. Biochem.* **80**, 193 (1977).
[6] S. Fukui, A. Tanaka, S. Kawamoto, S. Yasuhara, Y. Teranishi, and M. Osumi, *J. Bacteriol.* **123**, 317 (1975).
[7] J. P. van Dijken, M. Veenhuis, N. J. W. Kreger-van Rij, and W. Harder, *Arch. Microbiol.* **102**, 41 (1975).
[8] H. Sahm, R. Roggenkamp, F. Wagner, and W. Hinkelmann, *J. Gen. Microbiol.* **88**, 218 (1975).
[9] S. K. Srivastava and N. H. Ansari, *Biochem. Biophys. Acta* **633**, 317 (1980).

Reagents

For measurement of catalatic activity
   Potassium phosphate buffer, 50 mM, pH 7.2
   Hydrogen peroxide, 40 mM, dissolved in deionized water
For measurement of peroxidatic activity
1. Potassium phosphate buffer, 100 mM, pH 7.2
   Hydrogen peroxide, 20 mM, dissolved in deionized water
   β-(3,4-Dihydroxyphenyl)-L-alanine, 20 mM, dissolved in deionized water
2. Potassium phosphate buffer, 100 mM, pH 7.2
   Hydrogen peroxide, 1 μl of 31% (w/w) hydrogen peroxide solution/ml, in deionized water
   o-Methoxyphenol (guaiacol), 2 μl/ml, in deionized water
3. Potassium phosphate buffer, 66.7 mM, pH 6.0
   Glucose, 33 mM, dissolved in deionized water
   Methanol, 100 mM, dissolved in deionized water
   Glucose oxidase (31.1 U/mg protein, from *Aspergillus niger*), 30 μg/ml, in 66.7 mM potassium phosphate buffer (pH 6.0)
   HCl, 10 M
   Tryptophan, 0.1% (w/w), in 50% (v/v) ethanol
   Sulfuric acid, 90% (v/v)
   FeCl$_3$, 1% (w/w), in deionized water

Procedure for Assay of Catalatic Activity. In a quartz cuvette (10-mm light path) are added 1.7 ml of potassium phosphate buffer and 1.2 ml of hydrogen peroxide, and the mixture is incubated for 2.5 min at 30°. After incubation, the reaction is initiated by the addition of 0.1 ml of appropriately diluted enzyme solution. The decrease in absorbance at 240 nm is followed spectrophotometrically.

Procedure for Assay of Peroxidatic Activity. Method 1. In a glass cuvette are added 1.7 ml of potassium phosphate buffer (pH 7.2), 0.1 ml of hydrogen peroxide, and 0.1 ml of β-(3,4-dihydroxyphenyl)-L-alanine, and the solution is incubated for 10 min at 30°. After incubation, the reaction is initiated by the addition of 0.1 ml of appropriately diluted enzyme solution. The increase in absorbance at 470 nm is followed spectrophotometrically.

Method 2. In a glass cuvette are added 3 ml of potassium phosphate buffer (pH 7.2), 50 μl of o-methoxyphenol 40 μl of hydrogen peroxide, and 20 μl of enzyme. After incubation for 10 min at 30°, the change in absorbance at 436 nm is measured.

Method 3. The reaction mixture contains 1.3 ml of potassium phosphate buffer (pH 6.0), 0.1 ml of glucose, and 0.1 ml of methanol. After the purified catalase is added to the mixture, followed by incubation at 30° for

10 min, 50 μl of glucose oxidase is added, and the mixture is incubated at 30° for 30 min. The reaction is stopped by adding 10 μl of HCl at 30°. After centrifugation at 1000 g for 10 min, the supernatant is used to measure the amount of formaldehyde formed. One milliliter of the supernatant is mixed with 1 ml of tryptophan. One milliliter of sulfuric acid is added to the reaction mixture and mixed thoroughly. Then, 0.2 ml of FeCl$_3$ is added and mixed thoroughly, and the mixture is incubated at 70° for 90 min. After cooling at 30° for 15 min, the development of the violet color is monitored at 575 nm.[10]

Growth of Microorganism

Candida boidinii 2201 is cultivated aerobically at 30° in a medium containing 4 g/liter NH$_4$Cl, 1 g/liter KH$_2$PO$_4$, 1 g/liter K$_2$HPO$_4$, 0.5 g/liter MgSO$_4 \cdot$ 7H$_2$O, 1 ml/liter corn steep liquor, and a carbon source (10 g/liter glucose, 10 ml/liter methanol, or 10 ml/liter ethanol). The initial pH of the medium is adjusted to pH 6.0 with 1 M HCl before sterilization.[5]

Purification Procedure

All operations are carried out at 0–4°.

Step 1: Preparation of Cell-Free Extracts. Candida boidinii cells harvested at the midexponential growth phase by centrifugation at 1000 g for 5 min are washed twice with deionized water and suspended in 50 mM potassium phosphate buffer (pH 7.2) containing 0.2 mM phenylmethylsulfonyl fluoride to give a concentration of 0.5 g wet cells/ml. An aliquot of the cell suspension (20 ml) is disrupted with a Braun cell homogenizer for 2.5 min. The cell homogenate is centrifuged at 20,000 g for 30 min. The supernatant obtained is used as the cell-free extract.

Step 2: Ammonium Sulfate Precipitation. The cell-free extract (191 ml) prepared from methanol-grown cells (90 g wet cells) is fractionated with ammonium sulfate between 40 and 70% of saturation. The precipitate obtained is dissolved in a minimal volume of 50 mM potassium phosphate buffer (pH 7.2) containing 0.2 mM phenylmethylsulfonyl fluoride, followed by dialysis against the same buffer for 16 hr at 4°.

Step 3: Column Chromatographies. The dialyzed solution is applied to a hydroxyapatite column (4.0 × 14.5 cm) equilibrated with 50 mM potassium phosphate buffer (pH 7.2) containing 0.2 mM phenylmethylsulfonyl fluoride, and the enzyme is eluted with a linear concentration gradient of

---

[10] J. Chrastil and J. T. Wilson, *Anal. Biochem.* **63,** 202 (1975).

potassium phosphate buffer prepared from 20 and 500 mM potassium phosphate buffers (pH 7.2). The enzyme solution, concentrated with a Diaflo membrane PM10 (Amicon Far East, Tokyo), is applied on a Sephacryl S-300 column (2.2 × 85 cm), and the enzyme is eluted with 50 mM potassium phosphate buffer (pH 7.2). After concentration, the enzyme solution is applied to a DEAE-Sepharose column (2.2 × 22.5 cm), and elution is carried out with a linear concentration gradient of KCl prepared from 50 mM potassium phosphate buffer (pH 7.2) and the same buffer containing 0.4 M KCl. The fractions exhibiting catalase activity are combined, concentrated, and used as the purified enzyme.

A summary of the purification procedure is given in Table I.

Properties

Induction. The activity of catalase in methanol-grown cells of *C. boidinii* is more than 10 times higher than those in glucose- and ethanol-grown cells.

Molecular Masses of Native and Subunit Forms. The molecular mass of the native enzyme from *C. boidinii* is determined to be 240 kDa by ultracentrifugation analysis and gel filtration. Sodium dodecyl sulfate–polyacrylamide slab-gel electrophoresis shows that the subunit of the enzyme has a molecular mass of 62 kDa, indicating that the enzyme consists of four identical subunits.

Optimal Reaction pH. The optimal pH is observed at 7.2, although the enzyme shows catalatic activity from pH 5 to 9 at 30°.

Absorption Spectrum of Catalase. The absorption spectrum of the purified enzyme is characteristic of a heme protein having two major peaks at 280 and 405 nm (Soret band) and three minor peaks at 500, 530, and 630 nm. The A_{405}/A_{280} value is calculated to be 1.06, indicating that this enzyme is a T-type catalase.

Crystallization and Electron Microscopical Observation. The enzyme crystallized by the addition of ammonium sulfate is brown-colored and needlelike. The minimum size of the enzyme is about 6 × 10 nm when the enzyme is stained negatively with 4% uranyl acetate and observed electron microscopically with a Hitachi H-800 electron microscope at 200 kV.

Catalatic and Peroxidatic Activity. The enzyme exhibits the highest catalatic activity at 50 mM hydrogen peroxide, with activity decreasing significantly above this substrate concentration. The K_m value of the enzyme for hydrogen peroxide is 25 mM when calculated from the data obtained at low concentrations of the substrate. The V_{max} value is 286 mmol/min/mg protein. When hydrogen peroxide is directly added to the reaction mixture at different concentrations, the enzyme shows a low

TABLE I
PURIFICATION OF CATALASE FROM METHANOL-GROWN *Candida boidinii* 2201[a]

Fraction	Volume (ml)	Total protein (mg)	Total activity (mmol/min)	Yield (%)	Specific activity (mmol/min/mg)	Purification (-fold)
Cell-free extract	191	2290	10,500	100	4.6	1
Ammonium sulfate (40–70%) saturation	44	972	6660	63.4	6.9	1.5
Hydroxyapatite	8.2	70.6	3950	37.6	55.9	12.2
Sephacryl S-300	4.5	19.3	2480	23.6	128	27.8
DEAE-Sepharose	4.3	6.3	1775	16.9	282	61.3

[a] Catalatic activities of catalase are shown.

peroxidatic activity on β-(3,4-dihydroxyphenyl)-L-alanine, but not on o-methoxyphenol or methanol. However, the peroxidatic activity of *C. boidinii* catalase is observed on methanol when hydrogen peroxide is supplied continuously with a combination of glucose and glucose oxidase. Under these conditions, the K_m value of the enzyme for methanol is 83 mM and the V_{max} value is 1.79 μmol/min/mg protein.

Immunochemical Difference. Immunochemical titration with antiserum against *C. boidinii* catalase shows a weak cross-reaction to *n*-alkane-grown yeast *(C. tropicalis)* catalase but no cross-reaction to bovine liver catalase.

Subcellular Localization. In *C. boidinii,* catalase activity is detected in the cytosolic fraction as well as in peroxisomes. The results of purification of peroxisomal and cytosolic catalases from this yeast reveal that both enzyme molecules are indistinguishable in terms of molecular masses of native molecules and subunits, peptide maps, kinetic constants (K_m for H_2O_2: 25 mM), and the amino-terminal residue (alanine).[11]

[11] M. Ueda, S. Mozaffar, and A. Tanaka, *Mem. Fac. Eng., Kyoto Univ.* **50,** 154 (1988).

Author Index

Numbers in parentheses are footnote reference numbers and indicate that an author's work is referred to although the name is not cited in the text.

A

Abbott, B. J., 3, 4
Abeles, R. H., 425, 426
Adachi, O., 271
Adkins, H., 266
Ahmed, A. E., 355
Aitken, A., 60
Aldrich, T. L., 127, 130(3)
Alefounder, P. R., 243
Allais, J. J., 455, 459, 462(5)
Amano, Y., 286
Ambler, R. P., 286, 297
Ameyama, M., 271
Aminova, L. R., 445
Anders, M. W., 355
Anderson, R. F., 146
Ando, M., 398, 456
Anilionis, A., 169
Ansari, N. H., 463
Anthony, C., 14, 191, 202, 210, 212, 213, 214, 215, 216, 218, 219(2), 220(1, 2), 221(2, 4), 222, 227, 241, 247, 250(1), 284, 285(8, 9), 286, 287(9), 288(8), 289, 298, 299, 301, 302, 303, 304, 305(1), 307(6), 309, 331, 335, 336(2), 346, 365, 373, 386
Aperghis, P. N. G., 346, 397
Arciero, D. M., 82, 89, 90(4), 93(4), 94, 95, 105, 107, 109(1), 110(1)
Arfman, N., 223, 226, 391, 393
Arima, K., 144
Armstrong, J. M., 253
Armstrong, J. McD., 204, 324
Armstrong, J., 247
Artymiuk, P. J., 137
Asada, Y., 169
Asther, M., 166
Atheron, J. P., 421

Attwood, M. M., 204, 223, 226, 314, 323(3), 325, 363, 367, 391
Atwood, M. M., 250
Aubert, L., 171, 176
Aurich, H., 14, 18
Auton, K. A., 284, 285(8, 9), 287(9), 288(8)
Avilova, T. V., 459, 462(6)
Axcell, B. C., 52, 53, 56, 57(1, 5), 59(5), 134
Azoulay, E., 171, 172, 176

B

Babel, W., 346, 401, 405, 447, 449(11), 450(11), 451, 453, 455
Baeyer, A., 70, 71(1)
Ballou, D. P., 61, 62, 63(4), 64(1, 4), 67(1, 13), 68(6, 7, 8, 10, 11, 12), 69(1, 9), 70, 71, 75(11), 76(11), 139, 143, 144, 145
Balny, C., 207, 208(13), 209(13), 304
Bamforth, C. W., 435
Baratti, J., 420, 421(3), 422(3), 425(3), 455, 459, 462(5)
Barber, M. J., 260
Barbour, M. G., 102
Barist, I., 21, 22(1), 28
Barnabe, N., 21, 22(1), 28, 459, 462(4)
Barnsley, E. A., 46, 49(6), 115
Barry, S., 24
Bartsch, R. G., 297
Basu, D., 10
Batie, C. J., 61, 62, 64(1), 67(1, 8, 13), 68(7, 10, 11, 12), 69(1, 9), 70
Bayer, E., 194
Bayly, R. C., 102
Beardmore-Gray, M., 14, 212, 213(4), 215(4), 298, 299(3), 302, 303, 307(6)
Beardsmore, A. J., 346, 392, 397, 445, 451

Beer, R., 189
Beg, F., 60
Beinert, H., 251, 258, 260
Beintema, J. J., 143, 146
Belew, M. A., 169
Bellamy, H. D., 259
Belyaeva, P. Kh., 116
Ben-Bassat, A., 346
Bendayan, M., 412, 420(6)
Benson, A., 3
Bentsen, J. G., 189
Beppu, T., 144
Bergmeyer, H. U., 272
Bernhardt, F.-H., 54, 60
Bes, B. S. M., 423
Best, D. J., 191
Bethge, P. H., 259
Betts, R. P., 318
Bezemer, R. P., 266
Bill, E., 54
Binnema, D. J., 227
Black, S., 330
Blackkolb, F., 350
Blackmore, M. A., 362
Blake, C. C. F., 137
Blakley, E. R., 111
Böhme, H. J., 50, 352, 453
Bolbot, J. A., 216
Bonitz, S., 123
Bonting, K., 436
Borah, K. R., 4
Borriss, R., 350
Bos, J. W., 424
Bosma, G., 299
Botkin, J. H., 262
Boulikas, T., 411
Bradford, M. M., 224, 228, 242, 328, 393, 429, 452
Bradford, M., 85, 96, 103, 332
Brand, K., 406
Braster, M., 299
Brazil, H., 350
Briggs, R. T., 415
Bringer, S., 420
Britton, L. N., 70, 77
Bromley, J. W., 82, 95
Brooke, A. G., 223, 391
Bruice, T. C., 146
Buckel, W., 384

Buechi, G., 262
Bull, C., 61, 63(4), 64(4), 82, 161, 162(21), 170(21)
Bundock, P., 146
Burton, J. M., 397
Buswell, J. A., 159, 160(2), 169(2)
Byron, C. M., 309, 313(4)
Bystrykh, L. V., 342, 413, 425, 443, 445, 446, 447, 450(7), 451

C

Cammarck, R., 53, 54(6), 60
Cantoni, G. J., 365
Capdevila, C., 166
Carlemalm, E., 412
Cecchini, G., 208
Cerniglia, C. E., 148, 149(2, 3), 150(3), 151, 153(1, 6)
Chakrabarty, A. M., 122, 127, 130(3, 4)
Champion, P. M., 94
Chandrasekar, R., 284
Chang, J., 3
Chapman, P. J., 94, 102, 121, 143
Chapman, P., 82
Chari, R. V. J., 130, 133
Chatterjee, D. K., 122
Chen, Y. P., 122
Chen, Y.-C., 71
Chibisova, E. S., 336
Chistoserdov, A. Y., 335, 336(1), 346
Chiu, A. A., 160, 166(12)
Chrastil, J., 465
Claiborne, A., 145
Claus, D., 73, 79
Clement, W., 223, 226, 391
Cleton-Jansen, A.-M., 274
Cline, J. F., 62, 68(6)
Cohen-Bazire, G., 103
Colbert, J. E., 9
Colby, J., 181, 182(2), 183, 184, 188, 191, 250, 251, 252, 253(7), 254(1), 258(7), 310
Coleman, J. P., 21, 22(2), 24(2), 29
Collins, C., 169
Cook, A. M., 355
Coon, J. J., 3, 5(2)

Coon, M. J., 10, 41, 48, 171
Cooney, C. L., 441, 447
Cooper, R. M., 30
Corey, E. J., 262, 264(6), 265(6), 266(6)
Cornforth, J. W., 384
Correll, C. C., 62, 67(13)
Corrieu, G., 166
Coté, C. E., 234
Couderc, R., 420, 421(3), 422(3), 425(3)
Covert, S., 169
Cox, R. B., 364
Crawford, R. C., 102
Crawford, R. L., 82, 95, 160, 164(13), 167(13)
Cripps, R. E., 31
Croan, S., 160, 164(15), 165(15), 167(15), 169(15)
Cromartie, Th. H., 426
Cross, A. R., 218, 299, 309
Crutcher, S. E., 52, 53(2), 56, 57(2), 59(2), 60
Cullen, D., 169

D

Dagley, S., 82, 89, 94, 102
Dalton, H., 22, 40, 181, 182(2), 183, 184, 186, 187, 188, 189, 190, 191, 202(4), 289, 297, 331
Davidson, V. L., 208, 241, 242(1), 244, 245, 246, 247, 250(2), 251, 286, 299, 309, 311(1), 312(1), 313(4), 314(1)
Davies, G. E., 361
Davis, B. J., 336
Davis, R. E., 39
Day, D., 214, 215, 222, 289, 301
Day, E. P., 67, 68(14)
de Beer, R., 207, 209(14)
de Jong, R. J., 143
de Koning, W., 412, 414(1), 420(1), 436, 445
de Moor, P., 344, 442
de Vries, G. E., 223, 226, 391
de Vries, J. G., 262
deBoer, H. A., 169
deBont, J. A. M., 29, 30
Dege, J. E., 200
Derelanko, P., 420, 455, 459(4)
Dickson, D. P. E., 53, 54(4), 58

Dijkhuizen, L., 223, 226, 227, 314, 318, 391, 393, 405, 409, 410(2, 4), 436, 445
Dijkstra, M., 202, 207, 208(13), 209(13), 212, 215, 269, 299, 303, 304, 305(3), 307(1)
Dikinson, F. M., 171
Dilworth, M. J., 122
DiSpirito, A., 292
Dixon, G. H., 365
Dokter, P., 271
Donahue, P. W., 421
Donoghue, N. A., 71, 75(10), 76(10), 77, 81(2)
Dooley, D. M., 234, 235
Dorn, E., 122
Doten, R. C., 131
Doudoroff, M., 37, 42, 103
Douma, A. C., 412, 414(1), 420, 426
Draht, B. D., 415
Drenth, J., 142, 143, 144, 241, 247, 250(5), 283
Duich, L., 277
Duine, J. A., 33, 35, 37(5), 38, 202, 203(3), 204(3), 205(3), 206, 207, 208(9, 10, 13), 209(13, 14), 212, 215, 227, 229(2), 230(2), 232(2), 233(2), 234, 235(2), 241, 247, 250(5), 260, 261, 262, 266, 268, 269, 270, 271, 273, 277, 279, 280, 281, 282, 283, 289, 299, 303, 304, 307(1, 3), 327, 329
Dunford, H. B., 162, 169(23), 170(23)
Dunham, W. R., 62, 67, 68(11, 14), 251
Duppel, W., 171
Dutton, P. L., 154
Duvnjak, Z., 172, 176

E

Eady, R. R., 242, 247
Eaton, R. W., 61
Ebersold, H.-R., 355, 357(1)
Eberspächer, J., 134
Ebina, Y., 122
Eckart, M., 424
Edens, L., 424
Eggeling, L., 24, 321, 323(11), 330

Eggerer, H., 379, 380, 381, 384, 387, 389, 390(7), 391
Egli, C., 355
Egli, Th., 441, 445, 448
Egorov, A. M., 459, 462(6)
Egorova, O. A., 459, 462(6)
Eitner, G., 18
Elbert, E., 400
Elinarsdottir, G. H., 138
Elliott, E. J., 215, 299
Elsden, S. R., 72, 73(13), 79
Ensley, B. D., 46
Entsch, B., 139, 142, 143, 144, 145, 146, 147(10)
Ericson, A., 189
Eriksson, K.-E., 165
Evans, W. C., 154
Evers, M. E., 412, 414(1), 420(1)

F

Farr, A. L., 12, 15, 19, 34, 37(7), 90, 252, 290, 383, 407
Farrell, R. L., 159, 160, 161(14), 164(15), 165(15), 167(14, 15), 169
Favre, K., 224, 411
Favre, M., 310
Fee, J. A., 62, 67, 68(6, 7, 10, 14), 161, 162(21), 170(21)
Feiser, L. F., 389
Feiser, M., 389
Felton, D. G., 253
Ferenci, T., 401
Ferguson, S. J., 243
Fiechter, A., 160, 164, 168, 169, 441, 448
Fieser, L. F., 267, 278
Fieser, M., 267, 278
Finazzi-Argo, A., 235
Findling, K. L., 67, 68(14)
Finnerty, W. R., 12, 13, 14, 15(5), 18, 19(3, 4)
Fisher, J., 76, 145
Fixter, L. M., 14
Flechtner, V. R., 350
Fletcher, P., 131
Floris, G., 234, 235
Flossdorf, J., 455, 459, 462(2)
Fluckiger, R., 277

Folkerts, K., 266
Ford, S., 214, 216, 221(4), 222(4)
Fox, B. G., 191, 194(9), 200, 201, 202(15), 331, 332(2)
Frank Jzn, J., 279, 281(42)
Frank, J., 33, 35, 202, 203(3), 204(3), 205(3), 206, 207, 208(9, 10, 13), 209(13, 14), 212, 215, 234, 241, 260, 261, 269, 277, 283, 289, 299, 303, 304, 305(1, 3), 329
Frank, M. D. J., 247, 250(5)
Franklin, W., 148, 149(2)
Frantz, B., 122, 127, 130(3, 4)
Freeman, J. P., 151, 153(6)
Fritsche, B., 169
Froland, W. A., 200, 331, 332(2)
Froud, S. J., 299, 304
Fuhrhop, J. H., 304
Fuhrop, H.-H., 290
Fujii, T., 139, 422, 463
Fujisawa, H., 82, 83(10), 88(9), 111, 122
Fukui, S., 171, 173, 175, 176, 177, 463, 465(5)
Fukumori, Y., 284

G

Gainor, J. A., 262
Galli, R., 355, 357(1)
Gallo, M., 171, 176
Gallop, P. M., 277
Gallup, M., 251, 255(4), 256(4), 257(4), 309, 311(6), 312(6), 313(6)
Garavito, R. M., 412
Garland, P. B., 380
Garza, A., 39
Gaskell, J., 169
Geary, P. J., 52, 53, 54(4, 6), 56, 57(1, 2, 5), 58, 59(2, 5), 60, 134, 137
Geiger, O., 271
Geissler, J. F., 154
Geissler, J., 426
Gelep, P., 169
Geller, A. M., 366
Gerhardt, P., 103
Gersonde, K., 60
Geuze, H. J., 420
Ghisla, S., 145
Ghosal, D., 122

Giatosio, A., 234
Gibson, D. T., 39, 40, 42, 43(4), 44(2, 4), 45(2, 3, 4, 10), 46, 82, 95, 134
Gibson, J., 154, 156, 158(4)
Gill, B., 127, 130(3)
Gishla, S., 426
Giuseppin, M. L. F., 423
Gleeson, M. A. G., 445
Glenn, A. R., 122
Glenn, J. K., 160, 166(12), 167
Gold, M. H., 160, 162, 166(12), 167, 169, 170(23)
Goldberg, I., 346
Goldman, A., 126, 129(2)
Golovleva, L. A., 116
Gomez, L. E., 169
Gonen, L., 350
Goodman, J. M., 421
Goodwin, P. M., 361, 364(3)
Goovsen, N., 272, 274
Gordon, S. L., 4, 5(18)
Gorelick, R. J., 312
Görisch, H., 33, 35(2), 271
Gorlatova, N. V., 116
Gosh, R., 203
Govorukhina, N. I., 215, 248, 336
Granick, S., 33, 203, 204, 273
Gray, K. A., 245, 299
Green, J., 187, 188, 189, 190, 191, 427, 429, 431(2), 433(2), 434(2), 435(2)
Griffin, M., 77, 81(1)
Griffith, G. R., 3
Groen, B. W., 33, 35(1), 37(5), 38, 273, 277, 280
Groendijk, H., 142, 143(9), 241, 247, 250(5), 283
Groeneveld, A., 202
Grunert, R. R., 32
Grünewälder, C., 379, 381, 387, 390(7), 391
Grunow, M., 14
Gurbiel, R. J., 62, 68(7)
Gutschow, Ch., 384

H

Hacking, A. J., 364, 379, 387, 391(6)
Haemmerli, S. D., 160, 162, 169
Haigler, B. E., 46

Halvorson, H., 111
Hamlin, R., 259
Hammel, K. E., 159, 160, 161, 163, 164(4), 169(4, 20)
Hammond, R. C., 191
Hamzah, R. Y., 111, 138
Hancock, R., 411
Hansen, R. E., 60, 61
Hanson, R. S., 350, 366, 370(3), 371(3)
Harada, Y., 284, 288(1)
Harder, W., 204, 223, 224, 226, 227, 228, 229(1, 8), 230(8), 232(8), 233(8), 250, 251, 314, 319, 325, 363, 391, 392, 405, 410(2), 412, 413, 414(1, 8, 9), 415(2, 3, 8, 9), 417(9), 420, 421, 422(6), 423(6), 425(6), 426, 436, 445, 451, 455, 459, 462(3), 463
Harnisch, H., 14
Harpel, M. R., 102, 109
Harris, A., 60
Harris, P. H., 31
Hartmann, C., 235
Hartmans, S., 30, 355
Hartnett, C., 123
Hartree, E. I., 380
Harwood, C. S., 154, 156, 158(4)
Hauser, T. R., 400
Hayaishi, O., 82, 88(9), 89, 116, 118, 122
Haywood, G. W., 241, 427, 429, 431(2), 433(1, 2), 434(1, 2), 435(1, 2, 4)
Hearshen, D. O., 67, 68(14)
Hedman, B., 189
Heinz, F., 451
Heitkamp, M. A., 148, 149(2, 3), 150(3), 151, 153(1, 6)
Hemmerich, P., 145, 234, 426
Hendricksen, J. B., 262
Heptinstall, J., 363, 366, 374, 377(4), 381
Hersh, L. B., 387, 390(4)
Heyn, G., 172
Heyns, W., 344, 442
Heywood, G. W., 247, 250(4)
Higginbotham, D. H., 279
Higgins, I. J., 191
Higuchi, C., 443, 444(10)
Higuchi, T., 159, 169(5)
Hill, C. L., 259
Hille, R., 67, 68(14)
Hinkelmann, W., 463
Hodgson, K. O., 189

Hoffman, B. M., 62, 68(6, 7)
Hoffman, K. H., 447
Hofmann, E., 50, 352, 453
Hofmann, K. H., 451, 453, 455
Hofsteenge, J., 143
Hol, W. G. J., 142, 143, 241, 247, 250(5), 283
Hollenberg, C. P., 421, 424
Holloway, B. W., 147
Holm, R. H., 259
Hopkins, T. R., 420, 422(2, 4)
Hopper, D. J., 102
Horecker, B. L., 405, 406, 410(3)
Horiike, K., 122
Hou, C. J., 28
Hou, C. T., 21, 22(1), 420, 455, 459, 462(4)
Houghton, J., 131
Hughes, J., 131
Huitema, F., 283
Hultquist, D. E., 195
Husain, M., 111, 138, 143, 144, 145, 241, 242(1), 244(1), 245, 247, 250(2), 251, 286, 299, 309, 311(1), 312, 313(4), 314(1)
Hutber, G. N., 154
Hutton, S. W., 102
Huynh, B. H., 89, 90(4), 93(4), 94(4), 105
Huynh, V.-B., 160, 164(13), 167(13)
Hyde, D., 277

I

Iateava, I. A., 116
Ilchenko, V. Ya., 116
Ingraham, J. L., 82
Inouye, S., 122
Ioanesyan, L. S., 459, 462(6)
Itada, N., 116
Itagaki, E., 71, 77
Iwaki, M., 122
Iyengar, R., 446
Izumi, Y., 366, 367(5), 372, 459, 462(7)

J

Jacobs, T. L., 108
Jakoby, W. B., 373, 376

Janowicz, Z., 421, 424
Janschke, N. S., 241, 247, 250(4)
Jansen, J. C., 277
Jarowek, D., 404
Javahenian, K., 169
Jayasuria, G. C. N., 381
Jekel, P. A., 146
Jenkins, O., 250, 253(2), 254(2)
Jenkins, R. O., 40
Johansson, T., 160
Johnson, P. A., 331
Johnsrud, S. C., 165
Jollie, D. R., 198, 331, 332(2, 5)
Jones, C. W., 397
Jones, D., 250, 253(2), 254(2), 350
Jones-Mortimer, M. C., 331
Jongejan, J. A., 207, 234, 241, 260, 262, 266, 268, 269, 270, 277, 279, 280, 281, 282, 283, 234
Jönsson, L., 160
Joris, L., 406
Joyce, M., 46

K

Kagamiyama, H., 116, 122
Kagan, H. M., 277, 433
Kalin, M., 169
Kalk, K. H., 143, 283
Kallio, R. E., 42
Kalyanaraman, B., 159, 160(4), 161, 162(19), 163, 164(4), 169(4, 20, 25)
Kamen, M. D., 297
Kamin, H., 138
Kaneda, T., 139, 340
Kanetsuna, F., 122
Kano, M., 459, 461(1)
Kantelinen, A., 160
Kanzaki, H., 459, 462(7)
Karnovsky, M. J., 415
Karnovsky, M. L., 415
Kasprazak, A. A., 244, 251, 257(10), 258(10), 259(10)
Kataeva, I. A., 116
Kato, N., 392, 397, 399, 401, 420, 422(5), 425(5), 443, 444(10), 447, 448(12), 449(11), 450(11, 12), 451, 455, 458, 459, 460, 461(1), 462

Katopodis, A. G., 4, 8(17), 9
Katti, S., 130, 132(2)
Katz, B., 130, 132(2)
Kawamoto, S., 171, 176, 463
Kearney, E. B., 251
Keddie, R. M., 243, 254, 309
Keddy, R. M., 340
Keilin, D., 380
Kemp, G. D., 171
Kennelly, P. J., 286
Kenney, W. C., 241, 245(3), 258
Kenny, W. C., 247, 250(3)
Kent, T. A., 67, 68(14), 83, 88(12), 89, 90(4), 93(4), 94(4), 105
Kersten, P. J., 94, 161, 162(19), 163(19), 169(20)
Keyser, P. K., 61
Kilbane, J. F., 127, 130(3)
Killgore, J., 277
Kim, Y., 147
Kimura, Y., 169
Kiriuchin, M. Y., 335, 336(1), 346
Kirk, T. K., 159, 160, 161, 162(19, 21), 163, 164(4, 15), 165, 166(18), 167(14, 15), 169(1, 4, 11, 15, 20, 25), 170(11, 14, 21)
Kita, H., 111
Kitatsuji, K., 463
Kitayama, Y., 399
Kiyohara, H., 134, 148
Klapper, M. H., 284
Kletsova, L. V., 248, 335, 336, 346
Klinman, J. P., 235
Knackmuss, H.-J., 121, 122, 127
Knaff, D. B., 245, 299
Knowles, J. K. C., 169
Knowles, P. F., 234
Kobal, V. M., 39
Koch, J. R., 42
Kohler-Staub, D., 355
Kohn, L. D., 373, 376
Koide, K., 169
Koivusalo, M., 444
Kojima, J., 116
Kojima, Y., 122
Kolattukuda, P. E., 172
Komagata, K., 250
Kopperschlager, G., 50, 352, 453
Kornberg, H. L., 242, 365
Kotani, S., 118, 121
Kotb, M. Y., 366

Kozarich, J. W., 130, 133
Kozulic, B., 168
Kreger-van, N. J. W., 463
Krema, C., 361, 368
Krieg, N. R., 103
Kroneck, P. M. H., 426
Kuhn, E., 406
Kuila, D., 62, 68(10)
Kula, M. R., 455, 459, 462(2)
Kula, M.-R., 391, 397, 401
Kunkel, T. A., 147
Kuo, J. Y., 4, 8(16, 17), 9(15)
Kurkela, S., 46
Kurz, W., 111
Kusunose, E., 10
Kusunose, M., 10
Kuwahara, M., 160, 166(12), 167, 169

L

La Roche, S., 355
Lack, L., 102
Laemmli, U. K., 59, 224, 249, 257, 291, 294(10), 310, 360, 376, 411
LaHaie, E., 61, 62, 64(1), 67(1), 68(6), 69(1)
Lamed, R., 340
Lamprecht, W., 451
Large, P. J., 227, 241, 242, 247, 250(4), 364, 386, 427, 429, 431(2), 433(1, 2), 434(1, 2), 435
Large, P., 251
Laskin, A. I., 21, 22(1), 28, 420, 459, 462(4)
Lawton, R., 145
Lawton, S. A., 241, 284, 299
Leak, D. J., 181, 191
Lebeault, J. M., 171, 172, 176
Ledeboer, A. M., 424
Lederer, F., 259
Lee, G. C. M., 262
Lee, L. G., 4, 8(17), 9(19)
Lehvšlaiho, H., 46
Leisinger, T., 355, 357(1)
Leisola, M. S. A., 160, 162, 164, 168, 169
Lekim, D., 172
Lerch, K., 432, 434(9)
Levering, P. R., 224, 227, 229(1), 318, 319, 405, 409, 410(2, 4)
Levine, D. W., 441, 447

Lidstrom, M. E., 222, 289, 361, 368
Lidstrom, M., 292
Light, D. R., 76
Lim, L. W., 259, 309
Lin, E. C. C., 451
Lindholm, N., 164
Lingens, F., 134
Linko, P., 164
Linko, Y.-Y., 164
Lippard, S. J., 189
Lipscomb, J. D., 82, 83, 88, 89, 90, 93(4), 94, 95, 96, 101, 102, 105, 107, 109, 111, 191, 193, 194(9), 198, 200, 201, 202(15), 331, 332(2, 5)
Liu, T.-N., 39, 43(4), 44(2, 4), 45(2, 3, 4)
Lobenstein-Verbeek, C. L., 234, 279, 281(42)
Lode, E. T., 3, 5(2), 10, 41, 48
Loffhagen, N., 451
Loginova, N. V., 339
Long, A. R., 213, 215(6), 286
Louktibi, A., 455, 459, 462(5)
Lowry, O. H., 12, 15, 19, 34, 37(7), 90, 252, 290, 383, 407
Lück, H., 228
Ludwig, M. L., 62, 67(13), 68(12)
Lui, T.-N., 40
Lund, J., 186
Lynen, F., 389

M

Maat, J., 424
MacKenzie, A. R., 262
Maione, T. E., 169
Makula, R. A., 12
Mallinson, J., 251, 309
Marison, I. W., 314, 323(3)
Markovetz, A. J., 70, 77
Marquez, L., 162, 169(23), 170(23)
Martinkus, K., 286
Maruyama, K., 262
Mason, J. R., 52, 55(3), 56, 57(3), 59(3)
Massey, V., 111, 138, 139, 143, 144, 145, 234
Mateles, R. I., 346
Mathews, F. S., 259, 309

Matsueda, G., 3
Matsumoto, T., 241
Matsushita, K., 271
Maurer, H. R., 403
May, S. W., 3, 4, 5(18), 8(16, 17), 9
Mayer, F., 421
Mazumdar, S., 122, 124(6), 126(6)
McCordle, G. M., 128, 131
McCoy, C. J., 3
McCracken, J., 235
McDonnel, A., 290
McGuirl, M. A., 234, 235
McIntire, W. S., 234, 235(12), 253, 254, 255(24)
McIntire, W., 241, 245(3), 247, 250(3), 258, 260
McKenna, E. J., 3, 10
McNerney, T., 361, 366, 372(4)
Meagher, R. B., 127, 128, 130, 131, 132
Mehta, R. J., 321, 323(10)
Meiberg, J. B. M., 251
Melko, M., 277
Meussdoerffer, F., 168
Meyer, T. B., 297
Michaelis, L., 33, 203, 204, 273
Michal, G., 422
Miethe, D., 346
Mihalover, J. L., 111
Mildvan, A. S., 426
Miller, D. W., 151, 153(6)
Mincey, T., 426
Mitchell, D. J., 131
Miya, T., 420, 421(9), 422(9)
Miyamoto, N., 397, 401(6)
Miyazaki, S. S., 366, 367(5), 372
Mizzer, J. P., 312
Möllering, H., 272, 422
Moody, C. J., 262
Moog, R. S., 234
Morgan, A. F., 147
Morgan, M. A., 167
Morikawa, T., 173
Morita, S., 459, 462(7)
Morrice, N., 60
Morris, J. G., 242
Mozaffar, S., 463, 467
Mozuch, M. D., 160, 161(14), 163, 167(14), 169(25), 170(14)
Mudd, S. H., 365
Mulder, A. C., 271, 281, 282(48), 283(48)

Müller, F., 139, 142, 143(8), 145, 146, 420, 422(2)
Muller, R. H., 401, 405
Münck, E., 67, 68(14), 83, 88, 89, 90(4), 93(4), 94, 105, 201, 202(15)
Murrell, J. C., 22, 26
Murtagh, K. E., 160, 161(14), 164(15), 165(15), 167(14, 15), 169(15), 170(14)

N

Nagao, K., 134, 148
Nagi, M. N., 14
Nagy, J., 258
Nakai, C., 122
Nakazawa, A., 122
Nakazawa, T., 118, 122
Narro, M., 39, 44(2), 45(2)
Naruta, Y., 262
Nash, T., 315, 356, 392, 398, 402, 437, 438(4)
Nawa, H., 171, 176
Neher, J. W., 208, 245, 251, 309, 311(1), 312(1), 314(1)
Neidle, E. L., 123, 124(6, 11), 126(6)
Netrusov, A. I., 248, 336
Ngai, K.-L., 122, 126, 127, 128, 129(2), 130, 131, 132(5), 133
Nichols, C. S., 426
Nicolay, K., 426
Niku-Paavola, M.-L., 160
Nishikawa, H., 420, 421(9), 422(9), 448
Nishizawa, T., 443, 444(10), 458
Nonobe, M., 271
Norris, D. B., 71, 75(10), 76(10), 77, 81(2)
Nozaki, M., 89, 116, 118, 121, 122
Nunn, D. N., 215, 222, 302
Nyman, P. O., 160

O

O'Carra, P., 24
O'Connor, M. L., 361, 366, 370(3), 371(3), 372(4)
O'Keeffe, D. T., 212, 213(4), 214, 215(4), 298, 299, 302, 303, 307(1, 6)

Odier, E., 159, 160(2), 163, 169(2, 25)
Ogata, K., 392, 397, 401, 420, 421(9), 422(5, 9), 425(5), 448, 455, 459, 461(1)
Ogushi, S., 398, 456
Ohashi, H., 392, 397, 401
Ohlendorf, D. H., 82
Ohmann, E., 350
Ohsugi, M., 448
Ohta, S., 299
Oka, A., 169
Okada, H., 173, 177
Ollis, D. L., 126, 129(2), 130
Omori, Y., 420, 422(5), 425(5)
Ono, K., 89, 118, 121
Oostra-Demkes, G. J., 314, 455, 459, 462(3)
Ornstein, L., 336
Ornston, L. N., 115, 122, 123, 124(6, 11), 126, 127, 128, 130, 131, 132, 133, 147
Orville, A. M., 82, 90, 94, 95(12), 96, 102, 107, 109, 111, 193
Osborn, M., 404
Osborne, M., 257, 300
Osslund, T. D., 46
Osumi, M., 171, 173, 175, 176, 177, 463, 465(5)
Otto, R., 314, 420, 421(6), 422(6), 423(6), 425(6), 455, 459, 462(3), 463
Ougham, H. J., 70
Owens, J. D., 243, 254, 309, 340
Ozawa, T., 436

P

Page, M. D., 214, 215, 216, 218(2), 219(2), 220(2), 221(2, 4), 222(2, 4)
Palleroni, N. J., 37, 42, 89, 102, 103
Palva, E., 46
Pandeya, K. B., 234
Papas, E. J., 251, 257(10), 258(10), 259(10)
Parke, D., 133
Paszczynski, A., 160, 164(13), 167(13)
Patel, K. B., 146
Patel, R. H., 420
Patel, R. N., 122, 124(6), 126(6), 130, 191, 459, 462(4)
Patel, R. T., 134
Patel, R., 21, 22(1), 28, 455, 459(4)
Patel, T. R., 134

Patil, D., 53, 54(6), 60
Paz, M. A., 277
Peel, D., 364, 386
Peisech, J., 235
Penner-Hahn, J. E., 62, 68(8)
Peoples, O. P, 71
Perkins-Olson, P. E., 82, 95
Perry, J. J., 21, 22(2), 24(2), 29, 82
Peterson, J. A., 10
Petratos, K., 283
Phillips, P. H., 32
Pilkington, S. J., 191, 202(4), 331
Pilon, S. A. F., 413
Poels, P. A., 38, 327
Poh, C. L., 102
Pollock, V., 260
Presswood, R. P., 138
Prick, P. A. J., 143
Pringle, J. R., 404
Prior, S. D., 181, 189, 191
Pujar, B. G., 61

Q

Quayle, J. R., 203, 223, 331, 346, 362, 363,
    364, 366, 374, 377(4), 379, 380, 381,
    386, 387, 391(6), 392, 397, 401, 436,
    437, 443(5), 444(5),
445, 451
Que, L., Jr., 88

R

Racker, E., 406, 410(6), 411
Randall, R. J., 12, 15, 19, 34, 37(7), 90, 290,
    383, 407
Randall, R. L., 252
Ratledge, C., 171
Rea, T., 286
Reddy, C. A., 169
Redmond, J. W., 384
Rees, C. W., 262
Reganathan, V., 169
Reineke, W., 121
Reiner, A. M., 134
Reinhammar, B., 161, 169(20)
Reiser, J., 169

Reiser, K., 277
Reisfeld, R. A., 59
Reiske, J. S., 61
Remberger, U., 381, 391
Rey, H.-G., 428
Ribbons, D. W., 61, 111, 154
Rieske, J. S., 60, 304
Rikitake, K., 398, 456
Rinaldi, A., 234, 235
Rindt, K. R., 350
Roberts, J. D., 147
Robertson, G. R., 108
Robinson, J., 30
Roche, B., 171, 172, 176
Roche-Peneverne, B., 176
Roels, F., 415
Rogers, J. E., 134
Roggenhamp, R., 421, 463
Romanov, V. P., 425
Romero-Chapman, M., 277
Rose, I., 446
Rosebrough, N. H., 407
Rosebrough, N. J., 12, 15, 19, 34, 37(7), 90,
    252, 290, 383
Rosenberger, R. F., 72, 73(13), 79
Rotilio, G., 235
Rouche-Penverne, B., 171
Roxburgh, J. H., 340
Roy, M., 14
Rucker, R. B., 277
Ruettinger, R. T., 3
Ruis, F. X., 234
Rupp, M., 33, 35(2)
Ryerson, C. C., 71, 75(11), 76(11)

S

Saboowalla, F., 53, 54(6), 60
Saeki, Y., 122
Sahm, H., 24, 321, 323(11), 330, 346, 391,
    397, 401, 420, 421, 422(7), 427, 455,
    458, 459, 462, 463
Saitou, S., 436
Sakai, Y., 426
Sakazawa, C., 397, 399, 401(6), 443,
    444(10), 447, 448(12), 450(12), 451, 458
Salem, A. R., 379, 387
Salerno, J. C., 106

Saloheimo, M., 169
Sands, R. H., 251
Sanglard, D., 160
Santer, M., 224, 229, 271, 316, 325(6), 407, 423
Savas, J. C., 191
Sawada, H., 286
Sawicki, C. R., 356
Sawicki, E., 356, 400
Schalch, H., 169
Schindler, J., 350
Schlegel, H. G., 350
Schmidt, H. W. H., 162
Schoemaker, H. E., 162
Scholtz, R., 355
Schoonover, J. R., 62, 68(10)
Schopfer, L. M., 111, 138
Schopp, W., 14
Schreiner, R. P., 160
Schreuder, H. A., 143, 146
Schubert, M. P., 204
Schulz, G. E., 144
Schulz, J., 50, 352, 453
Schütte, H., 346, 391, 397, 401
Schutte, H., 455, 459, 462(2)
Schwartz, R. D., 3, 4
Scott, C. C. L., 13
Scott, C. W., 421
Scott, K. F., 142, 143(10), 147(10)
Segal, I. H., 258
Senoh, S., 111, 121
Serdar, C., 39, 45(3)
Sgin, K.-G., 420
Shaltiel, S., 358
Shamala, N., 259
Shamsuzzaman, K. M., 46, 49(6)
Shapiro, M., 131
Sherry, B., 425, 426(21)
Shibata, S., 398, 456
Shimao, M., 397, 399, 401(6), 447, 448(12), 450(12), 451
Shimizu, S., 420, 463
Shinagawa, E., 271
Shirai, S., 241
Shoun, H., 144
Siedel, J., 422
Siedow, J. N., 106
Simon, M. J., 46
Simpson, F. J., 111
Singer, M. E., 14, 15(5), 18, 19(3, 4)

Singer, T. P., 251, 252(16), 258, 259, 260
Sistrom, W. R., 103
Sivaraja, M., 62, 68(7)
Sjollema, K. A., 420
Sjöström, K., 160
Skerman, V. B. D., 148, 150(5)
Slot, J. W., 420
Smibert, R. M., 103
Smidt, C., 277
Smith, H. A., 4, 9(21)
Smith, K. M., 304
Smith, T. L., 169
Smyrniotis, P. Z., 406
Sokolov, A. P., 215, 339, 397, 443
Sorger, H., 18
Spence, J. T., 260
Spencer, R., 76, 145
Spenser, C. M., 234
Sprey, B., 420
Srere, P. A., 350
Srivastava, S. K., 463
Stadtman, E. R., 380
Staehelin, L. A., 290
Stainer, R. Y., 37, 42, 82, 89, 102, 103, 122, 126, 131, 147
Stanikowski, Z., 421
Stankovich, M. T., 138, 309, 313(4)
Stanley, S. H., 181, 191
Stanley, T. W., 400
Stark, G. R., 361
Steczko, J., 425
Steenkamp, D. J., 244, 251, 255, 256(4), 257(4,10), 258, 259, 260, 309, 311(6), 312, 313(6)
Steennis, P. J., 139
Steensma, M., 318
Steinbach, R. A., 346
Steitz, T. A., 126, 129(2)
Steltenkamp, M. S., 3, 4, 5(18)
Stephens, S. E., Jr., 160
Stevens, F. J., 279
Stevens, R. C., 251
Stirling, D. I., 22, 188
Stoffel, W., 172
Stone, S., 361, 364(3)
Stopher, D. A., 82, 89
Stouthamer, A. H., 299
Stratford, M. R. L., 146
Strøm, T., 401
Stucki, G., 355, 357(1)

Subramanian, V., 39, 42, 82, 95, 134
Subraminian, V., 39, 43(4), 44(2, 4), 45(2, 3, 4, 10)
Sullivan, K. A., 433
Surerus, K. K., 201, 202(15)
Suter, F., 169, 355
Swan, G. A., 253
Swarte, M. B. A., 142, 143(9), 241, 247, 250(5)

T

Takada, N., 286
Takanami, M., 169
Takeda, Y., 122
Takizawa, N., 134
Tamaoki, T., 455
Tanaka, A., 171, 173, 175, 176, 177, 178(7), 463, 465(5), 467
Tanaka, K., 447, 448(12), 450(12), 451
Tani, Y., 392, 397, 401, 420, 421(9), 422(5, 9), 425(5), 426, 443, 444(10), 455, 458, 459, 461(1)
Tardone, P. J., 160
Tarr, G. E., 67, 68(14)
Tauchert, H., 14
Taylor, D. G., 31, 70
Tayrien, G., 426
Teeri, T., 46, 169
Teranishi, Y., 171, 173, 176, 463
Terui, G., 286
Thorpe, C., 311, 312
Thowsen, J. R., 4, 8(17)
Tien, M., 159, 160, 161, 162(19, 21), 163, 164(15), 165, 166(18), 167(14, 15), 169, 170(11, 14, 21)
Tiesma, L., 318
Timkovitch, R., 286
Tinker, D., 277
Tobari, J., 241, 284, 286, 288(1), 299
Toki, S., 366, 367(5), 372
Tomita, I., 436
Tomoda, K., 3
Toniuchi, H., 122
Tonomura, K., 422, 463
Torrelio, B. M., 277
Tramontano, A., 262, 264(6), 265(6), 266(6)
Trautwein, A. X., 54

Trippett, S., 82, 89
Troller, J., 164
Trotsenko, Y. A., 215, 248, 336, 339, 342, 397, 425, 443, 445, 446, 450(7), 451
Trudgill, P. W., 31, 70, 71, 75(10), 76(10), 77, 79(3), 81(1, 2, 3)
True, A. E., 62, 68(7)
Tsang, H.-T., 62, 68(8, 12)
Tsolas, O., 405, 406, 410(3)
Tsukamoto, A., 169
Tsuru, D., 398, 456
Tsygankov, Y. D., 248, 335, 336, 346
Tu, C.-P. D., 169
Tu, S.-C., 111, 138, 147
Tubbs, P. K., 380
Tur, S. S., 432, 434(9)
Twilfer, H., 60

U

Ubiyvovk, V. M., 445
Ueda, M., 173, 177, 178(7), 463, 467
Ueda, T., 3, 10, 41, 48
Umezawa, T., 159, 169(5)
Uotila, L., 315, 444
Urakami, T., 250
Utting, J. M., 373

V

van Berkel, W. J. H., 139, 142, 143(8), 145, 146
Van Den Tweel, W. J. J., 29
van der Klei, I. J., 413, 426
van der Laan, J. M., 142, 143
van der Meer, R. A., 227, 229(2), 230(2), 232(2), 233(2), 234, 235(2), 277, 279, 280, 281, 282, 283
van der Putte, P., 272, 274
van der Zee, R., 408
van Dijken, J. P., 224, 227, 228, 229(1, 8), 230(8), 232(8), 233(8), 314, 319, 392, 405, 412, 413, 414(8, 9), 415(8, 9), 417(9), 420, 421, 422(6), 423(6), 425(6), 445, 451, 455, 459, 462(3), 463
van Eijk, H. M. L., 423

van Iersel, J., 207, 227, 229(2), 230(2), 232(2), 233(2), 234(2), 235(2), 280
van Kleef, M. A. G., 33, 37(5), 261, 268, 271
van Koningsveld, H., 277, 279(39)
van Oosterwijk, R. J., 223, 226, 391
van Rooijen, P. J., 139
van Verseveld, H. W., 299
van Vleits-Smits, M., 228, 229(8), 230(8), 232(8), 233(8)
van Wielink, J. E., 215, 303, 304(3), 307(3)
Veenhuis, M., 224, 227, 229(1), 319, 405, 412, 413, 414(1, 8, 9), 415(2, 3, 8, 9), 417(9), 420, 421, 426, 463
Vellieux, F. M. D., 241, 247, 250(5), 283
Venkataraman, R., 406, 410(6), 411
Vermaas, D. A. M., 272
Vermeulen, C. A., 412, 414(8), 415(8)
Verrips, C. T., 424
Villiger, V., 70, 71(1)
Villiger, W., 412
Vishniac, W., 224, 229, 271, 316, 325(6), 407, 423
Visser, A. J. W. G., 146
Visser, G., 424
Visser, S., 139
von Koningsveld, H., 266
Voordouw, G., 139
Vriend, G., 143

W

Wackett, L. P., 39, 45(3), 355
Wagner, F., 420, 422(7), 458, 459(8), 462, 463
Waites, M. J., 436, 437, 443(5), 444(5)
Waldner, R., 160
Walker, N., 73, 79
Wallis, J. M., 241, 247, 250(4)
Walsh, C. T., 71, 75(11), 76, 145
Walther, I., 169
Wang, L.-H., 111, 138
Wariishi, H., 162, 169(23), 170(23)
Waterham, H. R., 420
Watling, E. M., 223, 226, 391
Waxman, D. J., 76
Weaich, K., 142, 143(10), 147(10)
Weber, K., 257, 300, 404
Weber, P. C., 82

Weijer, W. J., 143, 146
Weinreb, S. M., 262
Weitzman, P. D. J., 350
Welling, G. W., 408
Werner, W., 428
Westerling, J., 202, 203(3), 204(3), 205(3), 207, 209(14)
Weyler, W., 254, 255(24)
Wheelis, M. L., 89, 102
White-Stevens, R. H., 138
Whitman, C. P., 130, 133
Whittaker, J. W., 83, 88, 90, 94, 96, 102, 111, 193
Whittenbury, R., 181, 289
Wicks, R. E., 146
Wielinger, H., 428
Wierenga, R. K., 143, 144, 146
Wijnands, R. A., 146
Wilchek, M., 340
Williams, D. E., 59
Williams, R. J., 154
Wilson, J. T., 465
Wilson, K. S., 143
Wimalasena, K., 4, 8(17), 9(19)
Woldendorp, J. P., 318
Wolgel, S. A., 82, 95, 101
Wood, J. M., 61, 88, 94
Woodland, M. P., 184, 186, 191, 202(4)
Woodruff, W. H., 62, 68(10)
Woods, N. R., 22, 26
Wray, V. P., 411
Wray, W., 411
Wyckoff, H. W., 130, 132(2)

X

Xia, Z.-X., 259
Xuong, N. H., 259

Y

Yakushijin, K., 262
Yamada, H., 366, 367(5), 372, 420, 443, 444(10), 458, 459, 462(7)
Yamada, T., 171, 176, 463
Yamagami, Y., 399

Yamamura, M., 171, 173, 176
Yamanaka, T., 284
Yamanoi, K., 173
Yang, N., 142, 143(10), 147(10)
Yano, K., 148
Yashima, Y., 446
Yasuhara, S., 463, 465(5)
Yasunobu, K. T., 3
Yeh, W. K., 131
Yeh, W.-K., 39, 40, 43(4), 44(2, 4), 45(2, 3, 4)
Yoshida, T., 67, 68(14)
Yoshikawa, H., 447, 448(12), 450(12), 451
Yoshimoto, T., 398, 456
Yoshimura, N., 446
You, I. S., 122

Yphantis, D. A., 383
Yu, Y., 111

Z

Zabinski, R., 94
Zaborsky, O. R., 4
Zager, J., 420
Zak, B., 334
Zakour, R. A., 147
Zamanian, M., 52, 55(3), 56, 57(3), 59(3)
Zatman, L. J., 250, 251, 252, 253(7), 254(1), 258(7), 310
Zaugg, W. S., 60, 61
Zhang, Y. Z., 169

Subject Index

A

Acetaldehyde dehydrogenase, soluble
NAD$^+$-dependent, from *Acinetobacter*,
properties, 20
Acetobacter methanolicus
citrate synthase, 352–354
3-hexulose-6-phosphate synthase from,
401–405
MB58
cell-free extract, preparation of, 403
growth of, 351, 402–403
3-hexulose-6-phosphate synthase from,
401–405
methanol dehydrogenase, 215
soluble cytochrome *c* of, 299
Acetol dehydrogenase, 27
assay, 31
Acetol monooxygenase, 27
assay, 30–31
Acetone monooxygenase, 27
assay, 30–31
Acetyl-CoA, production, in methylotrophs,
379, 386
Acinetobacter
catechol oxygenase from, 123, 126
ethanol dehydrogenase from, 17
hexadecanol dehydrogenase from, 16
HO1-N
alcohol dehydrogenase, 15–18
aldehyde dehydrogenase, 18–21
alkane oxidation in
assay
cell-free, 10, 13–14
principle, 10
reagents, 10
substrates, 10
whole cell, 10, 12–13
definition of specific activity, 12
pathway of, 10
cell extracts, preparation of, 12, 15–16,
18
cell fractionation, 12, 15–16, 18–19
cell-free extracts

alkane oxidation in, 10, 13–14
stability, 13
incorporation of [^{14}C]hexadecane into
cellular components, 13–14
whole cells, alkane oxidation by, 10,
12–13
NAD$^+$-dependent, membrane-bound fatty
aldehyde dehydrogenase from, 20
NCIMB 9871
cell-free extract, preparation of, 73
cyclohexanone 1,2-monooxygenase
from, 70–77
growth of, 72–73
trace element solution for, 72–73
soluble NAD$^+$-dependent acetaldehyde
dehydrogenase from, properties, 20
Acinetobacter calcoaceticus. See also
Pseudomonas putida
ADP1, growth of, 124
catechol oxygenase gene, 123
LMD 79.41, glucose dehydrogenase from,
274–276
PQQ$^-$ mutant
growth of, 272
PQQ assayed with, 271–272
69V, aldehyde dehydrogenase, 18
Acinetobacter lwoffii, 276
Acyl-CoA synthetase, 174–175
Alcaligenes eutrophus, dihydrodiol
dehydrogenase, 135
Alcaligenes faecalis, dihydrodiol dehydrogenase, 135
Alcohol dehydrogenase. *See also* Long-chain
alcohol dehydrogenase; Short-chain
alcohol dehydrogenase
from *Acinetobacter*, 15–18
soluble NAD$^+$- and NADP$^+$-dependent,
15–16
assay, 15
definition of specific activity, 15
electron acceptor specificity, 16
inhibitors, 16
kinetic constants, 16–17

properties, 16
reaction catalyzed, 15
substrate specificity, 16
assay, 29–30
NAD(P)+-independent, 33
propane-specific
reaction catalyzed, 21
from *Rhodococcus*, 21–26
assay, 21–23
inhibitors, 25
kinetics, 25–26
molecular weight, 26
pH optimum, 25
properties, 25
purification, 23–25
purity, 25
stability, 25
substrate specificity, 25
subunit structure, 26
units, 23
Alcohol oxidase, 463
assay, 421–422
from *C. boidinii*, 420, 426
distribution, 420
from *H. polymorpha*, 411–420, 420–427
from *H. polymorpha* CBS 4732, immunocytochemical detection of, 417–420
inactivators, 426–427
inhibitors, 426
molecular weight, 424
from *P. pastoris*, 420, 425–426
pH optimum, 427
properties, 424–427
purification, 422–424
purity, 424
reaction catalyzed, 420
stability, 425
structure, 424–425
substrate specificity, 425–426
synthesis, 421
from *T. candida*, 420
Aldehyde dehydrogenase. *See also* Long-
chain aldehyde dehydrogenase
from *Acinetobacter*, 18–21
isozymes, 21
kinetic constants, 20–21
soluble and membrane-bound NAD+-
and NADP+-dependent, 19–21
assay, 19

definition of specific activity, 19
reaction catalyzed, 19
assay, 29–30
from *Hyphomicrobium*, 327–330
activators, 330
assay, 328–329, 330
inhibitors, 330
properties, 329–330
purification, 329
purity, 329
stability, 329
substrate specificity, 330
NAD-linked, factor-independent, and
glutathione- independent, 327–330
NAD(P+), from bakers' yeast, 330
reaction catalyzed, 327
Alkane oxidation, 10–14
Amicyanin, 235, 241, 284
assay of, from absorption spectra, 285
from *M. extorquens* AM1, 288
from organism 4025, purification of,
286–288
properties, 288
Amine oxidase, 280
applications, 435
assay, 427–428
bovine plasma, 234
eukaryotic, 234–235
forms of, 427
inhibitors, 434–435
from methylotrophic yeasts, 427–435
enzyme mechanism, 433–434
Michaelis constants, 433–434
properties, 432–435
purification, 429–433
purity, 432
stability, 433
substrate specificity, 432–435
molecular weight, 435
from *P. pastoris*, 427
pH optimum, 435
plasma, quinone-type redox cofactor for,
235
reaction catalyzed, 427
subunit structure, 435
units, 428–429
Androstene-3,17-dione 13,17-monooxygen-
ase, from *C. radicicola* ATCC 11011, 71
Arthrobacter

catalase, 227
P1 NCIB 11625
  cell-free extract, preparation of, 319,
    407–408
  growth of, 229, 407
  methylamine oxidase from, 227–235
  transaldolase isoenzymes from,
    405–411
Arthrobacter globiformis
  citrate synthase, 352–354
  glucose-6-phosphate dehydrogenase,
    339–343
  growth of, 351
  6-phosphogluconate dehydrogenase,
    343–345
  VKM B-175, 339
  growth of, 340
Arthrobacter P1 NCIB 11625
  formaldehyde dehydrogenase from,
    318–321
  growth of, 319
Azurin, 284–285
  assay of, from absorption spectra, 285
  from *M. extorquens* AM1, 288–289
  from organism 4025, purification of, 286–
    288
  properties, 288–289

B

Bacillus
  C1
    cell-free extract, preparation of,
      224–225
    growth of, 224, 393–394
    3-hexulose-6-phosphate synthase from,
      391–397
    methanol dehydrogenase from,
      223–226
    phosphohexuloisomerase from, 393–396
    dihydrodiol dehydrogenase, 135
    thermotolerant methylotrophic strains,
      223
Bacillus circulans, protocatechuate
  2,3-dioxygenase from, 95
Bacillus macerans
  JJ1b, growth of, 97
  protocatechuate 2,3-dioxygenase from, 95
Bacterium W3A1 (NCIB 11348)

electron-transfer flavoproteins from, 251,
  309–314
growth of, 243–244, 253–255, 309–310
methylamine dehydrogenase, 242–245
trimethylamine dehydrogenase from,
  250–260
Baeyer–Villiger reaction, 70, 71
Baker's yeast. *See* Yeast
Benzene, bacterial oxidation of, 134–135
cis-Benzene dihydrodiol dehydrogenase,
  134–137
  activation by Fe^{2+}, 137
  assay, 134–136
  kinetics, 137
  molecular weight, 137
  from *P. putida*, 134–137
  properties of, 137
  purification of, 136–137
  specificity, 137
  X-ray crystallography, 137
Benzene dioxygenase
  antibody probing, 55–56
  assay, 53–55
  components, 52
  kinetic constants, 60
  molecular weight, 60
  from *P. putida*, 52–60
  properties, 59–60
  purification, 56–58
  purity, 59
  reaction catalyzed, 52
  stability, 60
  substrate specificity, 59–60
  subunit structure, 60
Benzoate, anaerobic metabolism of, 154
Benzoate-CoA ligase
  activity, 156–157
  assay, 154–156
  effect of growth conditions, 158–159
  kinetic parameters, 158
  molecular size, 158
  pH optimum, 158
  properties, 156–157
  purification, 156–158
  from *R. palustris*, 154–159
  stability, 156–157
  substrate specificity, 158
Benzylamine oxidase, from *C. boidinii*, 427,
  432–433

properties, 432–435
purification, 432–433
substrate specificity, 434
Benzylamine/putrescine oxidase, from *P.
pastoris*, 427, 432–435
substrate specificity, 434
Blue copper proteins, 284–289
assay of
from absorption spectra, 285
as electron acceptor for methylamine
dehydrogenase, 285
from organism 4025
properties of, 288–289
purification of, 285–288
Brevibacterium fuscum, 109
chlorocatechol 1,2-dioxygenase from, 122
growth of, 85–86, 111–112
protocatechuate 3,4-dioxygenase from,
82–88

C

Candida. See also Yeast
cell-free extracts, preparation of,
172–173, 177
long-chain alcohol dehydrogenase of,
171–175
subcellular fractionation, 173, 177
Candida boidinii. See also Yeast
2201
catalase from, 463–467
cell-free extract, preparation of,
456–457, 460, 465
growth of, 456, 460, 465
alcohol oxidase, 420, 426
amine oxidases, 427
ATCC 32195
formaldehyde dehydrogenase from,
455, 459
formate dehydrogenase from, 459–460,
462
benzylamine oxidase from, 427, 432–435
CBS 5777, growth of, 429
cell-free extract, preparation of, 429–430
KD1
dihydroxyacetone synthase from,
435–445
glycerone kinase from, 451
growth of, 441, 448
triokinase from, 445–451

(*Kloeckera* sp.) 2201
formaldehyde dehydrogenase (glutathione), 455
formate dehydrogenase from, 459–460
methylamine oxidase from, 427, 429–435
polyamine oxidase from, 427
Candida lipolytica
catalase from, 175
cytochrome oxidase from, 175
long-chain alcohol dehydrogenase from,
171–175
long-chain aldehyde dehydrogenase from,
176–178
NRRL Y-6795, growth of, 172, 177
short-chain alcohol dehydrogenase, subcellular distribution of, 175
Candida methanolica ATCC 26175,
formate dehydrogenase from, 459, 462
Candida methylica
cell-free extract, preparation of, 452
formate dehydrogenase from, 459–460,
462
glycerone kinase from, 451–455
growth of, 452
Candida tropicalis
Berkhout pK 233, growth of, 172, 177
catalase, 175, 463
cytochrome oxidase, 175
long-chain alcohol dehydrogenase from,
171–175
long-chain aldehyde dehydrogenase from,
176–178
short-chain alcohol dehydrogenase from,
175
Carboxyl oxygen exchange, for synthesis of
[^{17}O]- or [^{18}O]carboxyl-enriched
homoprotocatechuate, 110–111
Carboxymuconate cycloisomerase, 130
4-Carboxymuconolactone decarboxylase,
133
Catalase
from *Arthrobacter*, 227
assay, 463–465
from *C. boidinii*, 463–467
absorption spectrum, 466
catalytic activity, 466–467
crystallization, 466
electron microscopy, 466
immunochemical difference, 467

induction, 466
molecular masses of native and subunit
forms, 466
optimal reaction pH, 466
peroxidatic activity, 466–467
properties, 466–467
purification, 465–467
subcellular localization, 467
from *C. lipolytica*, subcellular distribution
of, 175
from *C. tropicalis*, 175, 463
subcellular distribution of, 175
from *Candida* N16, 463
catalytic activity, 463
from *H. polymorpha*, 411–420, 463
peroxidative activity, 463
reaction catalyzed, 463
Catechol, reactions of, 123
Catechol 1,2-dioxygenase, 122–126. *See
also* Pyrocatechase
assay, 123–124
distribution, 122
properties, 126
sources, 122
specific activity, 122
stability, 126
storage, 126
yield, 126
Catechol 2,3-dioxygenase, 115–121
from *P. aeruginosa*, 115–121
Catechol oxygenase
from *Acinetobacter*, 123, 126
catA, 122, 124
gene, from *A. calcoaceticus*, 123
from *Pseudomonas*, 122
Catechol 1,2-oxygenase, purification of,
124–125
Catechol oxygenase I. *See* Catechol
1,2-dioxygenase
Catechol oxygenase II. *See* Chlorocatechol
1,2- dioxygenase
Chlorocatechol 1,2-dioxygenase, 122–126
assay, 123–124
properties, 126
reactions of, 123
specific activity, 122
specificity, 122
stability, 126
storage, 126

yield, 126
Chlorocatechol oxygenase
gene, 122
from *Pseudomonas*, 122–123, 126
Chlorocatechol 1,2-oxygenase, purification
of, 125
2-Chlorodibenzo[*p*]dioxin, enzymatic
oxidation, 160
Chloromuconate cycloisomerase, 127, 130
Citrate synthase
from *A. globiformis*, 352–354
assay, 350–351
from *M. curvatus*, 352–354
from *Mc. aquaticus*, 352–354
from methylotrophs, 350–354
effectors, 353–354
homogeneity, 352–354
properties, 352–354
purification, 351–352
from *Pseudomonas* M27, 352–354
reaction catalyzed, 350
Clostridium pasteuranium, rubredoxin, 8
Comamonas testosteroni, 276
ATCC 15667, growth of, 37
PQQ assayed with, 273–274, 276
quinohemoprotein alcohol dehydrogenase
from, 36–39, 273–274, 276
Corynebacteria, acetone-utilizing, metabolism of acetone by, 31
Corynebacterium cyclohexanicum,
hydroxybenzoate monooxygenase from,
139
Cupredoxin, 284
Cyclindrocarpon radicicola ATCC 11011,
androstene- 3,17-dione 13,17-monoox-
ygenase from, 71
Cyclohexanone 1,2-monooxygenase, 70–77
absorption spectrum, 75
from *Acinetobacter* NCIMB 9871, 70–77
assay, 71–72
definition of units, 72
inhibitors, 76–77
molecular weight, 74
from *N. globerula* CL1, 71
native apoenzyme, preparation of, 75
pH optimum, 76
properties, 74–77
prosthetic group, 75
purification of, 74–75

purity, 74
reaction catalyzed, 71
reaction mechanism, 76
reconstitution of active holoenzyme,
    75–76
substrate specificity, 76
subunit structure, 74–75
Cyclopentanone 1,2-monooxygenase, 77–81
absorption spectrum, 80
assay, 77
definition of units, 78
inhibitors, 81
Michaelis constants, 81
molecular weight, 80
pH optimum, 81
properties, 80–81
prosthetic group, 80
purification, 79–81
purity, 80
reaction catalyzed, 77
substrate specificity, 81
subunit structure, 80
Cytochrome *c*, 284
assay, 299–300
of *M. methylotrophus*, 298, 307
of *Methylobacterium* AM1, 298–303, 307
molecular weights, by denaturing
    electrophoresis, 300
peroxidase assay, 300
soluble
    from *M. extorquens* AM1, purification
        of, 301–302
    of methanol-utilizing bacteria, 298–303
    from *Methylomonas*, 289–297
        assay, 290–291
        heme stain, 290–291
        properties of, 295–296
        purification, 291–295
        spectroscopy, 290
    in other methanotrophs, 297
    units, 299–300
Cytochrome *c*-550, 235, 241
Cytochrome *c*-551, purification of, 292–294
Cytochrome *c*-552, 235
    purification of, 294–295
Cytochrome *c*-553, purification of, 294–295
Cytochrome *c*-554, purification of, 292–294
Cytochrome *c*-555, in *M. capsulatus*, 297
Cytochrome c_H, 288, 303
    extinction coefficient, 299

from *Hyphomicrobium* X, 304–308
    purification, 305–307
of *M. extorquens* AM1, 298–303
from *M. methylotrophus*, 304
properties, 298
Cytochrome c_L, 209, 212–213, 216, 222,
    284–285, 303
    extinction coefficient, 299
    from *Hyphomicrobium* X, 304–308
        autoreduction, 307
        properties, 307–308
        purification, 305–307
        purity, 307
        reaction with methanol dehydrogenase,
            307–308
    of *M. extorquens* AM1
        properties of, 302
        purification, 301–302
    from methylamine-grown bacteria,
        purification of, 287–288
    properties, 298
    stopped-flow spectrophotometric assay
        with methanol dehydrogenase, 304
Cytochrome *co*, 288–289
Cytochrome oxidase
    from *C. lipolytica*, subcellular distribution
        of, 175
    from *C. tropicalis*, subcellular distribution
        of, 175
    of organism 4025, 288

D

Desulfovibrio gigas, rubredoxin, 8
Desulfovibrio vulgaris, rubredoxin, 8
Diamine oxidase
    plant, 234–235
    porcine kidney, 234–235
Diazonium reaction, for synthesis of [^{17}O]-
    or [^{18}O]hydroxyl-labeled 4-hydroxy-
    phenyl acetate, 108–110
Dichloromethane
    metabolism of, 355
    utilization of, by methylotrophic bacteria,
        355
Dichloromethane dehalogenase, 355–361
    assay, 355–357
    from *Hyphomicrobium* DM2, 355–361
    from *Hyphomicrobium* GJ21, 359

inhibitors, 360
from *Methylobacterium* DM4, 355, 359
molecular weight, 360–361
pH optimum, 360
properties, 360–361
purification, 357–359
reaction catalyzed, 355
specific activity, 357
stability, 360
structural gene, 355
structure, 360–361
substrates, 360
unit, 357
Dihydrodiol dehydrogenase, 134
from *A. eutrophus*, 135
from *A. faecalis*, 135
from *Bacillus*, 135
cis- forms of, comparison of, 135
from *P. putida*, 135
Dihydroxyacetone kinase, 445. *See also*
Glycerone kinase
Dihydroxyacetone synthase, 435–445
assay, 435–441
from *C. boidinii* KD1, 435–445
effect of metal cations, 445
from *H. polymorpha*, 411–420, 441,
443–445
inhibitors, 444
molecular weight, 435
pH optimum, 445
properties, 443–445
purification, 441–443
purity, 435
reaction catalyzed, 435
stability, 435–436
structure, 435
substrate specificity, 444
Dihydroxy aromatic compounds, ^{17}O- or
^{18}O-enriched, synthesis of, 107–115
cis-1,2-Dihydroxycyclohexa-3,5-diene
(NAD) oxidoreductase. *See cis*-Benzene
dihydrodiol dehydrogenase
3,4-Dihydroxyphenyl acetate, isotopically
enriched, synthesis of, 107–115
2,5-Diketocamphane 1,2-monooxygenase,
from *P. putida*, 70
7,9-Dimethoxycarbonyl-2-ethoxycarbonyl-
5-methoxy-1*H*-pyrrolo[2,3-*f*]quinoline,
264–265
7,9-Dimethoxycarbonyl-2-ethoxycarbonyl-

1*H*-pyrrolo[2,3-*f*]quinoline-4,5-dione,
265
Dimethylamine dehydrogenase, from
Hyphomicrobium X, 251
Dioxygenases, physiological role of,
120–121
Dodecanethiol, 441–442
Dopa decarboxylase, 280

E

Electron-transfer flavoproteins, 309–314
immunological and enzymatic cross-reactivity, 313–314
from *M. methylotrophus*, 309–314
isoelectric point, 311
oxidation states of, 313–314
mammalian, 312
from *P. denitrificans*, 312
pig liver, 314
properties of, 312
properties, 311–312
purification of, 309–311
purity, 311
role of, 309
spectral and catalytic properties, 311–313
stability, 311
subunits, 311
Entner–Doudoroff, 2-keto-3-deoxy-6-phos-
phogluconate aldolase/sedoheptulose
biphosphate (ribose monophosphate)
cycle. *See* Hexulose phosphate cycle
3-Epimerase, 436
Escherichia coli, recombinant, ISP$_{NAP}$, 46
Ethanol dehydrogenase, from *Acinetobacter*,
kinetic constants of, 17
2-Ethoxycarbonyl-5-methoxy-1*H*-pyr-
rolo[2,3-*g*]indole-7,8-dione, 267
Ethyl 6-amino-5-methoxyindole-2-carboxyl-
ate, 264
Ethyl 6-formylamino-5-methoxyindole-2-
carboxylate, 264
Ethyl pyruvate 3-formylamino-4-methoxy-
phenylhydrazone, 263–264

F

Fatty aldehyde dehydrogenase, NAD$^+$-de-
pendent, membrane-bound, from
Acinetobacter, properties, 20

Ferredoxin(ben), 52, 59
  amino acid sequence of, 60
  iron-sulfur clusters, 60
  redox potentials, 60
Ferredoxin$_{NAP}$, 46
  assay, 49
  purification of, 50–51
Ferredoxin$_{TOL}$, 39, 40
Formaldehyde dehydrogenase, 314–327,
    339
  from *Arthrobacter* P1, 318–321
    inhibitors, 320
    Michaelis constants, 320–321
    molecular weights, 321
    pH optimum, 320
    properties, 320–321
    purity, 320
    stability, 320
    substrate specificity, 320
    subunit structure, 321
    thiol reagents, 320
  assay, 318–319
  from *C. boidinii* ATCC 32195, 459
  definition of unit, 456
  effect of pH, 458
  inhibition, 458
  kinetics, 459
  from *M. methylotrophus*, 318
  from methylotrophic yeasts, 455–459
  molecular weight, 458
  from *P. putida*, 328, 330
  properties, 457–459
  purification, 319–320, 456–458
  purity, 457
  from *R. erythropolis*, 321–323
  reaction catalyzed, 314, 318
  reaction product, 458
  reverse reaction, 458
  specific activity, 456
  substrate specificity, 458
  subunit structures, 458
Formaldehyde dehydrogenase (acceptor),
    323–327
  assay, 323–325
  electron acceptors, 326
  enzyme inhibitors, 327
  Michaelis constants, 326
  molecular weight, 327
  pH optimum, 326
  properties, 326–327

  purification, 325–327
  purity, 326
  stability, 326
  substrate specificity, 326
Formaldehyde dehydrogenase (glutathione)
  assay, 315–316, 455–456
  from *C. boidinii* ATCC 32195, 455
  from *C. boidinii* (*Kloeckera* sp.) 2201, 455
  from *H. polymorpha*, 455
    purification of, 316–318
  Michaelis constants, 318
  from *P. pastoris* IFP 206, 455
  pH optimum, 318
  from *Pichia* NRRL-Y-11328, 455
  properties, 317–318
  purification, 316–318
  purity, 317
  reaction catalyzed, 455
  stability, 317
  substrate, 455
  substrate specificity, 317
  units, 316
Formaldehyde dehydrogenase (novel
    cofactor)
  assay, 321
  Michaelis constants, 323
  molecular weight, 323
  pH optimum, 323
  properties, 322–323
  purification, 322–323
  purified cofactor, 323
  purity, 322
  substrate specificity, 322–323
  subunit structure, 323
Formaldehyde:NAD$^+$ oxidoreductase
    (glutathione formylating). *See*
    Formaldehyde dehydrogenase (glutathi-
    one)
Formaldehyde transketolase. *See* Dihydrox-
    yacetone synthase
Formate dehydrogenase, 331–334, 339
  activity, 333
  alternative substrates, 334
  assay, 331–332, 459–460
  from *C. boidinii* ATCC 32195, 459–460
    properties, 462
  from *C. boidinii* (*Kloeckera* sp.) 2201,
    459–460
  from *C. methanolica* ATCC 26175, 459
    properties, 462

from *C. methylica*, 459–460
  properties, 462
cofactors, 333–334
definition of unit, 460
effects of pH, 461
effects of temperature, 461
[Fe–S]$_x$ centers, 334
from *H. polymorpha* CBS 4732, 459–460
  properties, 462
inhibition, 334, 462
kinetics, 462
from *M. trichosporium* OB3b, 331–334
from methylotrophic yeasts, 459–463
molecular weight, 333, 461
from *P. pastoris* IFP 206, 459–460
  properties, 462
from *P. pastoris* NRRL-Y-7556, 459–460
  properties, 462
properties, 333–334, 461–462
purification, 332–334, 460–462
purity, 333, 461
reaction catalyzed, 331, 459
specific activity, 460
stability, 333
substrate specificity, 461
subunit structure, 461
S-Formylglutathione, preparation of,
  315–316

G

Gentisate 1,2-dioxygenase, 101–107, 109
assay, 102–103
concentration, 103
holoenzyme molecular mass, 106
inhibition, 106–107
kinetic properties, 106
from *M. osloensis*, 102
molecular weight, 106
from *P. acidovorans*, 101–107
physical properties, 106
properties, 106–107
purification, 103–106
reaction catalyzed, 101
stability, 106
substrate specificity, 106–107
subunit structure, 106
units, 103
Gentisate:oxygen oxidoreductase. *See*
  Gentisate 1,2- dioxygenase

Gentisic acid, structure, 101
Glucose dehydrogenase, from *A. calcoace-*
  ticus LMD 79.41, 274–276
Glucose-6-phosphate dehydrogenase
from *Arthrobacter*, 339–343
  assay, 339–340
  effect of metal ions, 342
  effectors, 342–343
  Michaelis constants, 342
  molecular weight, 342
  pH optimum, 342
  properties, 342–343
  purification, 341–343
  purity, 342
  stability, 342
  substrate specificity, 342
  subunit structure, 342
  units, 340
assay, 346–347
from *Hydrogenomonas* H16, 350
from *M. flagellatum*, 335–337
  assay, 335
  forms of, 336–337
  inhibitors, 337
  Michaelis constants, 337
  polyacrylamide gel electrophoresis, 336
  properties, 336–337
  units, 335
from *M. flagellatum* KT, 346
from *M. methylotrophus*, 346
from *Methylomonas* M15, 346
from *P. fluorescens*, 350
from *Pseudomonas* C, 346
from *Pseudomonas* W6, 346–350
  inhibitors, 350
  kinetic properties, 348
  pH optimum, 348
  properties, 348–350
  purification, 347–349
  purity, 348
  stability, 348
  units, 347
from *R. sphaeroides*, 350
reaction catalyzed, 346
Glutathione reductase, 259
Glyceraldehyde-3-phosphate dehydrogenase,
  437
Glycerol kinase, 437
Glycerol-3-phosphate acyltransferase, 175
Glycerone kinase

assay, 451–452
from *C. boidinii* KD1, 451
from *C. methylica*, 451–455
  effect of metal ions, 453–455
  inhibitors, 455
  Michaelis constants, 453
  pH optimum, 455
  properties, 453–455
  purity, 453
  stability, 455
  substrate specificity, 453–454
from *H. polymorpha* CBS 4732, 451
purification, 452–454
reaction catalyzed, 451
units, 452

H

Halogenated aromatics, enzymatic
    oxidation, 160
Hansenula polymorpha. See also Yeast
alcohol oxidase, 411–420
catalase from, 411–420, 463
CBS 4732
  formate dehydrogenase from, 459–460,
      462
  glycerone kinase from, 451
de Marais et Maia, CBS 4732
  alcohol oxidase from, 420–427
  alcohol oxidase protein, immunocyto-
      chemical detection of, 417–420
  cell-free extract, preparation of, 423
  cytochemical staining procedures for,
      412
    localization of alcohol oxidase
        activity by, 414–417
    localization of catalase activity by,
        413–414
  formaldehyde dehydrogenase (glutathi-
      one) from, 316–318, 455
  growth of, 316–317, 413, 423
  protoplast formation, 413
dihydroxyacetone synthase from,
    411–420, 441, 443–445
DL1, 441, 448
  alcohol oxidase, 426
enzymes of methanol metabolism in,
    cytochemical staining methods for,
    411–420

triokinase from, 450–451
Heme-staining proteins, in methanotrophs
  compared to major monoheme cy-
      tochromes from *Methylomonas* A4,
      296–297
  cross-reacted with antisera from cy-
      tochromes *c*-554 and *c*-552, 296–297
Hexadecanol dehydrogenase, 171–172
from *Acinetobacter*, 16
  kinetic constants of, 17
  localization of, 18
  membrane-bound NAD^+-dependent,
      17–18
    assay, 17
    definition of specific activity, 17
    properties, 18
Hexadecanol:NAD^+ oxidoreductase. *See*
    Hexadecanol dehydrogenase
Hexulose phosphate cycle, 227
Hexulose-6-phosphate isomerase, 392
3-Hexulose-6-phosphate synthase, 391–397
from *A. methanolicus*, 401–405
  assay, 401–402
  effect of metal ions, 404
  kinetic properties, 405
  molecular weight, 404
  pH optimum, 404
  properties, 403–405
  purification, 403–404
  purity, 403–404
  stability, 404
  subunit structure, 404
  temperature optimum, 404–405
assay, 392–393
from *Bacillus* C1, 391–392
  effect of metal ions, 396
  immunology, 397
from *M. aminofaciens*, 391–392, 397
from *M. capsulatus*, 391, 397
from *M. gastri*, 392, 397–401
  assay, 397–398
  definition of unit, 398
  effect of metal ions, 401
  effect of pH, 401
  kinetic constants, 400–401
  molecular weight, 400
  properties, 400–401
  purification, 399–400
  purity, 400
  specific activity, 398

substrate specificity, 400
subunit structure, 400
from *M. methylotrophus*, 392, 397
from *Methylomonas* M15, 391, 397
Michaelis constants, 396–397
molecular weights, 396
pH optimum, 397
properties, 395–397
purification, 393–395
purity, 395
reaction catalyzed, 391, 397, 401
stability, 396
substrate specificity, 396
units of activity, 393
Homoprotocatechuate
isotopically enriched, synthesis of, 107,
107–115
[3- or 4-^{17}O]hydroxyl-enriched, 115
synthesis of, 113–115
[3- or 4-^{18}O]hydroxyl-enriched, 115
[^{18}O]hydroxyl-enriched, synthesis of,
113–115
Homoprotocatechuate 2,3-dioxygenase,
109, 112
Hydrocarbon monooxygenase, of *P.
oleovorans*, 3–9
cell-free production of oxygenated
products, 7–8
chemical mechanism of, 9
components, 3
isolation of, 7
oxygenation activities, 9
assay of, 4
oxygenation reactions, 9
reactions catalyzed, 3–4
Hydrogenomonas H16, glucose-6-phosphate
dehydrogenase from, 350
Hydroxybenzoate, isomers, hydroxylation,
enzymes for, 138
p-Hydroxybenzoate
analogs of, 143–144
metabolism of, by microorganisms, 147
p-Hydroxybenzoate hydroxylase
gene, 147
reaction catalyzed by, 138
3-Hydroxybenzoate 4-hydroxylase, 109
4-Hydroxybenzoate 3-hydroxylase, 109
purification of, 112–113
Hydroxybenzoate monooxygenase,
138–147

amino acid residues, 145–146
assay, 139
from *C. cyclohexanicum*, 139
crystallization, 141–142
flavin cofactor, 144–145
genetics, 146
molecular properties, 143
oxygen reactions, 146
from *P. aeruginosa*, 139–146
from *P. fluorescens*, 138–139, 143, 144,
147
properties, 142–146
purification, 139–142
stability, 142–143
storage, 142
substrate dependence, 143–144
4-Hydroxybenzoate 3-monooxygenase. *See
p*- Hydroxybenzoate hydroxylase
ω-Hydroxylase
of *P. oleovorans* hydrocarbon monooxygenase enzyme system, 3
in *P. putida*, 10
4-Hydroxyphenyl acetate, [4-^{17}O]- or
[4-^{18}O]hydroxyl-enriched, isolation of,
110
3-Hydroxyphenyl acetate 6-hydroxylase, 109
4-Hydroxyphenyl acetate 3-hydroxylase, 109
Hydroxypyruvate reductase, 373–378
assay, 373–374
forms of, 378
kinetics, 377–378
molecular weight, 377
from *P. acidovorans*, 373
production of, 374
properties, 377–378
purification, 374–377
reaction catalyzed, 373
specificity, 377
subunit structure, 377
unit of enzyme activity, 374
Hyphomicrobium
DM2
cell-free extract, preparation of, 358
dichloromethane dehalogenase from,
355–361
growth of, 357
GJ21, dichloromethane dehalogenase
from, 359
X
aldehyde dehydrogenase from, 327–330

cytochrome c_L and cytochrome c_H from, 303–308
dimethylamine dehydrogenase from, 251
dye linked formaldehyde dehydrogenase from, 325–327
growth of, 204–205, 325, 329
methanol dehydrogenase from, 202–209, 215, 305–307
serine hydroxymethyltransferase from, 372
soluble cytochromes c of, 299
trimethylamine dehydrogenase from, 250–251
Hyphomicrobium methylovorum, serine hydroxymethyltransferase from, 366, 372

I

Intradiol oxygenases, 126
Iron-sulfur protein (ben) [ISP(ben)], 52, 58
  α and β subunits of, 52
  preparation of, 59
  effect of metal ions, 60
  iron-sulfur clusters, 60
  redox potentials, 60
Iron-sulfur protein$_{NAP}$ [ISP$_{NAP}$], 46, 47–48
  purification of, 51–52
Iron-sulfur protein$_{TOL}$ [ISP$_{TOL}$], 39, 40
ISP. *See* Iron-sulfur protein

K

β-Ketoadipate enol-lactone, 130
β-Ketoadipate enol-lactone hydrolase, 131–132
β-Ketoadipate pathway, 126, 128, 130, 147

L

Lactate dehydrogenase, 259
Lignin, biodegradation, 159–160
Lignin peroxidase
  fungal, 159–171
  assay, 160–163
  oxidation products, electron spin resonance spectroscopy, 163–164
isoenzyme LiP1, kinetic parameters for, 170
isoenzymes, 160, 168–169
nomenclature, 160
from *Ph. chrysosporium*, 159–171
from *Phl. radiata*, 160
pH optimum, 170
production, 164–166
properties, 168–170
purification, 166–168
pyrene oxidation, demonstration of, 163–164
reaction catalyzed, 159
spectral properties, 169–170
substrate specificity, 169
from *T. versicolor*, 160
Long-chain alcohol dehydrogenase
  in *C. lipolytica*, 171–175
    subcellular distribution of, 175
  from *C. tropicalis*, 171–175
    subcellular distribution of, 175
  yeast, 171–175
    assay, 172
    function, 174–175
    induction, 174
    properties, 174–175
    solubilization, 174
    stability, 174
    subcellular localization, 174–175
    substrate specificity, 174
Long-chain aldehyde dehydrogenase, 171, 174–175
  from *C. lipolytica*, subcellular distribution of, 178
  from *C. tropicalis*, 176–178
    cellular levels of, 177
    subcellular distribution of, 178
    function, 178
    induction, 177
    properties, 177–178
    solubilization, 178
    stability, 178
    subcellular localization, 178
    substrate specificity, 177–178
  from yeast, 176–178
    assay, 176
    reaction catalyzed, 176
Long-chain-fatty-acid–CoA ligase. *See* Acyl-CoA synthetase
Lysyl oxidase, 433

M

Maleylpyruvic acid, structure, 101
L-Malic acid chloralide, 387–388
L-Malic acid chloralide chloride, 387–389
L-Malylchloralidyl-*N*-octanoylcysteamine, 387, 389–390
Malyl-CoA lyase, 379–386, 386–387
  assay, 379–381
  equilibrium constant, 386
  inhibition, 384
  metal ion requirement, 384–385
  Michaelis constants of, 385–386
  molecular weight, 383
  pH optimum, 384
  properties, 383–386
  purification, 382–383
  reaction catalyzed, 379
  source of, 381
  specificity, 384
  stability, 384
L-4-Malyl coenzyme A, 387–388, 390–391
  synthesis of, 386–391
4-Malyl coenzyme A, synthesis, 381
Mandelate racemase, 130
  amino acid sequence, 126–127
  structure, 126
Megasphaera elsdenii, rubredoxin, 8
Metapyrocatechase, activity, determination of, 116
Metapyrocatechase 1, 116
  amino acid composition, 118–119
  induction of, 120
  molecular mass of, 118–119
  physiochemical properties, 118
  physiological role of, 121
  properties of, 118–119
  purification, 116–117
  purity, 118
  spectral characteristics, 118
  substrate specificity, 118, 120, 121
    effect of aromatic inducer structure on, 121
Metapyrocatechase 2, 116
  amino acid composition, 118–119
  induction of, 120
  kinetics, 118–120
  molecular mass of, 118–119
  physiochemical properties, 118
  physiological role of, 121

properties of, 118–119
purification, 117–119
purity, 118
spectral characteristics, 118
substrate specificity, 118, 120
  effect of aromatic inducer structure on, 121
Methane monooxygenase, 331
  activity calculations, 193–194
  assay, 191–192
  definition of unit of activity, 193, 194
  gas chromatographic assays, procedure for, 193–194
  hydroxylase activity, 191
  from *M. trichosporium*, 191–202
  membrane-bound (particulate), 181, 191
  from *Methylobacterium* CRL-26, 191
  polarographic assays, procedure for, 192–193
  protein components, 191
    procedure for optimizing concentrations of, 192
  purification of, 197–200
  purification of, 197–200
  reaction catalyzed, 181, 191
  reductase NADH oxidoreductase activity assay, 195
    calculation of activity, 195
    definition of unit of activity, 195
  reductase NAD(P)H oxidoreductase activity, 191
  soluble, 191
    assay, 181–182
    from *M. capsulatus*, 181–191
      component proteins, 182
      kinetic properties, 189–190
      properties, 186–190
      protein C, 183
      purification, 183–187
    from *M. capsulatus* Bath, 181–190
      inhibitors, 188–189
      mechanism of, 186–188
      products, 188
      substrate specificity, 188
    from *M. trichosporium*, properties, 200–202
    from *M. trichosporium* OB3b
      cofactor of hydroxylase, 201–202
      cofactor of reductase, 202
      kinetic properties, 201

physical properties of, 201
purity, 200–201
stability, 200–201
protein A
   purification of, 184–185
   stability, 190
protein B
   purification of, 185–186
   stability, 190
protein C
   purification of, 185–187
   stability, 190
proteins, physicochemical characteristics of, 189
Methanol (alcohol) oxidase, 427
Methanol dehydrogenase, 260, 312
   from *A. methanolicus*, 215
   assay, 210–213, 223–224
   electron acceptor, 298, 303, 304
   formaldehyde reductase activity, assay,
      223–224
   from *Hyphomicrobium*, 202–209, 215
      absorption coefficients, 207
      activators, 206–207, 209
      apoenzyme, 208
      assay, 203–204
      cofactor content, 209
      inhibitors, 207
      kinetic parameters, 209
      molecular weight, 209
      natural electron acceptor, 209
      pH optimum, 209
      primary electron acceptors, 206
      properties, 206–209
      purification, 204–207
      purity, 206
      redox forms, 207–208
      stability, 209
      substrate specificity, 206
      subunit structure, 209
   from *Hyphomicrobium* X, 202–209, 215
      purification, 305–307
   from *M. glucoseoxidans*, 215
   from *M. methylotrophus*, 215
   from *Methylobacterium*, 210–216
      absorption spectrum, 215–216
      activators, 216
      assay, with oxygen electrode, 211
      electron acceptors, 216

methanol:cytochrome *c* oxidoreductase
      activity, 212–213
   microtiter assay, 212
   molecular weight, 216
   pH optimum, 216
   properties, 215–216
   purification, 213–215
   purity, 215–216
   spectrophotometric assay, 211–212
   substrate specificity, 216
   subunit composition, 216
   subunits, preparation of, 215
modifier protein of, 216–222
   assay, 216–217
   effect on alcohol oxidation, 222
   effect on formaldehyde oxidation, 222
   effect on methanol dehydrogenase in
      cytochrome-linked system, 222
   isoelectric point, 222
   molecular weight, 221
   periplasmic location, 222
   properties, 221–222
   purification, 218–221
   stability, 221
   subunit structure, 221
   units, 216–217
   from *P. denitrificans*, 215
      purification of, 220–221
   reaction catalyzed, 202, 210
   substrate specificity, 202, 216
   from thermotolerant *Bacillus*, 223–226
      Michaelis constants, 226
      molecular weights, 225–226
      properties, 225–226
      purification, 223–226
      purity, 225
      specificity, 226
      stability, 226
   unit of enzyme activity, 224
2-Methoxy-5-aminoformanilide, 262–263
4-Methoxybenzoate monooxygenase, 60
2-Methoxy-5-nitroformanilide, 262
Methylamine dehydrogenase, 260, 312
   amino acid composition, 241
   assay, 242, 247–248
   from bacterium W3A1, 242–245
      properties of, 245–246
      purification, 243–245
   definition of unit, 242

electron acceptor, 284
induction, 247
isoelectric point, 241–242
kinetic properties, 241, 246
from *M. flagellatum*, 247–250
  isoelectric focusing, 249
  Michaelis constants, 249–250
  molecular weight, 250
  pH optimum, 249
  properties, 249–250
  purification, 248–250
  purity, 249
  stability, 249
  substrate specificity, 249
  subunit structure, 250
from methylamine-grown bacteria,
  purification of, 287–288
from methylotrophic bacteria, 241–246
molecular weights, 241
from *P. denitrificans*, 242
  properties of, 245–246
purification, 242–245
reaction catalyzed, 241, 247
specific activity, 242
spectral properties, 241
subunits
  physical properties, 245
  properties, 245–246
  reconstitution, 245
  resolution, 245
  stability, 245–246
from *T. versutus*, 235–241
  assay, 236
  cofactor, 239
  properties, 239–241
  purification, 237–240
  purity, 239
  reaction with natural electron acceptors,
    241
  redox cycle of, 240–241
  spectral properties, 240
  structure, 239
units, 248
Methylamine oxidase, 260, 280
from *Arthrobacter*, 227–235
  assay, 227–229
  definition of unit, 228
  inhibitors, 233
  molecular weight, 233–234

oxidation–reduction properties, 234
properties, 232–235
purification, 229–232
purity, 232–233
specific activity, 228
spectral properties, 234
stability, 233
steady-state kinetic parameters, 233
subunit composition, 233–234
from *C. boidinii*, 427, 429–435
  properties, 432–435
  purification, 429–431
  substrate specificity, 434
cofactor, 235
isoelectric point, 235
reaction catalyzed, 227
Methylglyoxal dehydrogenase, 27
Methylobacillus flagellatum
  cell-free extracts, preparation of, 248, 336
  glucose-6-phosphate dehydrogenase from,
    335–337
  growth of, 248, 336
  methylamine dehydrogenase from,
    247–250
  6-phosphogluconate dehydrogenase,
    337–339
  temperature-sensitive mutants T623 and
    T903, growth of, 336
Methylobacterium
  3A2, serine hydroxymethyltransferase
    from, 372
  AM1. See *Methylobacterium extorquens*,
    AM1
  CRL-26, methane monooxygenase from,
    191
  DM4, dichloromethane dehalogenase
    from, 359
    structural gene for, 355
Methylobacterium extorquens
  AM1
    amicyanin from, 288
    azurin from, 288–289
    blue copper proteins, 286
    cell-free extract, preparation of, 382
    citrate synthase, 352–354
    c-type cytochromes of, 298–303, 307
    cytochrome c_H
      properties of, 303
      purification, 301–302

growth of, 213–214, 351, 381
hydroxypyruvate reductase from, 373–378
malyl-CoA lyase from, 379–386
methanol dehydrogenase from, 210–216
M proteins from, 221
mutant lacking malyl-CoA lyase, 379
serine hydroxymethyltransferase from, 372
crude extract, preparation of, 374–375
culture growth, 374
Methylobacterium organophilum XX
growth of, 368
serine hydroxymethyltransferases from, 365–372
Methylococcus capsulatus (Bath)
growth of, 183
heme-staining protein from, 296
methane monooxygenase from, 181–191
soluble, 181–190
Methylocystis aquaticus, citrate synthase from, 352–354
Methylocystis curvatus
citrate synthase from, 352–354
growth of, 351
Methylocystis parvus OBBP, heme-staining protein, 296
Methylomonas
A1, heme-staining protein, 296
A4
cell lysis, 291
growth of, 289
harvesting of cells, 289–290
heme-staining protein, 296
soluble cytochromes *c* from, 289–297
C1, heme-staining protein, 296
J, soluble cytochromes *c* of, 299
M15
glucose-6-phosphate dehydrogenase from, 346
3-hexulose-6-phosphate synthase from, 391, 397
MM2, heme-staining protein, 296
MN, heme-staining protein, 296
soluble cytochrome *c* from, 289–297
Methylomonas albus BG8, heme-staining protein, 296
Methylomonas aminofaciens

77a, 3-hexulose-6-phosphate synthase from, 391–392
3-hexulose-6-phosphate synthase from, 397
Methylomonas capsulatus
Bath, soluble *c*-type cytochromes, 297
3-hexulose-6-phosphate synthase from, 391, 397
Methylophilus, 253
Methylophilus glucoseoxidans, methanol dehydrogenase from, 215
Methylophilus methylotrophus
cytochrome c_L from, 298, 307
electron-transfer flavoproteins from, 309–314
formaldehyde dehydrogenase from, 318
glucose-6-phosphate dehydrogenase from, 346
3-hexulose-6-phosphate synthase, 392, 397
methanol dehydrogenase from, 215
M protein from, purification of, 219–220
NCIB 10515, growth of, 218, 309–310
soluble cytochromes *c* of, 299, 304
Methylosinus trichosporium OB3b
formate dehydrogenase from, 331–334
growth of, 195–197
heme-staining protein, 296
methane monooxygenase from, 191–202
Methylotrophic bacterium strain DM11, dichloromethane dehalogenase from, 359
Micrococcus aerogenes, rubredoxin, 8
Modifier protein, for methanol dehydrogenase of methylotrophs, 216–222
Moraxella osloensis, gentisate 1,2-dioxygenase from, 102
M protein, for methanol dehydrogenase of methylotrophs, 216–222
Muconate cycloisomerase, 126–130, 131
amino acid sequence, 126
assay, 127
from *P. putida*, 126
subunits, 129
properties of, 129–130
purification, 128–129
reaction catalyzed, 126
structure, 126
Muconolactone, enzymatic isomerization of, 133

Muconolactone isomerase, 128, 130–133
  amino acid sequence, 130
  assay, 131–132
  crystal structure of, 132
  from *P. putida*, 131–133
  properties of, 132–133
  purification, 132–133
  reaction catalyzed, 130
  structure, 130
  subunits, 130
Mycobacterium, polycyclic aromatic
    hydrocarbon- degrading, 148–153
  culture conditions, 149
  experiments with, 149–151
  growth of, 149
  isolation of, 148–149
Mycobacterium gastri MB19
  cell-free extract, preparation of, 399
  growth of, 398–399
  3-hexulose-6-phosphate synthase from,
    392, 397–401

N

NADH:ferredoxin oxidoreductase, 52
NADH:rubredoxin reductase
  of *P. oleovorans*, 3
    isolation of, 6–7
  of *P. putida*, 10
NAD(P)$^+$-independent formaldehyde
    dehydrogenase (dye- linked). *See* For-
    maldehyde dehydrogenase (acceptor)
Naphthalene dioxygenase
  amino acid sequence, 46
  components, 46
  genes, 46
  purification, 49
  reaction catalyzed, 46
Naphthalene oxygenase, assay, 46–49
Nocardia globerula CL1, 71

O

Octanol dehydrogenase, in *Saccharomyces
    cerevisiae*, 171
Octanol:NAD$^+$ oxidoreductase. *See* Octanol
    dehydrogenase
N-Octanoylcysteamine, 387, 389
L-*N*-Octanoylcysteamine, 387–388, 390

Organism 4025
  blue copper proteins from, 285–289
  growth of, 286–287
  soluble cytochromes *c* of, 299
2-Oxo-Δ^3-4,5,5-
    trimethylcyclopentenylacetyl-CoA
    monooxygenase, from *P. putida*, 70–71
Oxygen atoms, of phenolic and catecholic
    compounds, isotopic labeling of,
    107–108
Oxygen polarograph, 83–85
  calibration of, 83–84

P

Paracoccus denitrificans
  blue copper proteins, 286
  electron-transfer flavoproteins from, 312
  growth of, 242–243
  methanol dehydrogenase, 215
    purification of, 220–221
  methylamine dehydrogenase from,
    242–246
  M protein from, purification of, 219–221
  NCIB 8944
    growth of, 218
    periplasmic fractions, preparation of,
      218–219
    spheroplasts, preparation of, 218–219
  soluble cytochromes *c* of, 299
Phanerochaete chrysosporium
  BKM-F-1767, culture, 164–166
  lignin peroxidase from, 159–171
Phlebia radiata, lignin peroxidase from, 160
6-Phosphogluconate dehydrogenase
  from *Arthrobacter*, 343–345
    assay, 343
    effect of metal ions, 345
    effectors, 345
    Michaelis constants, 345
    molecular weight, 345
    pH optimum, 345
    properties, 344–345
    purification, 343–345
    purity, 344
    stability, 345
    substrate specificity, 345
    subunit structure, 345
    units, 343

from *M. flagellatum*, 337–339
  assay, 337–338
  inhibitors, 338
  isoelectrofocusing, 338
  isoenzymes, 338
    separation of, 338
  partial purification, 338
  properties, 338–339
  reaction catalyzed, 337
  stability, 338
  units, 338
Phosphohexuloisomerase
  assay, 393
  from *Bacillus* C1, purification of,
    394–396
Phosphoriboisomerase. *See* Ribose-5-phos-
  phate isomerase
Phthalate dioxygenase, 61–70
  assay, 62
  catalytic properties, 69
  molecular weight, 67
  mononuclear site, spectroscopic investiga-
    tion of, 68–69
  from *P. cepacia*, 61
  physical properties, 66–69
  prosthetic groups, 67
  purification, 63–65, 65–66
  reaction catalyzed, 61
  Rieske [2Fe–2S] center, structure of, 68
  specificity, 69
  spectral properties, 67–68
  stability, 66–67
  steady-state kinetics, 69–70
Phthalate dioxygenase reductase, 61
  assay, 62–63
  purification of, 64–66
Pichia NRRL-Y-11328, formaldehyde
  dehydrogenase (glutathione) from, 455
Pichia pastoris. See also Yeast
  alcohol oxidase from, 420, 425–426
  amine oxidases from, 427
  benzylamine/putrescine oxidase from,
    427, 432–435
    purification, 432–433
  CBS 704, growth of, 429
  IFP 206
    formaldehyde dehydrogenase (glutathi-
    one) from, 455
    formate dehydrogenase from, 459–460,
    462

NRRL-Y-7556, formate dehydrogenase
  from, 459–460, 462
Polyamine oxidase, from *C. boidinii*, 427
Polycyclic aromatic hydrocarbons
  bioremediation of, 148
  degradation, by *Mycobacterium*, 148–153
  enzymatic oxidation, 160
Propanal dehydrogenase, 27
Propane monooxygenase, 27
Propane oxygenase, assay, 28–29
Propane utilization, pathways of, 27
Propane utilization enzyme assays, cell-free
  methods, 26–32
1-Propanol dehydrogenase, 27
2-Propanol dehydrogenase, 27
Propionate-CoA ligase. *See* Propionyl-CoA
  synthetase
Propionyl-CoA synthetase, 27
  assay, 31–32
Protocatechuate dioxygenase
  extradiol, 82
  intradiol, 82
  reaction catalyzed, 82
  subclassifications, 82
Protocatechuate 2,3-dioxygenase, 95–101
  assay, 95–96
  from *B. circulans*, 95
  from *B. macerans*, 95
  definition of unit of activity, 96
  determination of protein concentration,
    96
  inhibitors, 100–101
  kinetic constants, 100
  molecular weight, 99–100
  properties, 99–100
  purification, 97–99
  purity, 99
  removal of adventitious Fe(III), 99
  spectroscopic properties, 101
  stability, 99
  substrate range, 100
  subunit structure, 99–100
Protocatechuate 3,4-dioxygenase, 82–88,
  109, 112, 116, 121
  assay, 83–84
  from *B. fuscum*, 82–88
  calculation of activity, 85
  definition of unit of activity, 84
  kinetic properties, 87
  molecular weight, 87

from *P. aeruginosa*, 82
from *P. putida*, 126
properties, 87–88
purification, 85–88
purity, 87
spectroscopic properties, 88
stability, 87
substrate range, 87–88
subunit structure, 87
Protocatechuate 4,5-dioxygenase, 109
assay, 89–90
definition of unit, 90
kinetic properties, 94
metal requirement, 93
molecular weight, 93–94
from *P. testosteroni*, 89–95
properties, 92–95
protein determination, 90
purification, 90–93
purity, 92–93
reaction catalyzed, 89
spectroscopic properties, 94–95
stability, 92–93
substrate range, 94
subunit structure, 93
Pseudoazurin, 284
Pseudomonads, fluorescent, 146
Pseudomonas
AM1. *See Methylobacterium extorquens,*
AM1
aromatic ring-cleaving enzymes, 115
B13
chlorocatechol 1,2-dioxygenase from,
122
crude extracts, preparation of, 125
growth of, 125
C, glucose-6-phosphate dehydrogenase
from, 346
catechol oxygenase from, 122
chlorocatechol oxygenase from, 122–123,
126
DM1, dichloromethane dehalogenase
from, 359
M27
citrate synthase from, 352–354
growth of, 351
MA
citrate synthase from, 352–354
growth of, 351
MS

citrate synthase from, 352–354
growth of, 351
naphthalene metabolism, 46
NCIB 9816
crude cell extracts, preparation of, 49
growth of, 49
naphthalene dioxygenase from, 46–52
NCIMB 9872
cyclopentanone 1,2-monooxygenase
from, 77–81
growth of, 78–79
W6
glucose-6-phosphate dehydrogenase
from, 346–350
growth of, 347
Pseudomonas acidovorans
ATCC 17438, 109
gentisate 1,2-dioxygenase from, 101–107
growth, 103–104
hydroxypyruvate reductase from, 373
Pseudomonas aeruginosa, 147, 276
hydroxybenzoate monooxygenase from,
139–146
LMD 76.39, growth of, 273
LMD 80.53, growth of, 34
PAO1C (ATCC 15692), growth of, 140
PQQ assayed with, 272–273
protocatechuate 3,4-dioxygenase from, 82
quinoprotein alcohol dehydrogenase
from, 33–36
Pseudomonas cepacia
DBO 1, growth of, 63
phthalate dioxygenase from, 61
salicylate monooxygenase from, 147
Pseudomonas fluorescens
glucose-6-phosphate dehydrogenase from,
350
hydroxybenzoate monooxygenase from,
138–139, 143–144, 147
Pseudomonas methylica
citrate synthase from, 352–354
growth of, 351
Pseudomonas oleovorans
growth of, 5
3-hexulose-6-phosphate synthase from,
397
hydrocarbon monooxygenase system, 3–9
maintenance, 5
NADH:rubredoxin reductase from, 3, 6–7
Pseudomonas putida, 33, 147

dihydrodiol dehydrogenase from, 135
2,5-diketocamphane 1,2-monooxygenase
    from, 70
F1
    cell-free extract, preparation of, 42
    growth of, 42
    toluene dioxygenase from, 39–45
formaldehyde dehydrogenase from, 328
ML2 (NCIB 12190), 136
    cis-benzene dihydrodiol dehydrogenase
        from, 134–137
    benzene dioxygenase from, 52–60
    growth of, 56
    muconate cycloisomerase from, 126, 129
    muconolactone isomerase from, 131–133
    NADH:rubredoxin reductase from, 10
    2-oxo-Δ^3-4,5,5-
        trimethylcyclopentenylacetyl-CoA
        monooxygenase from, 70–71
    protocatechuate 3,4-dioxygenase from,
        126
PRS2000
    crude extract, preparation of, 128
    growth of, 128
Pseudomonas testosteroni
    protocatechuate 4,5-dioxygenase from,
        89–95
PtL-5, 109
Pyrene
    initial ring oxidation of, enzymatic
        mechanism for, 151–153
    oxidation by Mycobacterium, metabolic
        pathways of, 152–143, 152–153
Pyrocatechase, 116
Pyrroloquinoline quinone, 260–283
    assay, 270–277
    chemical synthesis, 262–267
    concentration, 268
    covalently bound, determination of,
        277–284
    hydrazine method, 277–281, 283
    free, determination of, 268–270
    [8-^{2}H], preparation of, 266–267
    3-^{2}H- and 2-^{13}C-labeled, preparation of,
        266
    hexanol extraction procedure, 281–283
    isolation
        from biological materials, 261–262
        from culture medium, 261
        from whole bacteria, 262

prepurification, 268
redox cycling assay, 277
reversed-phase HPLC, 269–270
sample preparation, 268

Q

Quinohemoprotein alcohol dehydrogenase,
    33, 260
    from C. testosteroni, 36–39, 273–274,
        276
    absorption spectra, 38–39
    activators, 39
    assay, 36–37
    catalytic properties, 39
    cofactor content, 39
    electron acceptors, 39
    inhibitors, 39
    properties, 38–39
    purification, 37–38
    purity, 38
    structure, 39
Quinoprotein alcohol dehydrogenase,
    33–39, 260
    from P. aeruginosa, 33–36
    assay, 33–34
    catalytic properties, 36
    cofactor content, 35–36
    properties, 35–36
    purification, 34–36
    purity, 35
    structure, 35
Quinoproteins, 33, 260

R

R. sphaeroides, glucose-6-phosphate
    dehydrogenase from, 350
Reductase(ben), 52, 58
Reductase$_{NAP}$, 46
    assay, 48
    purification of, 49–50
Reductase$_{TOL}$, 39, 40
Rhodococcus erythropolis, formaldehyde
    dehydrogenase (novel cofactor) from,
        321–323
Rhodococcus rhodochrous PNKb1
    cell-free extracts, preparation of, 26–28
    enzymes of propane utilization, cell-free
        assay methods for, 26–32

growth of, 23–24, 26
propane-specific alcohol dehydrogenase
from, 21–26
Rhodopseudomonas palustris
benzoate-CoA ligase from, 154–159
cell extract, preparation of, 156–157
growth of, 156
Ribose-5-phosphate isomerase, 392, 397,
401, 436
Ribulose monophosphate cycle, ketodeoxyphosphogluconate aldolase/transaldo-
lase variant of, 335
Ribulose-phosphate, 436
Rieske proteins, respiratory, 67
Rieske-type [2Fe–S] center, 60, 61
Rubredoxin
from anaerobic bacteria, 8
cobalt, chemical modification, 8–9
C-terminal CNBr fragment, monooxygenation activity, 8–9
modified, monooxygenation activity, 9
from *P. oleovorans*, 3
extra sulfhydryls, 8–9
isolation of, 5–6
from *P. putida*, 10
from *Pseudomonas*
metal-binding sites, 8
properties, 8
sequence data, 8

S

Saccharomyces cerevisiae, octanol dehydrogenase from, 171
Salicylate monooxygenase, from *P. cepacia*,
gene for, 147
Salicylate monooxygenase (hydroxylase),
138
Serine hydroxymethyltransferase, 365–372
activities, in *Methylobacterium*, 365–366
assay, 366–368
cation stimulation, 372
cofactor requirements, 371
glyoxylate-activated (C_1-specific),
365–366, 370–372
purification of, 368–369, 371
glyoxylate-insensitive (heterotrophic),
365–366, 370–372
purification of, 369–371
from *H. methylovorum*, 366, 372

from *Hyphomicrobium* X, 372
kinetic properties, 370–371
from *M. extorquens* AM1, 372
from *M. organophilum* XX, 365–372
from *Methylobacterium* 3A2, 372
molecular weight, 371–372
pH optimum, 371
physiological role of, 365
properties, 370–372
purity, 370
reaction catalyzed, 365
specific activity, 368
stability, 370
stimulation by small molecules, 372
unit, 368
Serine pathway, for bacterial assimilation of
formaldehyde, 386–387
Serine transhydroxymethylase. *See* Serine
hydroxymethyltransferase
Short-chain alcohol dehydrogenase
from *C. lipolytica*, subcellular distribution
of, 175
from *C. tropicalis*, subcellular distribution
of, 175

T

Thermus thermophilus, oxidized Rieske
protein from, 67
Thiobacillus versutus
growth of, 236
methylamine dehydrogenase from,
235–241
Toluene dioxygenase
components, 39, 40
from *P. putida* F1, 39–45
cofactor content, 45
cytochrome c reduction assay, 40–41
molecular weights, 45
properties, 45
purification, 42–45
purity, 45
radiometric assay, 40–41
stability, 45
reaction catalyzed, 39, 40
reconstitution of, 39
Torulopsis candida. *See also* Yeast
alcohol oxidase from, 420
Trametes versicolor, lignin peroxidase from,
160

Transaldolase, 405–411
  from *Arthrobacter*, 405–411
  from *Arthrobacter* P1
    assay, 406–407
    Michaelis constants, 410–411
    molecular weights, 411
    pH optima, 410
    properties, 410–411
    purification, 407–410
    purity, 410
    stability, 411
    unit of activity, 407
  functions, 405
  reaction catalyzed, 405
  from yeast, 406
Transketolase, 435, 436
2,4,7-Tricarboxy-1*H*-pyrrolo[2,3-*f*]quino-
  line-4,5-dione. *See* Pyrroloquinoline
  quinone
Tricarboxylic acid cycle, 350
[8-²*H*]-2,7,9-Tricarboxy-5-methoxy-1*H*-pyr-
  rolo[2,3-*f*]quinoline, 267
2,4,6-Trichlorophenol, enzymatic oxidation,
  160
Trimethylamine dehydrogenase
  from bacterium W3A1, 250–260
    assay, 251–253
    inhibitors, 258
    molecular weight, 258–259
    properties, 257–260
    purification, 253–257
    purity, 257
    spectral and oxidation–reduction
      properties, 259–260
    stability, 258
    steady-state kinetic parameters, 258
    substrate specificity, 257–258
    subunit structure, 258–259
  electron acceptor for, 309
  from *Hyphomicrobium* X, 250–251
  reaction catalyzed, 250, 309
  reduction of *M. methylotrophus* electron-
    transfer flavoprotein by, 312–313

Triokinase, 445–451
  assay, 445–447
  from *C. boidinii* KD1, 445–451
    properties, 450–451
    purification of, 449–450
  from *C. methylica*, 450
  effect of metal ions, 451
  from *H. polymorpha*, 450, 451
    properties, 450–451
  isoelectric point, 450
  molecular weight, 450
  pH optimum, 451
  purity, 450
  reaction catalyzed, 445
  stability, 450
  structure, 450
  substrate specificity, 450
Triose-phosphate isomerase, 437

U

UTP-Agarose, synthesis of, 340–341

Y

Yeast. *See also Candida; Candida boidinii;*
    Hansenula polymorpha; Pichia pastoris;
    Torulopsis candida
  aldehyde dehydrogenase [NAD(P⁺)], 330
  cell-free extract, preparation of, 441
  flavocytochrome b_2, 259
  long-chain alcohol dehydrogenase,
    171–175
  long-chain aldehyde dehydrogenase from,
    176–178
  methylotrophic
    amine oxidases from, 427–435
    formaldehyde dehydrogenase from,
      455–459
    formate dehydrogenase from, 459–463
  peroxisomes, alcohol oxidase activity, 421
  transaldolase from, 406